D1250600

251 00794

OPERATIONAL METHODS
FOR
LINEAR SYSTEMS

This book is in the
ADDISON-WESLEY SERIES IN MATHEMATICS

OPERATIONAL METHODS
FOR
LINEAR SYSTEMS

by

WILFRED KAPLAN

Department of Mathematics
University of Michigan

ADDISON-WESLEY PUBLISHING COMPANY, INC.

READING, MASSACHUSETTS · PALO ALTO · LONDON

WILLIAM MADISON RANDALL LIBRARY UNC AT WILMINGTON

Copyright © 1962

ADDISON-WESLEY PUBLISHING COMPANY, INC.

Printed in the United States of America

ALL RIGHTS RESERVED. THIS BOOK, OR PARTS THERE-
OF, MAY NOT BE REPRODUCED IN ANY FORM WITH-
OUT WRITTEN PERMISSION OF THE PUBLISHERS.

Library of Congress Catalog Card No. 62–9401

QA432
.K3

PREFACE

The present volume is intended as a text for a course treating the mathematical methods employed in the design and analysis of linear systems, that is, physical systems whose behavior is described by a set of simultaneous ordinary linear differential or integrodifferential equations. The principal topics are Fourier series, Fourier transforms, Laplace transforms, and their application to ordinary linear differential equations. The concepts of stability, transfer function, frequency response, and weighting function are also considered extensively. The basic approach to the problems considered arose in modern engineering in the design of complicated control systems. However, the ideas have proven to be of value in many physical, biological, and social sciences. Furthermore, the mathematical problems, which concern the properties of certain linear operators, are of major interest in their own right.

Chapter 1 presents a brief review of linear differential equations and also extends the theory to integrodifferential equations and generalized functions (impulse functions), which are treated in a simplified manner. Chapter 2 introduces linear operators and relates them to linear differential equations; the superposition principle and its applications are given much weight.

Chapter 3 presents a concise treatment of analytic functions of a complex variable. For students already possessing a background in this field, the chapter can serve as a review; for others, it provides a brief introduction and covers those topics (especially residues) essential to the study of linear systems.

Chapter 4 develops the theory of Fourier series. After the basic theorems on convergence have been established, the finite Fourier transform is introduced and is shown to be an excellent tool for determining response to periodic forcing functions. Chapter 5 then presents a somewhat parallel development for infinite Fourier integrals and transforms; they are shown to be of value in finding the response of linear systems to forcing functions which are absolutely integrable from $-\infty$ to ∞ (and to more general forcing functions). The very closely related Laplace transform is then treated in detail in Chapter 6; here there are advantages, concerning on the one hand, the forcing functions allowed and, on the other, the ease with which initial conditions can be satisfied. The chapter closes with sections on the z-transform, sampled-data systems, and the Hilbert transform.

Chapter 7 concerns stability; the Routh-Hurwitz criterion, the Nyquist criterion, the root-locus method are discussed.

139320

The final chapter treats time-variant linear systems, i.e., systems described by linear differential equations with variable coefficients. The concepts of weighting function and transfer function are given precise generalizations. A variety of methods for explicit solution, at least in approximate form, are developed. Equations with periodic coefficients are studied. Some principal theorems on stability are established; in particular, stability is related to the type of response to bounded inputs.

Appendix I describes briefly the formalism developed by Mikusiński as an alternative to that based on Laplace transforms. A second appendix brings together for ready reference the principal tables of transforms and convolutions. Appendix III lists the more important symbols used throughout the book.

Numerous examples and problem sets, with answers, are distributed throughout the text. References for further reading are listed after each chapter.

The book as a whole provides enough material for a year's course. At the University of Michigan it has been used as a text for a one-semester course meeting four times a week, with the following outline: 1–1 to 1–3, 1–4 to 1–5, 1–6, 1–7 to 1–8, 1–9 to 1–11, 1–12, 1–13, 1–14, 2–1 to 2–2, 2–3 to 2–4, 2–5 to 2–6, 2–7 to 2–8, 2–9 to 2–10, 3–1 to 3–2, 3–3, 3–4, 3–5, 3–6, 3–7, 3–8, 3–9, 3–10 to 3–11, 3–12, 3–13 to 3–14, 3–15 to 3–16, 4–1 to 4–3, 4–4 to 4–5, 4–6 to 4–7, 4–8 to 4–9, 4–10 to 4–11, 5–1 to 5–5, 5–6 to 5–8, 5–9 to 5–10, 5–11, 5–12, 5–13 to 5–15, 5–16, 5–17 to 5–19, 5–20 to 5–22, 6–1 to 6–3, 6–4 to 6–5, 6–6, 6–7, 6–8, 6–9, 6–12, 6–13, 6–14, 6–15, 6–16, 7–1 to 7–3, 7–4, 7–5, 7–6 to 7–8. Chapter 8 by itself has formed the basis for a one-semester course, meeting twice a week, on time-variant systems.

The author wishes to express his appreciation for the valuable advice given by his colleagues at the University of Michigan, especially Professors R. C. F. Bartels, F. J. Beutler, R. V. Churchill, C. L. Dolph, E. G. Gilbert, E. O. Gilbert, R. M. Howe, A. B. Macnee, G. J. Minty, L. L. Rauch, R. K. Ritt, E. H. Rothe, and B. Williams.

To the Addison-Wesley Publishing Company he expresses his thanks for their cooperation. To his wife he expresses his gratitude for helping in ever so many ways and for her patience throughout the long and difficult period of preparation of the manuscript.

W. K.

Chardonne, Switzerland
October, 1961

CONTENTS

CHAPTER 4.   FOURIER SERIES AND FINITE FOURIER TRANSFORM

CHAPTER 7. STABILITY

CHAPTER 8. TIME-VARIANT LINEAR SYSTEMS

CHAPTER 1

LINEAR DIFFERENTIAL EQUATIONS

In this chapter the basic theorems concerning linear differential equations are stated and methods for obtaining solutions explicitly are reviewed. Integrodifferential equations are defined and shown to be equivalent to differential equations. The value of complex functions of a real variable for study of such equations is pointed out. The theory is extended to cover equations involving discontinuous functions and "generalized" functions such as the impulse function (delta function).

1–1 Existence theorem for linear differential equations. By an *ordinary linear differential equation of order n* is meant an equation

$$a_0(t) \frac{d^n x}{dt^n} + a_1(t) \frac{d^{n-1} x}{dt^{n-1}} + \cdots + a_{n-1}(t) \frac{dx}{dt} + a_n(t)x = f(t), \quad (1\text{--}10)$$

in which $a_0(t), \ldots, a_n(t), f(t)$ are defined on some interval of t, and $a_0(t)$ is not identically 0. By a *solution* of the equation is meant a function $x(t)$ which satisfies the equation identically over some interval; in particular, $x'(t), x''(t), \ldots, x^{(n)}(t)$ must exist over the interval.

The general existence theorem for a differential equation of order n guarantees existence of a unique solution satisfying given initial conditions, provided appropriate continuity conditions are satisfied. However, the solution may exist over only a part of the t-interval considered. For linear equations one has a stronger assertion:

THEOREM 1. *Let $a_0(t), a_1(t), \ldots, a_n(t), f(t)$ be defined and continuous on the interval $\alpha < t < \beta$ and let $a_0(t) \neq 0$ on this interval. Let t_0 be a point of the interval. Then there exists one and only one function $x(t)$, $\alpha < t < \beta$, such that $x(t)$ is a solution of Eq. (1–10) and such that $x(t)$ satisfies the prescribed initial conditions:*

$$x(t_0) = x_0, \ x'(t_0) = x'_0, \ \ldots, \ x^{(n-1)}(t_0) = x_0^{(n-1)}. \quad (1\text{--}11)$$

Remark. The theorem has been stated for an *open* interval: $\alpha < t < \beta$. It could equally well have been stated for a *closed* interval $\alpha \leqq t \leqq \beta$, with the understanding that derivatives at $t = \alpha$ are evaluated as derivatives *to the right* and those at $t = \beta$ as derivatives *to the left*. The theorem also applies to infinite intervals: $\alpha \leqq t < \infty$; $-\infty < t \leqq \beta$; $-\infty < t < \infty$; $\alpha < t < \infty$; $-\infty < t < \beta$.

1

The n numbers $x_0, x_0', \ldots, x_0^{(n-1)}$, which can be assigned arbitrarily, are termed the *initial values* of the solution at $t = t_0$. Since there is one solution for each set of initial values at t_0, we can state that the general solution depends on n *arbitrary constants*.

When $f(t) \equiv 0$, Eq. (1–10) is termed *homogeneous*. For the homogeneous equation one has further information:

THEOREM 2. *Under the conditions of Theorem 1, if $f(t) \equiv 0$, the general solution of Eq. (1–10) can be expressed in the form*

$$x = c_1 x_1(t) + \cdots + c_n x_n(t), \qquad (1\text{–}12)$$

where $x_1(t), \ldots, x_n(t)$ are solutions of Eq. (1–10) with $f(t) \equiv 0$, and $x_1(t), \ldots, x_n(t)$ are linearly independent for $\alpha < t < \beta$. Furthermore, the general solution can be constructed in the form (1–12) from any n solutions $x_1(t), \ldots, x_n(t)$ which are linearly independent for $\alpha < t < \beta$.

Linear independence of $x_1(t), \ldots, x_n(t)$ means that no one of the functions is expressible as a linear combination of the others over the given interval. Equivalently, it means that

$$k_1 x_1(t) + \cdots + k_n x_n(t) \equiv 0 \quad (\alpha < t < \beta)$$

can hold, with constant $k_1, \ldots, k_n$, only if

$$k_1 = 0, k_2 = 0, \ldots, k_n = 0.$$

The general solution of the nonhomogeneous equation is related to that of the *corresponding homogeneous equation*, obtained by replacing $f(t)$ by 0:

THEOREM 3. *Under the conditions of Theorem 1, the general solution of Eq. (1–10) can be expressed in the form*

$$x = x^*(t) + x_c(t), \qquad (1\text{–}13)$$

where $x^*(t)$ is a particular solution and $x_c(t)$ denotes the general solution (1–12) of the corresponding homogeneous equation.

For proofs of the theorems refer to Chapters 4 and 12 of Reference 4 listed at the end of the chapter.

1–2 Solution of the homogeneous equation with constant coefficients. We consider a homogeneous linear equation with constant coefficients $a_0, \ldots, a_n$, where $a_0 \neq 0$:

$$a_0 \frac{d^n x}{dt^n} + \cdots + a_n x = 0 \quad (-\infty < t < \infty). \qquad (1\text{–}20)$$

Its solution is reduced to an algebraic problem as follows. We seek a solution of the form

$$x = e^{st} \quad (s = \text{const}). \tag{1-21}$$

Substitution in the differential equation leads to the equation

$$(a_0 s^n + \cdots + a_{n-1} s + a_n) e^{st} = 0.$$

Hence Eq. (1–21) is a solution, provided s is chosen to satisfy the *characteristic equation*

$$a_0 s^n + \cdots + a_{n-1} s + a_n = 0.$$

If this equation has n distinct real roots (called *characteristic roots*) $s_1, \ldots, s_n$, we at once obtain n linearly independent solutions

$$x = e^{s_1 t}, \ldots, x = e^{s_n t}$$

and the corresponding general solution

$$x = c_1 e^{s_1 t} + \cdots + c_n e^{s_n t}.$$

If the characteristic equation has repeated roots, for example

$$s_1 = s_2 = \cdots = s_k,$$

one obtains linearly independent solutions by replacing the corresponding k functions by $e^{s_1 t}, te^{s_1 t}, \ldots, t^{k-1} e^{s_1 t}$. If the characteristic equation has a pair of complex roots (conjugate complex numbers) $s_1 = a + bi$, $s_2 = a - bi$, the corresponding functions are $e^{at} \cos bt$, $e^{at} \sin bt$; if these roots are repeated k times, there are $2k$ linearly independent solutions $t^m e^{at} \cos bt$, $t^m e^{at} \sin bt$ $(m = 0, 1, \ldots, k - 1)$.

EXAMPLE. $\dfrac{d^5 x}{dt^5} - 2 \dfrac{d^4 x}{dt^4} + 2 \dfrac{d^3 x}{dt^3} - 2 \dfrac{d^2 x}{dt^2} + \dfrac{dx}{dt} = 0.$

The characteristic equation is

$$s^5 - 2s^4 + 2s^3 - 2s^2 + s = 0,$$

and the characteristic roots are $0, 1, 1, \pm i$. The general solution is

$$x = c_1 + c_2 e^t + c_3 t e^t + c_4 \cos t + c_5 \sin t.$$

Operator notation. It will be convenient to write Dx for dx/dt, $D^2 x$ for $d^2 x/dt^2$, and so on. The example can then be written as

$$D^5 x - 2D^4 x + 2D^3 x - 2D^2 x + Dx = 0,$$

or more simply as

$$(D^5 - 2D^4 + 2D^3 - 2D^2 + D)x = 0.$$

The expression in parentheses is an example of a *differential operator*.

PROBLEMS

1. Verify that $x = e^t \sin t$ is a solution of

$$\frac{d^2x}{dt^2} - 2\frac{dx}{dt} + 2x = 0.$$

2. Verify that $x = 1/t$ is a solution of

$$2t^2\frac{d^2x}{dt^2} + 2t\frac{dx}{dt} - 2x = 0 \quad (t > 0).$$

3. Find the general solution of each of the following:

(a) $(D^2 - 3D - 4)x = 0$ (b) $(D^2 + 9)x = 0$
(c) $(D^2 + 6D + 9)x = 0$ (d) $(3D^3 + 5D)x = 0$
(e) $(D^4 - D^3)x = 0$ (f) $(D^4 + 2D^2 + 1)x = 0$
(g) $(D^3 - 6D^2 + 11D - 6)x = 0$ (h) $(D^2 + D + 1)x = 0$
(i) $(D^4 + 1)x = 0$ (j) $(5D + 6)x = 0$
(k) $(aD + b)x = 0$ (l) $(pD^2 + 2qD + r)x = 0$

In (k) and (l) a, b, p, q, r are constants and $a \neq 0$, $p \neq 0$.

4. Find a solution satisfying the given initial conditions:

(a) $(D^2 + 4)x = 0$; $x = 1$ and $dx/dt = 0$ when $t = 0$
(b) $(D^2 + 2D + 5)x = 0$; $x = 0$ and $dx/dt = -1$ when $t = 0$

5. For each of the following functions find a homogeneous linear differential equation with constant coefficients of which the function is a solution:

(a) $x = 2e^t + t$ (b) $x = te^t$
(c) $x = \cos t - 3\sin t$ (d) $x = t^4 - 1$
(e) $x = e^t \cos 5t + e^t$ (f) $x = \cos^2 t$

6. Choose a, b, c so that $x = t^3 - t$ is a solution of

$$(aD^2 + bD + c)x = 6t^3 + 3t^2 - 1.$$

ANSWERS

3. (a) $c_1e^{4t} + c_2e^{-t}$ (b) $c_1\cos 3t + c_2\sin 3t$
(c) $c_1e^{-3t} + c_2te^{-3t}$
(d) $c_1 + c_2\cos\sqrt{5/3}\,t + c_3\sin\sqrt{5/3}\,t$
(e) $c_1 + c_2t + c_3t^2 + c_4e^t$
(f) $c_1\cos t + c_2\sin t + c_3t\cos t + c_4t\sin t$

(g) $c_1 e^t + c_2 e^{2t} + c_3 e^{3t}$
(h) $e^{-t/2}[c_1 \cos(\sqrt{3}\,t/2) + c_2 \sin(\sqrt{3}\,t/2)]$
(i) $e^{\alpha t}(c_1 \cos \alpha t + c_2 \sin \alpha t) + e^{-\alpha t}(c_3 \cos \alpha t + c_4 \sin \alpha t)$, $(\alpha = \sqrt{2}/2)$
(j) $ce^{-6t/5}$ \hspace{2cm} (k) $ce^{-bt/a}$
(l) if $q^2 - pr > 0$, $x = c_1 e^{s_1 t} + c_2 e^{s_2 t}$, where

$$ s_1 = \frac{-q + \sqrt{q^2 - pr}}{p}, \qquad s_2 = \frac{-q - \sqrt{q^2 - pr}}{p}; $$

if $q^2 - pr = 0$, $x = e^{-qt/p}(c_1 + c_2 t)$; if $q^2 - pr < 0$,

$$ x = e^{-qt/p}(c_1 \cos \beta t + c_2 \sin \beta t), \text{ where } \beta = \sqrt{pr - q^2}/p. $$

4. (a) $x = \cos 2t$ \hspace{2cm} (b) $x = -\tfrac{1}{2}e^{-t}\sin 2t$
5. Equation of lowest order:

(a) $(D^3 - D^2)x = 0$ \hspace{2cm} (b) $(D^2 - 2D + 1)x = 0$
(c) $(D^2 + 1)x = 0$ \hspace{2.4cm} (d) $D^5 x = 0$
(e) $(D^3 - 3D^2 + 28D - 26)x = 0$ \hspace{0.5cm} (f) $(D^3 + 4D)x = 0$

6. $a = 1, b = 1, c = 6$

1–3 Linear differential equations with constant coefficients; nonhomogeneous case. By Theorem 3, solution of a nonhomogeneous equation

$$ (a_0 D^n + \cdots + a_n)x = f(t) \tag{1–30} $$

is reduced to finding $x_c(t)$ (the *complementary* function), the general solution of the corresponding homogeneous equation

$$ (a_0 D^n + \cdots + a_n)x = 0, \tag{1–31} $$

and to finding a particular solution $x^*(t)$ of Eq. (1–30). For example, the equation

$$ (D^2 - 4)x = 3e^t \tag{1–32} $$

has a particular solution $x^* = -e^t$, as can be verified. The corresponding homogeneous equation

$$ (D^2 - 4)x = 0 $$

has the general solution $x_c = c_1 e^{2t} + c_2 e^{-2t}$. Hence the general solution of Eq. (1–32) is

$$ x = -e^t + c_1 e^{2t} + c_2 e^{-2t}. \tag{1–33} $$

The principal difficulty here is determination of the particular solution $x^*(t)$. In the standard course on differential equations, a variety of methods to this end are developed: (a) undetermined coefficients, (b) variation of parameters, (c) operator methods. Since one of the main goals of this text is to give a thorough treatment of this very problem, we consider here briefly only the first of the standard methods: undetermined coefficients. The method is restricted to equations with constant coefficients.

In the case of Eq. (1–32) the method of undetermined coefficients would lead us to inspect the right-hand member, e^t, and then use as trial function $x = ke^t$, where k is a constant to be determined. If we replace x by ke^t in Eq. (1–32), we obtain the equation

$$ke^t - 4ke^t = 3e^t.$$

This is satisfied for all t if $-3k = 3$ or $k = -1$. Hence $x^* = -e^t$ is a solution of Eq. (1–32).

If the right-hand member of Eq. (1–32) had been e^{-t}, we would have tried $x^* = ke^{-t}$. If it had been be^{at} (a and b constant, $b \neq 0$), we would have tried

$$x^* = ke^{at},$$

except when $a = 2$ or $a = -2$. For e^{2t} and e^{-2t} are solutions of the corresponding homogeneous equation, and the substitution $x^* = ke^{2t}$ or $x^* = ke^{-2t}$ leads to the impossible equation $k \cdot 0 = b$. In these exceptional cases we use, respectively,

$$x^* = kte^{2t}, \qquad x^* = kte^{-2t}.$$

For the general equation (1–30), let the right-hand member $f(t)$ have the form

$$e^{at} \cos bt \, (\alpha_1 + \alpha_2 t + \cdots + \alpha_m t^{m-1})$$
$$+ \, e^{at} \sin bt \, (\beta_1 + \beta_2 t + \cdots + \beta_m t^{m-1}), \quad (1\text{–}34)$$

where $a, b, \alpha_1, \ldots, \beta_1, \ldots$ are constants; let $s_1, \ldots, s_n$ be the characteristic roots of the corresponding homogeneous equation (1–31). If no one of $s_1, \ldots, s_n$ equals $a \pm bi$, the trial function can be chosen to have a form similar to Eq. (1–34):

$$x^* = e^{at} \cos bt \, (k_1 + k_2 t + \cdots + k_m t^{m-1})$$
$$+ \, e^{at} \sin bt \, (k_{m+1} + k_{m+2} t + \cdots + k_{2m} t^{m-1}). \quad (1\text{–}35)$$

This expression is to be substituted in the differential equation (1–30), and $k_1, \ldots, k_{2m}$ are to be chosen so that the equation is identically satisfied. If $a \pm bi$ occurs p times among the roots $s_1, \ldots, s_n$, the expression (1–35) must be multiplied by t^p before substitution in the differential equation.

If the right-hand member of Eq. (1–30) is a sum of blocks of terms, each block having form (1–34), we find the particular solution for each block separately and then add the results; this procedure is an application of the *superposition principle* (see Section 2–3). Equivalently, we can form one trial function which is the sum of the trial functions of the separate blocks.

In the expression (1–34) many of the constants may be 0. For example, $te^t \cos t$ is a special case, with $a = 1$, $b = 1$, $m = 2$; $t^4 + t^3$ is a special case, with $a = 0$, $b = 0$, $m = 5$.

EXAMPLE 1. $(D^2 + 1)x = 5e^t \sin t$. The characteristic roots are $\pm i$; since $a = 1$ and $b = 1$, $a \pm bi$ does not occur among the characteristic roots. Accordingly, the trial function is

$$x^* = e^t(k_1 \cos t + k_2 \sin t).$$

Substitution in the differential equation leads to the relation

$$e^t[(k_1 + 2k_2) \cos t + (k_2 - 2k_1) \sin t] = 5e^t \sin t.$$

This is to be an identity. Therefore,

$$k_1 + 2k_2 = 0, \quad k_2 - 2k_1 = 5; \quad k_1 = -2, \quad k_2 = 1.$$

Hence $x^* = e^t(\sin t - 2 \cos t)$ is the particular solution sought. The general solution is

$$x = c_1 \cos t + c_2 \sin t + e^t(\sin t - 2 \cos t).$$

EXAMPLE 2. $(D^2 + D)x = 12t^2 + 4t^3$. Here the characteristic roots are 0, -1, so that $x_c = c_1 + c_2 e^{-t}$. For the right-hand side $a = b = 0$, so that $a \pm bi$ occurs once among the characteristic roots. Accordingly, the trial function is

$$x^* = t(k_1 + k_2 t + k_3 t^2 + k_4 t^3) = k_1 t + k_2 t^2 + k_3 t^3 + k_4 t^4.$$

Substitution in the differential equation leads to simultaneous equations for $k_1, \ldots, k_4$, from which we find $k_1 = 0$, $k_2 = 0$, $k_3 = 0$, $k_4 = 1$.

If the right-hand member of Eq. (1–30) is not a sum of terms of the form (1–34), the method of undetermined coefficients is *inapplicable*. The other methods mentioned above must then be used.

PROBLEMS

1. Find the general solution for each of the following:

(a) $(D^2 - 4)x = e^t$ (b) $(D^3 + D^2 + D + 1)x = 4e^t$
(c) $(D^2 - 2D - 3)x = 6e^{2t}$ (d) $(D^3 - 4D^2 + 5D - 2)x = 2e^t - 3$
(e) $(D^3 - 5D^2 + 8D - 4)x = 12e^{2t}$
(f) $(D^2 + 4)x = \sin 2t$
(g) $(D^3 - 2D - 4)x = e^{-t} \sin t$
(h) $(D^2 + 2hD + \lambda^2)x = \sin \omega t,\ h \geqq 0,\ \lambda > 0,\ \omega > 0$

2. Find a solution satisfying the stated initial conditions:

(a) $(D^2 + 4)x = 3e^t$; $x = 1$ and $Dx = 0$ for $t = 0$

(b) $(D^2 - 9)x = e^t \sin t$; $x = 0$ and $Dx = 1$ for $t = 0$

3. Find a particular solution of each of the following equations:

[*Hint:* It is not necessary to find the characteristic roots but merely to verify whether each number $a \pm bi$ arising from the right-hand member is a characteristic root.]

(a) $(D^3 + 3D^2 + 3D + 1)x = 6e^{-t}$

(b) $(D^2 + 17D + 11)x = e^{3t}$

(c) $(D^3 + 5D^2 + 2D + 1)x = \cos 2t + 9e^{-t}$

(d) $(D^7 + D + 1)x = t^2 + 1$

(e) $(D^4 - D^3 + D^2 - 1)x = e^t + 1$

ANSWERS

1. (a) $c_1 e^{-2t} + c_2 e^{2t} - e^t/3$

(b) $c_1 e^{-t} + c_2 \cos t + c_3 \sin t + e^t$

(c) $c_1 e^{-t} + c_2 e^{3t} - 2e^{2t}$

(d) $c_1 e^t + c_2 t e^t + c_3 e^{2t} - t^2 e^t + \frac{3}{2}$

(e) $c_1 e^{2t} + c_2 t e^{2t} + c_3 e^t + 6t^2 e^{2t}$

(f) $c_1 \cos 2t + c_2 \sin 2t - \frac{1}{4}t \cos 2t$

(g) $c_1 e^{2t} + e^{-t}(c_2 \cos t + c_3 \sin t) + \frac{1}{20}te^{-t}(3 \cos t - \sin t)$

(h) If $h = 0$ and $\lambda = \omega$, gen. sol. is $-[(t \cos \omega t)/(2\omega)] + x_c(t)$; otherwise, gen. sol. is $[(\lambda^2 - \omega^2)^2 + 4h^2\omega^2]^{-1/2} \cdot \sin (\omega t - \alpha) + x_c(t)$,

     where $\alpha = \arctan [2h\omega(\lambda^2 - \omega^2)^{-1}]$

2. (a) $0.4 \cos 2t - 0.3 \sin 2t + 0.6e^t$

(b) $\frac{1}{85}[17e^{3t} - 15e^{-3t} - e^t(2 \cos t + 9 \sin t)]$

3. (a) $t^3 e^{-t}$ (b) $e^{3t}/71$

(c) $-[(19 \cos 2t + 4 \sin 2t)/377] + 3e^{-t}$

(d) $t^2 - 2t + 3$ (e) $\frac{1}{3}t e^t - 1$

1–4 Simultaneous linear differential equations. The basic form for simultaneous linear equations is the following:

$$\frac{dx_k}{dt} = a_{k1}x_1 + \cdots + a_{kn}x_n + f_k(t) \quad (k = 1, \ldots, n). \quad (1\text{–}40)$$

It will be assumed that the coefficients a_{kl} and the functions $f_k(t)$ are continuous functions of t over some interval. For example, with $n = 2$, the equations

$$\frac{dx_1}{dt} = a_{11}x_1 + a_{12}x_2 + f_1(t),$$

$$\frac{dx_2}{dt} = a_{21}x_1 + a_{22}x_2 + f_2(t)$$

are in the basic form, with x_1 and x_2 as unknown functions of t. More general systems can be reduced to this form. In particular, a set of equations

$$\sum_{l=1}^{n}\left[p_{kl}(t)\frac{dx_l}{dt} + q_{kl}(t)x_l\right] = F_k(t) \quad (k = 1, \ldots, n) \qquad (1\text{--}41)$$

is reducible to the form (1–40) by solving for the derivatives, provided the determinant

$$\Delta = \begin{vmatrix} p_{11} & p_{12} & \cdots & p_{1n} \\ p_{21} & p_{22} & \cdots & p_{2n} \\ \vdots & & & \vdots \\ p_{n1} & p_{n2} & \cdots & p_{nn} \end{vmatrix}$$

is never zero. An example of (1–41) is provided by the set of equations

$$(2D + 3)x + (4D - 1)y = \sin t,$$
$$(5D + 2)x - (D - 1)y = 1,$$

for which

$$\Delta = \begin{vmatrix} 2 & 4 \\ 5 & -1 \end{vmatrix} = -22 \neq 0.$$

Upon solving for Dx and Dy, we find

$$Dx = \tfrac{1}{22}(-11x - 3y + \sin t + 4),$$
$$Dy = \tfrac{1}{22}(-11x + 7y + 5\sin t - 2).$$

When the determinant Δ is identically zero, the system is degenerate in some sense. For example, the equations

$$(D + 2)x + (D - 1)y = 0,$$
$$(D - 1)x + (D + 2)y = 3$$

have the determinant

$$\Delta = \begin{vmatrix} 1 & 1 \\ 1 & 1 \end{vmatrix} = 0.$$

Subtracting the equations, we obtain a relation: $3x - 3y = -3$. Hence $y = x + 1$. Elimination of y leads to the equation $(2D + 1)x = 1$, so that $x = 1 + ce^{-t/2}$, $y = 2 + ce^{-t/2}$; we can verify that (for each choice of the constant c) these functions satisfy the equations. On the other hand, for the equations

$$Dx + Dy = 1, \qquad Dx + Dy = 2,$$

again $\Delta = 0$, but the equations are contradictory.

Equations involving derivatives of second or higher order can also be reduced to the form (1–40), as the following example illustrates:

$$(D^2 - D + 2)x + (D - 1)y = \sin 2t,$$
$$Dx + (D^2 + 2D - 3)y = e^t.$$

An equivalent system of the form (1–40) is obtained by writing

$$Dx = u, \quad Dy = v,$$
$$Du = -2x + y + u - v + \sin 2t,$$
$$Dv = 3y - u - 2v + e^t.$$

In particular, the single linear equation of nth order can be written as a set of equations of the form (1–40). The equations

$$Dx_1 = x_2, \; Dx_2 = x_3, \; \ldots, \; Dx_{n-1} = x_n,$$
$$Dx_n = -a_n x_1 - a_{n-1} x_2 - \cdots - a_1 x_n + f(t)$$

are equivalent to

$$(D^n + a_1 D^{n-1} + \cdots + a_n)x_1 = f(t).$$

More generally, a set of equations

$$(a_0 D^n + \cdots + a_n)x + (b_0 D^m + \cdots + b_m)y = f,$$
$$(A_0 D^n + \cdots + A_n)x + (B_0 D^m + \cdots + B_m)y = F \quad (1–42)$$

can be reduced to basic form, provided

$$\Delta = \begin{vmatrix} a_0(t) & b_0(t) \\ A_0(t) & B_0(t) \end{vmatrix} \neq 0. \quad (1–43)$$

For then we can solve equations (1–42) for $D^n x$, $D^m y$ to obtain equations

$$D^n x = (r_0 D^{n-1} + \cdots)x + (s_0 D^{m-1} + \cdots)y + \phi(t),$$
$$D^m y = (R_0 D^{n-1} + \cdots)x + (S_0 D^{m-1} + \cdots)y + \Phi(t).$$

These equations are in turn equivalent to the set of equations

$$Dx_1 = x_2, \; Dx_2 = x_3, \; \ldots, \; Dx_{n-1} = x_n,$$
$$Dy_1 = y_2, \; Dy_2 = y_3, \; \ldots, \; Dy_{m-1} = y_m,$$
$$Dx_n = r_0 x_n + \cdots + s_0 y_m + \cdots + \phi(t),$$
$$Dy_m = R_0 x_n + \cdots + S_0 y_m + \cdots + \Phi(t),$$

in which $x_1 = x$, $y_1 = y$.

The general form of the determinant condition (1–43) is the requirement that it must be possible to solve the given equations for the highest derivatives of all dependent variables.

By a *solution* of equations (1–40) is meant a set of functions $x_1(t), \ldots,$ $x_n(t)$ which satisfy the equations identically over some interval of t. Existence of solutions is guaranteed by the following theorem:

THEOREM 4. *Let the functions $a_{kl}(t)$ and $f_k(t)$ be continuous for $\alpha < t < \beta$ and let t_0 be a point of this interval. Then there is one and only one solution $x_k = x_k(t)$ $(k = 1, \ldots, n)$ of equations (1–40) for $\alpha < t < \beta$ which satisfies prescribed initial conditions,*

$$x_1(t_0) = x_1^0, \quad \ldots, \quad x_n(t_0) = x_n^0. \tag{1–44}$$

The general solution of equations (1–40) for $\alpha < t < \beta$ can be written in the form

$$x_k = c_1 x_{k1}(t) + \cdots + c_n x_{kn}(t) + x_k^*(t), \tag{1–45}$$

where $c_1, \ldots, c_n$ are arbitrary constants.

The n functions $c_1 x_{k1}(t) + \cdots + c_n x_{kn}(t)$ $(k = 1, \ldots, n)$ together can be considered as a "complementary-function set"; they provide the general solution of the *corresponding homogeneous system:*

$$Dx_k = a_{k1} x_1 + \cdots + a_{kn} x_n. \tag{1–46}$$

The n functions $x_k = x_k^*(t)$ $(k = 1, \ldots, n)$ form a "particular-function set"; they provide a particular solution of equations (1–40).

For a proof of Theorem 4 and a more detailed discussion, refer to Chapters 6 and 12 of Reference 4 listed at the end of the chapter.

1–5 Simultaneous linear equations with constant coefficients. *Method of elimination.* Let

$$\phi_1(D) = a_0 D^n + a_1 D^{n-1} + \cdots + a_n,$$
$$\phi_2(D) = b_0 D^m + b_1 D^{m-1} + \cdots + b_m$$

be differential operators. We define their sum $\phi_1(D) + \phi_2(D)$ and product $\phi_1(D) \cdot \phi_2(D)$ as the operators such that

$$[\phi_1(D) + \phi_2(D)]x(t) = \phi_1(D)x(t) + \phi_2(D)x(t),$$
$$[\phi_1(D) \cdot \phi_2(D)]x(t) = \phi_1(D)[\phi_2(D)x(t)]$$

whenever $x(t)$ has derivatives of the orders required. It can be verified that, *when the coefficients are constant,* addition and multiplication of operators obey the rules for polynomials. For example,

$$[(2D + 1) + (3D + 5)]x = (2D + 1)x + (3D + 5)x$$
$$= 2Dx + x + 3Dx + 5x = (5D + 6)x,$$
$$[(D + 1)(D - 1)]x = (D + 1)(Dx - x)$$
$$= D(Dx - x) + Dx - x$$
$$= D^2x - Dx + Dx - x$$
$$= D^2x - x = (D^2 - 1)x.$$

(See Section 4–11 of Reference 4 for a general discussion.)

EXAMPLE 1.

$$(2D - 1)x + (D + 13)y = 14e^{3t}, \tag{i}$$

$$(2D - 3)x - (D - 7)y = 0. \tag{ii}$$

The solution will be obtained by elimination. We subtract the equations to obtain a new equation

$$2x + (2D + 6)y = 14e^{3t}. \tag{i'}$$

We consider Eq. (i') to be a replacement for Eq. (i) and now consider Eqs. (i') and (ii). We can eliminate x between these equations by multiplying Eq. (ii) by -2, Eq. (i') by $(2D - 3)$, and adding. We find, after combining the operators as above,

$$(4D^2 + 8D - 32)y = 42e^{3t}. \tag{ii'}$$

The new equation is numbered (ii'), and not (i''), to indicate that it replaces Eq. (ii). The reason for this choice is given below. From Eq. (ii') we now find

$$y = c_1e^{2t} + c_2e^{-4t} + \tfrac{3}{2}e^{3t}, \tag{1–50}$$

and hence from Eq. (i'),

$$x = 7e^{3t} - (D + 3)y = -5c_1e^{2t} + c_2e^{-4t} - 2e^{3t}. \tag{1–50'}$$

The last two equations provide the general solution sought.

In general, the elimination procedure consists of multiplying one equation by a nonzero constant k, multiplying a second equation by an operator $\phi(D)$, and adding the two resulting equations. The new equation is considered as a *replacement for the one which was multiplied by k*. In symbols, the two equations

$$f_1(D)x + f_2(D)y + \cdots = F(t), \tag{I}$$

$$g_1(D)x + g_2(D)y + \cdots = G(t) \tag{II}$$

are replaced by the equations

$$[kf_1(D) + \phi(D)g_1(D)]x + [kf_2(D) + \phi(D)g_2(D)]y$$
$$+ \cdots = kF(t) + \phi(D)G(t), \qquad \text{(I')}$$

$$g_1(D)x + g_2(D)y + \cdots = G(t). \qquad \text{(II)}$$

It can easily be verified that the two sets of equations are equivalent; that is, if $x(t)$, $y(t)$, ... satisfies the first pair, then $x(t)$, $y(t)$, ... satisfies the second pair, and conversely. It is also not difficult to show that repeated application of the procedure, in appropriate fashion, will lead to an equivalent set of equations in "triangular" form:

$$\psi_1(D)x = H_1(t),$$
$$\psi_2(D)x + \psi_3(D)y = H_2(t),$$
$$\psi_4(D)x + \psi_5(D)y + \cdots = H_3(t),$$
$$\vdots$$

The first equation gives x; with x known, the second gives y, and so on. If the given set of equations is nondegenerate, the coefficient $\psi_1(D)$ of x in the first equation, $\psi_3(D)$ of y in the second equation, etc., will all be different from zero, so that solution is possible.

If Eq. (I) were retained instead of Eq. (II), the new equations would not in general be equivalent to the old ones, and one would obtain *extraneous* solutions.

In many physical problems, we are given nondegenerate simultaneous linear differential equations in several unknowns x, y, z, ..., but are interested only in one of them, for example, x. The elimination procedure enables us to obtain an equation for x alone. If the given equations have forms (I), (II), ..., the resulting equation in x has the form

$$\psi_1(D)x = p_1(D)F(t) + p_2(D)G(t) + \cdots$$

We may, in particular, study only the response, x, to the forcing function $F(t)$, with $G(t)$, ... all being zero. Thus we are led to an equation of the form

$$(a_0D^n + a_1D^{n-1} + \cdots + a_n)x = (b_0D^m + \cdots + b_m)F(t). \qquad \text{(1–51)}$$

Here it can be verified that m must be less than n. If $F(t)$ does not have derivatives through order m, Eq. (1–51) appears to be meaningless; this apparent defect of the method of elimination can, in most cases, be overcome with the aid of *generalized functions* (Sections 1–13 and 1–14).

Method of exponential substitution. We illustrate this method for the homogeneous equations corresponding to Example 1.

EXAMPLE 2.

$$(2D - 1)x + (D + 13)y = 0,$$

$$(2D - 3)x - (D - 7)y = 0.$$

$$(1\text{--}52)$$

We seek a solution of the form $x = \alpha e^{st}$, $y = \beta e^{st}$, with α, β not both zero. Substitution in the equations leads to

$$(2s - 1)\alpha + (s + 13)\beta = 0,$$

$$(2s - 3)\alpha - (s - 7)\beta = 0;$$

$$(1\text{--}53)$$

these equations have a solution for α, β other than 0, 0, provided that the determinant of the coefficients is zero:

$$\begin{vmatrix} 2s - 1 & s + 13 \\ 2s - 3 & -s + 7 \end{vmatrix} = -4s^2 - 8s + 32 = 0. \qquad (1\text{--}54)$$

Equation (1–54) is the *characteristic equation* associated with equations (1–52). We find the *characteristic roots* to be $s_1 = 2$, $s_2 = -4$. When $s = s_1 = 2$, equations (1–53) become

$$3\alpha + 15\beta = 0, \qquad \alpha + 5\beta = 0;$$

one solution is $\alpha = -5$, $\beta = 1$, so that

$$x = -5e^{2t}, \qquad y = e^{2t}$$

is a solution of (1–52). Similarly, when $s = s_2 = -4$, we obtain the solution

$$x = e^{-4t}, \qquad y = e^{-4t}.$$

An arbitrary linear combination of the two solutions found is also a solution of equations (1–52) and indeed provides the *general solution*

$$x = -5c_1 e^{2t} + c_2 e^{-4t}, \qquad y = c_1 e^{2t} + c_2 e^{-4t}. \qquad (1\text{--}55)$$

The general solution for Example 1 is formed of the "complementary-function pair" (1–55) plus a particular solution. When the right-hand member is formed of exponential functions, as in Example 1, a method of *undetermined coefficients* can be used. We seek a solution of form $x = k_1 e^{3t}$, $y = k_2 e^{3t}$. Substitution in the equations (i) and (ii) of Example 1 leads to the equations

$$5k_1 + 16k_2 = 14, \qquad 3k_1 + 4k_2 = 0,$$

which have the solution $k_1 = -2$, $k_2 = \frac{3}{2}$. We thus obtain the particular solution $x = -2e^{3t}$, $y = \frac{3}{2}e^{3t}$. Adding this to the pair (1–55), we again obtain the general solution (1–50), (1–50′).

For further discussion of the method of exponential substitution, see Problem 4 below, Section 2–9, and Chapter 6 of Reference 4 listed at the end of the chapter.

PROBLEMS

1. Determine whether each of the following sets of equations is nondegenerate and, if it is, replace it by a set in basic form.

(a) $(D - 1)x + (2D - 3)y = 0$
    $(2D + 2)x + (3D - 1)y = e^t$
(b) $(D + 1)x + Dy - 2Dz = 0$
    $2Dx + (D - 1)y + (3D - 2)z = t$
    $(4D - 1)x + 3Dy - (D + 1)z = 0$
(c) $(D^2 + D - 1)x + (D + 2)y = t$
    $(D^2 - D + 1)x + (2D - 1)y = e^t$

2. Find the general solution of each of the following:

(a) $(D - 3)x + (D + 5)y = 0$
    $(D - 5)x - (D - 5)y = 0$
(b) $(2D - 3)x + (D + 1)y = 0$
    $(D - 5)x + (3D + 3)y = 0$
(c) $3Dx + y - 2z = 0$
    $3Dy - 4y + 5z = 0$
    $3Dz - 5y + 4z = 0$
(d) $(D - 4)x + (D - 2)y + (D + 2)z = 0$
    $(D - 2)x - (D + 4)y + (D + 4)z = 0$
    $(D - 3)x + (2D + 1)y - (D + 1)z = 0$
(e) $6Dx - 17x + y = e^t$
    $6Dy + 5x - 13y = 0$
(f) $Dx = x + y + 2z + 1 + 2t$
    $Dy = -y - 2z$
    $Dz = y + z$

3. Find the solution of the following equations satisfying the initial conditions given:

(a) Equations of Problem 2(a), $x = 1$ and $y = 3$ for $t = 0$
(b) Equations of Problem 2(f), $x = 0$, $y = 0$, $z = 0$ for $t = 0$

4. Show that $x = pe^{st}$, $y = qe^{st}$, $z = re^{st}$ (p, q, r, s constant) is a solution of the equations

$$Dx = a_1x + b_1y + c_1z,$$
$$Dy = a_2x + b_2y + c_2z, \qquad (a)$$
$$Dz = a_3x + b_3y + c_3z,$$

(a_1, b_1, ... constant) provided s is a root of the *characteristic equation*

$$\begin{vmatrix} a_1 - s & b_1 & c_1 \\ a_2 & b_2 - s & c_2 \\ a_3 & b_3 & c_3 - s \end{vmatrix} = 0 \qquad \text{(b)}$$

and p, q, r are not all zero and satisfy the equations

$$\begin{aligned}
(a_1 - s)p + b_1 q + c_1 r &= 0, \\
a_2 p + (b_2 - s)q + c_2 r &= 0, \\
a_3 p + b_3 q + (c_3 - s)r &= 0.
\end{aligned} \qquad \text{(c)}$$

Show that, if Eq. (b) has distinct real roots s_1, s_2, s_3 and if p_k, q_k, r_k are chosen to be not all zero and to satisfy (c) with $s = s_k$ ($k = 1, 2, 3$), then

$$\begin{aligned}
x &= C_1 p_1 e^{s_1 t} + C_2 p_2 e^{s_2 t} + C_3 p_3 e^{s_3 t}, \\
y &= C_1 q_1 e^{s_1 t} + C_2 q_2 e^{s_2 t} + C_3 q_3 e^{s_3 t}, \\
z &= C_1 r_1 e^{s_1 t} + C_2 r_2 e^{s_2 t} + C_3 r_3 e^{s_3 t},
\end{aligned}$$

with C_1, C_2, C_3 arbitrary constants, is the general solution of equations (a). [*Hint:* The fact that the equations do provide solutions is verified by substitution. To show that they provide the general solution, one must show that C_1, C_2, C_3 can be chosen to satisfy arbitrary initial conditions at $t = t_0$. Show that these conditions lead to simultaneous linear equations which can be solved unless a certain determinant $d = 0$; show that $d = 0$ implies that C_1, C_2, C_3 can be chosen, not all zero, so that $x(t_0) = 0$, $y(t_0) = 0$, $z(t_0) = 0$. Apply Theorem 4 to conclude that $x(t) \equiv 0$, $y(t) \equiv 0$, $z(t) \equiv 0$. Show that this is impossible if s_1, s_2, s_3 are distinct.]

ANSWERS

1. (a) $Dx = -7x - 7y + 2e^t$, $Dy = 4x + 5y. - e^t$
(b) degenerate
(c) $Dx = u$, $Dy = -2x + 3y + 2u + e^t - t$
     $Du = 3x - 5y - 3u + 2t - e^t$

2. (a) $x = c_1 e^{5t} + c_2 e^{-t}$, $y = -\frac{1}{5}c_1 e^{5t} + c_2 e^{-t}$
(b) $x = c_1 e^{4t/5}$, $y = \frac{7}{9}c_1 e^{4t/5} + c_2 e^{-t}$
(c) $x = c_1 + c_2 \cos t + c_3 \sin t$
     $y = (c_3 - 2c_2) \cos t - (2c_3 + c_2) \sin t$
     $z = (2c_3 - c_2) \cos t - (c_3 + 2c_2) \sin t$
(d) $x = c_1 e^{2t} + c_2 e^{-2t}$
     $y = c_3 + \frac{1}{2}c_1 e^{2t} - \frac{3}{2}c_2 e^{-2t}$
     $z = c_3 + \frac{1}{2}c_1 e^{2t} + \frac{1}{2}c_2 e^{-2t}$

(e) $x = c_1 e^{2t} + c_2 e^{3t} - \frac{7}{72} e^t$
$\quad y = 5c_1 e^{2t} - c_2 e^{3t} - \frac{5}{72} e^t$

(f) $x = c_1 e^t + c_2 \cos t + c_3 \sin t - 3 - 2t$
$\quad y = (c_2 - c_3) \sin t - (c_2 + c_3) \cos t$
$\quad z = c_3 \cos t - c_2 \sin t$

3. (a) $x = (8e^{-t} - 5e^{5t})/3,\ y = (8e^{-t} + e^{5t})/3$
(b) $x = 3e^t - 3 - 2t,\ y = 0,\ z = 0$

1-6 Integrodifferential equations. An equation such as the following,

$$\frac{dx}{dt} + 3x + 2\int x\, dt = f(t), \qquad (1\text{-}60)$$

is called a *linear integrodifferential equation.* Equation (1–60) is equivalent to a set of two differential equations:

$$\frac{dx}{dt} + 3x + 2y = f(t), \qquad \frac{dy}{dt} = x. \qquad (1\text{-}61)$$

Thus y is simply $\int x\, dt$. We can eliminate y between the two equations (1–61) by differentiating the first equation and replacing dy/dt by x:

$$\frac{d^2 x}{dt^2} + 3\frac{dx}{dt} + 2x = f'(t). \qquad (1\text{-}62)$$

This equation could also have been obtained directly from Eq. (1–60) by differentiation and applying the rule

$$\frac{d}{dt}\int x(t)\, dt = x(t).$$

It should be remarked that the general solution of either the pair of equations (1–61) or the single equation (1–62) depends on *two arbitrary constants.* Thus Eq. (1–60) should be considered as a *second order* equation. On the basis of equations (1–61) we would use the initial values of x and $y = \int x\, dt$ to determine the solution. We can regard y as an *antiderivative* of x and denote y_0 by $x_0^{(-1)}$; then x_0 and $x_0^{(-1)}$ are the initial values. On the basis of Eq. (1–62), x_0 and x_0' would be used as initial values. The first of equations (1–61), or the given equation (1–60), shows that x_0', x_0 and $y_0 = x_0^{(-1)}$ are related:

$$x_0' + 3x_0 + 2y_0 = f(t_0).$$

As an example, we seek the solution of Eq. (1–60) such that $x_0 = 1$, $x_0^{(-1)} = 2$ with $t_0 = 0$, $f(t) = \cos t$. Then $x_0 = 1, y_0 = 2$. By the method of Section 1–5 we find the general solution of equations (1–61) to be

$$x = c_1 e^{-t} + c_2 e^{-2t} + (3 \cos t - \sin t)/10,$$
$$y = -c_1 e^{-t} - \tfrac{1}{2} c_2 e^{-2t} + (\cos t + 3 \sin t)/10.$$

Imposing the initial conditions, we find

$$1 = c_1 + c_2 + \tfrac{3}{10}, \qquad 2 = -c_1 - \tfrac{1}{2} c_2 + \tfrac{1}{10},$$

from which we find $c_1 = -4.5$, $c_2 = 5.2$.

Remark. We could also have solved equations (1–61) by replacing x by dy/dt in the first equation, thereby eliminating x:

$$D^2 y + 3Dy + 2y = f(t).$$

After y has been found, x is obtained as dy/dt. This procedure has the advantage of avoiding differentiation of $f(t)$ and may be essential if $f'(t)$ does not exist.

Operator notation. We write

$$D^{-1} x = \int x \, dt, \quad D^{-2} x = \int \left(\int x \, dt \right) dt, \ \ldots,$$

and in general

$$D^{-k} x = \int\!\!\int \ldots \int x \, dt \ldots dt,$$

where k integrations are performed. Accordingly, $y = D^{-k}x$ is a function of t such that

$$D^k y = x.$$

Therefore, y is determined uniquely when we give the initial values y_0, y_0', ..., $y_0^{(k-1)}$. If y is a choice of $D^{-k}x$, then Dy is a choice of $D^{-(k-1)}x$, $D^2 y$ is a choice of $D^{-(k-2)}x$, ..., and $D^{k-1}y$ is a choice of $D^{-1}x$.

A general linear integrodifferential equation has the form

$$a_0 D^n x + \cdots + a_n x + a_{n+1} D^{-1} x + \cdots + a_{n+k} D^{-k} x = f(t), \quad (1\text{–}63)$$

where $a_0 \neq 0$, $n > 0$. We can introduce integrodifferential operators and write Eq. (1–63) as follows:

$$(a_0 D^n + \cdots + a_n + a_{n+1} D^{-1} + \cdots + a_{n+k} D^{-k}) x = f(t). \quad (1\text{–}64)$$

Equations (1–63) or (1–64) are each equivalent to the simultaneous equations

$$a_0 D^n x + \cdots + a_n x + a_{n+1} D^{k-1} y + \cdots + a_{n+k} y = f(t),$$
$$D^k y = x; \qquad\qquad\qquad (1\text{–}65)$$

accordingly, a solution is uniquely determined by giving initial values

$$x_0, \quad x_0', \quad \ldots, \quad x_0^{(n-1)}, \quad y_0 = x_0^{(-k)}, \quad y_0' = x_0^{(-k+1)}, \quad \ldots,$$
$$y_0^{(k-1)} = x_0^{(-1)} \tag{1-66}$$

for $t = t_0$. We can eliminate x between equations (1–65) to obtain an equation of order $n + k$ for y; when y is known, x is obtained as $D^k y$.

For many applications t_0 is fixed as zero and the initial values $x_0^{(-1)}$, $x_0^{(-2)}$, ... are all zero. In this case, we can write

$$D^{-1}x = \int_0^t x(t) \, dt = D_0^{-1}x,$$
$$D^{-2}x = \int_0^t D_0^{-1}x \, dt = D_0^{-2}x,$$
$$\vdots$$

For example, if $x = e^t$, then

$$D_0^{-1}x = \int_0^t e^t \, dt = e^t - 1,$$
$$D_0^{-2}x = \int_0^t (e^t - 1) \, dt = e^t - 1 - t,$$
$$\vdots$$

If $x_0^{(-1)}$, $x_0^{(-2)}$, ... are not all zero, we can write

$$D^{-1}x = \int_0^t x(t) \, dt + x_0^{(-1)} = D_0^{-1}x + x_0^{(-1)},$$
$$D^{-2}x = \int_0^t [D_0^{-1}x + x_0^{(-1)}] \, dt + x_0^{(-2)}$$
$$= D_0^{-2}x + x_0^{(-1)}t + x_0^{(-2)},$$
$$\vdots$$
$$D^{-k}x = D_0^{-k}x + x_0^{(-1)} \frac{t^{k-1}}{(k-1)!}$$
$$+ x_0^{(-2)} \frac{t^{k-2}}{(k-2)!} + \cdots + x_0^{(-k+1)}t + x_0^{(-k)}.$$

EXAMPLE. Find the solution of the equation

$$(D + 1 + D_0^{-1} + D_0^{-2})x = 1$$

such that $x = 1$ when $t = 0$. Only one initial value is required since the notations D_0^{-1}, D_0^{-2} indicate that $x_0^{(-1)}$ and $x_0^{(-2)}$ are zero. An equivalent system is

$$(D + 1)x + (D + 1)y = 1, \qquad D^2 y = x,$$

with $y = 0$, $y' = 0$ for $t = 0$, so that $y = D_0^{-2}x$, $Dy = D_0^{-1}x$. We find

$$x = c_1 e^{-t} - c_2 \cos t - c_3 \sin t,$$
$$y = c_1 e^{-t} + c_2 \cos t + c_3 \sin t + 1.$$

The initial conditions give

$$1 = c_1 - c_2, \qquad 0 = c_1 + c_2 + 1, \qquad 0 = -c_1 + c_3,$$

so that

$$c_1 = 0, \qquad c_2 = -1, \qquad c_3 = 0 \quad \text{and} \quad x = \cos t, \qquad y = 1 - \cos t.$$

Simultaneous integrodifferential equations. The concepts described can be extended at once to sets of simultaneous equations. The general set of two simultaneous linear integrodifferential equations has the form

$$\phi_1(D)x + \phi_2(D)y = f_1(t),$$
$$\Phi_1(D)x + \Phi_2(D)y = f_2(t),$$
(1-67)

where ϕ_1, ϕ_2, Φ_1, Φ_2 are general integrodifferential operators. For example, the equations

$$(2D + 1 + D_0^{-1})x + (D - D_0^{-1})y = \cos 2t,$$
$$(D - D_0^{-1})x + (D + D_0^{-1})y = \sin 2t$$

are of this form. We obtain an equivalent set of simultaneous *differential* equations by setting $u = D_0^{-1}x$, $v = D_0^{-1}y$:

$$(2D + 1)x + Dy + u - v = \cos 2t,$$
$$Dx + Dy - u + v = \sin 2t,$$
$$Du = x, \qquad Dv = y,$$

and $u = 0$, $v = 0$ for $t = 0$.

PROBLEMS

1. Find the solution of each of the following with given initial values for $t = 0$:

(a) $(D + 6 + 8D^{-1})x = 0$, $x_0 = 8$, $x_0^{(-1)} = -1$
(b) $(D + 6 + 8D^{-1})x = 12e^{2t}$, $x_0 = 0$, $x_0^{(-1)} = 0$
(c) $(D + 1 - D^{-1} - D^{-2})x = 0$, $x_0 = 1$, $x_0^{(-1)} = 1$, $x_0^{(-2)} = -1$
(d) $(D + 1 - D_0^{-1} - D_0^{-2})x = t^2$, $x_0 = 0$
(e) $(D^2 + D - 1 - D_0^{-1})x = 2t$, $x_0 = 0$, $x_0' = -2$

2. (a) Show that an integrodifferential equation containing a term in x but no terms in derivatives:

$$a_0 x + a_1 D_0^{-1} x + \cdots + a_k D_0^{-k} x = f(t)$$

has a unique solution, provided the coefficients and $f(t)$ are continuous and $a_0 \neq 0$.

(b) Show that an integrodifferential equation having only terms in negative powers of D:

$$(a_0 D_0^{-k} + a_1 D_0^{-k-1} + \cdots + a_m D_0^{-k-m})x = f(t) \quad (k \geqq 1),$$

with constant coefficients and right-hand member continuous for all t and with $a_0 \neq 0$, has a continuous solution for all t if and only if $f(t)$ has continuous derivatives through the kth order and $f(0) = 0, f'(0) = 0, \ldots, f^{(k-1)}(0) = 0$.

3. Find the solution of the following [see Problem 2(a)]:

(a) $(1 + 3D_0^{-1} + 2D_0^{-2})x = e^{-t}$
(b) $(1 + D_0^{-1} - D_0^{-2} - D_0^{-3})x = t^2$ [Hint: See Problem 1(d)]

4. Find the solution, if there is one, of the following [see Problem 2(b)]:

(a) $(D_0^{-1} + 2D_0^{-2})x = 1 + t$
(b) $(D_0^{-2} + 2D_0^{-3})x = 1 - \cos t$

5. Prove each of the following rules:

(a) $D_0^{-2}[D^2 x] = x - x_0 - x_0' t,$
(b) $D^2[D_0^{-2} x] = x,$
(c) $D_0^{-3}[D^2 x] = D_0^{-1} x - x_0 t - \frac{1}{2} x_0' t^2,$
(d) $D^2[D_0^{-3} x] = D_0^{-1} x.$

6. Prove each of the following rules:

(a) $D_0^{-k}[D^m x] = D^{m-k} x - x_0^{(m-k)} - x_0^{(m-k+1)} t - \cdots$

$$- x_0^{(m-1)} \frac{t^{k-1}}{(k-1)!} \quad (0 < k \leqq m)$$

(b) $D_0^{-k}[D^m x] = D_0^{m-k} x - x_0 \frac{t^{k-m}}{(k-m)!} - x_0' \frac{t^{k-m+1}}{(k-m+1)!} - \cdots$

$$- x_0^{(m-1)} \frac{t^{k-1}}{(k-1)!} \quad (0 < m < k)$$

(c) $D^m[D_0^{-k} x] = D^{m-k} x \quad (0 < k \leqq m)$
(d) $D^m[D_0^{-k} x] = D_0^{m-k} x \quad (0 < m < k)$

7. Find the solution of the following satisfying the given initial conditions:

(a) $(D + 1)x - 2y = 0, D_0^{-1} x + (1 - D_0^{-1})y = 0, x_0 = 2$
(b) $(3D + 5)x + (2D - 2)y = 1, (D + 1)x + (D - D_0^{-1})y = 0,$
     $x_0 = 0, y_0 = 1$

(c) $(2D + 1 - 3D_0^{-1})x + (D + 5 - 6D_0^{-1})y = 0$
$(D - D_0^{-1})x + (D + 1 - 2D_0^{-1})y = 0,\ x_0 = 1,\ y_0 = 2$

8. Find the solution of the following:

$$(2 + 3D_0^{-1})x + (1 + 6D_0^{-1})y = t^3,$$

$$(1 + D_0^{-1})x + (1 + 2D_0^{-1})y = 0$$

Answers

1. (a) $x = 12e^{-4t} - 4e^{-2t},\ y = D^{-1}x = -3e^{-4t} + 2e^{-2t}$
   (b) $x = 3e^{-2t} + e^{2t} - 4e^{-4t},\ y = D^{-1}x = e^{-4t} + \frac{1}{2}(e^{2t} - 3e^{-2t})$
   (c) $x = \frac{1}{2}(e^t + e^{-t} - 2te^{-t}),\ y = D^{-2}x = \frac{1}{2}(e^t - 3e^{-t} - 2te^{-t})$
   (d) $x = \frac{1}{2}(e^t + 3e^{-t} + 2te^{-t} - 4)$
   $y = D_0^{-2}x = \frac{1}{2}(7e^{-t} + 2te^{-t} + e^t - 2t^2 + 4t - 8)$
   (e) $x = 2e^{-t} - 2,\ y = D_0^{-1}x = 2 - 2t - 2e^{-t}$

3. (a) $4e^{-2t} - 3e^{-t} + te^{-t}$ (b) $\frac{1}{2}(e^t - e^{-t} - 2te^{-t})$

4. (a) no solution   (b) $(\cos t - 2 \sin t + 4e^{-2t})/5$

7. (a) $x = 2 \cos t - 2 \sin t,\ y = -2 \sin t$
   (b) $x = (2 + 5e^{-t} - 7e^{-5t})/10,\ y = (3e^{-t} + 7e^{-5t} + 2e^t)/12$
   (c) $x = 10 - 9e^t,\ y = -5 + 7e^t$

8. $x = (t^4 + 2t^3)/2,\ y = -(t^4 + 4t^3)/4$

1–7 Complex functions of a real variable. It will be seen that functions of a real variable taking on complex values are of importance both for linear differential equations and for the operational theory. The function $e^{i\omega t}$, where ω is a real constant, is an example, and is the function most frequently encountered. We note that

$$e^{i\omega t} = \cos \omega t + i \sin \omega t,$$

by the familiar identity. A general complex-valued function of a real variable is a function

$$F(t) = f(t) + ig(t) \quad (\alpha < t < \beta),$$

where f and g are real-valued functions of t for $\alpha < t < \beta$.

Familiarity with the basic algebraic properties of complex numbers will be assumed. We shall denote complex numbers by $z = x + iy$, $w = u + iv$ (and in later chapters by $s = \sigma + i\omega$). Complex numbers can be represented geometrically in a complex plane, as in Fig. 1–1. The polar coordinates of $z = x + iy$ are $r = |z| =$ *absolute value* of z and $\theta = \arg z =$ *argument* of z (determined up to multiples of 2π). The *conjugate* of z is

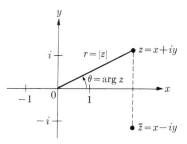

FIG. 1-1. Complex plane.

$\bar{z} = x - iy$. The quantities obey the following rules:

$$z = r(\cos\theta + i\sin\theta) = x + iy,$$

$$z\bar{z} = |z|^2 = x^2 + y^2,$$

$$z + \bar{z} = 2x = 2\,\mathrm{Re}\,z, \qquad z - \bar{z} = 2iy = 2i\,\mathrm{Im}\,z,$$

$$\overline{z_1 + z_2} = \bar{z}_1 + \bar{z}_2, \qquad \overline{z_1 \cdot z_2} = \bar{z}_1 \cdot \bar{z}_2, \qquad |z_1 \cdot z_2| = |z_1| \cdot |z_2|,$$

$$\arg(z_1 \cdot z_2) = \arg z_1 + \arg z_2 \quad \text{(up to multiples of } 2\pi\text{)},$$

$$|z_1 + z_2| \leq |z_1| + |z_2| \quad \text{(triangle inequality)},$$

$$|z_1 - z_2| = \text{distance from } z_1 \text{ to } z_2,$$

$$z^n = r^n(\cos n\theta + i\sin n\theta) \quad (n = 0, \pm 1, \pm 2, \ldots).$$

(1-70)

Complex-valued functions of z. If to each value of the complex number $z = x + iy$, with certain exceptions, there is assigned a value of the complex number $w = u + iv$, then w is given as a *complex-valued function* of z and we write $w = f(z)$. For example,

$$w = z^2, \qquad w = z^3 + 5z + 7, \qquad w = \frac{z+1}{z-2} \quad (z \neq 2)$$

are such functions. Important functions of this type are

polynomials: $w = a_0 z^n + \cdots + a_{n-1} z + a_n,$

rational functions: $w = \dfrac{a_0 z^n + \cdots + a_n}{b_0 z^m + \cdots + b_m},$

exponential function: $\exp z = e^z = e^{x+iy} = e^x(\cos y + i\sin y),$

trigonometric functions: $\sin z = \dfrac{e^{iz} - e^{-iz}}{2i}, \qquad \cos z = \dfrac{e^{iz} + e^{-iz}}{2},$

hyperbolic functions: $\sinh z = \dfrac{e^z - e^{-z}}{2}, \qquad \cosh z = \dfrac{e^z + e^{-z}}{2},$

nth root function: $\quad z^{1/n} = r^{1/n} \left[\cos\left(\dfrac{\theta}{n} + \dfrac{2k\pi}{n} \right) + i \sin\left(\dfrac{\theta}{n} + \dfrac{2k\pi}{n} \right) \right],$

$$(k = 0, 1, \ldots, n - 1, \quad n \text{ a positive integer}).$$

The last function is multiple-valued: it assigns n values to each z (except zero). The properties of these functions are similar to those of the analogous functions of a real variable (Problem 2 of Section 1–8); we note that in each case, the function reduces to the corresponding real function when z is real, that is, when $z = x$. For example,

$$\sin x = \frac{e^{ix} - e^{-ix}}{2i}, \qquad \cos x = \frac{e^{ix} + e^{-ix}}{2},$$

as follows from the identity: $e^{ix} = \cos x + i \sin x$.

Complex-valued functions of a real variable. We now restrict attention to functions of z for which z is *restricted to real values.* The values of the function will in general be complex, so that we are dealing with complex-valued functions of a real variable. To emphasize that z is to be real, we replace z by t in our notations and consistently require that t is to be a real number, $-\infty < t < \infty$. Typical functions of the class considered are the following:

$$w = 1 + it, \qquad w = e^{it}, \qquad w = \frac{1}{t + i}.$$

If we write $w = u + iv = F(t)$, then u and v are real-valued functions of t:

$$u = f(t), \qquad v = g(t). \tag{1–71}$$

Each function is therefore equivalent to a pair of real functions of t. To graph the complex function $w = F(t)$, we can graph the corresponding real functions (1–71). This is most conveniently done by regarding t as a parameter and graphing in the uv-plane; the result is a curve, on which the t-values can be marked.

EXAMPLE 1. $w = e^{it} = \cos t + i \sin t$. Here $u = \cos t$, $v = \sin t$, so that the graph is a circle in the uv-plane and t gives the polar angle of a point moving on the circle.

EXAMPLE 2. $w = t + it^2$. Here $u = t$, $v = t^2$, so that the graph is a parabola, as in Fig. 1–2.

The calculus can be developed for complex-valued functions of t in strict analogy with the development for real-valued functions. We write

$$\lim_{t \to t_0} F(t) = c = a + ib \tag{1–72}$$

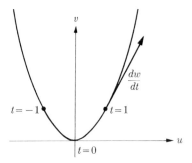

FIG. 1–2.   The function $w = t + it^2$.

if, given $\epsilon > 0$, we can choose $\delta > 0$ so that $|F(t) - c| < \epsilon$ when $0 < |t - t_0| < \delta$. It is assumed here that $F(t)$ is defined for t sufficiently close to t_0, but not necessarily for $t = t_0$; if $F(t)$ is defined only in an interval $t_0 < t < \beta$, the limit is interpreted as a limit *to the right* and is written

$$\lim_{t \to t_0+} F(t).$$

Limits to the left are defined similarly. Limits as $t \to \infty$ or $t \to -\infty$ are defined as for real functions. [However,

$$\lim_{t \to t_0} F(t) = \infty$$

is defined to mean

$$\lim_{t \to t_0} |F(t)| = \infty;$$

there is no concept of $+\infty$ or $-\infty$ for complex numbers; see Section 3–12.] If $F(t)$ is defined for $\alpha < t < \beta$ and t_0 lies in this interval, then $F(t)$ is said to be *continuous* at t_0 if

$$\lim_{t \to t_0} F(t) = F(t_0).$$

If $F(t)$ is also defined at $t = \alpha$ and

$$\lim_{t \to \alpha+} F(t) = F(\alpha),$$

then $F(t)$ is said to be continuous to the right at $t = \alpha$. Continuity to the left and continuity in an interval are defined as for real functions.

If $F(t) = f(t) + ig(t)$ then Eq. (1–72) is equivalent to the two equations

$$\lim_{t \to t_0} f(t) = a, \qquad \lim_{t \to t_0} g(t) = b. \tag{1–73}$$

For Eq. (1–72) signifies that $F(t)$ is as close to c as desired, for t sufficiently close to t_0; by a geometric argument we see that this is equivalent to the requirement that $f = \text{Re } F$ be as close to a as desired and $g = \text{Im } F$ as close to b as desired, for t sufficiently close to t_0. Similarly, continuity of $F(t)$ at t_0 is equivalent to continuity of $f(t)$ and $g(t)$ at t_0. Accordingly, such functions as $t^2 + it^3$ and e^{it} are continuous for all t. Furthermore, the rules for limits of sums, products and quotients, and the analogous continuity theorems must carry over to the complex case. For example, if $F_1(t) = f_1(t) + ig_1(t)$ and $F_2(t) = f_2(t) + ig_2(t)$ are continuous in an interval, then so also is

$$F_1(t) \cdot F_2(t) = [f_1(t) + ig_1(t)] \cdot [f_2(t) + ig_2(t)]$$
$$= f_1(t)f_2(t) - g_1(t)g_2(t) + i[f_1(t)g_2(t) + f_2(t)g_1(t)];$$

for $f_1(t)$, $g_1(t)$, $f_2(t)$, $g_2(t)$ must be continuous, so that the real and imaginary parts of $F_1(t) \cdot F_2(t)$ are continuous, and hence $F_1(t) \cdot F_2(t)$ is continuous.

1–8 Differentiation and integration of complex-valued functions of t. *Derivatives* can be defined as for real functions:

$$F'(t_0) = \lim_{\Delta t \to 0} \frac{F(t_0 + \Delta t) - F(t_0)}{\Delta t}. \tag{1–80}$$

By taking real and imaginary parts and applying (1–73), we conclude that

$$F'(t_0) = f'(t_0) + ig'(t_0); \tag{1–81}$$

that is, $F(t)$ has a derivative at t_0 precisely when $f(t)$, $g(t)$ have derivatives at t_0, and the derivatives are related by Eq. (1–81). Derivatives to the left or right are defined by requiring that $\Delta t < 0$ or $\Delta t > 0$, respectively, in Eq. (1–80); Eq. (1–81) also applies to these derivatives.

The rules for derivative of sum, product, quotient, constant, and constant times function all carry over, and the proofs for real functions can be repeated. We also note the rules

$$\frac{d}{dt}[F(t)]^n = n[F(t)]^{n-1}F'(t) \quad (n = 1, 2, \ldots), \tag{1–82}$$

$$\frac{d}{dt}e^{(a+bi)t} = (a + bi)e^{(a+bi)t} \quad (a, b \text{ real}). \tag{1–83}$$

Furthermore, if $F'(t) \equiv 0$ for $\alpha < t < \beta$, then $F(t)$ is identically constant for $\alpha < t < \beta$. The proofs are left as exercises (Problems 12 and 13 below).

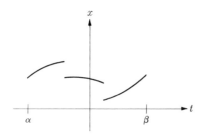

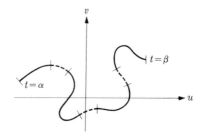

FIG. 1–3. Piecewise continuous real
function $f(t)$.

FIG. 1–4. Piecewise continuous
complex function $F(t)$.

Higher derivatives are obtained by repeated differentiation:

$$F''(t) = [F'(t)]' = D^2 F, \ F'''(t) = [F''(t)]' = D^3 F, \ldots$$

The first derivative of $F(t)$ can be thought of as the *velocity vector* of
the point (u, v) as it moves on the path $u = f(t)$, $v = g(t)$, with t as *time*.
In Example 2 of Section 1–7, $F'(t) = 1 + 2it$; the velocity vector $1 + 2i$
at $t = 1$ is shown in Fig. 1–2. The second derivative can be interpreted
as *acceleration*.

The *definite integral* of $F(t)$ over an interval $\alpha \leqq t \leqq \beta$ is defined as a
limit of a sum $\sum F(t_k^*) \, \Delta_k t$ as for real functions. However, again the limit
theorem permits us to take real and imaginary parts:

$$\int_\alpha^\beta F(t) \, dt = \int_\alpha^\beta f(t) \, dt + i \int_\alpha^\beta g(t) \, dt. \tag{1–84}$$

If $F(t)$ is continuous over the interval, then $f(t)$ and $g(t)$ are continuous
so that the integral exists. More generally, the integral exists if $f(t)$ and
$g(t)$ are *piecewise continuous* for $\alpha \leqq t \leqq \beta$; that is, they are continuous
except for a finite number of *jump discontinuities*, as suggested in Fig. 1–3.
At each jump discontinuity t_0, $f(t)$ is required to have limits from the left
and right. When f and g are piecewise continuous, we term $F = f + ig$
piecewise continuous; the graph of $F(t)$ in the uv-plane is then broken
into several pieces, as in Fig. 1–4.

The integral of $F(t)$ can also be given a meaning when $f(t)$ and $g(t)$ have
more complicated discontinuities, so long as the integrals on the right
of Eq. (1–78) have meaning, perhaps as improper integrals. In the same
spirit, we can allow integrals with infinite limits of integration; for example,

$$\int_0^\infty F(t) \, dt = \int_0^\infty f(t) \, dt + i \int_0^\infty g(t) \, dt,$$

provided both integrals on the right have meaning. Equivalently, we can define

$$\int_0^\infty F(t)\, dt = \lim_{b \to \infty} \int_0^b F(t)\, dt = \lim_{b \to \infty} \int_0^b f(t)\, dt + i \lim_{b \to \infty} \int_0^b g(t)\, dt.$$

Integration from $-\infty$ to ∞ is defined as

$$\int_{-\infty}^\infty F(t)\, dt = \int_{-\infty}^0 F(t)\, dt + \int_0^\infty F(t)\, dt; \qquad (1\text{–}85)$$

both terms on the right are required to have finite values. When this fails, it may happen that

$$\lim_{b \to \infty} \int_{-b}^b F(t)\, dt$$

exists. If it does, and the limit equals c, we write

$$c = (P)\int_{-\infty}^\infty F(t)\, dt = \lim_{b \to \infty} \int_{-b}^b F(t)\, dt, \qquad (1\text{–}86)$$

where c is the *principal value* of the integral (Cauchy principal value).

An *indefinite integral* of $F(t)$ is defined as a function $G(t)$ whose derivative is $F(t)$. As in ordinary calculus, we find that, if $G(t)$ is one indefinite integral, then $G(t) + c$ provides all indefinite integrals:

$$\int F(t)\, dt = G(t) + c,$$

c being an arbitrary complex constant. If an indefinite integral G of F is known, then it can be used to evaluate definite integrals of F as in calculus:

$$\int_\alpha^\beta F(t)\, dt = \int_\alpha^\beta G'(t)\, dt = G(\beta) - G(\alpha). \qquad (1\text{–}87)$$

The proof is left as an exercise (Problem 15 below).

From Eq. (1–84) we can verify the familiar rules for the integral of a sum, the integral of constant times function, the combination of integrals from α to β and from β to γ, and integration by parts. We also have the basic inequality

$$\left| \int_\alpha^\beta F(t)\, dt \right| \leq \int_\alpha^\beta |F(t)|\, dt \leq M(\beta - \alpha); \qquad (1\text{–}88)$$

this is valid if $\alpha < \beta$, if $F(t)$ is, for example, piecewise continuous for $\alpha \leq t \leq \beta$, and if $|F(t)| \leq M$ on this interval. The inequality is most easily obtained from the definition of the integral as limit of a sum; for we have, by repeated application of the triangle inequality [see Eq. (1–70) above],

$$\left| \sum_{k=1}^{n} F(t_k^*) \, \Delta_k t \right| \leqq \sum_{k=1}^{n} |F(t_k^*)| \, \Delta_k t \leqq M(\beta - \alpha),$$

and passage to the limit gives (1–88). Definite integrals and indefinite integrals are related by the rule

$$\frac{d}{dt} \int_{\alpha}^{t} F(u) \, du = F(t) \tag{1-89}$$

(see Problem 15 below).

EXAMPLE 1. $\displaystyle\int_{1}^{2} (t + it^2) \, dt = \left(\frac{t^2}{2} + i \frac{t^3}{3} \right) \Big|_{1}^{2} = \frac{3}{2} + \frac{7}{3} i.$

EXAMPLE 2. $\displaystyle\int_{0}^{1} e^{(a+bi)t} \, dt = \frac{e^{(a+bi)t}}{a + bi} \Big|_{0}^{1} = \frac{e^{a+bi} - 1}{a + bi} \quad (a + bi \neq 0).$

EXAMPLE 3. $\displaystyle\int_{0}^{\infty} e^{(-1+i)t} \, dt = \lim_{b \to \infty} \int_{0}^{b} e^{(-1+i)t} \, dt$

$$= \lim_{b \to \infty} \frac{e^{(-1+i)t}}{-1 + i} \Big|_{0}^{b} = \frac{-1}{-1 + i} = \frac{1 + i}{2}.$$

For

$$\lim_{b \to \infty} e^{(-1+i)b} = \lim_{b \to \infty} e^{-b} (\cos b + i \sin b) = 0 + i0 = 0.$$

EXAMPLE 4. $\displaystyle\int_{-\infty}^{\infty} \frac{1}{1 - it} \, dt = \int_{-\infty}^{\infty} \frac{1 + it}{1 + t^2} \, dt$

$$= \int_{-\infty}^{\infty} \frac{1}{1 + t^2} \, dt + i \int_{-\infty}^{\infty} \frac{t}{1 + t^2} \, dt.$$

Here the integral of the real part converges and

$$\int_{-\infty}^{\infty} \frac{1}{1 + t^2} \, dt = \int_{-\infty}^{0} \frac{1}{1 + t^2} \, dt + \int_{0}^{\infty} \frac{1}{1 + t^2} \, dt = \frac{\pi}{2} + \frac{\pi}{2} = \pi.$$

However, neither of the integrals

$$\int_{-\infty}^{0} \frac{t}{1 + t^2} \, dt = \frac{1}{2} \log (1 + t^2) \Big|_{-\infty}^{0}, \qquad \int_{0}^{\infty} \frac{t}{1 + t^2} \, dt = \frac{1}{2} \log (1 + t^2) \Big|_{0}^{\infty}$$

converges. But the principal value has meaning:

$$(P)\int_{-\infty}^{\infty} \frac{t}{1+t^2}\, dt = \lim_{b\to\infty} \frac{1}{2}\log(1+t^2)\Big|_{-b}^{b}$$

$$= \lim_{b\to\infty} \frac{1}{2}\log\frac{1+b^2}{1+b^2} = 0,$$

since the function $t/(1+t^2)$ is *odd*. Accordingly,

$$(P)\int_{-\infty}^{\infty} \frac{dt}{1-it} = \pi.$$

EXAMPLE 5.

$$\int p(t)e^{-at}\, dt = -e^{-at}\left[\frac{p(t)}{a} + \frac{p'(t)}{a^2} + \cdots + \frac{p^{(n)}(t)}{a^{n+1}}\right] + C,$$

where $p(t)$ is a polynomial of degree n and a is a complex constant, not 0. The equation is established by integration by parts [Problem 9(f)].

PROBLEMS

1. (a) Let $z_1 = r_1(\cos\theta_1 + i\sin\theta_1)$, $z_2 = r_2(\cos\theta_2 + i\sin\theta_2)$. Show that

$$z_1 \cdot z_2 = r_1 r_2[\cos(\theta_1 + \theta_2) + i\sin(\theta_1 + \theta_2)]$$

and deduce the rules given in Eq. (1–70) for $|z_1 \cdot z_2|$, $\arg(z_1 \cdot z_2)$.

(b) Show that $e^{i(\theta_1 + \theta_2)} = e^{i\theta_1} \cdot e^{i\theta_2}$ [see part (a)].

(c) Graph z_1, z_2, and $z_1 + z_2$ (parallelogram law), and show that $|z_1|$, $|z_2|$, $|z_1 + z_2|$ are sides of a triangle; deduce the triangle inequality in (1–70).

(d) Graph z_1, z_2, $z_2 - z_1$, and show that $|z_2 - z_1|$ is the distance from z_1 to z_2.

2. Prove the following identities:

(a) $e^{z_1} \cdot e^{z_2} = e^{z_1 + z_2}$ [see Problem 1(b)]

(b) $(e^z)^n = e^{nz}$ $(n = 0, \pm1, \pm2, \ldots)$

(c) $\sin^2 z + \cos^2 z = 1$

(d) $\sin(z_1 + z_2) = \sin z_1 \cos z_2 + \cos z_1 \sin z_2$

(e) $\operatorname{Re}(\sin z) = \sin x \cosh y$, $\operatorname{Im}(\sin z) = \cos x \sinh y$

(f) $\sin iz = i\sinh z$, $\cos iz = \cosh z$

(g) $\overline{e^z} = e^{\bar z}$, $\overline{\sin z} = \sin \bar z$, $\overline{\cos z} = \cos \bar z$

(h) $e^{ix} = 1 + (ix) + \frac{(ix)^2}{2!} + \cdots + \frac{(ix)^n}{n!} + \cdots$ (x real)

[*Hint:* Take real and imaginary parts.]

3. (a) Prove that $e^z \neq 0$ for all z.

(b) Prove that $\sin z$ and $\cos z$ are 0 only for appropriate real value of z.

4. Represent the following functions graphically:

(a) $w = (1 + t) + i(1 - t)$ (b) $w = t^4 + i(t^2 + 1)$

(c) $w = e^{3it}$ (d) $w = 2e^{(-1+2i)t}$

(e) $w = te^{(-1+2i)t}$ (f) $w = e^{-t} - ie^{it}$

5. Find the derivatives of the functions of Problem 4.

6. Graph $w = 3e^{2it}$ and indicate the first and second derivatives graphically for $t = 0$, $t = \pi/2$, $t = \pi$.

7. Integrate the functions of Problem 4 from 0 to 1.

8. Evaluate the following integrals wherever they have meaning, using the principal value where necessary:

(a) $\displaystyle\int_0^\infty e^{it}\, dt$ (b) $\displaystyle\int_0^\infty te^{(-1-i)t}\, dt$

(c) $\displaystyle\int_{-\infty}^\infty \frac{e^{it} - e^{-it}}{1 + it^2}\, dt$ (d) $\displaystyle\int_{-\infty}^\infty \frac{1}{c + t}\, dt$ (c not real)

(e) $\displaystyle\int_{-\infty}^\infty \frac{i}{(i - t)^2}\, dt$ (f) $\displaystyle\int_{-\infty}^\infty \frac{1}{t^2 + 3it - 2}\, dt$

9. Use integration by parts to evaluate each of the following:

(a) $\displaystyle\int (1 + it)^2 \sin t\, dt$ (b) $\displaystyle\int t^n e^{-at}\, dt$ ($n = 1, 2, \ldots$)

(c) $\displaystyle\int t^n \sin bt\, dt = \frac{1}{2i} \int t^n (e^{bit} - e^{-bit})\, dt$ ($n = 1, 2, \ldots$)

(d) $\displaystyle\int t^n \cos at\, dt = \operatorname{Re} \int t^n e^{iat}\, dt$ (a real, $n = 1, 2, \ldots$)

(e) $\displaystyle\int t^n \cos at \cos bt \cos ct\, dt$ ($n = 1, 2, \ldots$)

(f) $\displaystyle\int p(t)e^{-at}\, dt$, where $p(t)$ is a polynomial of degree n (Example 5 in text)

10. Assuming the truth of the theorem for real functions, namely, if $f(t)$ is continuous, $0 \le t < \infty$, and $\int_0^\infty |f(t)|\, dt$ converges, then $\int_0^\infty f(t)\, dt$ converges; prove the analogous theorem for complex functions $F(t)$. That is, show that for improper integrals, absolute convergence implies convergence.

11. Apply the theorem of Problem 10 to prove convergence of the following, but do not evaluate:

(a) $\displaystyle\int_0^\infty \frac{e^{it}}{1 + t^2}\, dt$ (b) $\displaystyle\int_0^\infty \frac{dt}{t^3 + it + 1}$

(c) $\displaystyle\int_0^\infty \frac{t}{(1 + it)^4}\, dt$ (d) $\displaystyle\int_0^\infty \frac{\log t}{(t - i)^2}\, dt$

12. Prove Eq. (1–82) by induction (repeated application of rule for differentiation of a product).

13. Prove (1–83) with the aid of (1–81).

14. Prove that if $F'(t) \equiv 0$, $\alpha < t < \beta$, then $F(t) \equiv$ constant for $\alpha < t < \beta$.

15. (a) Prove (1–89) by taking real and imaginary parts.

(b) Prove (1–87) either directly or as a consequence of (1–89).

Answers

5. (a) $1 - i$ (b) $4t^3 + 2it$ (c) $3ie^{3it}$ (d) $(-2 + 4i)e^{(-1+2i)t}$
(e) $e^{(-1+2i)t}[1 + t(-1 + 2i)]$ (f) $-e^{-t} + e^{it}$

6. $w' = 6ie^{2it}$, $w'' = -12e^{2it}$

7. (a) $(3 + i)/2$ (b) $(3 + 20i)/15$ (c) $(e^{3i} - 1)/3i$
(d) $2(e^{-1+2i} - 1)/(-1 + 2i)$ (e) $[1 + (2i - 2)e^{-1+2i}]/(-3 - 4i)$
(f) $2 - e^{-1} - e^i$

8. (a) no value (b) $-\frac{1}{2}i$ (c) 0 (d) $\pm\pi i$ (principal value), $+$ if Im $(c) < 0$, $-$ if Im $(c) > 0$ (e) 0 (f) 0

9. (a) $(t^2 - 2it - 3)\cos t + (2i - 2t)\sin t + c$

(b) $-e^{-at}\left(\dfrac{t^n}{a} + \dfrac{nt^{n-1}}{a^2} + \cdots + \dfrac{n!}{a^{n+1}}\right) + c$

(c) $\cos bt\left[-\dfrac{t^n}{b} + \dfrac{n(n-1)t^{n-2}}{b^3} - \dfrac{n(n-1)(n-2)(n-3)t^{n-4}}{b^5} + \cdots\right]$
$+ \sin bt\left[\dfrac{nt^{n-1}}{b^2} - \dfrac{n(n-1)(n-2)t^{n-3}}{b^4} + \cdots\right] + c$

(d) $\mathrm{Re}\left\{-e^{ait}\left[\dfrac{t^n}{-ai} + \dfrac{nt^{n-1}}{(-ai)^2} + \cdots + \dfrac{n!}{(-ai)^{n+1}}\right]\right\} + c$

(e) $-\dfrac{1}{8}\sum_{k=1}^{8}\left[e^{-a_k t}\left(\dfrac{t^n}{a_k} + \dfrac{nt^{n-1}}{a_k^2} + \cdots + \dfrac{n!}{a_k^{n+1}}\right)\right] + C$,

where the a_k are the 8 numbers $(\pm a \pm b \pm c)i$.

1–9 Complex solutions of linear differential equations. The theory of linear differential equations, as summarized in Sections 1–1 through 1–5, can be extended to complex functions. In the equation

$$(a_0 D^n + \cdots + a_n)x = F(t), \qquad (1\text{–}90)$$

the coefficients are allowed to be complex (constants or functions of t) and $F(t)$ is allowed to be a complex function of t; a solution is a complex function $x(t) = f(t) + ig(t)$ which satisfies the equation identically.   For example,

$$x = -\frac{1 + 2i}{5}e^{it} = \frac{1}{5}[2\sin t - \cos t - i(2\cos t + \sin t)]$$

is a solution of the equation

$$[(1 + i)D^2 + 2D + i]x = e^{it}.$$

The existence theorem of Section 1–1 applies to the complex case; the initial values $x_0, x_0', \ldots, x_0^{(n-1)}$ are now complex constants. [The theorem can be proved by taking real and imaginary parts and obtaining a set of two real equations for $u = f(t) = \mathrm{Re}\,x$, $v = g(t) = \mathrm{Im}\,x$; see Reference 4, pp. 499–501.]

The discussion of equations with constant coefficients can also be repeated as follows: when $a_0, \ldots, a_n$ are complex constants ($a_0 \neq 0$), the general solution of Eq. (1–90) has the form

$$x = x^*(t) + x_c(t), \tag{1-91}$$

where $x^*(t)$ is a particular complex solution and $x_c(t)$ is the general solution of the corresponding homogeneous equation. To obtain $x_c(t)$, we solve the characteristic equation

$$a_0 s^n + \cdots + a_n = 0.$$

If the roots $s_1, \ldots, s_n$ are distinct, then

$$x_c = C_1 e^{s_1 t} + \cdots + C_n e^{s_n t},$$

where $C_1, \ldots, C_n$ are arbitrary complex constants; when roots are repeated, the solution is modified as in the real case (Reference 4, pp. 142–144).

When the coefficients are not assumed constant, we can still write the general solution in form (1–91), but

$$x_c(t) = C_1 x_1(t) + \cdots + C_n x_n(t),$$

where $x_1(t), \ldots, x_n(t)$ are n complex solutions which are linearly independent with respect to complex coefficients; that is,

$$k_1 x_1(t) + \cdots + k_n x_n(t) \equiv 0$$

can hold, with $k_1, \ldots, k_n$ complex constants, only if $k_1 = 0, \ldots, k_n = 0$.

Equations having real form. If $a_0, \ldots, a_n$, and $F(t)$ are all real in Eq. (1–90), then we say that the equation has *real form*. An equation in real form must have some solutions which are complex and not real; for example, $x = e^{it}$ is a solution of the equation

$$(D^2 + 1)x = 0.$$

For an equation in real form we can form the *general complex solution* and the *general real solution*. These are related as follows: Let the general complex solution be

$$x = x^*(t) + C_1 x_1(t) + \cdots + C_n x_n(t), \qquad (1\text{--}92)$$

as above. Then the general real solution is given by

$$x = \operatorname{Re} [x^*(t) + C_1 x_1(t) + \cdots + C_n x_n(t)]; \qquad (1\text{--}93)$$

that is, the real solutions are simply the real parts of the complex solutions. Next, let the general real solution be given by

$$x = x^*(t) + c_1 x_1(t) + \cdots + c_n x_n(t), \qquad (1\text{--}94)$$

where $x^*(t)$, $x_1(t), \ldots, c_1, \ldots$ are real. Then the general complex solution is

$$x = x^*(t) + C_1 x_1(t) + \cdots + C_n x_n(t), \qquad (1\text{--}95)$$

where $C_1, \ldots C_n$ are arbitrary complex constants.

To justify these statements, we remark that since the coefficients and $F(t)$ are real, the real part of every complex solution is a real solution. Hence (1–93) does provide real solutions. But all real solutions are included in the general complex solution; hence (1–93) gives *all* the real solutions, that is, the general real solution. Next we reason that in the general real solution (1–94), $x^*(t)$ can be considered as a complex solution of (1–90) and $x_1(t), \ldots, x_n(t)$ can be considered as complex solutions of the corresponding homogeneous equation. Moreover, $x_1(t), \ldots, x_n(t)$ are linearly independent with respect to complex coefficients; for if

$$(k_1 + il_1)x_1(t) + \cdots + (k_n + il_n)x_n(t) \equiv 0,$$

then by taking real and imaginary parts we conclude that

$$k_1 x_1(t) + \cdots + k_n x_n(t) \equiv 0,$$
$$l_1 x_1(t) + \cdots + l_n x_n(t) \equiv 0.$$

Since $x_1(t), \ldots, x_n(t)$ are linearly independent as real functions, we conclude that $k_1 = 0, \ldots, k_n = 0, l_1 = 0, \ldots, l_n = 0$; hence $x_1(t), \ldots, x_n(t)$ are linearly independent with respect to complex coefficients. Therefore Eq. (1–95) gives the general complex solution.

EXAMPLE 1. $(D^2 + 1)x = \sin 2t$. The general complex solution is

$$x = C_1 e^{it} + C_2 e^{-it} - \tfrac{1}{3} \sin 2t.$$

Hence the general real solution is

$$\begin{aligned}
x &= \text{Re}\,[C_1 e^{it} + C_2 e^{-it} - \tfrac{1}{3} \sin 2t] \\
&= \text{Re}\,[(c_1' + ic_1'')e^{it} + (c_2' + ic_2'')e^{-it} - \tfrac{1}{3} \sin 2t] \\
&= (c_1' + c_2') \cos t + (c_2'' - c_1'') \sin t - \tfrac{1}{3} \sin 2t \\
&= c_1 \cos t + c_2 \sin t - \tfrac{1}{3} \sin 2t,
\end{aligned}$$

where $c_1, c_2, c_1', c_2', c_1'', c_2''$ are real. Starting from the last equation for the general real solution, we obtain the general complex solution as

$$x = K_1 \cos t + K_2 \sin t - \tfrac{1}{3} \sin 2t,$$

where K_1, K_2 are complex. Equivalently,

$$\begin{aligned}
x &= K_1 \frac{e^{it} + e^{-it}}{2} + K_2 \frac{e^{it} - e^{-it}}{2i} - \frac{1}{3} \sin 2t \\
&= C_1 e^{it} + C_2 e^{-it} - \tfrac{1}{3} \sin 2t,
\end{aligned}$$

where $C_1 = \tfrac{1}{2}(K_1 - iK_2)$, $C_2 = \tfrac{1}{2}(K_1 + iK_2)$.

Real coefficients, complex right-hand member. If $a_0, \ldots, a_n$ are real but $F(t)$ is complex in Eq. (1-90), then the solutions are in general complex. However, the real and imaginary parts of each solution $x(t)$ satisfy the differential equations

$$\begin{aligned}
(a_0 D^n + \cdots + a_n)\, \text{Re}\,[x(t)] &= \text{Re}\,[F(t)] = F_1(t), \\
&\qquad\qquad\qquad\qquad\qquad\qquad (1\text{-}96) \\
(a_0 D^n + \cdots + a_n)\, \text{Im}\,[x(t)] &= \text{Im}\,[F(t)] = F_2(t);
\end{aligned}$$

equations (1-96) are obtained from Eq. (1-90) by taking real and imaginary parts on both sides.

EXAMPLE 2. Find particular solutions of the equations

$$\begin{aligned}
(D^3 + 2D^2 + 3D + 1)x &= 53 \cos 2t, \\
(D^3 + 2D^2 + 3D + 1)x &= 53 \sin 2t.
\end{aligned}$$

Solution. We consider the equation

$$(D^3 + 2D^2 + 3D + 1)x = 53 e^{2it}.$$

If $x(t)$ satisfies this equation, then $\text{Re}\,[x(t)]$ and $\text{Im}\,[x(t)]$, respectively, will satisfy the given equations. By undetermined coefficients we find $x = (2i - 7)e^{2it}$, so that

$$\text{Re}\,[x(t)] = -7 \cos 2t - 2 \sin 2t,$$
$$\text{Im}\,[x(t)] = 2 \cos 2t - 7 \sin 2t$$

are the solutions sought.

Simultaneous differential equations. The results described carry over to sets of simultaneous linear differential equations with complex coefficients and complex forcing functions (Reference 4, pp. 500–501). An example is given in Problem 6 below. Since integrodifferential equations are equivalent to such simultaneous equations, the results also carry over to integrodifferential equations.

PROBLEMS

1. Find the general complex solution of each of the following equations:

(a) $(D^2 - 9)x = 0$ (b) $(D^2 - D + 1)x = 0$
(c) $(D^4 + 1)x = t^2$ (d) $(D^4 + 4D^2)x = 6i \sin t$

2. Find the general real solution of each of the following in both a form involving real constants and a form involving complex constants:

(a) $(D^2 + 9)x = 0$ (b) $(D^4 - 9)x = 0$
(c) $(D^2 + 4D + 8)x = e^t$ (d) $(D^3 + 1)x = \sin t$

3. Find a particular solution of each of the following (see Example 2 above):

(a) $(D + 3)x = \sin 3t$ (b) $(D^2 + 2D + 2)x = \cos t + 2 \sin t$
(c) $(D^2 + 4)x = \sin 2t$ (d) $(D^2 + 4)x = e^{2t} \sin t$
(e) $(D^2 + 2D + 2)x = e^{-t} \cos t$
(f) $(D^2 + 4)x = \sin^3 t = \left(\dfrac{e^{it} - e^{-it}}{2i} \right)^3$

4. Find a particular solution of the following:

$$[D^2 + (1 + i)D + 2]x = \cos 4t = \frac{e^{4it} + e^{-4it}}{2}$$

(Note that, since the coefficients are not real, the real part of a solution corresponding to right-hand member e^{4it} does not give a desired solution.)

5. Let an equation $(a_0 D^4 + a_1 D^3 + a_2 D^2 + a_3 D + a_4)x = 0$ be given, in which the coefficients are real constants and $a_0 \neq 0$. Let the characteristic roots be distinct complex numbers a, $\bar{a}$, b, $\bar{b}$.

(a) Show that the general real solution is

$$x = \text{Re}\,[C_1 e^{at}] + \text{Re}\,[C_2 e^{bt}],$$

where C_1 and C_2 are arbitrary complex constants.

(b) Show that the general real solution is

$$x = C_1 e^{at} + \overline{C}_1 e^{\bar{a}t} + C_2 e^{bt} + \overline{C}_2 e^{\bar{b}t},$$

where C_1 and C_2 are arbitrary complex constants.

(c) Modify the rules of parts (a) and (b) to cover the case of the following two pairs of equal roots: $a = b$, $\bar{a} = \bar{b}$.

6. Find a particular solution of the following:

$$(D - 2)x + (2D + 1)y = 3 \cos 2t, \quad (D + 1)x + (D + 2)y = 5 \cos 2t$$

[*Hint:* Replace the right-hand members by $3e^{2it}$, $5e^{2it}$, seek a solution of form $x = ae^{2it}$, $y = be^{2it}$, and then take real parts.]

7. Find a particular solution of the following:

$$(D + 5 + 10D^{-1})x = \cos 5t$$

ANSWERS

Throughout, C, $C_1, \ldots$ denote complex constants and c, $c_1, \ldots$ denote real constants.

1. (a) $C_1 e^{3t} + C_2 e^{-3t}$

(b) $C_1 e^{\alpha t} + C_2 e^{\bar{\alpha} t} \quad (\alpha = (1 + \sqrt{3}\,i)/2)$

(c) $t^2 + C_1 e^{(\alpha+i\alpha)t} + C_2 e^{(\alpha-i\alpha)t} + C_3 e^{(-\alpha+i\alpha)t} + C_4 e^{(-\alpha-i\alpha)t} \quad (\alpha = \sqrt{2}/2)$

(d) $C_1 + C_2 t + C_3 e^{2it} + C_4 e^{-2it} - 2i \sin t$

2. (a) $c_1 \cos 3t + c_2 \sin 3t, \qquad \mathrm{Re}\,[C_1 e^{3it} + C_2 e^{-3it}]$

(b) $c_1 \cos kt + c_2 \sin kt + c_3 e^{kt} + c_4 e^{-kt},$
$\qquad \mathrm{Re}\,[C_1 e^{kit} + C_2 e^{-kit} + C_3 e^{kt} + C_4 e^{-kt}] \quad (k = \sqrt{3})$

(c) $(e^t/13) + e^{-2t}(c_1 \cos 2t + c_2 \sin 2t),$
$\qquad (e^t/13) + \mathrm{Re}\,[C_1 e^{(-2+2i)t} + C_2 e^{(-2-2i)t}]$

(d) $\frac{1}{2}(\cos t + \sin t) + c_1 e^{-t} + e^{t/2}[c_2 \cos (\sqrt{3}\,t/2) + c_3 \sin (\sqrt{3}\,t/2)],$
$\frac{1}{2}(\cos t + \sin t) + \mathrm{Re}\,[C_1 e^{-t} + C_2 e^{at} + C_3 e^{\bar{a}t}] \quad (a = (1 + \sqrt{3}\,i)/2)$

3. (a) $\frac{1}{6} \,\mathrm{Im}\,[(1 - i)e^{3it}] = \frac{1}{6}\,(\sin 3t - \cos 3t)$

(b) $\frac{1}{5}\,\{\mathrm{Re}\,[(1 - 2i)e^{it}] + 2\,\mathrm{Im}\,[(1 - 2i)e^{it}]\}$

(c) $\mathrm{Im}\,[(-\tfrac{1}{4})ite^{2it}] = -\tfrac{1}{4}t \cos 2t$

(d) $\mathrm{Im}\left[\dfrac{7 - 4i}{65}\,e^{(2+i)t}\right]$

(e) $\mathrm{Re}\left[\dfrac{t}{2i}\,e^{(-1+i)t}\right]$

(f) $\dfrac{i}{40}\,(-e^{3it} - 5e^{it} + 5e^{-it} + e^{-3it}) = \mathrm{Im}\left[\dfrac{e^{3it} + 5e^{it}}{20}\right]$

4. $[(-9 - 2i)/340]e^{4it} + [(-5 + 2i)/116]e^{-4it}$

5. (c) $\mathrm{Re}\,[C_1 e^{at}] + \mathrm{Re}\,[C_2 t e^{at}],$
$\qquad C_1 e^{at} + \overline{C}_1 e^{\bar{a}t} + C_2 t e^{at} + \overline{C}_2 t e^{\bar{a}t}$

6. $x = \text{Re}\left[\dfrac{83 + 20i}{37} e^{2it}\right], \qquad y = \text{Re}\left[\dfrac{-11 - 82i}{37} e^{2it}\right]$

7. $\text{Re}\,[(5 - 3i)e^{5it}/34]$

1–10 Forced vibrations. To illustrate the theory of the preceding sections, we apply complex numbers to analyze the vibration problem

$$m\frac{d^2x}{dt^2} + 2q\frac{dx}{dt} + k^2 x = A\cos\omega t, \qquad (1\text{–}100)$$

where m, q, k, A, ω are constants, $m > 0$, $q \geqq 0$, $k > 0$, $A \geqq 0$, $\omega > 0$. We first replace Eq. (1–100) by the equation

$$(mD^2 + 2qD + k^2)x = Ae^{i\omega t}. \qquad (1\text{–}101)$$

Since $\text{Re}\,[Ae^{i\omega t}] = A\cos\omega t$, the real part of each solution of Eq. (1–101) will satisfy Eq. (1–100), while the imaginary part will satisfy the equation

$$(mD^2 + 2qD + k^2)x = A\sin\omega t.$$

We first discuss the *homogeneous case*, $A = 0$:

$$(mD^2 + 2qD + k^2)x = 0.$$

The characteristic equation has roots s_1 and s_2 given by

$$\frac{-q \pm \sqrt{q^2 - k^2 m}}{m}.$$

Depending on the nature of these roots, there are three cases:

I. *Overdamped case:* $q^2 > k^2 m$. The roots are real and negative; the real complementary function is

$$x_c(t) = c_1 e^{s_1 t} + c_2 e^{s_2 t}, \quad \text{and } x_c(t) \to 0 \text{ as } t \to +\infty.$$

II. *Underdamped case:* $q^2 < k^2 m$. The roots have form $\alpha \pm \beta i$, where α is negative if $q > 0$, and $\alpha = 0$ if $q = 0$. The real complementary function is (as in Problem 5 in Section 1–9)

$$\begin{aligned}
x_c(t) &= e^{\alpha t}(c_1 \cos\beta t + c_2 \sin\beta t) \\
&= Ce^{(\alpha+\beta i)t} + \overline{C}e^{(\alpha-\beta i)t} \\
&= \text{Re}\,[C_1 e^{(\alpha+\beta i)t}], \\
C_1 &= c_1 - ic_2 = 2C.
\end{aligned}$$

The solutions are vibrations of decreasing amplitude, approaching zero as $t \to \infty$, provided $q > 0$.

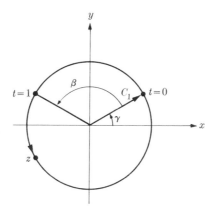

Fig. 1–5.   Graph of $z = C_1 e^{\beta i t}$.

III. *Critical damping:* $q^2 = k^2 m$.   The roots are real, negative, and equal. The real complementary function is

$$x_c(t) = c_1 e^{s_1 t} + c_2 t e^{s_1 t}.$$

Again $x_c(t) \to 0$ as $t \to \infty$.

It is instructive to consider the complex form of the solutions in Case II. Let us first assume that $q = 0$. The solution can then be written in the form

$$x_c(t) = \operatorname{Re}[C_1 e^{\beta i t}].$$

The curve $z = e^{\beta i t} = \cos \beta t + i \sin \beta t$ in the complex z-plane ($z = x + iy$) is a circle of radius 1 on which $z(t)$ moves with angular velocity β. Multiplication of z by the complex constant C_1 multiplies the radius by $|C_1|$ and adds $\arg C_1 = \gamma$ to the argument of z:

$$C_1 e^{\beta i t} = |C_1| e^{i\gamma} e^{\beta i t} = |C_1| e^{i(\gamma + \beta t)}.$$

Hence the curve $z = C_1 e^{\beta i t}$ is a circle of radius $|C_1|$, on which z moves with angular velocity β, and $\arg z = \arg C_1 = \gamma$ when $t = 0$; that is, $z = C_1$ when $t = 0$ (Fig. 1–5). The real solution $x_c(t)$ is $\operatorname{Re}[z(t)]$; hence $x_c(t)$ describes the motion of the point which is the *projection* of $z(t)$ on the real axis. Thus $x_c(t)$ oscillates back and forth between $|C_1|$ and $-|C_1|$, so that $|C_1|$ is the *amplitude* of the oscillation. If C_1 is real, $x_c = C_1 \cos \beta t$; if C_1 is pure imaginary, $C_1 = Bi$, then $x_c = -B \sin \beta t$. In general,

$$x_c = |C_1| \cos(\gamma + \beta t).$$

When $q > 0$, the solutions in Case II can be written in the form

$$x_c(t) = \operatorname{Re}[C_1 e^{\alpha t} e^{\beta i t}] \quad (\alpha < 0).$$

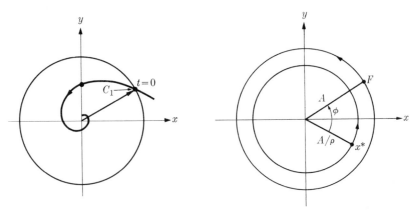

FIG. 1–6. Graph of $z = C_1 e^{(\alpha+i\beta)t}$.   FIG. 1–7. Response to $F(t) = Ae^{i\omega t}$.

The curve $z(t) = e^{\alpha t} \cdot C_1 e^{\beta it}$ is obtained from the curve $z = C_1 e^{\beta it}$ of Fig. 1–5 by multiplying z, at each t, by $e^{\alpha t}$. As t increases, $e^{\alpha t}$ decreases, approaching 0 as $t \to +\infty$. Hence $z(t)$ follows a spiral approaching the origin, as in Fig. 1–6. The function $x_c(t)$ is obtained by projecting this spiral motion on the x-axis. Hence $x_c(t)$ oscillates with decreasing amplitude, approaching 0.

Nonhomogeneous case. We seek a particular solution of Eq. (1–101). We write

$$V(s) = ms^2 + 2qs + k^2,$$

so that the characteristic equation is simply the equation $V(s) = 0$. The substitution $x^* = Ke^{i\omega t}$ in Eq. (1–101) gives the equations

$$K[m(i\omega)^2 + 2q\omega i + k^2]e^{i\omega t} = Ae^{i\omega t},$$

$$K = \frac{A}{k^2 - m\omega^2 + 2q\omega i} = \frac{A}{V(i\omega)},$$

so that

$$x^* = \frac{Ae^{i\omega t}}{V(i\omega)}, \qquad\qquad (1\text{–}102)$$

provided $V(i\omega) \neq 0$. We discuss the case $V(i\omega) = 0$ separately below and first assume that $V(i\omega) \neq 0$. Then, in terms of polar coordinates,

$$V(i\omega) = \rho e^{i\phi},$$

$$x^* = \frac{Ae^{i\omega t}}{V(i\omega)} = \frac{A}{\rho} e^{i(\omega t - \phi)}.$$

Again a graphical discussion is instructive. The complex forcing function $F(t) = Ae^{i\omega t}$ has as graph a circle of radius A on which the point $F(t)$ moves with angular velocity ω. The graph of the particular solution $x^*(t)$ is obtained from this graph by division by the complex constant $V(i\omega)$. Hence $x^*(t)$ moves on a circle of radius A/ρ and with angular velocity ω, but

$$\arg x^*(t) = \arg F(t) - \phi,$$

so that $x^*(t)$ *lags* the forcing function by ϕ radians (see Fig. 1–7). We note that

$$\rho = \sqrt{4q^2\omega^2 + (k^2 - m\omega^2)^2}, \qquad \tan \phi = \frac{2q\omega}{k^2 - m\omega^2}, \quad (1\text{–}103)$$

and $0 \leq \phi \leq \pi$, since by assumption $q \geq 0$ and $\omega > 0$.

To obtain a particular solution of our original real equation (1–100) we need only take real parts, that is, project on the x-axis. The forcing function becomes $A \cos \omega t$, and the particular solution becomes

$$x^* = \frac{A}{\rho} \cos (\omega t - \phi).$$

We now consider the exceptional case $V(i\omega) = 0$; that is,

$$k^2 - m\omega^2 + 2q\omega i = 0.$$

This equation can hold only if

$$k^2 - m\omega^2 = 0, \qquad 2q\omega = 0.$$

Hence $q = 0$ or $\omega = 0$; but by assumption $\omega > 0$. Accordingly, $q = 0$ and $\omega^2 = k^2/m$. The complementary function is obtained from the underdamped case discussed above, with $\alpha = 0$:

$$x = c_1 \cos \beta t + c_2 \sin \beta t, \qquad \beta^2 = \frac{k^2}{m} = \omega^2.$$

Thus the *forcing function has the same frequency as the natural vibrations.* The differential equation (1–101) has the form

$$m(D^2 + \beta^2)x = Ae^{i\beta t}.$$

By the method of undetermined coefficients we obtain a particular solution of form $x^* = Kte^{i\beta t}$:

$$x^* = \frac{-Ai}{2\beta m} te^{i\beta t}.$$

The function $e^{i\beta t}$ describes a circular motion, but the factor t forces the "radius" to increase as t increases, so that we have an oscillation of increasing amplitude, approaching ∞ as $t \to \infty$. This is the case of *resonance.*

From the discussion given, we can now describe the *general solution* of Eq. (1–100):

$$x = x_c(t) + x^*(t).$$

When $q > 0$, then $x_c(t)$ describes a *transient*; for large t only the particular solution, which is a sinusoidal oscillation, is of importance. When $q = 0$ but $V(i\omega) \neq 0$, then $x_c(t)$ is itself a sinusoidal oscillation of frequency β different from ω; since $x_c(t)$ does not approach 0 as $t \to \infty$, we have a superposition of two sinusoidal oscillations (of differing frequency) even for large t. When $q = 0$ and $V(i\omega) = 0$, then $x_c(t)$ is a sinusoidal oscillation but $x^*(t)$ is an oscillation of increasing amplitude, so that $x^*(t)$ dominates for large t.

1–11 The *LRC* circuit. As another illustration of the theory, we consider the electric circuit of Fig. 1–8, containing inductance L, resistance R, capacitance C, driving emf $E = E_0 \cos \omega t$, and current I. The current obeys the integrodifferential equation

$$\left(LD + R + \frac{1}{C}D^{-1}\right)I = E. \qquad (1\text{–}110)$$

We can introduce a new variable, equal to $D^{-1}I = \int I \, dt$. This can be interpreted as the charge Q stored in the capacitor:

$$Q = \int I \, dt = D^{-1}I.$$

The integrodifferential equation (1–110) is equivalent to the set of two differential equations

$$(LD + R)I + \frac{1}{C}Q = E,$$
$$DQ = I. \qquad (1\text{–}111)$$

We seek only a particular solution. We replace E by $E_0 e^{i\omega t}$ and then take real parts. We seek the particular solution in the form

$$I = ae^{i\omega t}, \qquad Q = be^{i\omega t}.$$

Substitution in equations (1–111), with $E = E_0 e^{i\omega t}$, gives the relations

$$(Li\omega + R)a + \frac{1}{C}b = E_0, \qquad bi\omega = a,$$

so that

$$a = \frac{E_0}{Li\omega + R + (Ci\omega)^{-1}}, \qquad b = \frac{a}{i\omega}.$$

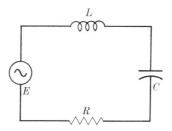

FIG. 1–8. LRC circuit.

Accordingly, for $E = E_0 \cos \omega t$,

$$I = \mathrm{Re}\left[\frac{E_0 e^{i\omega t}}{Li\omega + R + (Ci\omega)^{-1}}\right], \qquad (1\text{–}112)$$

and a similar expression is obtained for Q.

By analogy with Section 1–10 we write

$$Z(s) = Ls + R + \frac{s^{-1}}{C},$$

so that $Z(D)$ is the operator which appears in Eq. (1–110). Then the particular complex solution I corresponding to $E = E_0 e^{i\omega t}$ is

$$I = \frac{E_0 e^{i\omega t}}{Li\omega + R + (Ci\omega)^{-1}} = \frac{E_0 e^{i\omega t}}{Z(i\omega)}. \qquad (1\text{–}113)$$

This relation can be written concisely: $I = E/Z$; accordingly it is a generalization of *Ohm's law:*

$$I = \frac{E}{R}, \qquad (1\text{–}114)$$

which is valid when only a resistance R is present. The quantity Z, which replaces the resistance, is the *complex impedance.*

If I is eliminated from equations (1–111), we obtain a second order equation for Q:

$$\left(LD^2 + RD + \frac{1}{C}\right)Q = E = E_0 e^{i\omega t}. \qquad (1\text{–}115)$$

This equation has the same form as Eq. (1–101). Accordingly, the results described in Section 1–10 can be applied to the analogous electric circuit. It should be noted that, when Q has the form $Ke^{(\alpha+\beta i)t}$, the current I has a similar form:

$$I = \frac{dQ}{dt} = K(\alpha + \beta i)e^{(\alpha+\beta i)t},$$

but the factor $\alpha + \beta i$ will cause both the magnitude and argument of I to differ from that of Q; when $\alpha = 0$ and $\beta > 0$, *I leads Q by 90°.*

PROBLEMS

1. Consider the following as vibration problems of form (1–101). For each of the following, graph the forcing function and a particular solution in the complex plane:

(a) $(5D^2 + 2D + 40)x = 2e^{3it}$ (b) $(D^2 + D + 1)x = e^{2it}$
(c) $(D^2 + D + 10)x = 5e^{5it}$ (d) $(D^2 + 4)x = e^{it}$
(e) $(D^2 + 4)x = e^{2it}$ (f) $(D^2 + 20D + 1)x = 20e^{it}$

2. For each of the following, graph the right-hand member and a particular solution in the complex plane:

(a) $(2D^2 + D + 1)x = e^{(1+i)t}$ (b) $(4D^2 + 3D + 1)x = te^{it}$
(c) $(D^2 + 1)x = te^{(1+i)t}$

3. In the circuit of Fig. 1–8 let $L = 5$ henries, $R = 50$ ohms, $C = 10^{-4}$ farad, and $E = 100e^{i\omega t}$ volts (t in secs.). Obtain a particular solution $I = ae^{i\omega t}$, $Q = be^{i\omega t}$ and evaluate the phase differences between E, I and Q when (a) $\omega = 120\pi$ rad/sec, and (b) $\omega = 10$ rad/sec.

ANSWERS

1. (a) $(-10 - 12i)e^{3it}/61$ (b) $(-3 - 2i)e^{2it}/13$
(c) $(-3 - i)e^{5it}/10$ (d) $e^{it}/3$
(e) $-ite^{2it}/4$ (f) $-ie^{it}$

2. (a) $(2 - 5i)e^{(1+i)t}/29$ (b) $e^{it}[8 - 3i - (3 + 3i)t]/18$
(c) $e^{(1+i)t}[(5 - 10i)t + 14i - 2]/25$

3. (a) $I = (0.0014 - 0.054i)e^{120\pi it}$ amp

$$Q = (-0.00014 - 0.0000037i)e^{120\pi it} \text{ coul}$$
$$\arg E = \arg I + (\pi/2) = \arg Q + \pi \text{ (approx.)}$$

(b) $I = (0.0055 + 0.10i)e^{10it}$ amp

$$Q = (0.010 - 0.00055i)e^{10it} \text{ coul}$$
$$\arg E = \arg I - (\pi/2) = \arg Q \text{ (approx.)}$$

1–12 Response to unit function and other discontinuous inputs. In the differential equation

$$(a_0D^n + \cdots + a_n)x = f(t) \tag{1–120}$$

we have thus far assumed $f(t)$ to be continuous over a given interval. In many applications this condition is not satisfied, but $f(t)$ is *piecewise continuous;* that is, $f(t)$ is continuous except for jump discontinuities, as in Fig. 1–3. At most a finite number of discontinuities occur in each finite interval, and at each discontinuity $f(t)$ has limits from the left and from the right.

When $f(t)$ is piecewise continuous in an interval, we define a solution of Eq. (1–120) to be a function $x(t)$ which is continuous and has continuous derivatives through the $(n - 1)$st order in the interval and which satisfies the differential equation except at the discontinuities of $f(t)$.

On the basis of this definition it can be verified that the existence theorem of Section 1–1 continues to apply when f is piecewise continuous. Arbitrary initial values of x, Dx, ..., $D^{n-1}x$ can be assigned at t_0. For example, let $t_0 = 0$ and let α_1, α_2 be the first discontinuities of f to the right of 0. The existence theorem of Section 1–1 then applies to the interval $0 \leqq t \leqq \alpha_1$; since $f(t)$ has a limit as $t \to \alpha_1 -$, the solution exists in the closed interval, with derivatives at zero interpreted as derivatives to the right and derivatives at α_1 interpreted as derivatives to the left. The left-hand limits of x, Dx, ..., $D^{n-1}x$ at α_1 can now be used as *initial values* for the next interval $\alpha_1 \leqq t \leqq \alpha_2$; accordingly, the solution can be prolonged to the next interval and the derivatives up to order $n - 1$ remain continuous from zero to α_2; the nth derivative must have a jump discontinuity at α_1. Thus the solution can be extended to the right as far as desired, and similarly it can be extended to the left.

The argument just given remains valid if t_0 is a discontinuity point of f. It is also valid when the coefficients a_0, ..., a_n are functions of t which are either continuous or piecewise continuous; in the latter case, we obtain the solution piece by piece in each interval in which a_0, ..., a_n, f are continuous. We must also assume $a_0(t) \neq 0$ between discontinuities and that $a_0(t)$ does not have zero as a left- or right-hand limit at a discontinuity.

The *general solution* of Eq. (1–120) is given, as before, by $x = x_c(t) + x^*(t)$, where $x_c(t)$ is the general solution of the related homogeneous equation. For addition of a solution of the homogeneous equation to a particular solution $x^*(t)$ gives a function satisfying the nonhomogeneous equation, except at discontinuity points, and having the required continuity properties. If $x(t)$ is any solution of the nonhomogeneous equation, then $x(t) - x^*(t)$ will be a solution of the homogeneous equation; that is, $x(t) = x^*(t) + x_c(t)$, for some choice of $x_c(t)$.

EXAMPLE 1. $(D + 1)x = f(t)$, where $f(t) = 0$ for $t < 0$, $f(t) = e^t$ for $t > 0$. [The value of $f(t)$ at $t = 0$ need not be specified, since only the left and right limits affect the solution.] Let $x = x_0$ for $t = 0$. Then we first obtain a solution for $t \geqq 0$: $x = x_0 e^{-t} + \frac{1}{2}(e^t - e^{-t})$. We also use x_0 as initial value for $t \leqq 0$ and obtain the solution $x = \dot{x}_0 e^{-t}$ for $t \leqq 0$. Hence

$$x = x_0 e^{-t} \quad \text{for } t \leqq 0;$$
$$x = x_0 e^{-t} + \tfrac{1}{2}(e^t - e^{-t}) \quad \text{for } t \geqq 0.$$

This gives the general solution with x_0 as arbitrary constant.

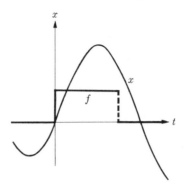

FIG. 1–9. Response of second order equation to discontinuous input.

EXAMPLE 2. $(D^2 + 1)x = f(t)$, where $f(t) = 0$ for $t < 0$, $f(t) = 1$ for $0 < t < \pi$, $f(t) = 0$ for $t > \pi$. Let $x = 0$, $Dx = 1$ for $t = 0$. In the interval $0 \leqq t \leqq \pi$ we solve the equation $(D^2 + 1)x = 1$, with the given initial values and find $x = 1 + \sin t - \cos t$; the limiting values of x, Dx at $t = \pi$ (from the left) are 2 and -1, respectively. We solve the equation $(D^2 + 1)x = 0$ for $t \geqq \pi$ with $x = 2$, $Dx = -1$ at $t = \pi$ and find $x = \sin t - 2 \cos t$; we solve the equation $(D^2 + 1)x = 0$ for $t \leqq 0$ with $x = 0$, $Dx = 1$ for $t = 0$ and find $x = \sin t$. Hence in all

$$x = \sin t \quad (t \leqq 0);$$
$$x = 1 + \sin t - \cos t \quad (0 \leqq t \leqq \pi);$$
$$x = \sin t - 2 \cos t \quad (t \geqq \pi).$$

Forcing function and response are shown in Fig. 1–9.

An important discontinuous function is the Heaviside *unit function* or *unit step*. This function, which we shall denote by $h(t)$, is defined as follows:

$$h(t) = 0, \quad t < 0; \qquad h(t) = 1, \quad t \geqq 0.$$

The solution of a differential equation (1–120) with forcing function $f = h(t)$ and initial values of x, Dx, ..., $D^{n-1}x$ at $t = 0$ equal to zero, is known as the *step response*.

The unit function is useful as a building block for other piecewise continuous functions. Thus $h(t - c)$, where c is constant, is a function equal to zero for $t < c$ and to 1 for $t \geqq c$; the function $1 - h(t)$ equals 1 for $t < 0$ and equals zero for $t \geqq 0$; the function $h(t - a) - h(t - b)$, where $a < b$, equals 1 for $a \leqq t < b$ and equals zero otherwise. By multiplying each of these functions by $g(t)$ we obtain a function equal to $g(t)$ in an interval, and equal to zero otherwise. By adding two such func-

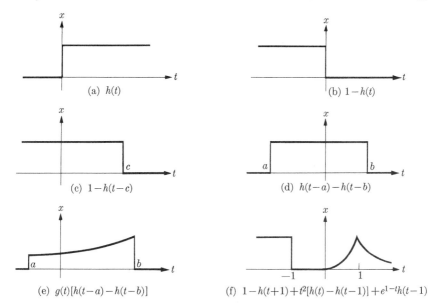

(a) $h(t)$

(b) $1-h(t)$

(c) $1-h(t-c)$

(d) $h(t-a)-h(t-b)$

(e) $g(t)[h(t-a)-h(t-b)]$

(f) $1-h(t+1)+t^2[h(t)-h(t-1)]+e^{1-t}h(t-1)$

Fig. 1-10.　Unit step function and functions constructed from it.

tions we obtain a function equal to $g_1(t)$ in one interval and to $g_2(t)$ in another.　The possibilities are summarized graphically in Fig. 1-10.

As illustrations we can write forcing functions f and response x for Examples 1 and 2 as follows:

$$f = e^t h(t), \quad x = x_0 e^{-t}[1 - h(t)] + [x_0 e^{-t} + \tfrac{1}{2}(e^t - e^{-t})]h(t);$$
$$f = h(t) - h(t - \pi),$$
$$x = [1 - h(t)] \sin t + (1 + \sin t - \cos t)[h(t) - h(t - \pi)]$$
$$+ (\sin t - 2 \cos t)h(t - \pi).$$

EXAMPLE 3.　$(D^2 + 3D + 2)x = h(t)$.　The step response is found to be

$$x = 0 \text{ for } t \leqq 0; \quad x = \tfrac{1}{2}(1 + e^{-2t} - 2e^{-t}) \text{ for } t \geqq 0;$$

that is, $x = \tfrac{1}{2}(1 + e^{-2t} - 2e^{-t})h(t)$.　The general solution is $x = c_1 e^{-t} + c_2 e^{-2t} + \tfrac{1}{2}(1 + e^{-2t} - 2e^{-t})h(t)$.

PROBLEMS

1. For each of the following find the solution with the given initial values at $t = 0$ and graph:

(a) $(D^2 + 3D + 2)x = f(t)$, $x_0 = 0$, $x_0' = 0$, where $f(t) = 0$ for $t < 0$, $f(t) = 4t$ for $0 \leqq t < 2$, $f(t) = 0$ for $t > 2$

(b) $(D^2 + 4)x = f(t)$, $x_0 = 1$, $x_0' = 0$, where $f(t) = 0$ for $t < 0$, $f(t) = e^{it}$ for $0 < t < \pi$, $f(t) = 0$ for $t > \pi$

(c) $(D + 2)x = f(t)$, $x_0 = 0$, where $f(t) = 0$ for $t < 0$, $f(t) = \cos t$ for $0 < t < \pi/2$, $f(t) = e^{(\pi/2)-t}$ for $t > \pi/2$

(d) $(D^2 - 2D + 2)x = f(t)$, $x_0 = 1$, $x_0' = -1$, where $f(t) = 4 - 2t$ for $t < 0$, $f(t) = 4$ for $0 < t < 2$, $f(t) = 0$ for $t > 2$

2. Graph the following functions:

(a) $h(t - 2)$

(b) $2[h(t - 3) - h(t - 6)]$

(c) $[h(t + \pi) - h(t - \pi)] \sin t$

(d) $\sum\limits_{k=0}^{\infty} (-1)^k h(t - k)$

(e) $2h(t) + 3h(t - 1) - 4h(t - 2)$

3. Find the step response for each of the following:

(a) $(2D + 3)x = f$

(b) $(D^2 + 4D + 5)x = f$

(c) $(D^3 - D)x = f$

4. Write the forcing functions of Problem 1 in terms of $h(t)$.

5. Obtain the general solution of each of the following:

(a) $(D + 1)x = 2h(t) - 2h(t - 3)$

(b) $Dx = h(t) - 2h(t + 1)$

(c) $(D^2 + 9)x = e^{3t}[h(t) - h(t - 3)]$

6. (a) Show that $2h[f(t)] - 1 = \operatorname{sgn} f(t)$, equal to $+1$ when $f(t) \geq 0$, equal to -1 when $f(t) < 0$;

b. Show that $\sum_{n=1}^{\infty} h(t - n) = [t]$, for $t \geq 0$, where $[t]$ is the integer part of t ($[1.23] = 1$, $[0.75] = 0$).

Answers

1. (a) $x = 0$ for $t < 0$; $x = 4e^{-t} - e^{-2t} + 2t - 3$ for $0 \leq t \leq 2$; $x = (4 + 4e^2)e^{-t} - (1 + 3e^4)e^{-2t}$ for $t \geq 2$

(b) $x = (e^{2it} + e^{-2it})/2$ for $t \leq 0$; $x = (3e^{2it} + 5e^{-2it} + 4e^{it})/12$ for $0 \leq t \leq \pi$; $x = e^{-2it}/3$ for $t \geq \pi$

(c) $x = 0$ for $t \leq 0$; $x = [2\cos t + \sin t - 2e^{-2t}]/5$ for $0 \leq t \leq \pi/2$; $x = e^{(\pi/2)-t} - [(4e^\pi + 2)e^{-2t}/5]$ for $t \geq \pi/2$

(d) $x = 1 - t$ for $t \leq 0$; $x = 2 - e^t \cos t$ for $0 \leq t \leq 2$; $x = e^{t-2}[(2\cos 2 + 2\sin 2 - e^2)\cos t + (2\sin 2 - 2\cos 2)\sin t]$ for $t \geq 2$

3. (a) $\frac{1}{3}(1 - e^{-3t/2})h(t)$

(b) $\frac{1}{5}[1 - e^{-2t}(\cos t + 2\sin t)]h(t)$

(c) $[\sinh t - t]h(t)$

4. (a) $4t[h(t) - h(t - 2)]$

(b) $e^{it}[h(t) - h(t - \pi)]$

(c) $[h(t) - h(t - \frac{1}{2}\pi)]\cos t + e^{(\pi/2)-t}h(t - \frac{1}{2}\pi)$

(d) $(4 - 2t)[1 - h(t)] + 4[h(t) - h(t - 2)]$

5. (a) $(2 - 2e^{-t})[h(t) - h(t - 3)] + (2e^3 - 2)e^{-t}h(t - 3) + ce^{-t}$

(b) $th(t) - 2(t + 1)h(t + 1) + c$

(c) $\frac{1}{18}(e^{3t} - \cos 3t - \sin 3t)[h(t) - h(t - 3)] + \frac{1}{18}\{[e^9(\cos 9 - \sin 9) - 1]$ $\cos 3t + [e^9(\cos 9 + \sin 9) - 1]\sin 3t\}h(t - 3) + c_1 \cos 3t + c_2 \sin 3t$

1-13 Impulse function and other generalized functions.   The Dirac delta function, or unit impulse function, is "defined" as a function $\delta(t)$ such that

$$\delta(t) = 0 \quad \text{for} \quad t \neq 0, \qquad \int_{-\infty}^{\infty} \delta(t)\, dt = 1; \qquad (1\text{--}130)$$

from these properties we "deduce" that

$$\int_{-\infty}^{\infty} \delta(t) f(t)\, dt = f(0) \qquad (1\text{--}131)$$

for each function $f(t)$ which is continuous at $t = 0$. Of course no ordinary function can have the properties mentioned. The situation is similar to that encountered in algebra: The equation $x^2 = -1$ can be satisfied by no real number x. Hence we invent an "imaginary" number i which has this property: $i^2 = -1$. In the same way we invent an "imaginary" or "ideal" function $\delta(t)$ to have the properties stated above.

We can consider $\delta(t)$ to be the limiting case of a pulse of area 1, as in Fig. 1–12, as the width of the pulse approaches zero. Below we shall interpret $\delta(t)$ as the derivative of $h(t)$; it is clear that the pulse of Fig. 1–12 can be considered as the derivative of the continuous approximation to $h(t)$ of Fig. 1–11. Differentiation of the pulse of Fig. 1–12 provides a useful model for $\delta'(t)$, as in Fig. 1–13. (The figures are on the next page.)

These interpretations suggest that $\delta(t)$ is to be an *even* function $[\delta(-t) = \delta(t)]$, that $\delta'(t)$ is to be *odd* $[\delta'(-t) = -\delta'(t)]$, $\delta''(t)$ is even, and so on. These properties can easily be verified to be in agreement with the definitions given below.

The delta function is but one of a class of ideal functions called *generalized functions* or *distributions*, whose theory has recently been developed by Schwartz, Temple, Korevaar, Lighthill, Mikusiński, and others (see References 5, 7, 9, 10 at the end of the chapter; see also Appendix I). It will be sufficient for our purposes to consider only certain members of this class which we can describe quite explicitly. Specifically, we introduce the following generalized functions:

(a) the delta function $\delta(t)$ and the translated functions $\delta(t - c)$;

(b) the first, second, . . . derivatives $\delta'(t)$, $\delta''(t)$, . . . and the translated functions $\delta'(t - c)$, $\delta''(t - c)$, . . . ;

(c) finite linear combinations, with constant (real or complex) coefficients of ordinary functions and the generalized functions (a) and (b). (Terms with coefficient zero may be omitted.)

The following are examples of the generalized functions considered:

$$3 \sin t + 2\, \delta(t - 1) - 5i\, \delta'(t + 2), \qquad 2h(t) - \delta''(t + 2).$$

The ordinary functions occurring will ordinarily be defined for all finite t;

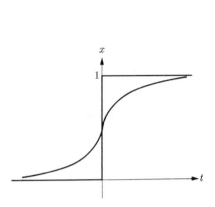

FIG. 1–11. Smooth approximation
of unit step function.

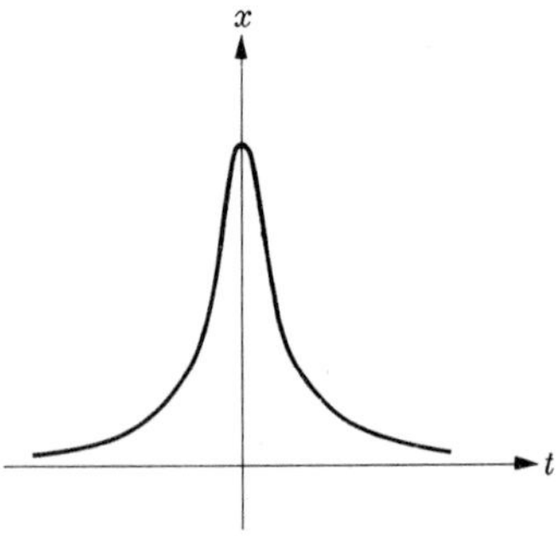

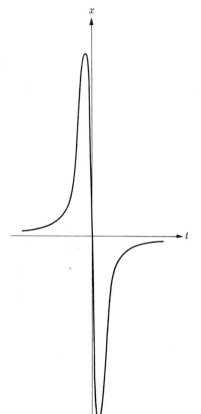

FIG. 1–12. Smooth approximation
of $\delta(t)$.

FIG. 1–13. Smooth approximation
of $\delta'(t)$.

an ordinary function should be considered as a special case of a generalized
function, in which the ideal terms are absent.

Algebraic operations. We define the sum of two generalized functions to
be the generalized function obtained by combining terms of similar form.
Thus

$$[2 \sin t + 2\,\delta(t) - 3\,\delta'(t-1)] + [3 \cos t + 5\,\delta(t) + 2\,\delta'(t-1) + \delta''(t)]$$
$$= (2 \sin t + 3 \cos t) + 7\,\delta(t) - \delta'(t-1) + \delta''(t).$$

Subtraction is defined similarly. The product of a generalized function
by a constant k is defined as the generalized function obtained by multiplying each term by k; a term $a\,\delta^{(n)}(t-c)$ becomes $(ka)\,\delta^{(n)}(t-c)$.

These operations obey the familiar algebraic rules: $f + g = g + f$, $f + (g + h) = (f + g) + h$, $k(f + g) = kf + kg, \ldots$

Multiplication of an ordinary function $f(t)$ by a generalized function is defined, under appropriate conditions. First, $f(t)\ \delta(t)$ is defined as $f(0)\ \delta(t)$, provided f is continuous at $t = 0$. This is easily justified in terms of the interpretation of $\delta(t)$ as limit of a pulse function; $f(t)\ \delta(t)$ is the limit of a similar pulse which, near $t = 0$, is an approximation to $f(0)\ \delta(t)$; the values of f away from $t = 0$ are of no importance in the limit. Next, $f(t)\ \delta'(t) = \delta'(t)f(t)$ is defined as $f(0)\ \delta'(t) - f'(0)\ \delta(t)$, provided f and f' are continuous at $t = 0$. This definition can be justified by an analysis of $[f(t)\ \delta(t)]'$ in terms of the limit process. The previous reasoning suggests that this derivative should be $f(0)\ \delta'(t)$; however, we also expect the product rule to hold:

$$f(0)\ \delta'(t) = [f(t)\ \delta(t)]' = f(t)\ \delta'(t) + f'(t)\ \delta(t);$$

the last term is $f'(0)\ \delta(t)$, so that

$$f(t)\ \delta'(t) = f(0)\ \delta'(t) - f'(0)\ \delta(t).$$

A similar reasoning leads to the general definition:

$$f(t)\ \delta^{(k)}(t - c) = \delta^{(k)}(t - c)f(t)$$
$$= (-1)^k \sum_{r=0}^{k} (-1)^r \binom{k}{r} f^{(k-r)}(c)\ \delta^{(r)}(t - c), \quad (1\text{--}132)$$

where $\binom{k}{r}$ is the binomial coefficient $k![r!(k - r)!]^{-1}$ and $f, f', \ldots, f^{(k)}$ are assumed to be continuous at $t = c$. Finally, an arbitrary generalized function can be multiplied by f by multiplying each term by f and evaluating $f \cdot [a\ \delta^{(k)}(t - c)]$ as $a[f(t)\ \delta^{(k)}(t - c)]$.

The rule (1–132) can be remembered simply as a rule to expand $(\delta - f)^k$ formally by the binomial formula and then replace powers by derivatives, those of f being evaluated at c and those of δ at $t - c$. For example,

$$f(t)\ \delta'''(t - c) = f(c)\ \delta'''(t - c) - 3f'(c)\ \delta''(t - c)$$
$$+ 3f''(c)\ \delta'(t - c) - f'''(c)\ \delta(t - c).$$

Multiplication of two generalized functions is not defined.

Differentiation. Not all ordinary functions have derivatives; in particular, the derivative fails to exist at a discontinuity point. This deficiency can be partly remedied by the generalized functions we have introduced. For example, we shall define

$$h'(t) = \delta(t), \qquad h'(t - c) = \delta(t - c).$$

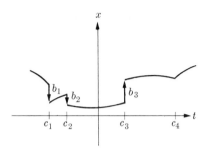

FIG. 1–14.   Function whose derivative is a generalized function.

The function $g(t)$ equal to zero for $t < 0$ and to $1 + t$ for $t \geqq 0$ has a jump discontinuity at $t = 0$. We can write $g(t) = th(t) + h(t)$. The term $th(t)$ is continuous, but its graph has a corner at $t = 0$; we agree to consider its derivative to be the discontinuous function $p(t)$ equal to zero for $t < 0$ and to 1 for $t > 0$, with $p(0)$ assigned arbitrarily; hence in particular we can choose $p(t) = h(t)$. The derivative of $g(t)$ is then defined as $[th(t) + h(t)]' = h(t) + \delta(t)$. We are thus defining the derivative as the usual one plus a contribution of $\delta(t)$ corresponding to the jump discontinuity at $t = 0$.

In general, we can define the derivative of every function $g(t)$ which has a derivative in the ordinary sense except at a finite number of points $c_1, \ldots, c_n$, at which $g(t)$ has limits from the left and right and at which corresponding left- and right-hand derivatives exist; that is,

$$\lim_{\Delta t \to 0+} \frac{g(c_k + \Delta t) - g(c_k+)}{\Delta t},$$

$$\lim_{\Delta t \to 0-} \frac{g(c_k + \Delta t) - g(c_k-)}{\Delta t}$$

must exist, where $g(c_k\pm) = \lim g(c_k + \Delta t)$ as $\Delta t \to 0\pm$. Such a function is suggested in Fig. 1–14. It may happen that g is continuous at c_k, but the left- and right-hand derivatives differ, so that the graph has a *corner* (as at c_4 in Fig. 1–14).

In order to define the derivative of $g(t)$, we let b_k be the jump in $g(t)$ at c_k:

$$b_k = g(c_k+) - g(c_k-).$$

Then the derivative is

$$g'(t) + \sum_{k=1}^{n} b_k \, \delta(t - c_k),$$

where $g'(t)$ is defined as the ordinary derivative except at the points

$c_1, \ldots, c_k$; here the value can be assigned arbitrarily as either the left- or the right-hand derivative (or the average of the two).

The derivative of $a\, \delta(t - c)$ is defined as $a\, \delta'(t - c)$, the derivative of $a\, \delta'(t - c)$ as $a\, \delta''(t - c)$, and so on. The derivative of an arbitrary generalized function can now be obtained by differentiating term by term; second and higher derivatives are obtained by repeating the process.

EXAMPLE 1. Let $f(t) = t^2[h(t) - h(t - 2)]$. There is a discontinuity at $t = 2$. Hence

$$f'(t) = 2t[h(t) - h(t - 2)] - 4\, \delta(t - 2),$$
$$f''(t) = 2[h(t) - h(t - 2)] - 4\, \delta(t - 2) - 4\, \delta'(t - 2),$$
$$f'''(t) = 2\, \delta(t) - 2\, \delta(t - 2) - 4\, \delta'(t - 2) - 4\, \delta''(t - 2).$$

Here the first term in $f'(t)$ has a corner at $t = 0$, while the first term in $f''(t)$ has a discontinuity at $t = 0$.

EXAMPLE 2. $f(t) = g(t)\, h(t)$, where $g(t)$ has derivatives of all orders for all t. Unless $g(0) = 0$, there is a discontinuity at $t = 0$ and we can write

$$f'(t) = g'(t)\, h(t) + g(0)\, \delta(t),$$
$$f''(t) = g''(t)\, h(t) + g'(0)\, \delta(t) + g(0)\, \delta'(t),$$
$$f'''(t) = g'''(t)\, h(t) + g''(0)\, \delta(t) + g'(0)\, \delta'(t)$$
$$+ g(0)\, \delta''(t),$$
$$\vdots$$

If any of the values $g(0)$, $g'(0)$, $g''(0), \ldots$ are zero, the corresponding terms can be omitted. It should be remarked that the formulas for $f'(t)$, $f''(t), \ldots$ can be obtained by differentiating $g(t)\, h(t)$ by the familiar product rule and then interpreting $g(t)\, \delta(t)$, $g'(t)\, \delta(t), \ldots$ as $g(0)\, \delta(t)$, $g'(0)\, \delta(t), \ldots$ in accordance with the definition of multiplication given above.

The derivative as defined here satisfies the rule for derivatives of a sum and of a constant times a function; also the derivative of $f(t - c)$ is equal to $f'(t)$ evaluated at $t - c$. It follows that every function of form

$$f = \sum_{k=1}^{n} A_k g_k(t - c_k) h(t - c_k)$$

can be differentiated as in Example 2 above; in particular, if each g_k has derivatives of all orders, f', f'', $f''', \ldots$ all are defined.

The product rule $(uv)' = u'v + uv'$ holds for differentiation of the product of a differentiable ordinary function and a generalized function differentiable as above. In view of the remark made above, this is true for

$u = f$, $v = g(t)h(t - c)$, where f and g are differentiable ordinary functions. Hence it is sufficient to consider $[f(t) \delta(t)]'$, $[f(t) \delta'(t)]'$, ... For example,

$$[f(t) \delta(t)]' = f'(t) \delta(t) + f(t) \delta'(t),$$

as we verify with the aid of the rule (1–132) for multiplication, and the general case is established in the same way.

Integration. From the definition of derivative, we conclude that the only generalized function whose derivative is identically zero is a constant. Hence, if we define an indefinite integral of a generalized function $f(t)$ as a function whose derivative is $f(t)$, we conclude that the indefinite integral is determined up to a constant. For example,

$$\int \delta(t) \, dt = h(t) + C,$$

$$\int [h(t) + 3i \, \delta'(t)] \, dt = th(t) + 3i \, \delta(t) + C.$$

It should be remarked that an ordinary function $g(t)$ which is piecewise continuous has an indefinite integral which is continuous. In particular, we can choose

$$\int g(t) \, dt = \int_0^t g(u) \, du.$$

The right-hand member is continuous and has a derivative equal to $g(t)$ except at the discontinuities of $g(t)$, at which the integral has a corner. This is illustrated by $\int h(t) \, dt = th(t)$.

If we define the definite integral of a generalized function $g(t)$ by the formula

$$\int_a^b g(t) \, dt = f(t)|_a^b,$$

where $f(t)$ is an indefinite integral of g, we are forced to evaluate $\delta(t)$, $\delta'(t)$, ... at numerical values of t. This leads to difficulties, in particular for $\delta(0)$, $\delta'(0)$, ... However, we can define definite integrals directly:

$$\int_a^b \delta(t - c) \, dt = h(c - a) - h(c - b),$$

$$\int_a^b \delta'(t - c) \, dt = 0, \tag{1–133}$$

$$\int_a^b \delta^{(k)}(t - c) \, dt = 0 \quad (a \neq c, b \neq c, k = 1, 2, \ldots).$$

We can now obtain integrals of products $g(t) f(t)$, where $g(t)$ is an ordinary function and $f(t)$ is a generalized function: if $a < c < b$, we find

$$\int_a^b g(t)\, \delta(t - c)\, dt = g(c),$$

$$\int_a^b g(t)\, \delta'(t - c)\, dt = -\int_a^b \delta(t - c) g'(t)\, dt = -g'(c),$$

$$\int_a^b g(t)\, \delta''(t - c)\, dt = -\int_a^b g'(t)\, \delta'(t - c)\, dt = g''(c),$$

with the aid of (1–132) and (1–133); in general

$$\int_a^b g(t)\, \delta^{(n)}(t - c)\, dt = (-1)^n g^{(n)}(c), \qquad (1\text{–}134)$$

provided $g^{(n)}(t)$ is continuous at $t = c$. The relations clearly suggest an integration by parts. From Eq. (1–134) we obtain a term-by-term definition for the integral of $g(t)f(t)$, where $f(t)$ is an arbitrary generalized function. For example,

$$\int_{-1}^1 e^{2t}[h(t) + 2\,\delta(t) - 3\,\delta'(t)]\, dt$$

$$= \int_{-1}^1 e^{2t} h(t)\, dt + 2\int_{-1}^1 e^{2t}\, \delta(t)\, dt - 3\int_{-1}^1 e^{2t}\, \delta'(t)\, dt$$

$$= \frac{e^2 - 1}{2} + 2 + 6.$$

In Eq. (1–134) the limits of integration may be infinite; for example,

$$\int_{-\infty}^\infty \sin 3t\, \delta'(t - \pi)\, dt = -3\cos 3\pi = 3.$$

In Eq. (1–134), if $c < a < b$ or $c > b > a$, the integral is found to be zero. This is consistent with considering $\delta^{(n)}(t - c)$ to be zero in any interval not containing c.

PROBLEMS

1. Graph and find the derivative for each of the following:
(a) $h(t) + 2h(t - 1) - 3h(t - 2)$
(b) $h(t - 1) - h(t - 2)$
(c) $3[h(t) - h(t - 1)] + 5[h(t - 2) - h(t - 3)]$
(d) $e^t[1 - h(t)]$

2. Find first and second derivatives for each of the following:
(a) $|t|$ (b) $t[1 - h(t)] + t^2 h(t)$
(c) $[h(t - \tfrac{1}{2}\pi) - h(t - \pi)]\sin t$

3. Evaluate each of the following integrals:

(a) $\displaystyle\int_{-\infty}^{\infty} e^t\, \delta(t)\, dt$
(b) $\displaystyle\int_{-\infty}^{\infty} t^2\, \delta(t-1)\, dt$

(c) $\displaystyle\int_{-\infty}^{\infty} \frac{1 + \delta(t-2)}{1 + t^2}\, dt$
(d) $\displaystyle\int_{-\infty}^{\infty} [e^t h(-t) + e^{-t} h(t)]\, dt$

(e) $\displaystyle\int_{-\infty}^{\infty} e^{-t}\, \delta'(t)\, dt$
(f) $\displaystyle\int_{-\infty}^{\infty} t^2 [2\, \delta'(t-1) + 3\, \delta''(t-2)]\, dt$

(g) $\displaystyle\int_{-\infty}^{\infty} e^{-t}[h(t) + i\, \delta(t-1) + 2\, \delta'(t-3)]\, dt$

4. Prove that if $g(t)$ is an ordinary function having a continuous derivative for all t and $g(t) \equiv 0$ for $|t|$ sufficiently large, then

$$\int_{-\infty}^{\infty} g(t)f'(t)\, dt = -\int_{-\infty}^{\infty} g'(t)f(t)\, dt$$

for every generalized function $f(t)$.

ANSWERS

1. (a) $\delta(t) + 2\, \delta(t-1) - 3\, \delta(t-2)$
(b) $\delta(t-1) - \delta(t-2)$
(c) $3[\, \delta(t) - \delta(t-1)] + 5[\, \delta(t-2) - \delta(t-3)]$
(d) $e^t[1 - h(t)] - \delta(t)$

2. (a) $2h(t) - 1$, $2\, \delta(t)$
(b) $(2t - 1)h(t) + 1$, $2h(t) - \delta(t)$
(c) $[h(t - \tfrac{1}{2}\pi) - h(t - \pi)] \cos t + \delta(t - \tfrac{1}{2}\pi)$,
$- [h(t - \tfrac{1}{2}\pi) - h(t - \pi)] \sin t + \delta(t - \pi) + \delta'(t - \tfrac{1}{2}\pi)$
3. (a) 1 (b) 1 (c) $\pi + \tfrac{1}{5}$ (d) 2 (e) 1 (f) 2 (g) $1 + ie^{-1} + 2e^{-3}$

1–14 Application of generalized functions to linear differential equations. We seek a solution of a linear differential equation

$$(a_0 D^n + a_1 D^{n-1} + \cdots + a_n)x = f(t), \qquad (1\text{–}140)$$

with constant coefficients, when $f(t)$ is a generalized function. By definition $f(t)$ is a finite linear combination of terms of form $g(t)$, $\delta(t - a)$, $\delta'(t - b)$, $\delta''(t - c)$, $\ldots$, where $g(t)$ is an ordinary function. We shall assume that $g(t)$ is defined and piecewise continuous for all t.

We first consider the special case of the equation

$$D^n x = f(t), \qquad (1\text{–}141)$$

where $n = 1, 2, \ldots$ Here the solution is obtained by repeated integration

EXAMPLE 1. $Dx = 2t + \delta(t - 1)$, $x = t^2 + h(t - 1) + c$.

EXAMPLE 2.

$$D^2x = \sin t + \delta(t - 2), \qquad Dx = -\cos t + h(t - 2) + c_1,$$
$$x = -\sin t + (t - 2)h(t - 2) + c_1 t + c_2.$$

Here we have reasoned:

$$\int h(t - 2) \, dt = \int_2^t h(t - 2) \, dt + \text{const} = (t - 2)h(t - 2) + \text{const}.$$

EXAMPLE 3.

$$D^3x = 120t^2 + 2\,\delta(t - 1) - 5\,\delta''(t - 2),$$
$$D^2x = 40t^3 + 2h(t - 1) - 5\,\delta'(t - 2) + c_1,$$
$$Dx = 10t^4 + 2(t - 1)h(t - 1) - 5\,\delta(t - 2) + c_1 t + c_2,$$
$$x = 2t^5 + (t - 1)^2 h(t - 1) - 5h(t - 2) + \tfrac{1}{2}c_1 t^2 + c_2 t + c_3.$$

We now turn to the general equation (1–140). Let m be the highest order of a derivative of a delta function appearing in $f(t)$. Thus in Example 1, $m = 0$; in Example 2, $m = 0$; in Example 3, $m = 2$. *Then $f(t)$ can be written as $D^{m+1}F(t)$, where $F(t)$ is piecewise continuous.* For we need only solve the equation

$$D^{m+1}u = f(t)$$

as in Examples 1 through 3. If $u = F(t)$ is a solution, then no delta or derivative of a delta will appear in $F(t)$, since each successive integration lowers the order of a derivative by 1. Accordingly, Eq. (1–140) can be written:

$$(a_0 D^n + \cdots + a_n)x = D^{m+1}F(t). \qquad (1\text{–}142)$$

We now replace Eq. (1–142) by the pair of equations

$$D^{m+1}y = x, \qquad (a_0 D^n + \cdots + a_n)y = F(t). \qquad (1\text{–}143)$$

If $y(t)$ is a solution of the second equation, then $(m + 1)$-fold differentiation of the second equation yields

$$D^{m+1}[(a_0 D^n + \cdots + a_n)y] = D^{m+1}F(t),$$
$$(a_0 D^n + \cdots + a_n)D^{m+1}y = D^{m+1}F(t),$$

so that $x = D^{m+1}y$ satisfies Eq. (1–142). In this way we obtain one solution of Eq. (1–140). The general solution is obtained by adding $x_c(t)$ in familiar fashion. [No generalized function can be a solution of a homogeneous equation (1–142); see Problem 2 below.]

EXAMPLE 4. $(D + 1)x = 2t + \delta(t - 1)$. Here $m = 0$ and we must integrate the right-hand side once. As in Example 1 above we find an integral to be $t^2 + h(t - 1)$. Thus

$$(D + 1)x = D\big(t^2 + h(t - 1)\big).$$

We now consider the pair

$$Dy = x, \qquad (D + 1)y = t^2 + h(t - 1).$$

The second equation has a particular solution

$$y = t^2 - 2t + 2 + (1 - e^{1-t})h(t - 1).$$

Hence

$$x = Dy = 2t - 2 + e^{1-t}h(t - 1).$$

The general solution is obtained by adding ce^{-t}. To check the particular solution we evaluate Dx:

$$Dx = 2 - e^{1-t}h(t - 1) + \delta(t - 1).$$

Then $Dx + x = 2t + \delta(t - 1)$ as required.

EXAMPLE 5. $(D^2 + 2D + 1)x = 120t^2 + 2\,\delta(t - 1) - 5\,\delta''(t - 2)$. With the aid of Example 3, we write

$$(D^2 + 2D + 1)x = D^3[2t^5 + (t - 1)^2 h(t - 1) - 5h(t - 2)]$$

and then

$$D^3 y = x, \qquad (D^2 + 2D + 1)y = 2t^5 + (t - 1)^2 h(t - 1) - 5h(t - 2).$$

A particular solution $y(t)$ is found to be

$$\begin{aligned}
y = {}& 2t^5 - 20t^4 + 120t^3 - 480t^2 + 1200t - 1440 \\
& + [(t - 1)^2 - 4(t - 1) + 6 - 6e^{1-t} - 2(t - 1)e^{1-t}]h(t - 1) \\
& - 5[1 - e^{2-t} - (t - 2)e^{2-t}]h(t - 2).
\end{aligned}$$

Hence

$$\begin{aligned}
x = D^3 y = {}& 120t^2 - 480t + 720 + 2(t - 1)e^{1-t}h(t - 1) \\
& - 5(t - 4)e^{2-t}h(t - 2) - 5\,\delta(t - 2).
\end{aligned}$$

The general solution is obtained by adding $c_1 e^{-t} + c_2 t e^{-t}$.

We note that if m is defined as above, then $x = D^{m+1}y$, where y is continuous, along with its derivatives up to order $n - 1$. Hence if $m + 1 \leqq n - 1$, then x is continuous, along with its derivatives up to order $n - m - 2$. If $m + 1 = n$, then x will have jump discontinuities; if $m + 1 = n + 1$, x will contain a term in $\delta(t - b)$; if $m + 1 > n + 1$, x will contain a term in $\delta^{(m-n)}(t - b)$. These rules can be remembered by considering the simplest case of the equation $D^n x = \delta^{(m)}(t)$; here x is obtained by integrating $\delta^{(m)}(t)$ n times; each successive integration lowers the order of derivative of $\delta(t)$; thus if $n = m$, $x = \delta(t)$; if $n = m + 1$, $x = h(t)$.

Solution determined by initial conditions. Although the general solution of Eq. (1–140) contains n arbitrary constants, it does not necessarily follow that a solution is uniquely determined by initial values of x, Dx, $\ldots$, $D^{n-1}x$ at t_0. This is indeed the case if $f(t)$ contains no term of form $\delta^{(k)}(t - t_0)$ $(k = 0, 1, 2, \ldots)$, provided we consider each term $a \, \delta^{(k)}(t - b)$ to be zero for $t \neq b$; for then the general solution is of form $x = x^*(t) + x_c(t)$, where $x^*(t)$ is formed of terms each of which has well-defined value and derivatives up to order $n - 1$ at t_0. Hence the constants in $x_0(t)$ can be adjusted to give the desired values of x, Dx, $\ldots$, $D^{n-1}x$ at t_0. For example, the particular solution of Example 5 above has initial values $x = 720$, $Dx = -480$ at $t = 0$. By choosing c_1, c_2 so that

$$c_1 e^{-t} + c_2 t e^{-t} + 720 = x_0,$$

$$-c_1 e^{-t} + c_2(e^{-t} - te^{-t}) - 480 = x_0',$$

when $t = 0$, the solution with initial values x_0, x_0' is obtained.

When $f(t)$ contains terms of form $\delta^{(k)}(t - t_0)$, there is difficulty in interpreting initial conditions at t_0. However, if $k < n$ for all such terms, we can impose initial conditions *from the right* (or from the left). We illustrate this for the equation

$$(D^2 + 3D + 2)x = 2 \, \delta(t) + 3 \, \delta'(t) \quad (t_0 = 0).$$

We seek a solution such that $x \to x_0 = a$ and $Dx \to x_0' = b$ as $t \to 0+$. The method of this section leads to the general solution

$$x = (4e^{-2t} - e^{-t})h(t) + c_1 e^{-t} + c_2 e^{-2t}.$$

For $t > 0$ we can write

$$x = 4e^{-2t} - e^{-t} + c_1 e^{-t} + c_2 e^{-2t},$$

$$Dx = -8e^{-2t} + e^{-t} - c_1 e^{-t} - 2c_2 e^{-2t}.$$

Hence as $t \to 0+$, our initial conditions give

$$a = 3 + c_1 + c_2, \qquad b = -7 - c_1 - 2c_2,$$

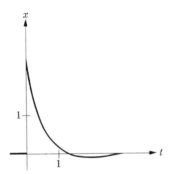

FIG. 1–15.   Response to $2\delta(t) + 3\delta'(t)$.

so that $c_1 = 2a + b + 1$, $c_2 = -a - b - 4$. Hence arbitrary initial conditions can be satisfied from the right. A similar discussion applies to initial conditions from the left. For $t < 0$, $x = c_1e^{-t} + c_2e^{-2t}$ and x and Dx have limiting values $c_1 + c_2$ and $-c_1 - 2c_2$, respectively, as $t \to 0-$. We note that x jumps by 3, Dx by -7 as t crosses through 0. The solution with c_1 and c_2 equal to 0 is shown in Fig. 1–15.

Simultaneous equations. Simultaneous linear differential equations with constant coefficients, with generalized functions as forcing functions, can be treated by the elimination method of Section 1–5. The elimination method leads to successive equations for one variable alone, and each of these can be solved by the method of this section. It is important to note that even though the given differential equations may contain no generalized functions, the elimination process may, as a result of differentiation, introduce generalized functions.

EXAMPLE 6. $(D - 1)x + 3y = 8h(t)$, $x + (D + 1)y = 16h(t - 1)$. Elimination of x requires applying the operator $D - 1$ to the second equation; we find

$$(D^2 - 4)y = 16\,\delta(t - 1) - 16h(t - 1) - 8h(t),$$
$$y = (4 + 2e^{2t-2} - 6e^{2-2t})h(t - 1) + (2 - e^{2t} - e^{-2t})h(t)$$
$$+ c_1e^{2t} + c_2e^{-2t},$$
$$x = 16h(t - 1) - (D + 1)y$$
$$= (12 - 6e^{2t-2} - 6e^{2-2t})h(t - 1)$$
$$+ (-2 + 3e^{2t} - e^{-2t})h(t) - 3c_1e^{2t} + c_2e^{-2t}.$$

It can be verified that x and y are continuous, but Dx and Dy have jump discontinuities.

PROBLEMS

1. For each of the following find the general solution and check by substitution in the equation:

(a) $(D + 2)x = \delta(t)$ (b) $(D + 2)x = \delta(t - 1)$
(c) $(D + 2)x = \delta'(t)$ (d) $(D + 2)x = \delta''(t)$
(e) $(D + 2)x = t + \delta(t - 1)$ (f) $(D^2 + 3D + 2)x = 3\,\delta''(t - 2)$
(g) $(D^2 + 3D + 2)x = \delta'''(t)$ (h) $(D^2 + 4)x = 4\,\delta(t) + 4\,\delta'(t)$

2. Show that a generalized function f, not an ordinary function, cannot be a solution of a homogeneous linear differential equation with constant coefficients. [*Hint:* Show that $D^n f$ contains a term in $\delta^{(k)}(t - c)$ not present in $D^{n-1}f$, $D^{n-2}f, \ldots$] Prove also that $D^{-m}f$ cannot be a solution, for $m \geq 1$.

3. Let $H(t)$ be the step response for Eq. (1–140); that is, $H(t)$ is the solution with initial values zero at $t = 0$ when $f = h(t)$. Show that $H^{(k)}(t - c)$ is a solution when $f = \delta^{(k-1)}(t - c)$. Obtain a solution when f is a generalized function $a_1\,\delta^{(k_1)}(t - b_1) + a_2\,\delta^{(k_2)}(t - b_2) + \cdots$

4. Obtain the step response H for the equation $(D + 2)x = f$ and apply the results of Problem 3 to verify the solutions of Problem 1 (a), (b), (c), (d).

5. For each of the following find a solution satisfying the given initial conditions:

(a) Equation of Problem 1(a), $x = 1$ for $t = 1$
(b) Equation of Problem 1(f), $x = 0$, $Dx = 0$ for $t = 0$
(c) Equation of Problem 1(a), $x \to 1$ as $t \to 0+$
(d) Equation of Problem 1(a), $x \to 1$ as $t \to 0-$
(e) Equation of Problem 1(h), $x \to 1$, $Dx \to 2$ as $t \to 0+$

6. Let a_0, a_1, a_2, k_1, k_2 be constants, $a_0 \neq 0$.

(a) Apply the method of Problem 3 to obtain the solution of the equation

$$(a_0 D^2 + a_1 D + a_2)x = k_1\,\delta(t) + k_2\,\delta'(t)$$

such that $x \to 0$, $Dx \to 0$ as $t \to 0-$.

(b) Show that, as $t \to 0+$, $H(t) \to 0$, $H'(t) \to 0$, $H''(t) \to a_0^{-1}$, $H'''(t) \to -a_1 a_0^{-2}$ and conclude that for the solution $x(t)$ of part (a) we have, as $t \to 0+$,

$$x \to x_{0+} = \frac{k_2}{a_0}, \qquad Dx \to \frac{k_1 a_0 - k_2 a_1}{a_0^2} = x'_{0+}$$

Show that k_1, k_2 can be chosen to make x_{0+}, x'_{0+} equal to arbitrary preassigned values. [*Hint:* For $t > 0$, $(a_0 D^2 + a_1 D + a_2)H(t) = 1$. Let $t \to 0+$ to find the limit of $H''(t)$. Differentiate and let $t \to 0+$ to find the limit of $H'''(t)$.]

Remark. The results of this problem show that if we are interested in solutions of a second order differential equation for $t \geq 0$, imposing initial conditions as $t \to 0+$ is equivalent to adding appropriate terms in $\delta(t)$, $\delta'(t)$ and then imposing *zero* initial conditions as $t \to 0-$. A similar statement applies to the general equation (1–140); we need terms in $\delta(t)$, $\delta'(t), \ldots, \delta^{(n-1)}(t)$.

7. Find the general solution of each of the following:

(a) $(D - 4)x + 3y = h(t), \; 5x - (D + 4)y = h(t - 2)$

(b) $(D - 4)x + (D - 2)y + (D + 2)z = h(t)$
$(D - 2)x - (D + 4)y + (D + 4)z = 0$
$(D - 3)x + (2D + 1)y - (D + 1)z = 0$

Answers

1. (a) $e^{-2t}h(t) + ce^{-2t}$ (b) $e^{2-2t}h(t - 1) + ce^{-2t}$

(c) $-2e^{-2t}h(t) + \delta(t) + ce^{-2t}$

(d) $4e^{-2t}h(t) - 2\,\delta(t) + \delta'(t) + ce^{-2t}$

(e) $(2t - 1)/4 + e^{2-2t}h(t - 1) + ce^{-2t}$

(f) $(3e^{2-t} - 12e^{4-2t})h(t - 2) + 3\,\delta(t - 2) + c_1e^{-t} + c_2e^{-2t}$

(g) $(8e^{-2t} - e^{-t})h(t) - 3\,\delta(t) + \delta'(t) + c_1e^{-t} + c_2e^{-2t}$

(h) $(2\sin 2t + 4\cos 2t)h(t) + c_1\cos 2t + c_2\sin 2t$

3. $x = a_1 H^{(k_1+1)}(t - b_1) + a_2 H^{(k_2+1)}(t - b_2) + \cdots$

4. $H = \frac{1}{2}(1 - e^{-2t})h(t)$

5. (a) $e^{-2t}[h(t) + e^2 - 1]$

(b) $(3e^{2-t} - 12e^{4-2t})h(t - 2) + 3\,\delta(t - 2)$

(c) $e^{-2t}h(t)$ (d) $e^{-2t}h(t) + e^{-2t}$

(e) $(2\sin 2t + 4\cos 2t)h(t) - 3\cos 2t - \sin 2t$

6. (a) $x = k_1 H'(t) + k_2 H''(t)$

(b) $k_1 = a_1 x_{0+} + a_0 x'_{0+}, \; k_2 = a_0 x_{0+}$

7. (a) $x = \frac{1}{2}(5e^t + 3e^{-t} - 8)h(t) + \frac{1}{2}(-6 + 3e^{t-2} + 3e^{2-t})h(t - 2) + c_1e^t + \frac{3}{5}c_2e^{-t}$,

$y = \frac{5}{2}(e^t + e^{-t} - 2)h(t) + \frac{1}{2}(-8 + 3e^{t-2} + 5e^{2-t})h(t - 2) + c_1e^t + c_2e^{-t}$

(b) $x = \frac{1}{16}(4 - 3e^{2t} - e^{-2t})h(t) + c_1e^{2t} + c_2e^{-2t}$

$y = \frac{1}{32}(28t - 3e^{2t} + 3e^{-2t})h(t) + \frac{1}{2}c_1e^{2t} - \frac{3}{2}c_2e^{-2t} + c_3$

$z = \frac{1}{32}(28t + 4 - 3e^{2t} - e^{-2t})h(t) + \frac{1}{2}c_1e^{2t} + \frac{1}{2}c_2e^{-2t} + c_3$

SUGGESTED REFERENCES

1. RALPH P. AGNEW, *Differential Equations*, 2nd ed. New York: McGraw-Hill, 1960.

2. DAVID K. CHENG, *Analysis of Linear Systems*. Reading, Mass.: Addison-Wesley, 1959.

3. CHARLES S. DRAPER, WALTER MCKAY and SIDNEY LEES, *Instrument Engineering*, vol. 2. New York: McGraw-Hill, 1953.

4. WILFRED KAPLAN, *Ordinary Differential Equations*. Reading, Mass.: Addison-Wesley, 1958.

5. JACOB KOREVAAR, "Distributions Defined by Fundamental Sequences," *Nederl. Akad. Wetensch. Proc. Ser. A*, vol. 58 (1955) pp. 368–389, 483–503, 663–674.

6. DEREK F. LAWDEN, *Mathematics of Engineering Systems*. New York: John Wiley, 1954.

7. M. J. LIGHTHILL, *An Introduction to Fourier Analysis and Generalized Functions*. Cambridge, Eng.: Cambridge University Press, 1958.

8. W. T. MARTIN and E. REISSNER, *Elementary Differential Equations*, 2nd ed. Reading, Mass.: Addison-Wesley, 1961.

9. JAN MIKUSIŃSKI, *Operational Calculus*, transl. from the second Polish ed. New York: Pergamon Press, 1959.

10. L. SCHWARTZ, *Théorie des Distributions*, vols. I and II. Paris: Hermann and Cie., 1950, 1951.

CHAPTER 2

BASIC CONCEPTS OF SYSTEMS ANALYSIS

2–1 Operators. Let a class of functions be given, all defined for the same values of the variable; for example, the class of all continuous complex functions of t, $-\infty < t < \infty$. By an *operator* or *transformation* we mean an assignment of a member F of a second class of functions to each function f in the given class.

For example, let the given class be, as above, formed of all functions f continuous for $-\infty < t < \infty$. To each such f we assign a function $F(t)$ by the rule

$$F(t) = \int_0^t f(u)\, du.$$

The functions $F(t)$ are precisely those functions which have a continuous derivative for $-\infty < t < \infty$ and which equal zero for $t = 0$.

As a second example we may define, for the same class of functions f,

$$F(t) = \max \operatorname{Re} [f(u)], \quad u \text{ on the interval from 0 to } t. \tag{2–10}$$

If $f = e^t$, then $F = 1 - h(t) + e^t h(t)$; if $f = t^2 + it^3$, then $F = t^2$.

One can easily construct more examples:

$$F(t) = |f(t)| \text{ (absolute value operator)}; \tag{2–11}$$

$$F(t) = [f(t)]^2 \text{ (squaring operator)}; \tag{2–12}$$

$$F(t) = f(t + 1) - f(t) \text{ (difference operator)}; \tag{2–13}$$

$$F(t) = (D^2 - 1)f \text{ (differential operator)}; \tag{2–14}$$

for the first three the functions f can be as above. For the fourth example, the functions f must have derivatives of second order on the interval considered.

The following examples are of much importance in later chapters:

(a)
$$F(\omega) = \int_{-\infty}^{\infty} f(t) e^{-i\omega t}\, dt, \tag{2–15}$$

where the operation is that of forming the *Fourier transform* of f.

(b)
$$F(s) = \int_0^{\infty} f(t) e^{-st}\, dt, \tag{2–16}$$

which is the *Laplace transform* of f.

(c)
$$F(t) = \int_{-\infty}^{\infty} f(u) W(t - u)\, du, \tag{2–17}$$

64

where $W(t)$ is a given function, $-\infty < t < \infty$; this is the operation of forming the *convolution* of f with W. The descriptions of the classes concerned and precise definitions are given in later chapters.

An operator can be regarded as a generalization of the concept of a function. A function assigns numerical values to numerical values; to each t there is assigned the number $f(t)$. An operator assigns functions to functions; to each f there is a corresponding F. For the present we write $F = T[f]$ to suggest the functionlike relationship between f and F. The analogy is brought out even more forcefully by tabular representation. Thus for the function $g(t) = \sin t$ and the operator $T[f] = F = f(t+1) - f(t)$ we have Tables 2–1 and 2–2.

<table>
<tr><td colspan="2">TABLE 2–1</td></tr>
<tr><td>t</td><td>$\sin t$</td></tr>
<tr><td>0</td><td>0</td></tr>
<tr><td>0.1</td><td>0.09983</td></tr>
<tr><td>0.2</td><td>0.19867</td></tr>
<tr><td>0.3</td><td>0.29552</td></tr>
</table>

<table>
<tr><td colspan="2">TABLE 2–2</td></tr>
<tr><td>f</td><td>$F = f(t+1) - f(t)$</td></tr>
<tr><td>t^2</td><td>$2t + 1$</td></tr>
<tr><td>t^3</td><td>$3t^2 + 3t + 1$</td></tr>
<tr><td>e^t</td><td>$e^t(e - 1)$</td></tr>
<tr><td>$\sin t$</td><td>$(\cos 1 - 1) \sin t + \sin 1 \cos t$</td></tr>
</table>

The tabulation of operators is now almost as widespread as the tabulation of functions. Consider, for example, the tables of integrals, of Laplace transforms, of Fourier transforms.

One-to-one transformations. Inverse transformation. Let $T[f]$ be a transformation. If $T[f_1] = T[f_2]$ implies $f_1 = f_2$, then T is said to be a *one-to-one transformation*. For example, if $T[f] = 2f + e^t$ in the class of continuous functions, then $T[f_1] = T[f_2]$ means $2f_1 + e^t = 2f_2 + e^t$, so that $f_1 = f_2$. On the other hand, if $T[f] = f' = Df$ in the class of continuously differentiable functions for $-\infty < t < \infty$, then T is not a one-to-one transformation, since for each f_1 all functions $f_2 = f_1 + \text{const}$ have the same derivative as f_1. When T is one-to-one, f is uniquely determined by $g = T[f]$ and we write $f = T^{-1}[g]$. We thereby define the *inverse* T^{-1} of transformation T: T^{-1} is a transformation defined in the class of all $g = T[f]$, for f in its given class. For example, if $T[f] = 2f + e^t$ as above, then $T^{-1}[g] = \frac{1}{2}(g - e^t)$ in the class of continuous functions.

The operator D^{-1}. As remarked above, the operator D is not one-to-one if we consider D in the class of continuously differentiable functions. Hence D^{-1} is not defined. However, as in Section 1–6, it is convenient to use D^{-1} as the indefinite integral, an ambiguously defined inverse of D. We can make D one-to-one by restricting ourselves to functions $f(t)$ with given value at $t = 0$; if, for example, we require $f(0) = 0$, then the inverse operator is D_0^{-1}.

2-2 Linear operators. A class of functions is said to be a *linear space* if, whenever f_1, f_2 belong to the class, $c_1f_1 + c_2f_2$ belongs to the class for every choice of the complex constants c_1, c_2. For example, the functions continuous on a given interval form a linear space.

Let a linear space of functions be given and let T be an operator defined in this class, with values in another class of functions. If T satisfies the identity

$$T[c_1f_1 + c_2f_2] = c_1T[f_1] + c_2T[f_2] \qquad (2\text{-}20)$$

for all f_1, f_2 in the given linear space and all complex constants c_1, c_2, then T is called a *linear operator* or *linear transformation*. We can restrict the constants c_1, c_2 to be real. We then speak of a *real* linear space and *real* linear operator. (Generally we assume functions and constants to be complex.)

The simplest linear operator is multiplication by a scalar; for example, $T[f] \equiv 2f$ defines a linear operator, since

$$2[c_1f_1 + c_2f_2] = c_1[2f_1] + c_2[2f_2].$$

Multiplication by the scalar 1 defines the *identity operator: $T[f] \equiv f$.* The identity operator is usually denoted by I.

More generally, multiplication by a *fixed* function is a linear operator; for example, let $f_0 = \sin t$ and let $F = T[f] = f_0 f$. Then

$$T[c_1f_1 + c_2f_2] = f_0(c_1f_1 + c_2f_2) = c_1(f_0f_1) + c_2(f_0f_2)$$
$$= c_1T[f_1] + c_2T[f_2].$$

However, multiplication of f by itself (i.e., squaring) is nonlinear. For example,

$$(2t + 3e^t)^2 \neq 2t^2 + 3e^{2t}.$$

Integration from 0 to t (in the class of continuous functions) is a linear operator:

$$T[c_1f_1 + c_2f_2] = \int_0^t [c_1f_1(u) + c_2f_2(u)]\, du$$
$$= c_1\int_0^t f_1(u)\, du + c_2\int_0^t f_2(u)\, du = c_1T[f_1] + c_2T[f_2].$$

From given linear operators, T_1, T_2, ... defined in a given linear space, with values in another linear space, we can construct new linear operators by addition and multiplication of operators and by multiplication by scalars. Thus $T_1 + T_2$ is defined to be the operator such that

$$(T_1 + T_2)[f] = T_1[f] + T_2[f]. \qquad (2\text{-}21)$$

The operator $T_1 \cdot T_2$, or simply $T_1 T_2$, is defined by the equation

$$(T_1 T_2)f = T_1[T_2[f]]; \qquad (2\text{–}22)$$

that is, T_1 is to be applied to the function $T_2[f]$. Accordingly, we must require that the values of T_2 be in the linear space in which T_1 is defined. As a special case, the powers of a linear operator T are defined by the equations

$$T^0 = I, \quad T^1 = T, \quad T^2 = T \cdot T, \quad T^3 = T \cdot T^2, \ldots \qquad (2\text{–}23)$$

Finally, if c is a scalar, cT is the linear operator defined by the equation

$$(cT)[f] = c(T[f]). \qquad (2\text{–}24)$$

Combining the operations, we can form a polynomial operator:

$$c_0 T^n + c_1 T^{n-1} + \cdots + c_{n-1} T + c_n I. \qquad (2\text{–}25)$$

The verification of linearity for the operators of the preceding paragraph is left as an exercise (Problems 9 and 10 below).

Inverse of a linear operator. A linear operator need not be one-to-one. For example, $T[f] = f'$, in the class of differentiable functions for $-\infty < t < \infty$, is linear but not one-to-one. If a linear operator T is one-to-one, then its inverse T^{-1} is defined in a linear space and is linear. For if $T^{-1}[g_1]$ and $T^{-1}[g_2]$ are defined, then $T[f_1] = g_1$, $T[f_2] = g_2$ for some (uniquely determined) f_1, f_2. Accordingly, $T[c_1 f_1 + c_2 f_2] = c_1 g_1 + c_2 g_2$, so that $T^{-1}[c_1 g_1 + c_2 g_2]$ is defined and

$$T^{-1}[c_1 g_1 + c_2 g_2] = c_1 f_1 + c_2 f_2 = c_1 T^{-1}[g_1] + c_2 T^{-1}[g_2].$$

When the power $(T^{-1})^k$ is defined, it is denoted by T^{-k} ($k = 0, 1, 2 \ldots$).

Equality of linear operators. We write $T_1 = T_2$ if T_1, T_2 are defined in the same linear space and $T_1[f] = T_2[f]$ for all f.

Problems

1. Determine whether each class described forms a linear space (real or complex):

(a) All functions defined for $-\infty < t < \infty$ and having real values;

(b) All functions defined for $-\infty < t < \infty$ and having positive real values;

(c) All complex functions defined for $-\infty < t < \infty$ and having a continuous first derivative;

(d) All complex functions f defined for $t \geqq 0$ and such that $|f(t)|$ is *bounded* (that is, $|f(t)| < M$ for some constant M, depending on f);

(e) All complex functions defined and continuous for $t \geq 0$ and such that $\int_0^\infty |f(t)|\, dt$ exists.

2. A linear space of functions is said to form an *algebra* if, whenever f and g are in the class, their product fg is in the class. Determine which of the classes of Problem 1 is an algebra [(e) is difficult].

3. (a) Show that the operator defined by (2–10) is nonlinear.

(b) Tabulate this operator for the following values of f:

$$t,\ t^2 + it^3,\ t + t^2 + it^3,\ \sin t,\ e^{-t}.$$

4. Show that the operator defined by (2–11) is nonlinear.

5. Show that the operators defined by (2–13) and (2–14) are linear.

6. (a) Show that the operators (2–15) and (2–16) are linear if we restrict f in each case to the class of piecewise continuous functions for which the integral exists for all real ω and for Re $s > 0$ respectively.

(b) Show that, in general, an *integral operator* of the form

$$T[f] \equiv F(t) = \int_a^b f(u)K(t, u)\, du$$

is linear if K is continuous for $a \leq t \leq b$, $a \leq u \leq b$, and $f(t)$ is piecewise continuous for $a \leq t \leq b$.

7. Let f be restricted to the class of functions continuous for $t \geq 0$. Let $T[f] = F$, where

$$F(t) = e^{-2t}\int_0^t e^{2u}f(u)\, du + f(0)e^{-2t}.$$

Show that T is linear and one-to-one. Find an expression for $f = T^{-1}[F]$.

8. Show that for a linear transformation, $T[0] = 0$. (Here 0 stands for a function which is identically zero.)

9. Let T_1, T_2 be linear operators defined in the same linear space and having values in another linear space. (a) Show that $T_1 + T_2$, as defined by Eq. (2–21), is a linear operator. (b) Show that cT_1, as defined by Eq. (2–24), is a linear operator.

10. (a) Let T_1, T_2 be linear operators such that $T_1[T_2[f]]$ is defined wherever $T_2[f]$ is defined. Show that T_1T_2, as defined by Eq. (2–22), is a linear operator.

(b) Let T be a linear operator, defined and having values in the same linear space. Show that T^n ($n = 0, 1, 2, \ldots$), as defined by Eq. (2–23), is a linear operator, and that $T^{m+n} = T^m \cdot T^n$ for $m = 0, 1, 2, \ldots, n = 0, 1, 2, \ldots$

(c) Show that, under the hypotheses of part (b), each polynomial operator (2–25) is linear.

11. Let T and T^{-1} both be linear operators defined in the same linear space, so that T^n is defined for $n = 0, \pm 1, \pm 2, \ldots$ Show that $T^{m+n} = T^m \cdot T^n$ ($m = 0, \pm 1, \ldots, n = 0, \pm 1, \ldots$).

12. Let T_1, T_2, T_1^{-1}, T_2^{-1} all be defined in the same linear space. Show that T_1T_2, $(T_1T_2)^{-1}$ are defined in the same space and

$$(T_1T_2)^{-1} = T_2^{-1}T_1^{-1}.$$

13. Let T_1, T_2, T_3 be linear operators defined in the same linear space, with values in a linear space. Prove the following:

(a) $T_1 + T_2 = T_2 + T_1$
(b) $T_1 + (T_2 + T_3) = (T_1 + T_2) + T_3$
(c) $c(T_1 + T_2) = (cT_1) + (cT_2)$
(d) $T_1(T_2T_3) = (T_1T_2)T_3$
(e) $c(T_1T_2) = (cT_1)T_2 = T_1(cT_2)$
(f) $T_1(T_2 + T_3) = T_1T_2 + T_1T_3$

For (d), (e), and (f) assume that the operators have values in the given linear space in which they are defined.

14. (a) Show that the commutative law for multiplication of linear operators is not generally valid by proving that $T_1T_2 \neq T_2T_1$, when $T_1[f] = f(0)t$, $T_2[f] = f(1)t^2$, where f ranges over the linear space of all continuous functions of t, $-\infty < t < \infty$.

(b) Show that $(cI)T = T(cI)$ for every linear operator T with values in the same linear space in which T is defined.

(c) For each of the following determine which of the operators T_1, T_2 commute (that is, $T_1T_2 = T_2T_1$), the operators being defined in the linear space of continuous functions of t, $-\infty < t < \infty$:

(i) $T_1[f] = 3f$, $\qquad T_2[f] = \int_0^t f(u)\,du$
(ii) $T_1[f(t)] = f(t-1)$, $\qquad T_2[f(t)] = f(-t)$
(iii) $T_1[f(t)] = tf(t)$, $\qquad T_2[f] = \int_0^t f(u)\,du$

15. For each $f(t)$ defined for $-\infty < t < \infty$ and for fixed real c, let $\Delta_c[f]$ be equal to $f(t-c)$. We call Δ_c the *translation* or *time-delay* operator.

(a) Show that Δ_c is linear.
(b) Show that $\Delta_a\Delta_b = \Delta_{a+b}$.
(c) Determine the functions f for which $\Delta_c[f] = f$ $(c \neq 0)$.

16. For each $f(t)$ defined for $-\infty < t < \infty$ let $\Theta[f]$ be equal to $f(-t)$. We call Θ the *reflection operator*.

(a) Show that Θ is a linear operator.
(b) Show that $\Theta^2 = I$.
(c) Determine the functions f for which $\Theta[f] = f$ and the functions f for which $\Theta[f] = -f$.

17. For the class of complex-valued $f(t)$, $-\infty < t < \infty$, let $\mathcal{C}$ be the operator such that $\mathcal{C}[f] = \bar{f}$; we call $\mathcal{C}$ the *conjugation* operator. Show that $\mathcal{C}[f + g] = \mathcal{C}[f] + \mathcal{C}[g]$, but that $\mathcal{C}$ is not linear, unless we restrict to real coefficients. Show that $\mathcal{C}^2 = I$.

18. We can permit the functions f and F to be *generalized functions*, in the definition of operators. Consider the class of generalized functions f of form $g(t)h(t) + a_1\,\delta^{(k_1)}(t - c_1) + a_2\,\delta^{(k_2)}(t - c_2) + \cdots$, where $g(t)$ is an ordinary function having continuous derivatives of all orders for $-\infty < t < \infty$.

(a) Show that D, the operation of differentiation, is a linear operator in this class, as is D^k $(k = 1, 2, 3, \ldots)$.

(b) Show that $D^k\Delta_c = \Delta_c D^k$, where Δ_c is the translation operator of Problem 15.

ANSWERS

1. (a) is a real linear space; (c), (d), (e) are complex linear spaces.
2. (a), (c), (d) are algebras.
3. (b) $T[t] = th(t)$

$T[t^2 + it^3] = t^2$

$T[t + t^2 + it^3] = (t + t^2) \cdot [1 - h(t + 1) + h(t)]$

$T[\sin t] = 1 - h\left(t + \dfrac{3\pi}{2}\right)$

$\qquad + \sin t \left[h\left(t + \dfrac{3\pi}{2}\right) - h(t + \pi) + h(t) - h\left(t - \dfrac{\pi}{2}\right) \right]$

$\qquad + h\left(t - \dfrac{\pi}{2}\right)$

$T[e^{-t}] = e^{-t}[1 - h(t)] + h(t)$

7. $T^{-1}[F] = F' + 2F$
14. (c) (i) and (ii) commute.
15. (c) f has period c.
16. (c) the even functions, the odd functions.

2–3 Operators associated with linear differential equations. With the differential equation

$$(a_0 D^n + \cdots + a_n)x = f(t), \qquad (2\text{–}30)$$

we associate first the *differential operator*

$$T = a_0 D^n + \cdots + a_n = a_0 D^n + \cdots + a_{n-1}D + a_n I. \quad (2\text{–}31)$$

The operator is defined in the linear space of functions having derivatives through the nth order on a given interval. The coefficients $a_0, \ldots, a_n$ may be constant or may be functions of t on the given interval. The operator is *linear* within the space described:

$$T[c_1 f_1 + c_2 f_2] = c_1 T[f_1] + c_2 T[f_2].$$

This follows from the linearity of the operator D and its powers $D^2, \ldots, D^n$. From the linearity we deduce the *superposition principle:* if $x_1(t)$ is a solution of Eq. (2–30) with $f = f_1(t)$, and $x_2(t)$ is a solution of Eq. (2–30) with $f = f_2(t)$, then $c_1 x_1(t) + c_2 x_2(t)$ is a solution of Eq. (2–30) with $f = c_1 f_1(t) + c_2 f_2(t)$ (c_1, c_2 being constants). For

$$T[x_1] = f_1, \qquad T[x_2] = f_2,$$

$$T[c_1 x_1 + c_2 x_2] = c_1 T[x_1] + c_2 T[x_2] = c_1 f_1 + c_2 f_2.$$

Another operator is obtained by assigning to each $f(t)$ (of a certain class) the solution of Eq. (2–30) satisfying given initial conditions. For example, let $T[f]$ be the solution of the equation

$$(D + 1)x = f(t)$$

such that $x = 1$ when $t = 0$. We find (Problem 1 below)

$$x = e^{-t}\int_0^t e^u f(u)\, du + e^{-t} = T[f].$$

This operator is *not* linear. In particular,

$$T[e^{-t}] = (t + 1)e^{-t},$$
$$T[2e^{-t}] = (2t + 1)e^{-t} \neq 2T[e^{-t}].$$

For the general first order equation

$$[a_0(t)D + a_1(t)]x = f(t), \quad (a_0(t) \neq 0) \tag{2–32}$$

the solution with $x = x_0$ for $t = 0$ is given by

$$x = q(t)\left[\int_0^t \{q(u)a_0(u)\}^{-1}f(u)\, du + x_0\right],$$
$$q(t) = \exp\left[-\int_0^t a_1(v)\{a_0(v)\}^{-1}\, dv\right] \tag{2–33}$$

(Problem 2 below). For each fixed x_0, this defines an operator T which is not linear, *except when* $x_0 = 0$. When $x_0 = 0$, we obtain an operator which we denote by $T_0[f]$:

$$T_0[f] = q(t)\int_0^t \{q(u)a_0(u)\}^{-1}f(u)\, du.$$

We can verify at once that T_0 is linear.

The conclusion for the first order equation can be generalized to the equation of order n:

THEOREM 1. *In Eq.* (2–30) *let the coefficients* $a_0, \ldots, a_n$ *and the right-hand member* f *be continuous functions of* t, $-\infty < t < \infty$, *with* $a_0(t) \neq 0$. *The solution with given initial values* $x_0, x_0', \ldots, x_0^{(n-1)}$ *at* $t = 0$ *can be written thus:*

$$x = T[f] = T_0[f] + x_0 g_1(t) + x_0' g_2(t) + \cdots + x_0^{(n-1)}g_n(t), \tag{2–34}$$

where $T_0[f]$ *is the solution when all initial values are zero, and* $g_1(t), \ldots, g_n(t)$ *are certain solutions of the related homogeneous equation. The operator* $T_0[f]$ *is linear.*

To prove the theorem we let $f(t)$ be given and denote by $F(t)$ the function $T_0[f]$; thus $x = F(t)$ is the solution of Eq. (2–30) such that $x = 0$, $Dx = 0, \ldots, D^{n-1}x = 0$ when $t = 0$. The function $g_1(t)$ is chosen as the solution of the related homogeneous equation

$$(a_0 D^n + \cdots + a_n)x = 0 \qquad (2\text{--}35)$$

such that $x = 1$, $Dx = 0$, $D^2x = 0, \ldots, D^{n-1}x = 0$ when $t = 0$; the function $g_2(t)$ is the solution of (2–35) such that $x = 0$, $Dx = 1$, $D^2x = 0$, $\ldots, D^{n-1}x = 0$ when $T = 0$. In general, $g_k(t)$ is the solution of (2–35) such that the initial value of $D^{k-1}x$ is 1, while the other initial values are zero. In the equation

$$x = F(t) + x_0 g_1(t) + \cdots + x_0^{(n-1)} g_n(t) \qquad (2\text{--}36)$$

we now set $t = 0$; we find

$$x(0) = F(0) + x_0 g_1(0) + \cdots + x_0^{(n-1)} g_n(0) = x_0,$$

by the choice of initial values. Similarly,

$$x^{(k-1)}(t) = F^{(k-1)}(t) + x_0 g_1^{(k-1)}(t) + \cdots + x_0^{(n-1)} g_n^{(k-1)}(t),$$

$$x^{(k-1)}(0) = x_0^{(k-1)} \qquad (k = 1, 2, \ldots, n),$$

so that all initial conditions are satisfied. Since $F(t)$ is a solution of the nonhomogeneous equation, while $g_1(t), \ldots, g_n(t)$ are solutions of the related homogeneous equation, it follows that Eq. (2–36) defines $x(t)$ as a solution of Eq. (2–30).

Finally we verify that T_0 is linear. Let $x_1(t) = T_0[f_1]$, $x_2(t) = T_0[f_2]$, so that $(a_0 D^n + \cdots + a_n)x_1(t) = f_1$, $(a_0 D^n + \cdots)x_2(t) = f_2$ and $x_1(t)$, $x_2(t)$ have all initial values zero at $t = 0$. By the linearity of the operator $(a_0 D^n + \cdots + a_n)$, we conclude that

$$(a_0 D^n + \cdots + a_n)[c_1 x_1(t) + c_2 x_2(t)] = c_1 f_1(t) + c_2 f_2(t);$$

furthermore, $c_1 x_1(t) + c_2 x_2(t)$ also has all initial values zero at $t = 0$. Hence

$$T_0[c_1 f_1 + c_2 f_2] = c_1 x_1(t) + c_2 x_2(t) = c_1 T_0[f_1] + c_2 T_0[f_2].$$

Thus T_0 is linear.

Remark 1. Equation (2–34) gives x as a linear expression in the forcing function f and initial values $x_0, x_0', \ldots, x_0^{(n-1)}$. We might say that x depends linearly on the $n + 1$ "quantities" $f, x_0, x_0', \ldots, x_0^{(n-1)}$. We can also consider the initial values as part of the input and say simply that x depends linearly on the input. Another procedure is to seek solutions only for $t > 0$ and impose initial conditions only as $t \to 0+$. For an equation (2–30) with constant coefficients, *all* solutions are obtained by adding to f all possible linear combinations of $\delta(t)$, $\delta'(t)$, ..., $\delta^{(n-1)}(t)$ and by choosing the solution of the new equation such that $x \to 0$, $Dx \to 0$, ..., $D^{(n-1)}x \to 0$ as $t \to 0-$. By proper choice of the coefficients of $\delta(t)$, $\delta'(t)$, ..., the solution will have desired initial values x_0, x_0', ..., as $t \to 0+$. In other words, we obtain all solutions for $t > 0$ as $T_0[f]$, where f is allowed to contain terms in $\delta(t)$, ..., $\delta^{(n-1)}(t)$, and $T_0[f]$ is the solution such that $x \to 0$, $Dx \to 0$, ... as $t \to 0-$; $T_0[f]$ is a linear operator in the linear space of generalized functions concerned. For a proof of these results see Problem 6 following Section 1–14; see also Section 6–18.

Remark 2. For differential equations (2–30) with constant coefficients, we can consider the solution x as the result of applying an *inverse differential operator* to f:

$$x = \frac{1}{a_0 D^n + \cdots + a_n} f.$$

We can consider all initial values to be zero, so that the new operator is $T_0[f]$, or we can leave the initial values arbitrary, so that the inverse differential operator contains n arbitrary constants. A special case of this operator is the operator D^{-1} considered in Section 1–6; the operator D_0^{-1} corresponds to T_0 for this case. The *Heaviside calculus* is a set of rules for manipulating differential operators and their inverses; see, for example, pp. 161–170 of Reference 4 listed at the end of the chapter. Equivalent rules are obtained by means of the Fourier and Laplace transforms studied in the present book.

2–4 Operators associated with simultaneous linear differential equations. Let us consider first a pair of equations

$$\frac{dx}{dt} - a_{11}x - a_{12}y = f(t), \qquad \frac{dy}{dt} - a_{21}x - a_{22}y = g(t), \quad (2\text{–}40)$$

as in Section 1–4. Our input is now a pair of functions f, g, and our output is a pair of functions $x(t)$, $y(t)$. The natural generalization of the transformation of the preceding sections leads one to the concept of *matrices*. (See Section 8–13 of this book and pp. 273–299 of Reference 4.) Here we follow the point of view stressed at the end of Section 1–5 and consider the dependence of each output function on each input function.

We consider first the dependence of x on f, g being zero. Let initial values x_0, y_0 be given for $t = 0$ and assume continuity conditions satisfied so that there is a unique solution $x(t)$, $y(t)$ with these initial values for $-\infty < t < \infty$. Considering only x, we can write $x = T[f]$. When x_0, y_0 are zero, T reduces to a linear transformation T_0. For given f_1, f_2

(continuous for $-\infty < t < \infty$), let $x_1(t)$, $y_1(t)$ and $x_2(t)$, $y_2(t)$ be the solutions of

$$(D - a_{11})x - a_{12}y = f_1, \qquad -a_{21}x + (D - a_{22})y = 0,$$
$$(D - a_{11})x - a_{12}y = f_2, \qquad -a_{21}x + (D - a_{22})y = 0,$$

respectively, with $x(0) = 0$, $y(0) = 0$. Then

$$(D - a_{11})(c_1x_1 + c_2x_2) - a_{12}(c_1y_1 + c_2y_2) = c_1f_1 + c_2f_2,$$
$$-a_{21}(c_1x_1 + c_2x_2) + (D - a_{22})(c_1y_1 + c_2y_2) = 0;$$

that is, $c_1x_1 + c_2x_2$, $c_1y_1 + c_2y_2$ is a solution pair, and clearly the initial values for $t = 0$ are zero. Considering only x, we conclude that

$$T_0[c_1f_1 + c_2f_2] = c_1x_1 + c_2x_2 = c_1T_0[f_1] + c_2T_0[f_2].$$

Thus T_0 is linear. By an argument parallel to that of the preceding section we conclude that for general initial values,

$$x = T_0[f] + x_0p_1 + y_0p_2, \tag{2-41}$$

where p_1, p_2 are functions of t (independent of f).

For the function y we obtain similar results:

$$y = S_0[f] + x_0q_1 + y_0q_2. \tag{2-41'}$$

Furthermore, when $f = 0$ but $g \neq 0$, a similar analysis applies; let $x = U_0[g]$, $y = V_0[g]$ be the corresponding solution with initial values zero. Then for arbitrary initial values and arbitrary continuous f, g,

$$x = T_0[f] + U_0[g] + x_0p_1 + y_0p_2,$$
$$y = S_0[f] + V_0[g] + x_0q_1 + y_0q_2.$$

The functions $p_1(t)$, $q_1(t)$ are the solution pair x, y for the related homogeneous equations

$$(D - a_{11})x - a_{12}y = 0,$$
$$-a_{21}x + (D - a_{22})y = 0 \tag{2-42}$$

with initial values $x(0) = 1$, $y(0) = 0$; the functions $p_2(t)$, $q_2(t)$ form the solution x, y of (2–42) with initial values $x(0) = 0$, $y(0) = 1$. (See Problem 9 below.)

The formulas (2–41) and (2–41′) generalize to the case of n equations of first order:

$$Dx_1 - a_{11}x_1 - a_{12}x_2 - \cdots - a_{1n}x_n = f_1(t),$$
$$Dx_2 - a_{21}x_1 - a_{22}x_2 - \cdots - a_{2n}x_n = f_2(t), \qquad (2\text{–}43)$$
$$\vdots$$
$$Dx_n - a_{n1}x_1 - a_{n2}x_2 - \cdots - a_{nn}x_n = f_n(t).$$

Here each solution is a set of n functions $x_1(t), \ldots, x_n(t)$. If $f_1(t), \ldots, f_n(t)$ are continuous for all t, the solution is uniquely determined by the initial values $x_{10}, \ldots, x_{n0}$ of $x_1, \ldots, x_n$. The solution can be written in the form

$$x_1 = T_{11}[f_1] + \cdots + T_{1n}[f_n] + x_{10}p_{11} + \cdots + x_{n0}p_{1n},$$
$$\vdots \qquad (2\text{–}44)$$
$$x_n = T_{n1}[f_1] + \cdots + T_{nn}[f_n] + x_{10}p_{n1} + \cdots + x_{n0}p_{nn},$$

where the T_{ij} are linear transformations and the p_{ij} are functions of t independent of $f_1, \ldots, f_n$.

Similar results apply to general simultaneous equations not in basic form. For example, the solutions of the equations

$$(a_0 D^n + \cdots)x + (b_0 D^m + \cdots)y = f(t),$$
$$(c_0 D^n + \cdots)x + (d_0 D^m + \cdots)y = g(t), \qquad (2\text{–}45)$$

assumed solvable for $D^n x$ and $D^m y$, are given as follows:

$$x = T_{11}[f] + T_{12}[g] + x_0 p_{11}(t) + \cdots + x_0^{(n-1)}p_{1n}(t)$$
$$+ y_0 q_{11}(t) + \cdots + y_0^{(m-1)}q_{1m}(t),$$
$$\qquad (2\text{–}46)$$
$$y = T_{21}[f] + T_{22}[g] + x_0 p_{21}(t) + \cdots + x_0^{(n-1)}p_{2n}(t)$$
$$+ y_0 q_{21}(t) + \cdots + y_0^{(m-1)}q_{2m}(t).$$

Integrodifferential equations are reducible to simultaneous linear differential equations, so that the conclusions can be extended to them. For example, the solutions of an equation

$$(a_0 D^2 + a_1 D + a_2 + a_3 D^{-1} + a_4 D^{-2})x = f \qquad (2\text{–}47)$$

are represented thus:

$$x = T_0[f] + x_0 p_1(t) + x_0' p_2(t) + x_0^{(-1)}p_3(t) + x_0^{(-2)}p_4(t). \qquad (2\text{–}48)$$

(See Problem 10 below.)

PROBLEMS

1. Find the solution of the equation

$$\frac{dx}{dt} + x = f(t)$$

such that $x = 1$ when $t = 0$. [*Hint:* e^t is an integrating factor.]

2. Verify directly that Eq. (2–33) defines a solution of Eq. (2–32) such that $x = x_0$ when $t = 0$. [*Hint:* Note that $q'(t) = -a_1(t)q(t)/a_0(t)$.]

3. For each of the following find the solution with initial value x_0 for $t = 0$ and show that it has the form $T_0[f] + x_0 g$, where T_0 is linear and g is independent of f and x_0:

(a) $\dfrac{dx}{dt} + 2tx = f(t)$ (b) $\dfrac{dx}{dt} + \dfrac{t}{1 + t^2} x = f(t)$

4. Verify that for each of the following equations with constant coefficients, the solution with initial values zero for $t = 0$ is given by the operator T_0 shown.

(a) $(D + a)x = f$, $T_0[f] = \displaystyle\int_0^t e^{-a(t-u)} f(u)\, du$

(b) $(D^2 + 3D + 2)x = f$, $T_0[f] = \displaystyle\int_0^t [e^{u-t} - e^{2(u-t)}] f(u)\, du$

(c) $[D^2 + (a + b)D + ab]x = f$, $(a \neq b)$

$$T_0[f] = \frac{1}{b - a} \int_0^t [e^{a(u-t)} - e^{b(u-t)}] f(u)\, du$$

(d) $[D^2 - 2aD + a^2 + b^2]x = f$, $(b \neq 0)$

$$T_0[f] = \frac{i}{2b} \int_0^t [e^{(a-bi)(t-u)} - e^{(a+bi)(t-u)}] f(u)\, du$$

$$= \frac{1}{b} \int_0^t e^{a(t-u)} \sin b(t - u)\, f(u)\, du$$

5. *Method of variation of parameters.* To find a particular solution of a second order equation

$$[a_0(t) D^2 + a_1(t) D + a_2(t)]x = f(t),$$

when the complementary function $x_c(t) = c_1 p(t) + c_2 q(t)$ is known, we write the equations

$$x = c_1 p(t) + c_2 q(t), \qquad x' = c_1 p'(t) + c_2 q'(t)$$

and replace the constants c_1, c_2 by u, v, unknown functions of t:

$$x = u p(t) + v q(t), \qquad x' = u p'(t) + v q'(t).$$

From these two equations it follows that

$$u'p(t) + v'q(t) = 0. \tag{a}$$

Substitution in the differential equation gives the relation

$$a_0(t)[u'p'(t) + v'q'(t)] = f(t). \tag{b}$$

Equations (a) and (b) can be solved for u', v'; integration then gives u, v up to two arbitrary constants, which can be chosen to make $u = 0, v = 0$ for $t = 0$; then $x = up(t) + vq(t) = T_0[f]$.

Apply the method described to find the solution $T_0[f]$ for the following equations:

(a) $[(t - 1)D^2 - tD + 1]x = f(t)$, $x_c = c_1 t + c_2 e^t$
(b) $[(t^2 - 1)D^2 - 2tD + 2]x = f(t)$, $x_c = c_1 t + c_2(1 + t^2)$
(c) The equation of Problem 4(c)   (d) The equation of Problem 4(d)

6. Find the solution $x(t)$, $y(t)$ of the simultaneous equations

$$(D + 4)x - 3y = 0, \qquad 2x + (D - 1)y = g$$

with initial values x_0, y_0 for $t = 0$. [*Hint:* Use the method of elimination of Section 1–5 to obtain an equation for x similar to that of Problem 4(b) above.]

7. *Method of variation of parameters for simultaneous equations.* For the equations

$$(D - a_{11})x - a_{12}y = f, \qquad -a_{21}x + (D - a_{22})y = g$$

let the general solution for $f = 0$, $g = 0$ be written as

$$x = c_1 p_1(t) + c_2 p_2(t), \qquad y = c_1 q_1(t) + c_2 q_2(t).$$

Now introduce new variables u, v by the equations

$$x = up_1(t) + vp_2(t), \qquad y = uq_1(t) + vq_2(t).$$

Substitution in the differential equations then leads to the equations

$$u'p_1(t) + v'p_2(t) = f, \qquad u'q_1(t) + v'q_2(t) = g,$$

from which u', v' can be found. Integration then gives u, v, up to two arbitrary constants, which can be chosen so that $u = 0$, $v = 0$ for $t = 0$; accordingly

$$x = T_0[f] + U_0[g], \qquad y = S_0[f] + V_0[g],$$

as in Eqs. (2–41) and (2–41′).

Apply the method to find the solution with initial values zero for $t = 0$ for the following pairs of equations. The method of elimination can be used to find the general solution for $f = 0$, $g = 0$.

(a) $(D + 4)x - 3y = f$, $2x + (D - 1)y = g$ (see Problem 6)

(b) $(D + 3)x - 5y = f,$ $x + (D - 3)y = g$
(c) $(D + 1)x - y = f,$ $2x + (D - 1)y = g$

8. A physical system is known to be governed by a linear differential equation of form $(a_0 D^n + \cdots + a_n)x = f$. With a certain fixed set of initial conditions the following observations are made: when $f(t) = t$, $x = g_1(t)$; when $f(t) = 1 + t$, $x = g_2(t)$; when $f(t) = t^2$, $x = g_3(t)$. Find $x(t)$ when all initial values are zero and $f(t)$ is as follows:

(a) $f(t) = 1$ (b) $f(t) = t^2 - t$ (c) $f(t) = 2 - 3t + 3t^2$

9. Verify that Eqs. (2-41) and (2-41') define the solution x, y of Eq. (2-40) such that $x(0) = x_0$, $y(0) = y_0$, provided $T_0[f]$, $U_0[g]$, $S_0[f]$, $V_0[g]$, $p_1(t)$, $p_2(t)$, $q_1(t)$, $q_2(t)$ are chosen as described in the text.

10. Show that the solution of Eq. (2-47) with initial values of x, Dx, $D^{-1}x$, $D^{-2}x$ for $t = 0$ equal to x_0, x_0', $x_0^{(-1)}$, $x_0^{(-2)}$, is representable in the form (2-48), where T_0 is a linear operator.

11. Given the integrodifferential equation $(D + 4 + 3D^{-1})x = f$, show that the solution with all initial values zero for $t = 0$ is given by

$$x = T_0[f] = \frac{1}{2}\int_0^t [3e^{3(u-t)} - e^{u-t}]f(u)\, du.$$

[Hint: Set $y = D^{-1}x$, so that $x = Dy$, and replace the given equation by the pair $(D + 4)x + 3y = f$, $Dy - x = 0$. Then apply variation of parameters as in Problem 7.]

ANSWERS

1. $e^{-t}\int_0^t e^u f(u)\, du + e^{-t}$
3. (a) $e^{-t^2}[\int_0^t e^{u^2} f(u)\, du + x_0]$
  (b) $(1 + t^2)^{-1/2}[\int_0^t (1 + u^2)^{1/2} f(u)\, du + x_0]$
5. (a) $-t\int_0^t (u - 1)^{-2} f(u)\, du + e^t \int_0^t (u - 1)^{-2} u e^{-u} f(u)\, du$
  (b) $-t\int_0^t (u^2 - 1)^{-2}(u^2 + 1)f(u)\, du + (1 + t^2)\int_0^t (u^2 - 1)^{-2} u f(u)\, du$
6. $x = 3T_0[g] + x_0(3e^{-2t} - 2e^{-t}) + y_0(3e^{-t} - 3e^{-2t})$
  $y = (D + 4)T_0[g] + x_0(2e^{-2t} - 2e^{-t}) + y_0(3e^{-t} - 2e^{-2t})$
  where $T_0[g] = \int_0^t [e^{u-t} - e^{2(u-t)}]g(u)\, du$
7. (a) Let $T_k[\phi] = e^{kt}\int_0^t e^{-k\alpha}\phi(\alpha)\, d\alpha$. Then

$$x = T_{-1}[3g - 2f] + T_{-2}[3f - 3g]$$
$$y = T_{-1}[3g - 2f] + T_{-2}[2f - 2g]$$

(b) With notations as in part (a),

$$x = \tfrac{1}{4}\{T_2[5g - f] + T_{-2}[5f - 5g]\}$$
$$y = \tfrac{1}{4}\{T_2[5g - f] + T_{-2}[f - g]\}$$

(c) With notations as in part (a),

$$x = \tfrac{1}{2}\{T_i[(1+i)f - ig] + T_{-i}[(1-i)f + ig]\}$$
$$y = \tfrac{1}{2}\{T_i[2if + (1-i)g] + T_{-i}[-2if + (1+i)g]\}$$

8. (a) $g_2 - g_1$ (b) $g_3 - g_1$ (c) $3g_3 - 5g_1 + 2g_2$

2–5 Superposition principle and interchange of operations. The *superposition principle* is encountered in one form in Section 1–3: if $x_1(t)$, $x_2(t)$ are solutions of a linear differential equation

$$(a_0 D^n + \cdots + a_n)x = f, \tag{2–50}$$

when $f = f_1, f_2$, respectively, then each linear combination $c_1 x_1 + c_2 x_2$ is a solution when $f = c_1 f_1 + c_2 f_2$. This rule is simply the statement that $a_0 D^n + \cdots + a_n$ is a linear operator:

$$(a_0 D^n + \cdots + a_n)[c_1 x_1 + c_2 x_2] = c_1(a_0 D^n + \cdots + a_n)[x_1]$$
$$+ c_2(a_0 D^n + \cdots + a_n)[x_2].$$

We can also formulate the rule for *inverse* differential operators:

$$\frac{1}{a_0 D^n + \cdots + a_n}[c_1 f_1 + c_2 f_2] = c_1\left[\frac{1}{a_0 D^n + \cdots}\right]f_1$$
$$+ c_2\left[\frac{1}{a_0 D^n + \cdots}\right]f_2.$$

Here, however, there is ambiguity because each term depends on n arbitrary constants. One can state that for each particular choice of the constants in two of the terms, there is a choice of the constants in the remaining term such that the relationship is correct.

Another method is to consider only solutions with fixed initial conditions. However, the sum of two solutions x_1, x_2 with given initial values is not a solution with the same initial values, *unless all initial values are zero.* Accordingly, we turn to the operator T_0 associated with Eq. (2–50); we have seen in Section 2–3 that T_0 is linear.

EXAMPLE 1. Given the differential equation

$$(D^2 + 5D + 4)x = 90e^t + 54e^{2t} - 168e^{3t},$$

find the solution with initial values zero when $t = 0$. We first find the solution with initial values zero when the right-hand member is e^{at}; that is, we evaluate $T_0[e^{at}]$:

$$T_0[e^{at}] = \frac{3e^{at} + (1+a)e^{-4t} - (4+a)e^{-t}}{3(a^2 + 5a + 4)}.$$

Then, successively,

$$T_0[e^t] = \frac{3e^t + 2e^{-4t} - 5e^{-t}}{30},$$

$$T_0[e^{2t}] = \frac{3e^{2t} + 3e^{-4t} - 6e^{-t}}{54}.$$

$$T_0[e^{3t}] = \frac{3e^{3t} + 4e^{-4t} - 7e^{-t}}{84}.$$

Accordingly, the solution sought is

$$T_0[90e^t + 54e^{2t} - 168e^{3t}] = 9e^t + 3e^{2t} - 6e^{3t} + e^{-4t} - 7e^{-t}.$$

The rule $T_0[cf] = cT_0[f]$ is equivalent to the statement that the linear operator T_0 commutes with the linear operator cI (see Problem 14 following Section 2–2). It is natural to ask whether other operators commute with T_0. We give some examples here. The coefficients in Eq. (2–50) will be assumed *constant*, with $a_0 \neq 0$.

Integration. We have the rule

$$T_0\left[\int_0^t f(u)\,du\right] = \int_0^t g(u)\,du, \tag{2–51}$$

where $g = T_0[f]$; that is, T_0 and D_0^{-1} commute:

$$T_0[D_0^{-1}[f]] = D_0^{-1}[T_0[f]]. \tag{2–51'}$$

We assume that f is piecewise continuous for $-\infty < t < \infty$. Then $F(t) = \int_0^t f(u)\,du$ is continuous and has the derivative $F'(t) = f(t)$ for $-\infty < t < \infty$ (except at the jumps of f). The functions $x = T_0[f]$ and $y = T_0[F]$ satisfy the equations

$$(a_0 D^n + \cdots + a_n)x = f, \qquad (a_0 D^n + \cdots + a_n)y = F \tag{2–52}$$

and x, y have initial values zero for $t = 0$. If we differentiate the second equation, we find

$$(a_0 D^n + \cdots + a_n)Dy = f, \tag{2–53}$$

except at the discontinuities of f. But $y, y', \ldots, y^{(n-1)}$ are continuous and, by the second part of (2–52), $y^{(n)}$ is also continuous. Hence Dy is a solution of Eq. (2–53) for all t. Also $Dy, D^2y, \ldots, D^{n-1}y$ are zero for $t = 0$; by the second part of (2–52), $a_0 D^n y = F(0)$ for $t = 0$; but $F(0) = 0$. Hence Dy is the solution of Eq. (2–53) with initial values zero; that is,

$Dy = x$. Since $y = 0$ for $t = 0$,

$$y = \int_0^t x(u)\, du = D_0^{-1}[x] = D_0^{-1}[T_0[f]].$$

Accordingly, the relation (2–51′) is valid.

EXAMPLE 2. $(D + 2)x = (4 + 8t + 16t^2)h(t)$.   The step response $H(t) = T_0[h(t)]$ is found to be $\frac{1}{2}(1 - e^{-2t})h(t)$. Accordingly,

$$T_0[th(t)] = T_0[D_0^{-1}[h]] = D_0^{-1}[\tfrac{1}{2}(1 - e^{-2t})h(t)]$$
$$= \tfrac{1}{4}(2t - 1 + e^{-2t})h(t),$$

$$T_0\left[\frac{t^2}{2}h(t)\right] = T_0[D_0^{-1}[th(t)]] = D_0^{-1}[T_0[th(t)]]$$
$$= D_0^{-1}[\tfrac{1}{4}(2t - 1 + e^{-2t})h(t)]$$
$$= \tfrac{1}{8}(2t^2 - 2t + 1 - e^{-2t})h(t).$$

Therefore, by superposition,

$$T_0[(4 + 8t + 16t^2)h(t)] = 2(1 - e^{-2t})h(t) + 2(2t - 1 + e^{-2t})h(t)$$
$$+ 4(2t^2 - 2t + 1 - e^{-2t})h(t) = (8t^2 - 4t + 4 - 4e^{-2t})h(t).$$

Differentiation. The rule

$$T_0[Df] = DT_0[f] \tag{2–54}$$

is valid only under special assumptions. It is valid, first of all, if f is continuous and has a piecewise continuous derivative f' for $-\infty < t < \infty$, *provided* $f(0) = 0$. For then we can write $f(t) = D_0^{-1}[g(t)]$, where $g(t) = Df$. By (2–51′), applied to g,

$$T_0[f] = T_0[D_0^{-1}[g]] = D_0^{-1}[T_0[g]],$$

so that

$$DT_0[f] = T_0[g] = T_0[Df].$$

If f, f' have the stated continuity, but $f(0) \neq 0$, we can write $f(t) = [f(t) - f(0) \cdot 1] + f(0) \cdot 1$, where 1 is thought of as the constant function always equal to 1. Then Eq. (2–54) applies to the first term, so that

$$T_0[Df] = DT_0[f(t) - f(0) \cdot 1] + 0 = DT_0[f] - f(0)DT_0[1]. \tag{2–55}$$

EXAMPLE 3. $(D + 1)x = \cos t$. Here $T_0[\cos t] = \frac{1}{2}(\cos t + \sin t - e^{-t})$, $T_0[1] = 1 - e^{-t}$, so that by (2–55)

$$T_0[D\cos t] = T_0[-\sin t] = \tfrac{1}{2}D(\cos t + \sin t - e^{-t}) - D(1 - e^{-t})$$
$$= \tfrac{1}{2}(\cos t - \sin t - e^{-t}).$$

Another case of importance is that in which $f(t)$ is of form $g(t)h(t)$, where $g(t)$ has a continuous derivative. Then Df is the generalized function $g'(t)h(t) + g(0)\,\delta(t)$. The rule (2–54) remains valid under these assumptions *provided we interpret $T_0[Df]$ as the solution whose initial values approach zero as $t \to 0-$.* To justify the rule, we first let $f(t) = h(t)$. Then our assertion is that $T_0[Dh(t)] = DT_0[h(t)]$ or that

$$T_0[\delta(t)] = H'(t), \qquad\qquad (2\text{–}56)$$

where $H(t)$ is the step response. But $x = T_0[\delta(t)]$ is found by solving the equations

$$(a_0 D^n + \cdots + a_n)y = h(t),$$

$$x = Dy.$$

These equations are clearly satisfied if $y = H(t)$, $x = H'(t)$. Furthermore, since $H(0) = 0$, $H'(0) = 0, \ldots, H(t)$ is identically zero for $t < 0$, so that $H'(0) = 0$ for $t < 0$; that is, all initial values of $H'(t)$ approach zero as $t \to 0-$. Accordingly, Eq. (2–56) is proved. For an arbitrary function $g(t)h(t)$ we can write $g(t)h(t) = [g(t) - g(0)]h(t) + g(0)h(t)$. The first term is continuous, has the value zero for $t = 0$, and has a piecewise continuous derivative $g'(t)h(t)$. Accordingly, Eq. (2–54) applies to this term; by (2–56), the rule also applies to the second term, so that it applies to $g(t)h(t)$.

Equation (2–56) can be generalized to derivatives of $\delta(t)$:

$$T_0\left[\delta^{(k)}(t)\right] = H^{(k+1)}(t), \qquad\qquad (2\text{–}57)$$

so that Eq. (2–54) applies to generalized functions of the form

$$g(t)h(t) + c_0\,\delta(t) + c_1\,\delta'(t) + \cdots + c_m\,\delta^{(m)}(t). \qquad (2\text{–}58)$$

If we interpret $D_0^{-1}[f]$ as $T_0[f]$ for the equation $Dx = f$, with $x \to 0$ as $t \to 0-$, then Eqs. (2–56) and (2–57) state that

$$D_0^{-1}[\delta(t)] = h(t),$$

$$D_0^{-1}\left[\delta^{(k)}(t)\right] = h^{(k)}(t) = \delta^{(k-1)}(t) \quad (k = 1, 2, \ldots). \qquad (2\text{–}59)$$

From these rules it follows that (2–51′) is valid when f is a generalized function of the form (2–58). For

$$T_0\left[D_0^{-1}\left[\delta^{(k)}(t)\right]\right] = T_0\left[\delta^{(k-1)}(t)\right] = H^{(k)}(t) = D_0^{-1}\left[H^{(k+1)}(t)\right],$$

since $DH^{(k)}(t) = H^{(k+1)}(t)$ and $H^{(k)}(t) \to 0$ as $t \to 0-$.

2–6 Translation and stationary systems.   Another operator which can be interchanged with T_0 is the translation operator Δ_c (Problem 15 following Section 2–2):

$$\Delta_c[f(t)] = f(t - c). \tag{2–60}$$

If $c > 0$, the graph of $f(t - c)$ is obtained from that of f by translating c units to the *right*, as in Fig. 2–1. We verify at once that $\Delta_c f$ is a *linear* operator. With reference to our differential equation (2–50) with constant coefficients, we can compare $T_0[f]$ and $T_0[\Delta_c[f]] = T_0[f(t - c)]$. If $x(t) = T_0[f]$, then $x(t - c)$ satisfies the differential equation (2–50) with f replaced by $f(t - c)$. However, $x(t - c)$ has initial values zero at $t = c$, instead of at $t = 0$; that is, in general, T_0 and Δ_c do not commute:

$$T_0[\Delta_c[f]] \neq \Delta_c[T_0[f]].$$

Despite this inequality it is important to know that $\Delta_c x$ is a solution of the equation with $\Delta_c f$ as input. We describe the property concerned by stating that the differential equation (2–50), with *constant coefficients*, is *stationary;* delay of the input by c units of time leads to a delay of output by c units of time. For equations with *variable coefficients*, the property in general fails.

Under further restrictions the inequality obtained above can be replaced by an equality: *if $f(t) \equiv 0$ for $t < 0$ and $c > 0$, then*

$$T_0[\Delta_c[f]] = \Delta_c[T_0[f]]. \tag{2–61}$$

We assume that $f(t)$ is piecewise continuous for $-\infty < t < \infty$. Then $\Delta_c[f] = f(t - c)$ is identically zero for $t \leqq 0$ and also for $0 \leqq t < c$. Hence $T_0[f(t - c)]$ is also identically zero for $t < c$ and, in particular, has all initial values zero at $t = c$. But this means that $T_0[f(t - c)]$ coincides with $x(t - c) = \Delta_c[T_0[f]]$; for we saw above that $x(t - c)$ is the solution of the equation with $\Delta_c f$ as input and having initial values zero at $t = c$. Thus (2–61) is proved. We remark that (2–61) also holds for generalized functions of form (2–58), provided always that $T_0[f]$ is

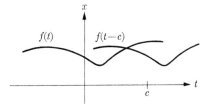

Fig. 2–1.   Translation.

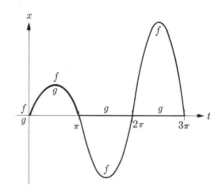

FIG. 2–2. Superposition and translation.

interpreted as the solution whose initial values approach zero as $t \to 0-$. We shall refer to Eq. (2–61), with the conditions stated, as the *translation principle*.

A combination of the superposition and translation principles leads to new procedures for evaluating $T_0[f]$.

EXAMPLE 1.   Let $f(t) = 0$ for $t < 0$, $f(t) = \sin t$ for $0 \leq t \leq \pi$, $f(t) = 2 \sin t$ for $\pi \leq t \leq 2\pi$, $f = 3 \sin t$ for $2\pi \leq t \leq 3\pi$, $f(t) = 0$ for $t > 3\pi$. To find $T_0[f]$ for the equation $(D + 1)x = f$, we first evaluate $G = T_0[g]$, where $g(t) = [h(t) - h(t - \pi)] \sin t$. We then remark that

$$f(t) = g(t) - 2g(t - \pi) + 3g(t - 2\pi),$$

as shown in Fig. 2–2. Accordingly, by the translation principle, $T_0[f] = G(t) - 2G(t - \pi) + 3G(t - 2\pi)$.   A calculation shows that $G(t) = \frac{1}{2}\{(\sin t - \cos t + e^{-t}) \cdot [h(t) - h(t - \pi)] + (e^\pi + 1)e^{-t}h(t - \pi)\}$, so that $T_0[f]$ is easily found.

EXAMPLE 2. Find $T_0[f]$ for the equation $(D^2 + 1)x = f$, if $f = h(t) - h(t - 1) + 2[h(t - 1) - h(t - 3)]$. Here f is a step function. We need only find the step response $H(t)$; then

$$T_0[f] = H(t) - H(t - 1) + 2[H(t - 1) - H(t - 3)].$$

Since $H(t) = (1 - \cos t)h(t)$, $T_0[f]$ is easily found.

The procedures can be extended to generalized functions. By Eqs. (2–56) and (2–57) we have, for $c > 0$,

$$T_0\,[\delta(t - c)] = H'(t - c),$$

$$T_0\,[\delta^{(k)}(t - c)] = H^{(k+1)}(t - c).$$

We note also that, by (2–51′) and by translation,

$$T_0[h(t - c)] = H(t - c),$$

$$T_0[(t - c)h(t - c)] = D_0^{-1}[H(t - c)],$$

$$T_0\left[\frac{(t - c)^2}{2} h(t - c)\right] = D_0^{-2}[H(t - c)],$$

$$\vdots$$

PROBLEMS

1. (a) For the equation $(3D + 1)x = f$, evaluate $T_0[e^{at}]$ $(a \neq -\frac{1}{3})$.

(b) Find the solution with initial value zero for $t = 0$ for each of the following equations:

(i) $(3D + 1)x = 2e^t - 7e^{2t}$,   (ii) $(3D + 1)x = \cos t$,
(iii) $(3D + 1)x = 2e^{-t} + \sin t + 3 \cos t$

2. Let $T_0[f]$ be the operator associated with an equation $(a_0 D^n + \cdots)x = f$, with constant coefficients, as defined in the text.

(a) If $T_0[1] = 1 - 2e^{-t} + e^{-2t}$, find

(i) $T_0[t]$,   (ii) $T_0[t^2]$,   (iii) $T_0[4t^2 - 3t + 2]$

(b) If $T_0[t^2 e^t] = \frac{1}{4}(e^t - 2te^t + 2t^2 e^t - e^{-t})$, find

(i) $T_0[e^t(t^2 + 2t)]$,   (ii) $T_0[e^t(t^2 + 4t + 2)]$,   (iii) $T_0[te^t]$,
(iv) $T_0[e^t]$,   (v) $T_0[1]$

(c) If $T_0[th(t)] = (t - \sin t)h(t)$, find

(i) $T_0[t^2 h(t)]$,   (ii) $T_0[h(t)]$,   (iii) $T_0[\delta(t)]$,   (iv) $T_0[\delta''(t)]$

(d) If $T_0[h(t)] = H(t)$, find

(i) $T_0[(t - 1)h(t - 1)]$,   (ii) $T_0[(t - 1)^2 h(t - 1)]$,
(iii) $T_0[th(t - 1)]$,   (iv) $T_0[t^2 h(t - 1)]$,
(v) $T_0[(at + b)\{h(t - c_1) - h(t - c_2)\}]$, $0 < c_1 < c_2$

(e) If $T_0[h(t)] = H(t)$, find

(i) $T_0[\delta(t - 1)]$,   (ii) $T_0[(t + 1)h(t - 1) + 3\,\delta(t - 2) + 2\,\delta'(t - 3)]$

(f) If $T_0[\{h(t) - h(t - \pi)\} \sin t] = g(t)$, find

(i) $T_0[h(t) \sin t]$,   (ii) $T_0[h(t) \cos t]$

3. Given the simultaneous equations $(D - a_{11})x - a_{12}y = f$, $-a_{21}x + (D - a_{22})y = g$ with constant coefficients, let $x = T_0[f] + U_0[g]$, $y = S_0[f] + V_0[g]$ be the solution with initial values zero at $t = 0$, as in Section 2–4. Let f, g be piecewise continuous for all t.

(a) Show that $T_0[D_0^{-1}f] = D_0^{-1}T_0[f]$.

(b) Show that $T_0[Df] = DT_0[f] - f(0)DT_0[1]$, provided f is continuous and Df is piecewise continuous.

(c) Show that $T_0[Df] = DT_0[f]$, provided

$$f = g(t)h(t) + b_0\,\delta(t) + b_1\,\delta'(t) + \cdots + b_k\,\delta^{(k)}(t),$$

where $g(t)$ is continuous and has a piecewise continuous derivative for all t, and the initial values zero apply as $t \to 0-$.

(d) Show that, if $f(t)$ is the same as in part (c), then for $c > 0$

$$T_0[\Delta_c f] = \Delta_c T_0[f].$$

4. Let T_0 be defined as in Problem 2 and let U_0 be defined similarly for an equation $(b_0 D^m + \cdots + b_m)x = f$. Show that

$$U_0[T_0[f]] = T_0[U_0[f]].$$

Show also that Eq. (2–51′) is a special case of this rule.

ANSWERS

1. (a) $(e^{at} - e^{-t/3})/(3a + 1)$
   (b) (i) $\frac{1}{2}(e^t - 2e^{2t} + e^{-t/3})$
       (ii) $\mathrm{Re}\,[(e^{it} - e^{-t/3})/(1 + 3i)]$
       (iii) $\frac{1}{2}\{4T_0[e^{-t}] + (3 - i)T_0[e^{it}] + (3 + i)T_0[e^{-it}]\}$
                   $= e^{-t/3} - e^{-t} + \sin t$

2. (a) (i) $t + 2e^{-t} - \frac{1}{2}(e^{-2t} + 3)$
       (ii) $t^2 - 4e^{-t} + \frac{1}{2}(e^{-2t} - 6t + 7)$
       (iii) $4t^2 - 15t - 26e^{-t} + \frac{1}{2}(41 + 11e^{-2t})$

   (b) (i) $\frac{1}{4}(-e^t + 2te^t + 2t^2e^t + e^{-t})$
       (ii) $\frac{1}{4}(e^t + 6te^t + 2t^2e^t - e^{-t})$
       (iii) $\frac{1}{4}(-e^t + 2te^t + e^{-t})$
       (iv) $\frac{1}{2}(e^t - e^{-t})$
       (v) $1 - e^{-t}$

   (c) (i) $(t^2 + 2\cos t - 2)h(t)$
       (ii) $(1 - \cos t)h(t)$
       (iii) $\sin t\, h(t)$
       (iv) $-\sin t\, h(t) + \delta(t)$

   (d) (i) $H_1(t - 1)$, where $H_1(t) = D_0^{-1}H$
       (ii) $2H_2(t - 1)$, where $H_2(t) = D_0^{-2}H$
       (iii) $H_1(t - 1) + H(t - 1)$
       (iv) $2H_2(t - 1) + 2H_1(t - 1) + H(t - 1)$
       (v) $aH_1(t - c_1) - aH_1(t - c_2) + (b + ac_1)H(t - c_1)$
                        $- (b + ac_2)H(t - c_2)$

   (e) (i) $H'(t - 1)$
       (ii) $H_1(t - 1) + 2H(t - 1) + 3H'(t - 2) + 2H''(t - 3)$, where
            $H_1 = D_0^{-1}[H]$

   (f) (i) $g(t) - g(t - \pi) + g(t - 2\pi) - \cdots$
       (ii) $g'(t) - g'(t - \pi) + g'(t - 2\pi) - \cdots$

2-7 Duhamel's integral. Let $f(t)$ be piecewise continuous for all t and let $f(t) \equiv 0$ for $t < 0$. We can then approximate $f(t)$ by a step function, as suggested in Fig. 2–3. We write

$$f(t) \sim f(t_1)[h(t) - h(t - t_1)] + f(t_2)[h(t - t_1) - h(t - t_2)] + \cdots$$

For an equation with step response $H(t)$, we then have, as in Section 2–6,

$$T_0[f] \sim f(t_1)[H(t) - H(t - t_1)] + f(t_2)[H(t - t_1) - H(t - t_2)] + \cdots$$

Now by the law of the mean,

$$H(t) - H(t - t_1) = t_1 H'(t - u_1) \quad (0 < u_1 < t_1),$$

$$H(t - t_1) - H(t - t_2) = (t_2 - t_1)H'(t - u_2) \quad (t_1 < u_2 < t_2),$$
$$\vdots$$

Accordingly,

$$T_0[f] \sim f(t_1)H'(t - u_1)t_1 + f(t_2)H'(t - u_2)(t_2 - t_1) + \cdots$$

For each fixed t, the series on the right breaks off at the kth term, where $t_k \geqq t$, since $H(t) \equiv 0$ for $t < 0$. If we now improve the accuracy of the approximation by increasing the number of subdivision points, it is reasonable to expect the right-hand side to approach a limit which is exactly $T_0[f]$. Because of the form of the right-hand side, this limit is an integral:

$$T_0[f] = \int_0^t f(u)H'(t - u)\,du. \tag{2–70}$$

The integral on the right is *Duhamel's integral*.

THEOREM 2. *Given a differential equation*

$$(a_0 D^n + \cdots + a_n)x = f \quad (a_0 \neq 0) \tag{2–71}$$

with constant coefficients, let $T_0[f]$ be the solution with initial values zero for $t = 0$. Let $H(t) = T_0[h(t)]$ be the step response. Then for arbitrary piecewise continuous f, equal to 0 for $t < 0$, Eq. (2–70) is valid.

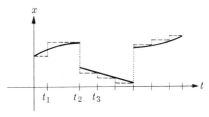

FIG. 2–3.   Approximation by a step function.

The derivation given above is intuitive. We now give a precise proof. By advanced calculus (see Reference 3, p. 220),

$$\frac{d}{dt} \int_0^t g(t, u)\, du = g(t, t) + \int_0^t \frac{\partial g}{\partial t}\, du, \tag{2–72}$$

provided g and $\partial g/\partial t$ are continuous for $0 \leq u \leq t$. The result is easily extended to piecewise continuous functions, as needed here. Let

$$x = \int_0^t f(u)H'(t - u)\, du,$$

and let f be continuous for all t. Then by (2–72), for $t \geq 0$,

$$\frac{dx}{dt} = f(t)H'(0) + \int_0^t f(u)H''(t - u)\, du.$$

But $H(0) = H'(0) = \cdots = H^{(n-1)}(0) = 0$. Hence

$$Dx = \int_0^t f(u)H''(t - u)\, du.$$

We note that $Dx \to 0$ as $t \to 0+$, and $x \equiv 0$ for $t < 0$, so that Dx is continuous for all t. Similarly,

$$D^2 x = f(t)H''(0) + \int_0^t f(u)H'''(t - u)\, du = \int_0^t f(u)H'''(t - u)\, du,$$

$$D^{n-1}x = \int_0^t f(u)H^{(n)}(t - u)\, du \quad (t \geq 0),$$

and $D^{n-1}x$ is continuous for all t, $D^{n-1}x = 0$ for $t = 0$. Now since

$$(a_0 D^n + \cdots + a_n)H(t) = h(t) = 1 \quad (t > 0) \tag{2–73}$$

and $H, DH, \ldots D^{n-1}H$ approach 0 as $t \to 0+$, we conclude that $D^n H \to 1/a_0$ as $t \to 0+$. Hence a further application of (2–72) gives

$$D^{(n)}x = f(t) \cdot \frac{1}{a_0} + \int_0^t f(u)H^{(n+1)}(t - u)\, du.$$

Therefore, for $t > 0$,

$$(a_0 D^n + \cdots + a_n)x = f(t)$$

$$+ \int_0^t f(u)[a_0 H^{(n+1)}(t - u) + a_1 H^{(n)}(t - u) + \cdots + a_n H'(t - u)]\, du.$$

By (2–73), for $t > 0$,

$$a_0 H^{(n+1)}(t) + \cdots + a_n H'(t) = h'(t) = 0,$$

so that $(a_0 D^n + \cdots + a_n)x = f(t)$. Since $x \equiv 0$ and $f(t) \equiv 0$ for $t < 0$, it follows that $x(t)$ is indeed $T_0[f]$. The proof is easily modified to cover the case of a piecewise continuous $f(t)$.

EXAMPLE 1. $(a_0 D + a_1)x = f(t)$, $a_1 \neq 0$, $f(t) = 0$ for $t < 0$.

$$H(t) = \frac{1 - e^{-a_1 t/a_0}}{a_1} h(t),$$

$$H'(t) = \frac{e^{-a_1 t/a_0}}{a_0} h(t),$$

so that

$$x = T_0[f] = \frac{1}{a_0} \int_0^t f(u) e^{-a_1(t-u)/a_0} h(t - u) \, du.$$

For $t > 0$, $0 \leq u \leq t$, $h(t - u) = 1$; for $t < 0$, $f(t) = 0$. Hence we can also write

$$x = \frac{e^{-a_1 t/a_0}}{a_0} \int_0^t f(u) e^{a_1 u/a_0} \, du$$

[compare with Eq. (2–33) above].

EXAMPLE 2. $[D^2 + (a + b)D + ab]x = f$, $a \neq b$, $ab \neq 0$, $f = 0$ for $t < 0$. Here

$$H(t) = \frac{1}{ab}\left[1 + \frac{be^{-at} - ae^{-bt}}{a - b}\right]h(t), \qquad H'(t) = \frac{e^{-bt} - e^{-at}}{a - b} h(t),$$

$$x = T_0[f] = \frac{1}{a - b} \int_0^t f(u)[e^{b(u-t)} - e^{a(u-t)}]h(t - u) \, du.$$

Again the factor $h(t - u)$ can be dropped, so that the result is in agreement with Problem 4(c) following Section 2–4.

2–8 Weighting function, impulse response, convolution.   We write $W(t) = H'(t)$. By Eq. (2–56), $W(t)$ is simply $T_0[\delta(t)]$ and is therefore termed the *impulse response*. The Duhamel formula therefore reads

$$x = T_0[f] = \int_0^t f(u) W(t - u) \, du. \qquad (2\text{–}80)$$

The expression on the right has meaning for any two piecewise continuous functions f, W. It arises in the theory of the Laplace transform as the *convolution* of f and W, and is denoted by $f * W$:

$$f * W = \int_0^t f(u)W(t - u) \, du. \qquad (2\text{--}81)$$

Thus the convolution of two functions is a third function. *We here restrict the functions f, W to be zero for $t < 0$, so that $f * W$ is also zero for $t < 0$.* Because of this assumption, we can also write

$$f * W = \int_{-\infty}^{\infty} f(u)W(t - u) \, du; \qquad (2\text{--}82)$$

for $f(u)$ is zero for $u < 0$; $W(t - u)$ is zero for $u > t$. The more general convolution (2–82) appears in the theory of Fourier transforms.

The convolution (2–81) has several important properties which we list:

$$f * (p + q) = f * p + f * q, \qquad (2\text{--}83)$$

$$f * (cg) = (cf) * g = c(f * g), \quad c = \text{const}, \qquad (2\text{--}84)$$

$$f * g = g * f. \qquad (2\text{--}85)$$

From these properties we deduce that

$$f * (c_1 p + c_2 q) = c_1 f * p + c_2 f * q,$$
$$(c_1 f + c_2 g) * p = c_1 f * p + c_2 g * p; \qquad (2\text{--}86)$$

that is, for fixed f, $f * g$ is a linear transformation $T[g]$ in the linear space of piecewise continuous functions g, equal to zero for $t < 0$; similarly, for fixed g, $f * g$ is a linear transformation $U[f]$. Accordingly, we call $f * g$ a *bilinear operator*. In the above equations, parentheses are omitted about the convolutions, as for multiplication; thus the right-hand side of (2–83) is understood as $(f * p) + (f * q)$. [The proofs of Eqs. (2–83) through (2–86) are considered in Problem 5 below.]

The convolution of two piecewise continuous functions f, g is a *continuous* function. For example, let $g(t) = h(t - c)$, $c \geqq 0$. Then

$$f * g = f * h(t - c) = \int_0^t h(u - c)f(t - u) \, du = q(t).$$

Now $q(t)$ is zero for $t \leqq c$, and for $t \geqq c$

$$q(t) = \int_c^t f(t - u) \, du = \int_0^{t-c} f(v) \, dv = F(t - c),$$

where $F(t) = D_0^{-1}f$, so that $F(t)$ is continuous for all t and $F(0) = 0$.

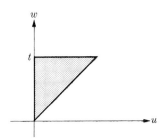

FIG. 2-4.   Region of integration.

Thus $q(t) = F(t - c)h(t - c)$, and $q(t)$ is continuous for all t. Since an arbitrary piecewise continuous function can be expressed as a continuous function plus a linear combination of functions of form $h(t - c)$, and since the convolution of two continuous functions is continuous, we conclude from the bilinearity of the convolution that $f * g$ is continuous whenever f and g are piecewise continuous (and equal to zero for $t < 0$).

The convolution also satisfies the *associative law:*

$$(f * p) * q = f * (p * q). \tag{2-87}$$

To prove this, we let $F = f * p$, $G = p * q$. Then we must show that $F * q = f * G$. Both sides are zero for $t \leqq 0$. For $t > 0$,

$$F(t) = \int_0^t f(t - u)p(u) \, du = \int_0^t f(t - v)p(v) \, dv,$$

$$F(t - u) = \int_0^{t-u} f(t - u - v)p(v) \, dv = \int_u^t f(t - w)p(w - u) \, dw,$$

by the substitution $w = u + v$, so that

$$F * q = \int_0^t F(t - u)q(u) \, du$$

$$= \int_0^t \int_u^t f(t - w)p(w - u)q(u) \, dw \, du.$$

The double integral is taken over the triangle of Fig. 2-4 in the uw-plane. We interchange the order of integration and obtain

$$F * q = \int_0^t \int_0^w p(w - u)q(u) \, du \, f(t - w) \, dw$$

$$= \int_0^t G(w)f(t - w) \, dw = f * G.$$

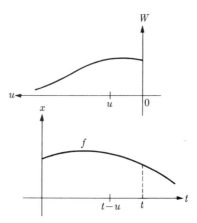

FIG. 2–5. Weighting function.

By (2–85) we can write (2–80) in another way:

$$x = T_0[f] = \int_0^t W(u)f(t - u)\, du. \qquad (2\text{–}80')$$

This relation leads to the interpretation of $W(t)$ as a *weighting function*, for the integral in (2–80′) can be considered as a weighted average of $f(t)$ over the interval zero to t. The value of $f(t)$ is given weight $W(0)$, the value $f(t - 1)$ is given weight $W(1)$, etc., and the values with appropriate weights are "summed" to yield the integral. The process can be carried out graphically as indicated in Fig. 2–5. Above the graph of f we place the graph of W, with abscissa pointing to the left and origin above $[t, f(t)]$. We then multiply the two functions, point by point, to obtain a new function in the interval zero to t; the integral of the new function from zero to t is the value $T_0[f]$ at time t. Each value of t necessitates a separate calculation.

The shape of the graph of $W(t)$ gives qualitative information on the behavior of the system. We can consider $T_0[f]$ at each t as a compromise of the different "memories" of $f(t)$ in the past; the value $W(u)$ indicates how well we remember $f(t - u)$. In the case (a) of Fig. 2–6, we remember essentially only the value $f(t - 2)$; in the case (b) we remember only the very recent past; in case (c) we remember all the interval $t - 3$ to t equally well, but forget all that precedes.

Equation (2–80) was deduced for the operator T_0 associated with an equation $(a_0 D^n + \cdots + a_n)x = f$, with constant coefficients. It can be justified in similar fashion for the linear operators associated with simultaneous equations and with integrodifferential equations. As the intuitive justification of Section 2–7 indicates, the formula is a consequence of

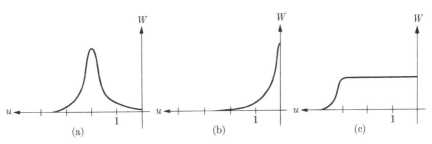

Fig. 2–6. Various forms of memory.

linearity and *stationarity*. It is important to note that Eq. (2–80) can be applied even though we do not know the precise form of the differential equation (or equations) from which it arose. In particular, we can determine $H(t)$ experimentally as the response to $h(t)$, with all initial values zero at $t = 0$; then $W(t) = H'(t)$. We can also obtain $W(t)$ approximately as the response to an impulse $\delta(t)$, that is, a pulse of very short duration (see Problem 4 below).

PROBLEMS

1. Given the equation $(D + 3)x = f$,

(a) find $H(t)$ and $W(t)$;
(b) find $T_0[e^{2t}h(t)]$ and verify that (2–80) holds;
(c) find $T_0[e^{-t^2}h(t)]$ graphically with the aid of (2–80').

2. A system has weighting function $W = te^{-t}h(t)$.

(a) Find $T_0[h(t) \sin \omega t] = \operatorname{Im} T_0[h(t)e^{i\omega t}]$.
(b) Find $T_0 [\delta(t)]$, $T_0[h(t)]$, $T_0 [\delta'(t)]$.
(c) Find $T_0[t(1 + t^2)^{-1}h(t)]$ graphically.

3. Represent $T_0[f]$ in integral form (2–80) for each of the following equations:

(a) $(D^2 + 1)x = f$ (b) $(D^3 + 5D + 4)x = f$
(c) $(D^2 + 2D + 2)x = f$ (d) $(D^3 + D^2 + D + 1)x = f$

4. Let $h_\epsilon(t) = \{t[h(t) - h(t - \epsilon)]/\epsilon\} + h(t - \epsilon)$.

(a) Graph $h_\epsilon(t)$ and show that for $t \neq 0 \lim h_\epsilon(t) = h(t)$ as $\epsilon \to 0$.
(b) Find $Dh_\epsilon(t)$ and graph. [Consider $Dh_\epsilon(t)$ as an approximation to $\delta(t)$.]
(c) For an equation $(a_0 D^n + \cdots + a_n)x = f$, with constant coefficients, find $T_0[Dh_\epsilon(t)]$ and show that $\lim T_0[Dh_\epsilon(t)] = W(t)$, as $\epsilon \to 0$.

5. (a) Prove (2–83). (b) Prove (2–84). (c) Prove (2–85). [*Hint:* Set $v = t - u$ in (2–80).] (d) Prove (2–86) from (2–83), (2–84), (2–85).

6. Evaluate the following convolutions:

(a) $e^t h(t) * e^{2t}h(t)$ (b) $h(t) * h(t)$
(c) $e^t h(t) * \sin t\, h(t)$ (d) $h(t) * f(t)$

7. Given a pair of equations $(D - a_{11})x - a_{12}y = f$, $-a_{12}x + (D - a_{22})y = 0$, with constant coefficients, let $x = T_0[f]$, $y = U_0[f]$ be the solution with $x = 0$, $y = 0$ for $t = 0$. Let $H(t) = T_0[h(t)]$, $H_0(t) = U_0[h(t)]$. Show that, for arbitrary f, continuous and equal to zero for $t < 0$,

$$T_0[f] = \int_0^t f(u)H'(t - u)\, du, \qquad U_0[f] = \int_0^t f(u)H_0'(t - u)\, du.$$

8. Given an equation $(a_0 D^n + \cdots + a_n)x = f$, with constant coefficients, let $W(t)$ be the corresponding weighting function.

(a) Show that $W(t)$ is a solution of the related homogeneous equation for $t > 0$ and $W(0) = 0$, $W'(0) = 0, \ldots, W^{(n-2)}(0) = 0$, $\lim_{t \to 0+} W^{(n-1)}(t) = 1/a_0$.

(b) Let $W_1(t)$ be defined as the solution of the homogeneous equation for *all* t with initial values $W_1(0) = 0$, $W_1'(0) = 0, \ldots, W_1^{(n-2)}(0) = 0$, $W_1^{(n-1)}(0) = 1/a_0$, so that $W_1(t) \equiv W(t)$ for $t > 0$. Show that, for arbitrary continuous f (not necessarily zero for $t < 0$),

$$T_0[f] = \int_0^t f(u)W_1(t - u)\, du = f * W_1 \quad (-\infty < t < \infty).$$

[*Hint:* Imitate the proof of the theorem of Section 2–7.]

Answers

1. (a) $H = [(1 - e^{-3t})/3]h(t)$, $W = e^{-3t}h(t)$ (b) $[(e^{2t} - e^{-3t})/5]h(t)$

2. (a) $(1 + \omega^2)^{-2}[-2\omega \cos \omega t + (1 - \omega^2) \sin \omega t + \omega(1 + \omega^2)te^{-t} + 2\omega e^{-t}]h(t)$
   (b) $T_0[\delta] = W$, $T_0[h] = D_0^{-1}W = [1 - (1 + t)e^{-t}]h(t)$,
   $T_0[\delta'] = W' = [e^{-t}(1 - t)]h(t)$

3. (a) $\int_0^t f(u) \sin (t - u)\, dt$ (b) $\frac{1}{3}\int_0^t f(u)[e^{u-t} - e^{4(u-t)}]\, du$
   (c) $\int_0^t f(u)e^{u-t} \sin (t - u)\, du$
   (d) $\frac{1}{2}\int_0^t f(u)[e^{u-t} + \sin (t - u) - \cos (t - u)]\, du$

4. (b) $[h(t) - h(t - \epsilon)]/\epsilon$
   (c) $T_0[Dh_\epsilon(t)] = [H(t) - H(t - \epsilon)]/\epsilon$

6. (a) $(e^{2t} - e^t)h(t)$ (b) $th(t)$
   (c) $\frac{1}{2}(e^t - \cos t - \sin t)$ (d) $D_0^{-1}f$

2–9 Characteristic function, transfer function, frequency response function. The operators such as D, D^{-1}, or, more generally, the integro-differential operators

$$a_0 D^n + \cdots + a_n + \cdots + a_{n+m}D^{-m}, \qquad (2\text{–}90)$$

have been treated thus far as formal expressions which describe particular linear transformations. Since these are constructed from the basic operator D by operations which resemble those of algebra, it is natural to think of them as functions of D, just as $x^2 + x + 1$ is a function of x. However, D is not a number, and the notation $f(D)$ which is sometimes used might give rise to confusion. Accordingly, we agree to associate with each operator (2–90) the ordinary function

$$a_0 s^n + \cdots + a_n + \cdots + a_{n+m} s^{-m},$$

where s is now allowed to take on arbitrary numerical values, real or complex.

For a differential equation with constant coefficients,

$$(a_0 D^n + \cdots + a_n)x = f(t), \tag{2–91}$$

the corresponding function

$$V(s) = a_0 s^n + \cdots + a_n$$

is called the *characteristic function* associated with Eq. (2–91). The equation $V(s) = 0$, that is,

$$a_0 s^n + \cdots + a_n = 0,$$

is the *characteristic equation;* its roots are the characteristic *roots*. The function

$$Y(s) = \frac{1}{V(s)} = \frac{1}{a_0 s^n + \cdots + a_n}$$

is called the *transfer function* associated with (2–91). We can think of $Y(s)$ as the function of s associated with the inverse differential operator

$$\frac{1}{a_0 D^n + \cdots + a_n}.$$

An important property of $Y(s)$ is the following: if $f(t) = A e^{st}$ in (2–91), $A = \text{const}$, then $x = Y(s) A e^{st}$ is a particular solution, provided s is not a characteristic root. We verify the property by substitution in (2–91). If $s = i\omega$ (ω real), our particular solution becomes

$$Y(i\omega) A e^{i\omega t};$$

as in Section 1–10, this describes a sinusoidal oscillation of circular frequency ω in complex form. If $A = R e^{i\alpha}$, where R and α are real, then the input $f(t)$ has amplitude R, while the output has amplitude $|Y(i\omega)| \cdot R$; that is, $|Y(i\omega)|$ is the *amplification* of the input. Similarly, $\arg Y(i\omega)$ gives the *phase shift* of the input. Because of these properties, $Y(i\omega)$ is termed the *frequency response function*.

These concepts can be extended to an integrodifferential equation:

$$(a_0 D^n + \cdots + a_n + \cdots + a_{n+m} D^{-m})x = f. \tag{2-92}$$

We assume $a_0 \neq 0$, $a_{n+m} \neq 0$, $m > 0$. The characteristic function is

$$V(s) = a_0 s^n + \cdots + a_n + \frac{a_{n+1}}{s} + \cdots + \frac{a_{n+m}}{s^m}$$

$$= \frac{a_0 s^{n+m} + \cdots + a_n s^m + a_{n+1} s^{m-1} + \cdots + a_{n+m}}{s^m}.$$

The characteristic equation is again the equation $V(s) = 0$; however, this is equivalent to the equation

$$a_0 s^{n+m} + \cdots + a_n s^m + \cdots + a_{n+m} = 0.$$

We note that since $a_{n+m} \neq 0$, $s = 0$ cannot be a characteristic root. The transfer function is

$$Y(s) = \frac{1}{V(s)} = \frac{s^m}{a_0 s^{n+m} + \cdots + a_n s^m + \cdots + a_{n+m}}.$$

The concepts can also be extended to simultaneous equations. For example, let the outputs x_1, x_2 be the solutions of nondegenerate equations of the form

$$T_{11}x_1 + T_{12}x_2 = f_1,$$

$$T_{21}x_1 + T_{22}x_2 = f_2, \tag{2-93}$$

where T_{11}, T_{12}, T_{21}, T_{22} are differential operators with constant coefficients. One can replace D by s in each T_{ij} to obtain a characteristic function $V_{ij}(s)$. The array of these characteristic functions:

$$\begin{bmatrix} V_{11}(s) & V_{12}(s) \\ V_{21}(s) & V_{22}(s) \end{bmatrix}$$

is called the *characteristic matrix* associated with (2-93). The *characteristic equation* is obtained by setting the corresponding determinant equal to zero:

$$\begin{vmatrix} V_{11}(s) & V_{12}(s) \\ V_{21}(s) & V_{22}(s) \end{vmatrix} = 0;$$

the roots of this equation are the *characteristic roots*.

The methods of Section 1–5 provide the general solution of equations (2–93) in the form

$$x_1 = x_1^*(t) + c_1\alpha_1 e^{s_1 t} + c_2\alpha_2 e^{s_2 t} + \cdots,$$

$$x_2 = x_2^*(t) + c_1\beta_1 e^{s_1 t} + c_2\beta_2 e^{s_2 t} + \cdots,$$

appropriately modified for multiple roots. The pair $\alpha_1 e^{s_1 t}$, $\beta_1 e^{s_1 t}$ is a solution of the homogeneous equations related to (2–93). Substitution in these equations yields the relations

$$V_{11}(s_1)\alpha_1 + V_{12}(s_1)\beta_1 = 0, \qquad V_{21}(s_1)\alpha_1 + V_{22}(s_1)\beta_1 = 0.$$

Since α_1, β_1 are not both zero, we conclude that

$$\begin{vmatrix} V_{11}(s_1) & V_{12}(s_1) \\ V_{21}(s_1) & V_{22}(s_1) \end{vmatrix} = 0;$$

that is, s_1 is a characteristic root. In general, the general solution of the homogeneous equations is built of terms of form const $\times\ t^k e^{st}$, where s is a characteristic root. [See Chapter 6 of Reference 4 for a full discussion.]

Associated with the characteristic matrix is the *transfer matrix*

$$\begin{bmatrix} Y_{11}(s) & Y_{12}(s) \\ Y_{21}(s) & Y_{22}(s) \end{bmatrix}.$$

In matrix terminology, this matrix is the *inverse* of the characteristic matrix. To obtain the $Y_{ij}(s)$, we solve the *algebraic* equations

$$V_{11}(s)x_1 + V_{12}(s)x_2 = f_1,$$

$$V_{21}(s)x_1 + V_{22}(s)x_2 = f_2 \tag{2–94}$$

for x_1, x_2. The results have the form

$$x_1 = Y_{11}(s)f_1 + Y_{12}(s)f_2, \qquad x_2 = Y_{21}(s)f_1 + Y_{22}(s)f_2.$$

For example,

$$x_1 = \frac{\begin{vmatrix} f_1 & V_{12}(s) \\ f_2 & V_{22}(s) \end{vmatrix}}{\begin{vmatrix} V_{11}(s) & V_{12}(s) \\ V_{21}(s) & V_{22}(s) \end{vmatrix}} = \frac{V_{22}(s)f_1 - V_{12}(s)f_2}{\begin{vmatrix} V_{11}(s) & V_{12}(s) \\ V_{21}(s) & V_{22}(s) \end{vmatrix}},$$

so that

$$Y_{11}(s) = \frac{V_{22}(s)}{\begin{vmatrix} V_{11}(s) & V_{12}(s) \\ V_{21}(s) & V_{22}(s) \end{vmatrix}}, \qquad Y_{12}(s) = \frac{-V_{12}(s)}{\begin{vmatrix} V_{11}(s) & V_{12}(s) \\ V_{21}(s) & V_{22}(s) \end{vmatrix}}.$$

We call $Y_{ij}(s)$ the *transfer function of x_i relative to f_j*. When $f_1 = Ae^{st}$, $f_2 = 0$, a particular solution is given by

$$x_1 = Y_{11}(s)Ae^{st}, \qquad x_2 = Y_{21}(s)Ae^{st},$$

provided that s is not a characteristic root. Indeed, the substitution $x_1 = k_1 e^{st}$, $x_2 = k_2 e^{st}$ in the differential equations (2–93) leads to the equations

$$V_{11}(s)k_1 + V_{12}(s)k_2 = A,$$
$$V_{21}(s)k_1 + V_{22}(s)k_2 = 0;$$

solution as above gives $k_1 = Y_{11}(s) \cdot A$, $k_2 = Y_{21}(s) \cdot A$. When $s = i\omega$, $Y_{ij}(s)$ becomes $Y_{ij}(i\omega)$, the *frequency response function* of x_i relative to f_j; as above, the frequency response function describes the response to sinusoidal inputs (in complex form).

The preceding definitions can be repeated for simultaneous integro-differential equations, in which the T_{ij} are integrodifferential operators with constant coefficients. As explained in Section 1–6, such equations are in fact equivalent to simultaneous differential equations, and all quantities can be evaluated with the aid of the equivalent differential equations. One can obtain the characteristic matrix directly by replacing D by s, D^{-1} by $1/s$ in the operators T_{ij}. However, the characteristic equation may not be the same as the equation obtained by setting the corresponding determinant equal to zero, *since zero roots can be lost in this process*. Instead, *we equate to zero the product of this determinant by $s^{m_1+m_2}$, where m_1, m_2 are the largest integers such that $D^{-m_1}x_1$ and $D^{-m_2}x_2$ appear in the T_{ij}*. The transfer matrix can be obtained by solving for x_1, x_2 as before.

EXAMPLE 1. $(D + D^{-1})x_1 + (D + D^{-1})x_2 = f_1,$

$$(2D + 1)x_1 + (3D + 1)x_2 = f_2.$$

The equations (2–94) become

$$\left(s + \frac{1}{s}\right)x_1 + \left(s + \frac{1}{s}\right)x_2 = f_1,$$

$$(2s + 1)x_1 + (3s + 1)x_2 = f_2.$$

Here $m_1 = 1$, $m_2 = 1$, since $D^{-1}x_1$, $D^{-1}x_2$ appear, and the characteristic equation is

$$s^2 \begin{vmatrix} s + \dfrac{1}{s} & s + \dfrac{1}{s} \\ 2s + 1 & 3s + 1 \end{vmatrix} = s^2(s^2 + 1) = 0,$$

so that the characteristic roots are 0, 0, $\pm i$. The characteristic matrix is

$$\begin{bmatrix} s + \dfrac{1}{s} & s + \dfrac{1}{s} \\ 2s + 1 & 3s + 1 \end{bmatrix}.$$

The transfer matrix is obtained by solving for x_1, x_2:

$$x_1 = \frac{\begin{vmatrix} f_1 & s + \dfrac{1}{s} \\ f_2 & 3s + 1 \end{vmatrix}}{s^2 + 1} = \frac{(3s + 1)f_1}{s^2 + 1} - \frac{f_2}{s},$$

$$x_2 = \frac{\begin{vmatrix} s + \dfrac{1}{s} & f_1 \\ 2s + 1 & f_2 \end{vmatrix}}{s^2 + 1} = \frac{-(2s + 1)f_1}{s^2 + 1} + \frac{f_2}{s}.$$

Hence $Y_{11}(s) = (3s + 1)/(s^2 + 1)$, $Y_{12}(s) = -1/s, \ldots$ If the factor s^2 had been omitted above, the double root of zero would have been lost. The fact that this root must be counted is seen by considering the equivalent differential equations

$$Dx_1 + Dx_2 + u + v = f_1,$$

$$(2D + 1)x_1 + (3D + 1)x_2 = f_2, \quad Du = x_1, \quad Dv = x_2,$$

for which the characteristic equation is $s^2(s^2 + 1) = 0$.

EXAMPLE 2. $(3D - 1)x + (2D + 1)y = 2 \cos 3t + 3 \sin 5t$, $(D + 2)x + (D - 1)y = 5 \sin 3t - 2 \cos 5t$. Here $x_1 = x$, $x_2 = y$. If f_1 were $\cos 3t$ alone and $f_2 = 0$, then a particular solution would be given by

$$x = \text{Re} \, [Y_{11}(3i)e^{3it}],$$

$$y = \text{Re} \, [Y_{21}(3i)e^{3it}].$$

Similar reasoning applies to the other terms. We find

$$Y_{11}(s) = \frac{s - 1}{s^2 - 9s - 1}, \qquad Y_{12}(s) = \frac{-(2s + 1)}{s^2 - 9s - 1},$$

$$Y_{21}(s) = \frac{-(s + 2)}{s^2 - 9s - 1}, \qquad Y_{22}(s) = \frac{3s - 1}{s^2 - 9s - 1}.$$

Accordingly, by superposition,

$$x = 2 \text{ Re } [Y_{11}(3i)e^{3it}] + 3 \text{ Im } [Y_{11}(5i)e^{5it}] + 5 \text{ Im } [Y_{12}(3i)e^{3it}]$$

$$- 2 \text{ Re } [Y_{12}(5i)e^{5it}] = 2 \text{ Re }\left[\frac{(3i-1)e^{3it}}{-10-27i}\right] + 3 \text{ Im }\left[\frac{(5i-1)e^{5it}}{-26-45i}\right]$$

$$+ 5 \text{ Im }\left[\frac{(-1-6i)e^{3it}}{-10-27i}\right] - 2 \text{ Re }\left[\frac{(-1-10i)e^{5it}}{-26-45i}\right]$$

$$= \frac{2}{829}(-71 \cos 3t + 57 \sin 3t) + \cdots,$$

$$y = 2 \text{ Re } [Y_{21}(3i)e^{3it}] + 3 \text{ Im } [Y_{21}(5i)e^{5it}] + 5 \text{ Im } [Y_{22}(3i)e^{3it}]$$

$$- 2 \text{ Re } [Y_{22}(5i)e^{5it}] = 2 \text{ Re }\left[\frac{(-2-3i)e^{3it}}{-10-27i}\right] + \cdots$$

Remark. It should be noted that initial conditions have played no part in the definitions of this section. In particular, the operators D_0^{-1}, D_0^{-2}, ..., which imply a choice of initial values, are not used. To apply the concepts of this section to equations containing these operators, we should first consider the equation without initial conditions, that is, with D^{-1}, $D^{-2}, \ldots$ replacing D_0^{-1}, $D_0^{-2}, \ldots$ For example, the transfer function for the equation

$$(D + 1 + D_0^{-1})x = f$$

is

$$Y(s) = \frac{s}{s^2 + s + 1}.$$

If $f = e^{st}$, then $Y(s)e^{st}$ provides a solution of the equation

$$(D + 1 + D^{-1})x = f,$$

but not a solution of the given equation, since

$$(D + 1 + D_0^{-1})e^{st} = (s + 1)e^{st} + \int_0^t e^{su}\, du$$

$$= \left(s + 1 + \frac{1}{s}\right)e^{st} - \frac{1}{s} \neq \frac{e^{st}}{Y(s)}.$$

PROBLEMS

1. Find the characteristic function, transfer function, frequency response function, and characteristic roots for the following equations:

(a) $(D^3 + 1)x = f$ (b) $(D + 4D^{-1})x = f$

(c) $(D + 1 + 3D^{-1})x = f$ (d) $(1 + D^{-1} + D^{-2})x = f$

2. For each of the following find a particular solution, with the aid of the results of Problem 1:

(a) $(D^3 + 1)x = e^{2t}$
(b) $(D + 4D^{-1})x = e^{3it}$
(c) $(D + 1 + 3D^{-1})x = 4 \cos 3t$
(d) $(1 + D^{-1} + D^{-2})x = 7e^{2t} - \sin t$

3. For the following sets of equations find the characteristic matrix, characteristic equation, characteristic roots, and transfer matrix:

(a) $(D + 2)x_1 + (D - 3)x_2 = f_1$
    $(2D - 1)x_1 + (D + 1)x_2 = f_2$
(b) $(D + D^{-1})x_1 + (3D - D^{-1})x_2 = f_1$
    $(D - D^{-1})x_1 + (D - D^{-1})x_2 = f_2$
(c) $(D + 1)x_1 + Dx_2 - Dx_3 = f_1$
    $Dx_1 + Dx_2 = f_2, \ Dx_2 - Dx_3 = f_3$
(d) $D^2x_1 + (1 + D^{-1})x_2 = f_1$
    $-Dx_1 + (1 - 3D^{-1})x_2 = f_2$

4. For the equations of Problem 3(b) find a particular solution when

(a) $f_1 = e^{2t}, f_2 = 0$ (b) $f_1 = 3e^{2t}, f_2 = e^{3t}$
(c) $f_1 = \cos t, f_2 = 0$ (d) $f_1 = \cos(t - 1), f_2 = 0$

5. (a) Show that if $x = \phi(t, \alpha)$ satisfies the equation

$$(a_0 D^n + \cdots + a_n + \cdots + a_{n+m}D^{-m})x = f(t, \alpha),$$

with constant coefficients (independent of α), then $x = \partial\phi/\partial\alpha$ satisfies the same equation with $f(t, \alpha)$ replaced by $\partial f/\partial\alpha$ (assume all differentiability conditions needed).

(b) Let $Y(s) = (a_0 s^n + \cdots + a_{n+m}s^{-m})^{-1}$ be the transfer function of part (a). Show that when $f = te^{st}$, a particular solution is $x = Y(s)te^{st} + Y'(s)e^{st}$, provided s is not a characteristic root. [*Hint:* Consider $x = e^{st}$ as $\phi(t, s)$ and apply the result of part (a).]

(c) Find a solution when $f = t^k e^{st}$ $(k = 1, 2, \ldots)$.

6. Let the following integrodifferential equations be given:

$$[\phi_{11}(D) + a_1 D^{-1}]x_1 + [\phi_{12}(D) + b_1 D^{-1}]x_2 = f_1,$$

$$[\phi_{21}(D) + a_2 D^{-1}]x_1 + [\phi_{22}(D) + b_2 D^{-1}]x_2 = f_2, \tag{a}$$

where the $\phi_{ij}(D)$ are polynomials in D with constant coefficients and a_1, b_1, a_2, b_2 are constants, $a_1^2 + a_2^2 \neq 0$, $b_1^2 + b_2^2 \neq 0$, with corresponding differential equations

$$\phi_{11}(D)x_1 + \phi_{12}(D)x_2 + a_1 x_3 + b_1 x_4 = f_1,$$

$$\phi_{21}(D)x_1 + \phi_{22}(D)x_2 + a_2 x_3 + b_2 x_4 = f_2, \tag{b}$$

$$Dx_3 - x_1 = 0, \quad Dx_4 - x_2 = 0.$$

Show that the characteristic equation of equations (b):

$$\begin{vmatrix} \phi_{11}(s) & \phi_{12}(s) & a_1 & b_1 \\ \phi_{21}(s) & \phi_{22}(s) & a_2 & b_2 \\ -1 & 0 & s & 0 \\ 0 & -1 & 0 & s \end{vmatrix} = 0$$

can be written in the form

$$s^2 \begin{vmatrix} \phi_{11}(s) + \dfrac{a_1}{s} & \phi_{12}(s) + \dfrac{b_1}{s} \\ \phi_{21}(s) + \dfrac{a_2}{s} & \phi_{22}(s) + \dfrac{b_2}{s} \end{vmatrix} = 0.$$

Show in particular that zero is a characteristic root if $a_1 b_2 - a_2 b_1 = 0$.

ANSWERS

1.

Part	char. func.	trans. func.	freq. resp.	roots
(a)	$s^3 + 1$	$\dfrac{1}{s^3 + 1}$	$\dfrac{1}{1 - i\omega^3}$	$-1, \ \dfrac{1 \pm \sqrt{3i}}{2}$
(b)	$s + 4s^{-1}$	$\dfrac{s}{s^2 + 4}$	$\dfrac{i\omega}{4 - \omega^2}$	$\pm 2i$
(c)	$s + 1 + 3s^{-1}$	$\dfrac{s}{s^2 + s + 3}$	$\dfrac{i\omega}{3 - \omega^2 + i\omega}$	$\dfrac{-1 \pm \sqrt{11}i}{2}$
(d)	$1 + s^{-1} + s^{-2}$	$\dfrac{s^2}{s^2 + s + 1}$	$\dfrac{-\omega^2}{1 - \omega^2 + i\omega}$	$\dfrac{-1 \pm \sqrt{3}i}{2}$

2. (a) $e^{2t}/9$ (b) $-3ie^{3it}/5$
   (c) $(8 \sin 3t + 4 \cos 3t)/5$ (d) $4e^{2t} - \cos t$

3. Characteristic matrices:

(a) $\begin{bmatrix} s + 2 & s - 3 \\ 2s - 1 & s + 1 \end{bmatrix}$ (b) $\begin{bmatrix} s + s^{-1} & 3s - s^{-1} \\ s - s^{-1} & s - s^{-1} \end{bmatrix}$

(c) $\begin{bmatrix} s + 1 & s & -s \\ s & s & 0 \\ 0 & s & -s \end{bmatrix}$ (d) $\begin{bmatrix} s^2 & 1 + s^{-1} \\ -s & 1 - 3s^{-1} \end{bmatrix}$

Characteristic equations and roots:
(a) $s^2 - 10s + 1 = 0$, $5 \pm \sqrt{24}$ (b) $s^4 - 2s^2 + 1 = 0$, $\pm 1, \pm 1$
(c) $s^3 + s^2 = 0$, $-1, 0, 0$ (d) $s^3 - 2s^2 + s = 0$, $1, 1, 0$

Transfer matrices:

(a)
$$\begin{bmatrix} \dfrac{s+1}{-s^2+10s-1} & \dfrac{3-s}{-s^2+10s-1} \\[2ex] \dfrac{-2s+1}{-s^2+10s-1} & \dfrac{s+2}{-s^2+10s-1} \end{bmatrix}$$

(b)
$$\begin{bmatrix} \dfrac{s}{-2(s^2-1)} & \dfrac{s-3s^3}{-2(s^2-1)^2} \\[2ex] \dfrac{s}{2(s^2-1)} & \dfrac{s^3+s}{-2(s^2-1)^2} \end{bmatrix}$$

(c)
$$\begin{bmatrix} \dfrac{1}{s+1} & 0 & \dfrac{-1}{s+1} \\[2ex] \dfrac{-1}{s+1} & \dfrac{1}{s} & \dfrac{1}{s+1} \\[2ex] \dfrac{-1}{s+1} & \dfrac{1}{s} & \dfrac{-1}{s(s+1)} \end{bmatrix}$$

(d)
$$\begin{bmatrix} \dfrac{s-3}{s(s-1)^2} & \dfrac{-s-1}{s(s-1)^2} \\[2ex] \dfrac{s}{(s-1)^2} & \dfrac{s^2}{(s-1)^2} \end{bmatrix}$$

4. (a) $x_1 = -\tfrac{1}{3}e^{2t}$, $x_2 = \tfrac{1}{3}e^{2t}$
   (b) $x_1 = -e^{2t} + \tfrac{39}{64}e^{3t}$, $x_2 = e^{2t} - \tfrac{15}{64}e^{3t}$
   (c) $x_1 = -\tfrac{1}{4}\sin t$, $x_2 = \tfrac{1}{4}\sin t$
   (d) $x_1 = -\tfrac{1}{4}\sin(t-1)$, $x_2 = \tfrac{1}{4}\sin(t-1)$

5. (c) $\dfrac{\partial^k}{\partial s^k}[Y(s)e^{st}] = e^{st}\left[t^k Y(s) + kt^{k-1}Y'(s) + \dfrac{k(k-1)}{2!}t^{k-2}Y''(s) + \cdots\right]$

2–10 Stability. For a system described by a linear differential equation, the output x is of form $T_0[f]$ plus terms arising from initial values; when all initial values are zero, x reduces to $T_0[f]$. For many systems, no matter how the initial values are chosen, the output x approaches $T_0[f]$ as t increases; more precisely, $x - T_0[f] \to 0$ as $t \to +\infty$. When this holds for all choices of f, the system is said to be *stable;* otherwise, the system is called *unstable.* Typical solutions of stable and unstable systems are shown in Fig. 2–7.

If a system is stable, then any two outputs $x_1(t)$, $x_2(t)$ approach each other (that is, $x_1(t) - x_2(t) \to 0$) as $t \to +\infty$. For $x_1(t) - T_0[f] \to 0$, $x_2(t) - T_0[f] \to 0$, so that

$$x_2(t) - x_1(t) = \{x_2(t) - T_0[f]\} + \{T_0[f] - x_1(t)\} \to 0$$

as $t \to +\infty$. Thus in a sense there is, for given input f, only one output, which all solutions approach as $t \to +\infty$.

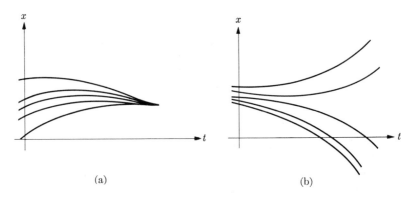

FIG. 2–7.   (a) Stable system, and (b) unstable system.

In the expression $x = T_0[f] + x_0 p_1(t) + \cdots$ for the output, the terms $x_0 p_1(t) + \cdots$ form the difference $x - T_0[f]$; these terms also form the *complementary function*, the general solution of the related homogeneous equation. Accordingly, stability is equivalent to the condition that *every solution of the homogeneous equation approaches zero as $t \to +\infty$*.

When the coefficients are constant, each term in the complementary function is of form $ct^k e^{\lambda t}$, where λ is a characteristic root. But

$$\lim_{t \to \infty} t^k e^{\lambda t} = 0 \quad \text{if Re} (\lambda) < 0$$

and the limit fails to exist otherwise. Accordingly, for an equation with constant coefficients stability is equivalent to the condition that *all characteristic roots have negative real parts*.

The concepts extend at once to systems described by simultaneous differential or integrodifferential equations. Each output x is of form $T_0[f_1] + U_0[f_2] + \cdots$ plus terms arising from initial values. The system is stable if for every choice of initial values the latter terms approach zero as $t \to \infty$, so that each output x has a unique limiting form.

When the coefficients are constant, it has been seen that the solutions of the homogeneous equations are constructed from terms of form const $\times$ $t^k e^{st}$, where s is a characteristic root. Therefore, stability is again reduced to the condition that all characteristic numbers have negative real parts. It can happen that we have stability for some outputs and not for others, as the following example shows:

$$(D + 1)x_1 = f_1, \qquad (D - 1)x_2 + x_1 = f_2.$$

Here the general solution of the homogeneous equations ($f_1 = f_2 = 0$) is given by

$$x_1 = 2c_1 e^{-t}, \qquad x_2 = c_1 e^{-t} + c_2 e^{t}.$$

Thus we have stability for x_1 and instability for x_2. For a system to be stable, we require that there be stability for *all* outputs.

Stability and resonance. The stability of a linear system can be related to *resonance.* We call a system *resonant* if some input f which is bounded ($|f| \leq$ const) for $t \geq 0$ has an output $x(t)$ which is not bounded for $t \geq 0$.

THEOREM 3. *Let a linear system be described by a differential equation with constant coefficients,*

$$(a_0 D^n + \cdots + a_n)x = f,$$

and let $W(t)$ be the corresponding weighting function. Then the following conditions are equivalent:
(a) *the system is stable;*
(b) *the system is not resonant;*
(c) $\int_0^\infty |W(t)|\, dt = B < \infty.$

Proof. Let us suppose the system is stable. Then the complementary function is formed of terms of form $t^k e^{st}$, where Re $(s) < 0$. Furthermore, $W(t)$ is a linear combination of such terms for $t \geq 0$ (Problem 8 following Section 2–8). Now for Re $(s) < 0$

$$\int_0^\infty |t^k e^{st}|\, dt < \infty \quad \text{for } k = 0, 1, \ldots \tag{2–100}$$

Hence $W(t)$ satisfies condition (c). If $f(t)$ is bounded for $t \geq 0$, then $|f(t)| \leq K$ for some constant K, for $t \geq 0$. Now, if $x^*(t)$ is the corresponding output $T_0[f]$, we have for $t \geq 0$

$$x^*(t) = \int_0^t f(t - u)W(u)\, du,$$

$$|x^*(t)| \leq \int_0^t |f(t - u)|\, |W(u)|\, du \leq K \int_0^t |W(u)|\, du \leq KB,$$

so that $x^*(t)$ is bounded. The general output is $x^*(t) + x_c(t)$. Since the system is stable, $x_c(t) \to 0$ as $t \to \infty$ for every choice of the arbitrary constants. Hence every output is bounded. Therefore condition (b) is proved.

Next let us suppose the system is unstable. Then at least one characteristic root s_0 has nonnegative real part. If Re $(s_0) = 0$, then we choose $f(t) = e^{s_0 t}$; there is an output $x(t)$ of form $t^k e^{s_0 t}$, $k \geq 1$, found by undetermined coefficients (Section 1–3). Now f is bounded for $t \geq 0$, but x is not. If Re $(s_0) > 0$, then the complementary function contains at least one unbounded term; thus for every input there may be some bounded output, but addition of a solution of the homogeneous equation yields an unbounded output. Hence the system is resonant. The proof that condition (c) fails is postponed to Section 6–18, where it is found as an immediate consequence of properties of Laplace transforms.

Remarks. From the proof we see that when the system is stable, $|f(t)| \leq K$ for $t \geq 0$ implies $|x(t)| \leq BK$ for $t \geq 0$, where $x = T_0[f]$. Thus in a sense the constant B is an upper estimate for the maximum amplification of the system.

The theorem can be generalized to equations with variable coefficients, though certain modifications are necessary (see Section 8–16). An extended discussion of stability is given in Chapter 7.

Problems

1. Test each of the following for stability:

(a) $(D^2 + 2D + 7)x = f$ (b) $(D^5 + 1)x = f$

(c) $(D + 2 + D^{-1})x = f$ (d) $(D^2 + D + 1 + D^{-1})x = f$

2. Test each of the following for stability:

(a) $(D + 2)x_1 + (D - 2)x_2 = f_1$, $(D - 1)x_1 + (2D + 3)x_2 = f_2$

(b) $(D + D^{-1})x_1 + (D - D^{-1})x_2 = f_1$, $(2D - D^{-1})x_1 + Dx_2 = f_2$

(c) $(D + 1 + D^{-1})x_1 + (D + D^{-1})x_2 = f_1$

   $(2D + 1)x_1 + (3D + 1)x_2 = f_2$

Answers

1. (a) stable   (b) unstable   (c) stable   (d) unstable
2. (a) stable   (b) unstable   (c) unstable

SUGGESTED REFERENCES

1. DAVID K. CHENG, *Analysis of Linear Systems*. Reading, Mass.: Addison-Wesley, 1959.

2. H. M. JAMES, N. B. NICHOLS and RALPH S. PHILLIPS, *Theory of Servomechanisms*. New York: McGraw-Hill, 1947.

3. WILFRED KAPLAN, *Advanced Calculus*. Reading, Mass.: Addison-Wesley, 1952.

4. WILFRED KAPLAN, *Ordinary Differential Equations*. Reading, Mass.: Addison-Wesley, 1958.

5. DEREK F. LAWDEN, *Mathematics of Engineering Systems*. New York: John Wiley, 1954.

6. JOHN H. TRUXAL, *Automatic Feedback Control System Synthesis*. New York: McGraw-Hill, 1955.

7. H. TSIEN, *Engineering Cybernetics*. New York: McGraw-Hill, 1954.

8. T. VON KÁRMÁN and M. A. BIOT, *Mathematical Methods in Engineering*. New York: McGraw-Hill, 1940.

9. CHARLES S. WILTS, *Principles of Feedback Control*. Reading, Mass.: Addison-Wesley, 1960.

CHAPTER 3

ANALYTIC FUNCTIONS OF A COMPLEX VARIABLE

Complex functions have played an important role in Chapters 1 and 2. For the most part, they have been complex functions of a *real* variable. The full power of the theory of complex functions of a complex variable has not been needed, nor is it essential for the chapters immediately following. The first important application appears in Section 5–14 (on the evaluation of inverse Fourier transforms by residues). An acquaintance with analytic functions is very helpful as background for the Laplace transform, and important applications are made in Chapter 6. Accordingly, the systematic study of the present chapter can be postponed if desired. However, there are advantages in studying it without delay: the material is necessary for most advanced topics; it is helpful as background for all that follows; by studying it early, interruptions in studying the chapters on Fourier and Laplace transforms can be avoided.

3–1 Functions of a complex variable. Limits and continuity. We consider complex functions $w = f(z)$, $w = u + iv$, as in Section 1–7. Thus u and v are real functions of x and y: $u = u(x, y) = \mathrm{Re}\,[f(z)]$, $v = v(x, y) = \mathrm{Im}\,[f(z)]$. Examples of such functions, given in Section 1–7, are polynomials, rational functions, the exponential function $e^z = \exp z$, and trigonometric functions.

In general, we assume $w = f(z)$ to be defined in a *domain* (open region) D in the z-plane, as suggested in Fig. 3–1. If z_0 is a point of D, we can then find a circular *neighborhood* $|z - z_0| < k$ about z_0 in D. If $f(z)$ is defined in such a neighborhood, except perhaps at z_0, then we write

$$\lim_{z \to z_0} f(z) = w_0 \qquad (3\text{–}10)$$

if, for every $\epsilon > 0$, we can choose $\delta > 0$, so that

$$|f(z) - w_0| < \epsilon \qquad (3\text{–}11)$$
$$\text{for } 0 < |z - z_0| < \delta.$$

If $f(z_0)$ is defined and equals w_0, and (3–10) holds, then we call $f(z)$ *continuous at z_0.*

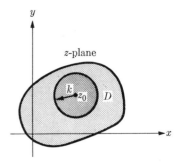

Fig. 3–1. Domain and neighborhood.

THEOREM 1. *The function $w = f(z)$ is continuous at $z_0 = x_0 + iy_0$ if and only if $u(x, y) = \mathrm{Re}\,[f(z)]$ and $v(x, y) = \mathrm{Im}\,[f(z)]$ are continuous at (x_0, y_0).*

Thus $w = z^2 = x^2 - y^2 + 2ixy$ is continuous for all z, since $u = x^2 - y^2$ and $v = 2xy$ are continuous for all (x, y). The proof of Theorem 1 is left as an exercise (Problem 5 below).

THEOREM 2. *The sum, product, and quotient of continuous functions of z are continuous, except for division by zero; a continuous function of a continuous function is continuous. Similarly, if the limits exist,*

$$\lim_{z \to z_0} [f(z) + g(z)] = \lim_{z \to z_0} f(z) + \lim_{z \to z_0} g(z), \quad \dots \qquad (3\text{--}12)$$

These properties are proved as for real variables. (It is assumed in Theorem 2 that the functions are defined in appropriate domains.)

It follows from Theorem 2 that polynomials in z are continuous for all z, and each rational function is continuous except where the denominator is zero. From Theorem 1 it follows that

$$e^z = e^x \cos y + ie^x \sin y$$

is continuous for all z. Hence, by Theorem 2, so also are the functions

$$\sin z = \frac{e^{iz} - e^{-iz}}{2i}, \qquad \cos z = \frac{e^{iz} + e^{-iz}}{2}.$$

We write

$$\lim_{z \to z_0} f(z) = \infty \quad \text{if } \lim_{z \to z_0} |f(z)| = +\infty;$$

that is, if for each real number K there is a positive δ such that $|f(z)| > K$ for $0 < |z - z_0| < \delta$. Similarly, if $f(z)$ is defined for $|z| > R$, for some R, then $\lim_{z \to \infty} f(z) = c$ if for each $\epsilon > 0$ we can choose a number R_0 such that $|f(z) - c| < \epsilon$ for $|z| > R_0$. All these definitions emphasize that there is but *one* complex number ∞ and that "approaching ∞" is equivalent to receding from the origin.

3–2 Derivatives and differentials. Let $w = f(z)$ be given in D and let z_0 be a point of D. Then w is said to have a derivative $f'(z_0)$ if

$$\lim_{\Delta z \to 0} \frac{f(z_0 + \Delta z) - f(z_0)}{\Delta z} = f'(z_0).$$

In appearance this definition is the same as that for functions of a real variable, and it will be seen that the derivative does have the usual properties. However, it will also be shown that if $w = f(z)$ has a continuous derivative in a domain D, then $f(z)$ has a number of additional properties; in particular, the second derivative $f''(z)$, third derivative $f'''(z), \ldots$, must also exist in D.

The reason for the remarkable consequences of possession of a derivative lies in the fact that the increment Δz is allowed to approach zero in any manner. If we restricted Δz so that $z_0 + \Delta z$ approached z_0 along a particular line, then we would obtain a "directional derivative." But here the limit obtained is required to be the *same for all directions*, so that the "directional derivative" has the same value in all directions. Moreover, $z_0 + \Delta z$ may approach z_0 in a quite arbitrary manner, for example along a spiral path. The limit of the ratio $\Delta w / \Delta z$ must be the same for all manners of approach.

We say that $f(z)$ has a *differential* $dw = c\,\Delta z$ at z_0 if $f(z_0 + \Delta z) - f(z_0) = c\,\Delta z + \epsilon\,\Delta z$, where ϵ depends on Δz and is continuous at $\Delta z = 0$, with value zero when $\Delta z = 0$.

THEOREM 3. *If $w = f(z)$ has a differential $dw = c\,\Delta z$ at z_0, then w has a derivative $f'(z_0) = c$. Conversely, if w has a derivative at z_0, then w has a differential at z_0: $dw = f'(z_0)\,\Delta z$.*

This is proved just as for real functions. We also write $\Delta z = dz$, as for real variables, so that

$$dw = f'(z)\,dz, \qquad \frac{dw}{dz} = f'(z). \tag{3-20}$$

From Theorem 3 we see that existence of the derivative $f'(z_0)$ implies continuity of f at z_0, for $f(z_0 + \Delta z) - f(z_0) = c\,\Delta z + \epsilon\,\Delta z \to 0$ as $\Delta z \to 0$.

THEOREM 4. *If w_1 and w_2 are functions of z which have differentials in D, then*

$$d(w_1 + w_2) = dw_1 + dw_2,$$

$$d(w_1 w_2) = w_1\,dw_2 + w_2\,dw_1, \tag{3-21}$$

$$d\,\frac{w_1}{w_2} = \frac{w_2\,dw_1 - w_1\,dw_2}{w_2^2} \qquad (w_2 \neq 0).$$

If w_2 is a differentiable function of w_1, and w_1 is a differentiable function of z, then wherever $w_2[w_1(z)]$ is defined

$$\frac{dw_2}{dz} = \frac{dw_2}{dw_1} \cdot \frac{dw_1}{dz}. \tag{3-22}$$

These rules are proved as in elementary calculus. We can now prove as usual the basic rule:

$$\frac{d}{dz} z^n = nz^{n-1} \quad (n = 1, 2, \ldots). \tag{3-23}$$

Furthermore, the derivative of a constant is zero.

PROBLEMS

1. For each of the following write the given function as two real functions of x and y and determine where the given function is continuous:

(a) $w = (1+i)z^2$

(b) $w = \dfrac{z}{z+i}$

(c) $w = \tan z = \dfrac{\sin z}{\cos z}$

(d) $w = \dfrac{e^{-z}}{z+1}$

2. Evaluate each of the following limits:

(a) $\displaystyle\lim_{z \to \pi i} \frac{\sin z + z}{e^z + 2}$

(b) $\displaystyle\lim_{z \to 0} \frac{z^2 - z}{2z}$

(c) $\displaystyle\lim_{z \to 0} \frac{\cos z}{z}$

(d) $\displaystyle\lim_{z \to \infty} \frac{z}{z^2 + 1}$

3. Differentiate each of the following complex functions:

(a) $w = z^3 + 5z + 1$

(b) $w = \dfrac{1}{z-1}$

(c) $w = [1 + (z^2 + 1)^3]^7$

(d) $w = \dfrac{z^2}{(z+1)^3}$

4. Prove the rule (3–23).

5. Prove Theorem 1.

ANSWERS

1. (a) $u = x^2 - y^2 - 2xy$, $v = x^2 - y^2 + 2xy$, all z
(b) $u = (x^2 + y^2 + y)[x^2 + (y+1)^2]^{-1}$
   $v = -x[x^2 + (y+1)^2]^{-1} \quad (z \neq -i)$
(c) $u = \tan x \operatorname{sech}^2 y [1 + \tan^2 x \tanh^2 y]^{-1}$
   $v = \tanh y \sec^2 x [1 + \tan^2 x \tanh^2 y]^{-1}$
   $z \neq (\pi/2) + n\pi, n = 0, \pm 1, \pm 2, \ldots$
(d) $u = e^{-x}[(1 + x) \cos y - y \sin y][(1 + x)^2 + y^2]^{-1}$
   $v = -e^{-x}[(1 + x) \sin y + y \cos y][(1 + x)^2 + y^2]^{-1} \quad (z \neq -1)$

2. (a) $i(\pi + \sinh \pi)$ (b) $-\frac{1}{2}$ (c) ∞ (d) 0

3. (a) $3z^2 + 5$ (b) $-(z-1)^{-2}$ (c) $42[1 + (z^2 + 1)^3]^6(z^2 + 1)^2 z$
(d) $(z+1)^{-4}(2z - z^2)$

3–3 Integrals. The complex integral $\int f(z)\, dz$ is defined as a line integral, and its properties are closely related to those of the integral $\int P\, dx + Q\, dy$ (see pp. 225–258 of Reference 5).

Let C be a path from A to B in the complex plane: $x = x(t)$, $y = y(t)$, $a \leq t \leq b$. We assume C to have a direction, usually that of increasing t. We subdivide the interval $a \leq t \leq b$ into n parts by $t_0 = a$, $t_1, \ldots,$ $t_n = b$. We let $z_j = x(t_j) + iy(t_j)$ and $\Delta_j z = z_j - z_{j-1}$, $\Delta_j t = t_j - t_{j-1}$. We choose an arbitrary value t_j^* in the interval $t_{j-1} \leq t \leq t_j$ and set $z_j^* = x(t_j^*) + iy(t_j^*)$. These quantities are all shown in Fig. 3–2. We then write

$$\int_C f(z)\, dz = \int_A^B f(z)\, dz = \lim_{\substack{n \to \infty \\ \max \Delta_j t \to 0}} \sum_{j=1}^n f(z_j^*)\, \Delta_j z. \qquad (3\text{–}30)$$

If we take real and imaginary parts in (3–30), we find

$$\int_C f(z)\, dz = \lim \sum (u + iv)\,(\Delta x + i\,\Delta y)$$
$$= \lim \{\sum(u\,\Delta x - v\,\Delta y) + i\sum(v\,\Delta x + u\,\Delta y)\};$$

that is,

$$\int_C f(z)\, dz = \int_C (u + iv)\,(dx + i\,dy)$$

$$= \int_C (u\, dx - v\, dy) + i \int_C (v\, dx + u\, dy). \qquad (3\text{–}31)$$

The complex line integral is thus simply a combination of two real line integrals. Hence we can apply all the theory of real line integrals. In the

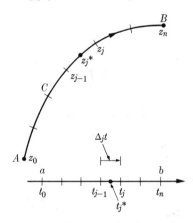

Fig. 3–2. Complex line integral.

following, each path is assumed to be *piecewise smooth;* that is, $x(t)$ and $y(t)$ are to be continuous with piecewise continuous derivatives.

THEOREM 5. *If $f(z)$ is continuous in domain D, then the integral (3–30) exists and*

$$\int_C f(z)\, dz = \int_a^b \left(u\frac{dx}{dt} - v\frac{dy}{dt} \right) dt + i\int_a^b \left(v\frac{dx}{dt} + u\frac{dy}{dt} \right) dt. \quad (3\text{–}32)$$

If we introduce the derivative

$$\frac{dz}{dt} = \frac{dx}{dt} + i\frac{dy}{dt}$$

of z with respect to the real variable t, and also use the theory of integrals of such functions (Section 1–8), we can write (3–32) more concisely:

$$\int_C f(z)\, dz = \int_a^b f[z(t)]\frac{dz}{dt}\, dt. \quad (3\text{–}33)$$

EXAMPLE 1.   Let C be the path $x = 2t$, $y = 3t$, $1 \leq t \leq 2$.   Let $f(z) = z^2$. Then

$$\int_C z^2\, dz = \int_1^2 (2t + 3it)^2(2 + 3i)\, dt = (2 + 3i)^3\int_1^2 t^2\, dt$$

$$= -107\tfrac{1}{3} + 21i.$$

EXAMPLE 2. Let C be the circular path $x = \cos t$, $y = \sin t$, $0 \leq t \leq 2\pi$. This can be written more concisely thus: $z = e^{it}$, $0 \leq t \leq 2\pi$. Since $dz/dt = ie^{it}$,

$$\int_C \frac{1}{z}\, dz = \int_0^{2\pi} e^{-it}ie^{it}\, dt = i\int_0^{2\pi} dt = 2\pi i.$$

Further properties of complex integrals follow from those of real integrals:

THEOREM 6. *Let $f(z)$ and $g(z)$ be continuous in a domain D. Let C be a piecewise smooth path in D. Then*

$$\int_C [f(z) + g(z)]\, dz = \int_C f(z)\, dz + \int_C g(z)\, dz,$$

$$\int_C kf(z)\, dz = k\int_C f(z)\, dz \quad (k = \text{const}),$$

$$\int_C f(z)\, dz = \int_{C_1} f(z)\, dz + \int_{C_2} f(z)\, dz,$$

where C is composed of a path C_1 from z_0 to z_1 and a path C_2 from z_1 to z_2, and

$$\int_C f(z)\, dz = -\int_{C'} f(z)\, dz,$$

where C' is obtained from C by reversing direction on C.

Upper estimates for the absolute value of a complex integral are obtained by the following theorem.

THEOREM 7. *Let $f(z)$ be continuous on C, let $|f(z)| \leqq M$ on C, and let*

$$L = \int_C ds = \int_a^b \sqrt{(dx/dt)^2 + (dy/dt)^2}\, dt$$

be the length of C. Then

$$\left| \int_C f(z)\, dz \right| \leqq \int_C |f(z)|\, ds \leqq M \cdot L. \tag{3–34}$$

Proof. The line integral $\int |f(z)|\, ds$ is defined as a limit:

$$\int_C |f(z)|\, ds = \lim \Sigma |f(z_j^*)|\, \Delta_j s,$$

where $\Delta_j s$ is the length of the jth arc of C. Now

$$|f(z_j^*)\, \Delta_j z| = |f(z_j^*)| \cdot |\Delta_j z| \leqq |f(z_j^*)| \cdot \Delta_j s,$$

for $|\Delta_j z|$ represents the *chord* of the arc $\Delta_j s$. Hence

$$|\Sigma f(z_j^*)\, \Delta_j z| \leqq \Sigma |f(z_j^*)\, \Delta_j z| \leqq \Sigma |f(z_j^*)|\, \Delta_j s$$

by repeated application of the triangle inequality (1–70). Passing to the limit, we conclude that

$$\left| \int_C f(z)\, dz \right| \leqq \int_C |f(z)|\, ds. \tag{3–35}$$

Also, if $|f| \leqq M = $ const,

$$\Sigma |f(z_j^*)|\, \Delta_j s \leqq \Sigma M\, \Delta_j s = M \cdot L.$$

Hence

$$\int_C |f(z)|\, ds \leqq M \cdot L. \tag{3–36}$$

Inequalities (3–34) follow from (3–35) and (3–36).

PROBLEMS

1. Evaluate the following integrals:

(a) $\displaystyle\int_0^{1+i} (x^2 - iy^2)\, dz$ on the straight line from 0 to $1 + i$.

(b) $\displaystyle\int_0^{\pi} z\, dz$ on the curve $y = \sin x$.

(c) $\displaystyle\int_1^{1+i} \frac{dz}{z}$ on the line $x = 1$.

2. Write each of the following integrals in the form $\int u\, dx - v\, dy + i\int v\, dx + u\, dy$; then show that each of the two real integrals is independent of path in the xy-plane.

(a) $\int (z + 1)\, dz$ (b) $\int e^z\, dz$
(c) $\int z^4\, dz$ (d) $\int \sin z\, dz$

3. (a) Evaluate

$$\oint \frac{1}{z}\, dz$$

on the circle $|z| = R$.

(b) Show that

$$\oint \frac{1}{z}\, dz = 0$$

on every simple closed path not meeting or enclosing the origin.

(c) Show that

$$\oint \frac{1}{z^2}\, dz = 0$$

on every simple closed path not passing through the origin.

ANSWERS

1. (a) $\frac{2}{3}$ (b) $\pi^2/2$ (c) $\frac{1}{2}\log 2 + i(\pi/4)$
3. (a) $2\pi i$

3–4 Analytic functions.   Cauchy-Riemann equations.   A function $w = f(z)$, defined in a domain D, is said to be an *analytic function* in D if w has a continuous derivative in D. Almost the entire theory of functions of a complex variable is confined to the study of such functions. Furthermore, almost all functions used in the applications of mathematics to physical problems are analytic functions or are derived from such.

It will be seen that possession of a continuous derivative implies possession of a continuous second derivative, third derivative, . . . , and, in fact, convergence of the Taylor series

$$f(z_0) + f'(z_0) \frac{(z - z_0)}{1!} + f''(z_0) \frac{(z - z_0)^2}{2!} + \cdots$$

in a neighborhood of each z_0 of D. Thus one could define an analytic function as one so representable by Taylor series, and this definition is often used. The two definitions are equivalent, for convergence of the Taylor series in a neighborhood of each z_0 implies continuity of the derivatives of all orders.

While it is possible to construct continuous functions of z which are not analytic (examples will be given below), it is impossible to construct a function $f(z)$ possessing a derivative, but not a continuous one, in D. In other words, if $f(z)$ has a derivative in D, the derivative is necessarily continuous, so that $f(z)$ is analytic. Therefore we could define an analytic function as one merely possessing a derivative in domain D, and this definition is also often used. For a proof that existence of the derivative implies its continuity, refer to Vol. I of Reference 6.

THEOREM 8. *If $w = u + iv = f(z)$ is analytic in D, then u and v have continuous first partial derivatives in D and satisfy the Cauchy-Riemann equations*

$$\frac{\partial u}{\partial x} = \frac{\partial v}{\partial y}, \qquad \frac{\partial u}{\partial y} = -\frac{\partial v}{\partial x} \tag{3-40}$$

in D. Furthermore,

$$\frac{dw}{dz} = \frac{\partial u}{\partial x} + i\frac{\partial v}{\partial x} = \frac{\partial v}{\partial y} + i\frac{\partial v}{\partial x} = \frac{\partial u}{\partial x} - i\frac{\partial u}{\partial y} = \frac{\partial v}{\partial y} - i\frac{\partial u}{\partial y}. \tag{3-41}$$

Conversely, if $u(x, y)$ and $v(x, y)$ have continuous first partial derivatives in D and satisfy the Cauchy-Riemann equations (3–40), then $w = u + iv = f(z)$ is analytic in D.

Proof. Let z_0 be a fixed point of D and let

$$\Delta w = \Delta u + i\,\Delta v = f(z_0 + \Delta z) - f(z_0), \qquad \Delta z = \Delta x + i\,\Delta y,$$

as in Fig. 3–3. We consider several equivalent formulations of the condition that $f'(z_0)$ exists. Throughout, ϵ, ϵ_1, ϵ_2, ϵ_3, ϵ_4 denote functions of $\Delta z = \Delta x + i\,\Delta y$, continuous and equal to zero at $\Delta z = 0$. By Theorem 3, existence of $f'(z_0)$ is equivalent to the statement

$$\Delta w = c \cdot \Delta z + \epsilon \cdot \Delta z, \qquad c = f'(z_0), \qquad c = a + ib; \tag{3-42}$$

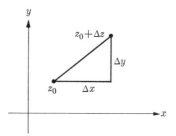

FIG. 3–3.   Complex derivative.

this is equivalent to

$$\Delta w = c\,\Delta z + \epsilon\,\Delta x + i\epsilon\,\Delta y \qquad (3\text{–}42')$$

and also to

$$\Delta w = c\,\Delta z + \epsilon_1\,\Delta x + \epsilon_2\,\Delta y + i(\epsilon_3\,\Delta x + \epsilon_4\,\Delta y), \qquad (3\text{–}42'')$$

where ϵ_1, ϵ_2, ϵ_3, ϵ_4 are real. For if (3–42′) holds, then (3–42″) holds with $\epsilon_1 = \mathrm{Re}\,(\epsilon)$, $\epsilon_2 = -\mathrm{Im}\,(\epsilon)$, $\epsilon_3 = \mathrm{Im}\,(\epsilon)$, $\epsilon_4 = \mathrm{Re}\,(\epsilon)$. Conversely, if (3–42″) holds, then (3–42′) holds with $\epsilon = 0$ for $\Delta z = 0$ and

$$\epsilon = (\epsilon_1 + i\epsilon_3)\frac{\Delta x}{\Delta z} + (\epsilon_2 + i\epsilon_4)\frac{\Delta y}{\Delta z} \quad (\Delta z \neq 0). \qquad (3\text{–}43)$$

As Fig. 3–3 shows,

$$\left|\frac{\Delta x}{\Delta z}\right| \leqq 1, \qquad \left|\frac{\Delta y}{\Delta z}\right| \leqq 1,$$

so that we deduce from (3–43) that $\epsilon \to 0$ as $\Delta z \to 0$. Thus (3–42), (3–42′) and (3–42″) are all equivalent to existence of $f'(z_0) = c = a + ib$. By taking real and imaginary parts in (3–42″), we obtain one more equivalent condition:

$$\begin{aligned} \Delta u &= a\,\Delta x - b\,\Delta y + \epsilon_1\,\Delta x + \epsilon_2\,\Delta y, \\ \Delta v &= b\,\Delta x + a\,\Delta y + \epsilon_3\,\Delta x + \epsilon_4\,\Delta y; \end{aligned} \qquad (3\text{–}42''')$$

these equations state that u, v have differentials $du = a\,dx - b\,dy$, $dv = b\,dx + a\,dy$ at (x_0, y_0), and hence at this point

$$\frac{\partial u}{\partial x} = a = \frac{\partial v}{\partial y}, \qquad \frac{\partial u}{\partial y} = -b = -\frac{\partial v}{\partial x}.$$

Thus differentiability of $f'(z)$ at any z is equivalent to differentiability of u, v along with validity of the Cauchy-Riemann equations. Furthermore, $f'(z)$ and $\partial u/\partial x, \ldots$ are related by (3–41). By Theorem 1, these equations show that continuity of $f'(z)$ in D is equivalent to continuity of $\partial u/\partial x, \ldots$ Thus the theorem is proved.

The theorem provides a perfect test for analyticity: if $f(z)$ is analytic, then the Cauchy-Riemann equations hold; if the equations hold (and the derivatives concerned are continuous), then $f(z)$ is analytic.

EXAMPLE 1. $w = z^2 = x^2 - y^2 + i \cdot 2xy$. Here $u = x^2 - y^2$, $v = 2xy$. Thus

$$\frac{\partial u}{\partial x} = 2x = \frac{\partial v}{\partial y}, \qquad \frac{\partial u}{\partial y} = -2y = -\frac{\partial v}{\partial x}$$

and w is analytic for all z.

EXAMPLE 2. $w = \dfrac{x}{x^2 + y^2} - \dfrac{iy}{x^2 + y^2}$. Here

$$\frac{\partial u}{\partial x} = \frac{y^2 - x^2}{(x^2 + y^2)^2} = \frac{\partial v}{\partial y}, \qquad \frac{\partial u}{\partial y} = \frac{-2xy}{(x^2 + y^2)^2} = -\frac{\partial v}{\partial x}.$$

Hence w is analytic except for $x^2 + y^2 = 0$, that is, for $z = 0$.

EXAMPLE 3. $w = x - iy = \bar{z}$. Here $u = x, v = -y$ and

$$\frac{\partial u}{\partial x} = 1, \qquad \frac{\partial v}{\partial y} = -1, \qquad \frac{\partial u}{\partial y} = 0 = \frac{\partial v}{\partial x}.$$

Thus w is not analytic in any domain.

EXAMPLE 4. $w = x^2 y^2 + 2x^2 y^2 i$. Here

$$\frac{\partial u}{\partial x} = 2xy^2, \qquad \frac{\partial v}{\partial y} = 4x^2 y, \qquad \frac{\partial u}{\partial y} = 2x^2 y, \qquad \frac{\partial v}{\partial x} = 4xy^2.$$

The Cauchy-Riemann equations give $2xy^2 = 4x^2 y, 2x^2 y = -4xy^2$. These equations are satisfied only along the lines $x = 0$, $y = 0$. There is *no domain* in which the Cauchy-Riemann equations hold, hence no domain in which $f(z)$ is analytic. One does not consider functions analytic only at certain points unless these points form a domain.

The terms "analytic at a point" or "analytic along a curve" are used, apparently in contradiction to the remark just made. However, we say that $f(z)$ is *analytic at the point* z_0 only if there is a domain containing z_0 within which $f(z)$ is analytic. Similarly, $f(z)$ is *analytic along a curve C* only if $f(z)$ is analytic in a domain containing C.

THEOREM 9. *The sum, product, and quotient of analytic functions is analytic (provided in the last case the denominator is not equal to zero at any point of the domain under consideration). All polynomials are analytic for all z. Every rational function is analytic in each domain containing no root of the denominator. An analytic function of an analytic function is analytic.*

This follows from Theorem 4.

We readily verify (Problem 1 below) that the Cauchy-Riemann equations are satisfied for $u = \text{Re } (e^z)$, $v = \text{Im } (e^z)$. Hence e^z is analytic for all z. It then follows from Theorem 9 that $\sin z$, $\cos z$, $\sinh z$, and $\cosh z$ are analytic for all z, while $\tan z$, $\sec z$, and $\csc z$ are analytic except for certain points (Problem 6 below). Furthermore, the usual formulas for derivatives hold:

$$\frac{d}{dz} e^z = e^z, \qquad \frac{d}{dz} \sin z = \cos z, \qquad \ldots \qquad (3\text{--}44)$$

(Problem 3).

Two basic theorems of more advanced theory are useful at this point. Proofs are given in Chapter IV of Reference 3.

THEOREM 10. *Given a function $f(x)$ of the real variable x, $a \leqq x \leqq b$, there is at most one analytic function $f(z)$ which reduces to $f(x)$ when z is real.*

THEOREM 11. *If $f(z)$, $g(z)$, ... are functions which are all analytic in a domain D which includes part of the real axis, and $f(z)$, $g(z)$, ... satisfy an algebraic identity when z is real, then these functions satisfy the same identity for all z in D.*

Theorem 10 implies that our definitions of e^z, $\sin z$, ... are the only ones which yield analytic functions and agree with the definitions for real variables.

Because of Theorem 11, we can be sure that all familiar identities of trigonometry, namely,

$$\sin^2 z + \cos^2 z = 1, \qquad \sin \left(\frac{\pi}{2} - z \right) = \cos z, \qquad \ldots \qquad (3\text{--}45)$$

continue to hold for complex z. A general algebraic identity is formed by replacing the variables $w_1, \ldots, w_n$ in an algebraic equation by functions $f_1(z), \ldots, f_n(z)$. Thus, in the two examples given, one has

$$w_1^2 + w_2^2 - 1 = 0 \quad (w_1 = \sin z, w_2 = \cos z),$$

$$w_1 - w_2 = 0 \quad \left[w_1 = \sin \left(\frac{\pi}{2} - z \right), w_2 = \cos z \right].$$

To prove identities such as

$$e^{z_1} \cdot e^{z_2} = e^{z_1 + z_2}, \qquad (3\text{--}46)$$

it may be necessary to apply Theorem 11 several times. (See Problems 4 and 5 below.)

It should be remarked that while e^z is written as a power of e, it is best not to think of it as such. Thus $e^{1/2}$ has only one value, not two, as would a usual complex root. To avoid confusion with the general power function, to be defined below, we often write $e^z = \exp z$ and refer to e^z as the *exponential function of z.*

To obtain the real and imaginary parts of $\sin z$, we use the identity

$$\sin(z_1 + z_2) = \sin z_1 \cos z_2 + \cos z_1 \sin z_2,$$

which holds, by the reasoning described above, for all complex z_1 and z_2. Hence $\sin(x + iy) = \sin x \cos iy + \cos x \sin iy$. Now from the definitions (Section 1–7),

$$\sinh y = -i \sin iy,$$

$$\cosh y = \cos iy.$$

(3–47)

Hence

$$\sin z = \sin x \cosh y + i \cos x \sinh y. \tag{3-48}$$

Similarly, we prove

$$\cos z = \cos x \cosh y - i \sin x \sinh y,$$

$$\sinh z = \sinh x \cos y + i \cosh x \sin y, \tag{3-49}$$

$$\cosh z = \cosh x \cos y + i \sinh x \sin y.$$

PROBLEMS

1. Verify that the following are analytic functions of z:

(a) $2x^3 - 3x^2 y - 6xy^2 + y^3 + i(x^3 + 6x^2 y - 3xy^2 - 2y^3)$
(b) $w = e^z = e^x \cos y + i e^x \sin y$
(c) $w = \sin z = \sin x \cosh y + i \cos x \sinh y$

2. Test each of the following for analyticity:

(a) $x^3 + y^3 + i(3x^2 y + 3xy^2)$
(b) $\sin x \cos y + i \cos x \sin y$
(c) $3x + 5y + i(3y - 5x)$

3. Prove the following properties directly from the definitions of the functions:

(a) $\dfrac{d}{dz} e^z = e^z$
     (b) $\dfrac{d}{dz} \sin z = \cos z$, $\dfrac{d}{dz} \cos z = -\sin z$

(c) $\sin(z + \pi) = -\sin z$
     (d) $\sin(-z) = -\sin z$, $\cos(-z) = \cos z$

4. Prove the identity $e^{z_1 + z_2} = e^{z_1} \cdot e^{z_2}$ by application of Theorem 11. [*Hint:* Let $z_2 = b$, a fixed real number, and $z_1 = z$, a variable complex number. Then $e^{z+b} = e^z \cdot e^b$ is an identity connecting analytic functions which is known to be true for z real. Hence it is true for all complex z. Now proceed similarly with the identity $e^{z_1 + z} = e^{z_1} \cdot e^z$.]

5. Prove the following identities by application of Theorem 11 (see Problem 4):

(a) $\cos (z_1 + z_2) = \cos z_1 \cos z_2 - \sin z_1 \sin z_2$
(b) $e^{iz} = \cos z + i \sin z$
(c) $(e^z)^n = e^{nz}$ $(n = 0, 1, 2, \ldots)$

6. Determine where the following functions are analytic (see Problem 3 following Section 1–8):

(a) $\tan z = \dfrac{\sin z}{\cos z}$

(b) $\cot z = \dfrac{\cos z}{\sin z}$

(c) $\tanh z = \dfrac{\sinh z}{\cosh z}$

(d) $\dfrac{\sin z}{z}$

(e) $\dfrac{e^z}{z \cos z}$

(f) $\dfrac{e^z}{\sin z + \cos z}$

ANSWERS

2. (a) Analytic nowhere, (b) analytic nowhere, (c) analytic for all z.

6. The functions are analytic except at the following points: (a) $\frac{1}{2}\pi + n\pi$; (b) $n\pi$; (c) $\frac{1}{2}\pi i + n\pi i$; (d) 0; (e) 0, $\frac{1}{2}\pi + n\pi$; (f) $-\frac{1}{4}\pi + n\pi$, where $n = 0$, ± 1, ± 2, ...

3–5 The functions log z, a^z, z^a, $\cos^{-1} z$, $\cos^{-1} z$. The function $w = \log z$ is defined as the inverse of the exponential function $z = e^w$. We write $z = re^{i\theta}$, in terms of polar coordinates r, θ, and $w = u + iv$, so that

$$re^{i\theta} = e^{u+iv} = e^u e^{iv},$$

$$e^u = r, \qquad v = \theta + 2k\pi \quad (k = 0, \pm 1, \ldots).$$

Accordingly,

$$w = \log z = \log r + i(\theta + 2k\pi) = \log |z| + i \arg z, \qquad (3\text{–}50)$$

where $\log r$ is the real logarithm of r Thus $\log z$ is a multiple-valued function of z, with infinitely many values except for $z = 0$. We can select one value of θ for each z and obtain a single-valued function, $\log z = \log r + i\theta$; however, θ cannot be chosen to depend continuously on z for all $z \neq 0$, since θ will increase by 2π each time one encircles the origin in the positive direction.

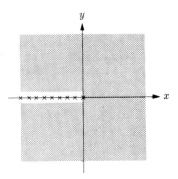

FIG. 3–4.   Domain for log z.

If we concentrate on an appropriate portion of the z-plane, we can choose θ to vary continuously within the domain. For example, the inequalities

$$-\pi < \theta < \pi, \qquad r > 0$$

together describe a domain (Fig. 3–4) and also tell how to assign the values of θ within the domain. With θ so restricted, $\log r + i\theta$ then defines a *branch* of log z in the domain chosen; this particular branch is called the *principal value* of log z and is denoted by Log z. The points on the negative real axis are excluded from the domain, but we usually assign the values Log $z = \log |x| + i\pi$ on this line. Within the domain of Fig. 3–4, Log z is *an analytic function of z* (Problem 4 below). Other branches of log z are obtained by varying the choice of θ or of the domain. For example, in the domain of Fig. 3–4, we might choose θ so that $\pi < \theta < 3\pi$, or so that $-3\pi < \theta < -\pi$. The inequalities $0 < \theta < 2\pi$, $\pi/2 < \theta < 5\pi/2, \dots$ also suggest other domains and choices of θ. We can verify that so long as θ varies continuously in the domain, $\log z = \log r + i\theta$ is analytic there. The most general domain possible here is an arbitrary simply-connected domain not containing the origin.

As a result of this discussion, it appears that log z is formed of many branches, each analytic in some domain not containing the origin. The branches fit together in a simple way; in general, we can get from one branch to another by moving around the origin a sufficient number of times, while varying the choice of log z continuously. We say that the branches form "analytic continuations" of each other.

We can further verify that for each branch of log z, the rule

$$\frac{d}{dz} \log z = \frac{1}{z} \tag{3–51}$$

remains valid. The familiar identities are also satisfied (Problems 4 and 5 below).

The *general exponential function* a^z is defined, for $a \neq 0$, by the equation

$$a^z = e^{z \log a} = \exp(z \log a). \tag{3–52}$$

Thus for $z = 0$, $a^0 = 1$. Otherwise, $\log a = \log |a| + i \arg a$, and we obtain many values: $a^z = \exp[z(\log |a| + i(\alpha + 2n\pi))]$, $(n = 0, \pm 1, \pm 2, \ldots)$, where α denotes one choice of arg a. For example,

$$(1 + i)^i = \exp\left[i\left\{\log \sqrt{2} + i\left(\frac{\pi}{4} + 2n\pi\right)\right\}\right]$$

$$= e^{-(\pi/4) - 2n\pi}(\cos \log \sqrt{2} + i \sin \log \sqrt{2}).$$

If z is a positive integer m, a^z reduces to a^m and has only one value. The same holds for $z = -m$, and we have

$$a^{-m} = \frac{1}{a^m}. \tag{3–53}$$

If z is a fraction p/q (in lowest terms), we find that a^z has q distinct values, which are the qth roots of a^p.

If a fixed choice of $\log a$ is made in (3–52), then a^z is simply e^{cz}, $c = \log a$, and is hence an analytic function of z for all z. Each choice of $\log a$ determines such a function.

If a and z are interchanged in (3–52), we obtain the *general power function*,

$$z^a = e^{a \log z}. \tag{3–54}$$

If an analytic branch of $\log z$ is chosen as above, then this function becomes an analytic function of an analytic function and is hence analytic in the domain chosen. In particular, the *principal value* of z^a is defined as the analytic function $z^a = e^{a \operatorname{Log} z}$, in terms of the principal value of $\log z$.
For example, if $a = \frac{1}{2}$, we have

$$z^{1/2} = e^{(1/2) \log z} = e^{(1/2)(\log r + i\theta)} = e^{(1/2) \log r} e^{(1/2)i\theta}$$

$$= \sqrt{r}\left(\cos \frac{\theta}{2} + i \sin \frac{\theta}{2}\right),$$

as in Section 1–7. If Log z is used, then $\sqrt{z} = f_1(z)$ becomes analytic in the domain of Fig. 3–4. A second analytic branch $f_2(z)$ in the same domain is obtained by requiring that $\pi < \theta < 3\pi$. These are the only two analytic branches which can be obtained in this domain. It should be remarked that these two branches are related by the equation $f_2(z) = -f_1(z)$. For f_2 is obtained from f_1 by increasing θ by 2π, which replaces $e^{(1/2)i\theta}$ by

$$e^{(1/2)i(\theta + 2\pi)} = e^{\pi i} e^{(1/2)i\theta} = -e^{(1/2)i\theta}.$$

The functions $\sin^{-1} z$ and $\cos^{-1} z$ are defined as the inverses of $\sin z$ and $\cos z$. We then find

$$\sin^{-1} z = \frac{1}{i} \log [iz \pm \sqrt{1 - z^2}],$$

(3–55)

$$\cos^{-1} z = \frac{1}{i} \log [z \pm i\sqrt{1 - z^2}].$$

The proofs are left to the exercises (Problem 2). It can be shown that analytic branches of both these functions can be defined in each simply-connected domain not containing the points ± 1. For each z other than ± 1, one has two choices of $\sqrt{1 - z^2}$ and then an infinite sequence of choices of the logarithm, differing by multiples of $2\pi i$.

PROBLEMS

1. Obtain all values of each of the following:

(a) $\log 2$ (b) $\log i$ (c) $\log (1 - i)$ (d) i^i (e) $(1 + i)^{2/3}$ (f) $i^{\sqrt{2}}$ (g) $\sin^{-1} 1$
(h) $\cos^{-1} 2$

2. Prove the formulas (3–55). [*Hint:* If $w = \sin^{-1} z$, then $2iz = e^{iw} - e^{-iw}$; multiply by e^{iw} and solve the resulting equation as a quadratic for e^{iw}.]

3. (a) Evaluate $\sin^{-1} 0$, $\cos^{-1} 0$.

(b) Find all roots of $\sin z$ and $\cos z$ [compare part (a)].

4. Show that each branch of $\log z$ is analytic in each domain in which θ varies continuously and that

$$(d/dz) \log z = 1/z.$$

[*Hint:* Show from the equations $x = r \cos \theta$, $y = r \sin \theta$ that $\partial\theta/\partial x = -y/r^2$, $\partial\theta/\partial y = x/r^2$. Show that the Cauchy-Riemann equations hold for $u = \log r$, $v = \theta$.]

5. Prove the following identities in the sense that, for proper selection of values of the multiple-valued functions concerned, the equation is correct for each allowed choice of the variables:

(a) $\log (z_1 \cdot z_2) = \log z_1 + \log z_2$ $(z_1 \neq 0, z_2 \neq 0)$
(b) $e^{\log z} = z$ $(z \neq 0)$
(c) $\log e^z = z$
(d) $\log z_1^{z_2} = z_2 \log z_1$ $(z_1 \neq 0)$

6. For each of the following determine all analytic branches of the multiple-valued function in the domain given:

(a) $\log z$, $x < 0$ (b) $\sqrt[3]{z}$, $x > 0$

7. Prove that for the analytic function z^a (principal value),

$$(d/dz)z^a = (az^a)/z = az^{a-1}.$$

8. Plot the functions $u = \text{Re}(\sqrt{z})$ and $v = \text{Im}(\sqrt{z})$ as functions of x and y and show the two branches described in the text.

ANSWERS

1. (a) $0.693 + 2n\pi i$ (b) $i(\tfrac{1}{2}\pi + 2n\pi)$ (c) $0.347 + i(\tfrac{7}{4}\pi + 2n\pi)$

(d) $\exp(-\tfrac{1}{2}\pi - 2n\pi)$ (e) $\sqrt[3]{2}\exp\left(\tfrac{1}{6}\pi i + \dfrac{4n\pi}{3}i\right)$

(f) $\exp\left(\dfrac{\sqrt{2}}{2}\pi i + 2\sqrt{2}\,n\pi i\right)$ (g) $\tfrac{1}{2}\pi + 2n\pi$ (h) $2n\pi \pm 1.317i$

The range of n is $0, \pm1, \pm2, \ldots$, except in (e), where it is $0, 1, 2$.

3. (a) and (b) $n\pi$ and $(\pi/2) + n\pi$ $(n = 0, \pm1, \pm2, \ldots)$

6. (a) $\log r + i\theta$, $\tfrac{1}{2}\pi + 2n\pi < \theta < \tfrac{3}{2}\pi + 2n\pi$ $(n = 0, \pm1, \pm2, \ldots)$

   (b) $\sqrt[3]{r}\exp(i\theta/3)$, $-(\pi/2) + 2n\pi < \theta < (\pi/2) + 2n\pi$ $(n = 0, 1, 2)$

3–6 Integrals of analytic functions. Cauchy integral theorem. All paths in the integrals concerned here, as elsewhere in the chapter, are assumed to be piecewise smooth.

The following theorem is fundamental for the theory of analytic functions:

THEOREM 12 (Cauchy integral theorem). *If $f(z)$ is analytic in a simply-connected domain D, then*

$$\oint_C f(z)\,dz = 0$$

on every simple closed path C in D (Fig. 3–5).

Proof. We have, by (3–31) above,

$$\oint_C f(z)\,dz = \oint_C u\,dx - v\,dy + i\oint_C v\,dx + u\,dy.$$

The two real integrals are equal to zero (see Section 5–6 of Reference 5)

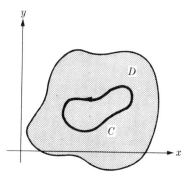

FIG. 3–5. Cauchy integral theorem.

provided u and v have continuous derivatives in D and

$$\frac{\partial u}{\partial y} = -\frac{\partial v}{\partial x}, \qquad \frac{\partial v}{\partial y} = \frac{\partial u}{\partial x}.$$

These are just the Cauchy-Riemann equations. Hence

$$\oint_C f(z)\, dz = 0 + i \cdot 0 = 0.$$

This theorem can be stated in an equivalent form:

THEOREM 12′. *If $f(z)$ is analytic in the simply-connected domain D, then $\int f(z)\, dz$ is independent of the path in D.*

For independence of path and equaling zero on closed paths are equivalent properties of line integrals. If C is a path from z_1 to z_2, we can now write

$$\int_C f(z)\, dz = \int_{z_1}^{z_2} f(z)\, dz,$$

the integral being the same for all paths C from z_1 to z_2.

THEOREM 13. *Let $f(z) = u + iv$ be defined in domain D and let u and v have continuous partial derivatives in D. If*

$$\oint_C f(z)\, dz = 0 \qquad\qquad (3\text{--}60)$$

on every simple closed path C in D, then $f(z)$ is analytic in D.

Proof. The condition (3–60) implies that

$$\oint_C u\, dx - v\, dy = 0, \qquad \oint_C v\, dx + u\, dy = 0$$

on all simple closed paths C; that is, the two real line integrals are independent of path in D. Therefore (p. 248 of Reference 5)

$$\frac{\partial u}{\partial y} = -\frac{\partial v}{\partial x}, \qquad \frac{\partial v}{\partial y} = \frac{\partial u}{\partial x};$$

since the Cauchy-Riemann equations hold, f is analytic.

This theorem can be proved with the assumption that u and v have continuous derivatives in D replaced by the assumption that f is continuous in D; it is then known as *Morera's* theorem. For a proof, see Chapter 5 of Vol. I of Reference 6.

THEOREM 14. *If $f(z)$ is analytic in D, then*

$$\int_{z_1}^{z_2} f'(z)\, dz = f(z)\Big|_{z_1}^{z_2} = f(z_2) - f(z_1) \qquad (3\text{–}61)$$

on every path in D from z_1 to z_2. In particular,

$$\oint f'(z)\, dz = 0$$

on every closed path in D.

Proof. By (3–41) above,

$$\int_{z_1}^{z_2} f'(z)\, dz = \int_{z_1}^{z_2} \left(\frac{\partial u}{\partial x} + i\frac{\partial v}{\partial x} \right)(dx + i\, dy)$$

$$= \int_{z_1}^{z_2} \frac{\partial u}{\partial x}\, dx + \frac{\partial u}{\partial y}\, dy + i\int_{z_1}^{z_2} \frac{\partial v}{\partial x}\, dx + \frac{\partial v}{\partial y}\, dy$$

$$= \int_{z_1}^{z_2} du + i\, dv = (u + iv)\Big|_{z_1}^{z_2} = f(z_2) - f(z_1).$$

This rule is the basis for evaluation of simple integrals, just as in elementary calculus. Thus we have

$$\int_i^{1+i} z^2\, dz = \frac{z^3}{3}\Big|_i^{1+i} = \frac{(1+i)^3 - i^3}{3} = -\tfrac{2}{3} + i,$$

$$\int_i^{-i} \frac{1}{z^2}\, dz = -\frac{1}{z}\Big|_i^{-i} = -i - i = -2i.$$

In the first of these any path can be used; in the second, any path not through the origin.

THEOREM 15. *If $f(z)$ is analytic in D and D is simply-connected, then*

$$F(z) = \int_{z_1}^{z} f(z)\, dz \quad (z_1 \text{ fixed in } D) \qquad (3\text{–}62)$$

is an indefinite integral of $f(z)$; that is, $F'(z) = f(z)$. Thus $F(z)$ is itself analytic.

Proof. Since $f(z)$ is analytic in D and D is simply-connected, $\int_{z_1}^{z} f(z)\, dz$ is independent of path and defines a function F which depends only on the upper limit z. We have, further, $F = U + iV$, where

$$U = \int_{z_1}^{z} u\, dx - v\, dy, \qquad V = \int_{z_1}^{z} v\, dx + u\, dy$$

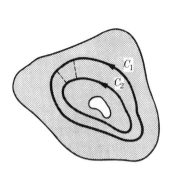
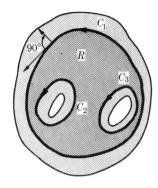

FIG. 3–6. Cauchy theorem for doubly-connected domain.

FIG. 3–7. Cauchy theorem for triply-connected domain.

and both integrals are independent of path. Hence $dU = u\,dx - v\,dy$, $dV = v\,dx + u\,dy$. Thus U and V satisfy the Cauchy-Riemann equations, so that $F = U + iV$ is analytic and

$$F'(z) = \frac{\partial U}{\partial x} + i\frac{\partial V}{\partial x} = u + iv = f(z).$$

Cauchy's theorem for multiply-connected domains. If $f(z)$ is analytic in a multiply-connected domain D, then we cannot conclude that

$$\oint f(z)\,dz = 0$$

on every simple closed path C in D. Thus, if D is the doubly-connected domain of Fig. 3–6 and C is the curve C_1 shown, then the integral around C need not be zero. However, by introducing cuts, we can reason that

$$\oint_{C_1} f(z)\,dz = \oint_{C_2} f(z)\,dz; \tag{3–63}$$

that is, the integral has the same value on all paths which go around the inner "hole" once in the positive direction. For a triply-connected domain, as in Fig. 3–7, we obtain the equation

$$\oint_{C_1} f(z)\,dz = \oint_{C_2} f(z)\,dz + \oint_{C_3} f(z)\,dz. \tag{3–64}$$

This can be written in the form

$$\oint_{C_1} f(z)\,dz + \oint_{C_2} f(z)\,dz + \oint_{C_3} f(z)\,dz = 0; \tag{3–65}$$

Eq. (3–65) states that the integral around the complete boundary of a certain region in D is equal to zero. More generally, we have the following theorem:

THEOREM 16 (Cauchy's theorem for multiply-connected domains). *Let $f(z)$ be analytic in a domain D and let $C_1, \ldots, C_n$ be n simple closed curves in D which together form the boundary B of a region R contained in D. Then*

$$\int_B f(z)\, dz = 0,$$

where the direction of integration on B is such that the outer normal is $90°$ behind the tangent vector in the direction of integration.

3–7 Cauchy's integral formula. Now let D be a simply-connected domain and let z_0 be a fixed point of D. If $f(z)$ is analytic in D, the function $f(z)/(z - z_0)$ will fail to be analytic at z_0. Hence

$$\oint \frac{f(z)}{z - z_0}\, dz$$

will in general not be zero on a path C enclosing z_0. However, as above, this integral will have the same value on all paths C about z_0. To determine this value, we reason that if C is a very small circle of radius R about z_0, then $f(z_0)$ has, by continuity, approximately the constant value $f(z_0)$ on the path. This suggests that

$$\oint_C \frac{f(z)}{z - z_0}\, dz = f(z_0) \cdot \oint_{|z-z_0|=R} \frac{dz}{z - z_0} = f(z_0) \cdot 2\pi i,$$

since we find

$$\oint_{|z-z_0|=R} \frac{dz}{z - z_0} = \int_0^{2\pi} \frac{Rie^{i\theta}}{Re^{i\theta}}\, d\theta = i\int_0^{2\pi} d\theta = 2\pi i,$$

with the aid of the substitution: $z - z_0 = Re^{i\theta}$. The correctness of the conclusion reached is the content of the following fundamental result:

THEOREM 17 (Cauchy integral formula). *Let $f(z)$ be analytic in a domain D. Let C be a simple closed curve in D, within which $f(z)$ is analytic and let z_0 be inside C. Then*

$$f(z_0) = \frac{1}{2\pi i} \oint_C \frac{f(z)}{z - z_0}\, dz. \tag{3–70}$$

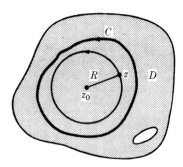

FIG. 3-8.   Cauchy integral formula.

Proof. The domain D is not required to be simply-connected, but since f is analytic within C, the theorem concerns only a simply-connected part of D, as shown in Fig. 3-8. We reason as above to conclude that

$$\oint_C \frac{f(z)}{z - z_0}\, dz = \oint_{|z-z_0|=R} \frac{f(z)}{z - z_0}\, dz.$$

It remains to show that the integral on the right is indeed $f(z_0) \cdot 2\pi i$. Now, since $f(z_0) = $ const,

$$\oint \frac{f(z_0)}{z - z_0}\, dz = f(z_0) \oint \frac{dz}{z - z_0} = f(z_0) \cdot 2\pi i,$$

where we integrate always on the circle $|z - z_0| = R$. Hence, on the same path,

$$\oint \frac{f(z)}{z - z_0}\, dz - f(z_0) \cdot 2\pi i = \oint \frac{f(z) - f(z_0)}{z - z_0}\, dz. \qquad (3\text{-}71)$$

Now $|z - z_0| = R$ on the path, and since $f(z)$ is continuous at z_0, $|f(z) - f(z_0)| < \epsilon$ for $R < \delta$, for each preassigned $\epsilon > 0$. Hence, by Theorem 7,

$$\left| \oint \frac{f(z) - f(z_0)}{z - z_0}\, dz \right| < \frac{\epsilon}{R} \cdot 2\pi R = 2\pi\epsilon.$$

Thus the absolute value of the integral can be made as small as desired by choosing R sufficiently small. But the integral has the same value for all choices of R. This is possible only if the integral is zero for all R. Hence the left side of (3-71) is zero and (3-70) follows.

The integral formula (3-70) is remarkable in that it expresses the values of the function $f(z)$ at points z_0 inside the curve C in terms of the values along C alone. If C is taken as a circle $z = z_0 + Re^{i\theta}$, then (3–70) reduces to the following:

$$f(z_0) = \frac{1}{2\pi} \int_0^{2\pi} f(z_0 + Re^{i\theta})\, d\theta. \tag{3–72}$$

Thus the *value of an analytic function at the center of a circle equals the average (arithmetic mean) of the values on the circumference.*

Just as with the Cauchy integral theorem, the Cauchy integral formula can be extended to multiply-connected domains. Under the hypotheses of Theorem 16,

$$f(z_0) = \frac{1}{2\pi i} \int_B \frac{f(z)}{z - z_0}\, dz = \frac{1}{2\pi i} \left(\oint_{C_1} \frac{f(z)}{z - z_0}\, dz + \oint_{C_2} \frac{f(z)}{z - z_0}\, dz + \cdots \right),$$

$$\tag{3–73}$$

where z_0 is any point inside the region R bounded by C_1 (the outer boundary), $C_2, \ldots, C_n$. The proof is left as an exercise (Problem 6 below).

PROBLEMS

1. Evaluate the following integrals:

(a) $\oint z^2 \sin z\, dz$ on the ellipse $x^2 + 2y^2 = 1$

(b) $\oint \dfrac{z^2}{z+1}\, dz$ on the circle $|z - 2| = 1$

(c) $\displaystyle\int_1^{2i} ze^z\, dz$ on the line segment joining the endpoints

(d) $\displaystyle\int_{1+i}^{1-i} \dfrac{1}{z^2}\, dz$ on the parabola $2y^2 = x + 1$

2. (a) Evaluate $\int_{-i}^{i} (dz/z)$ on the path $z = e^{it}$, $-\pi/2 \le t \le \pi/2$, with the aid of the relation $(\log z)' = 1/z$, for an appropriate branch of $\log z$.

(b) Evaluate $\int_i^{-i} (dz/z)$ on the path $z = e^{it}$, $\pi/2 \le t \le 3\pi/2$, as in part (a).

(c) Why does the relation $(\log z)' = 1/z$ not imply that the sum of the two integrals of parts (a) and (b) is zero?

3. A certain function $f(z)$ is known to be analytic except for $z = 1$, $z = 2$, $z = 3$, and it is known that

$$\oint_{C_k} f(z)\, dz = a_k \quad (k = 1, 2, 3),$$

where C_k is a circle of radius $\frac{1}{2}$ with center at $z = k$. Evaluate

$$\oint f(z)\,dz$$

on each of the following paths:

(a) $|z| = 4$ (b) $|z| = 2.5$ (c) $|z - 2.5| = 1$

4. A certain function $f(z)$ is analytic except for $z = 0$, and it is known that

$$\lim_{z \to \infty} zf(z) = 0.$$

Show that

$$\oint f(z)\,dz = 0$$

on every simple closed path not passing through the origin. [*Hint:* Show that the value of the integral on a path $|z| = R$ can be made as small as desired by making R sufficiently large.]

5. Evaluate each of the following with the aid of the Cauchy integral formula:

(a) $\oint \dfrac{z}{z-3}\,dz$ on $|z| = 5$ (b) $\oint \dfrac{e^z}{z^2 - 3z}\,dz$ on $|z| = 1$

(c) $\oint \dfrac{z+2}{z^2-1}\,dz$ on $|z| = 2$ (d) $\oint \dfrac{\sin z}{z^2+1}\,dz$ on $|z| = 2$

[*Hint for* (c) *and* (d): Expand the rational function in partial fractions.]

6. Prove (3–73) under the hypotheses stated.

7. Prove that if $f(z)$ is analytic in domain D and $f'(z) \equiv 0$, then $f(z) \equiv$ constant. [*Hint:* Apply Theorem 14.]

ANSWERS

1. (a) 0   (b) 0   (c) $(2i - 1)e^{2i}$ (d) $-i$
2. (a) πi (b) πi
3. (a) $a_1 + a_2 + a_3$ (b) $a_1 + a_2$ (c) $a_2 + a_3$
5. (a) $6\pi i$ (b) $-2\pi i/3$ (c) $2\pi i$ (d) $2\pi i \sinh 1$

3–8 Power series as analytic functions. We now proceed to enlarge the class of specific analytic functions still further by showing that every power series

$$\sum_{n=0}^{\infty} c_n(z - z_0)^n = c_0 + c_1(z - z_0) + \cdots + c_n(z - z_0)^n + \cdots$$

converging for some values of z other than $z = z_0$ represents an analytic function.

For the theory of real power series, refer to Chapter 6 of Reference 5. The following fundamental theorem for complex power series is proved just as for real series (p. 350 of Reference 5):

THEOREM 18. *Every power series $\sum_{n=0}^{\infty} c_n(z - z_0)^n$ has a radius of convergence r^* such that the series converges absolutely when $|z - z_0| < r^*$, and diverges when $|z - z_0| > r^*$. The series converges uniformly for $|z - z_0| \leqq r_1$, provided $r_1 < r^*$.*

The number r^* can be zero, in which case the series converges only for $z = z_0$, a positive number, or ∞, in which case the series converges for all z.

The number r^* can be evaluated as follows:

$$r^* = \lim_{n \to \infty} \left| \frac{c_n}{c_{n+1}} \right|, \quad \text{if the limit exists,}$$

$$r^* = \lim_{n \to \infty} \frac{1}{\sqrt[n]{|c_n|}}, \quad \text{if the limit exists,}$$

(3–80)

and in any case by the formula

$$r^* = \frac{1}{\overline{\lim_{n \to \infty}} \sqrt[n]{|c_n|}}.$$

(3–81)

As for real variables, no general statement can be made about convergence on the boundary of the domain of convergence. This boundary (when $r^* \neq 0$, $r^* \neq \infty$) is a circle $|z - z_0| = r^*$, termed the *circle of convergence* (Fig. 3–9). The series may converge at some points, all points, or no points of this circle.

EXAMPLE 1. $\sum_{n=1}^{\infty} (z^n/n^2)$. The first formula (3–80) gives

$$r^* = \lim_{n \to \infty} \frac{(n + 1)^2}{n^2} = 1.$$

The series converges absolutely on the circle of convergence, for when $|z| = 1$, the series of absolute value is the convergent series $\sum (1/n^2)$.

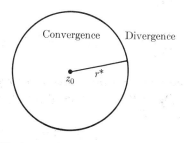

FIG. 3–9.   Circle of convergence of a power series.

EXAMPLE 2. $\sum_{n=0}^{\infty} z^n$. This complex geometric series converges for $|z| < 1$, as (3–80) shows. We have further

$$\sum_{n=0}^{\infty} z^n = \frac{1}{1 - z}, \quad (|z| < 1),$$

as for real variables. On the circle of convergence, the series diverges everywhere, since the nth term fails to converge to zero.

The following theorems are proved as for real variables.

THEOREM 19. *A power series with nonzero convergence radius represents a continuous function within the circle of convergence.*

THEOREM 20. *A power series can be integrated term by term within the circle of convergence; that is, if $r^* \neq 0$ and*

$$f(z) = \sum_{n=0}^{\infty} c_n(z - z_0)^n, \quad (|z - z_0| < r^*),$$

then, for every path C inside the circle of convergence

$$\int_{\substack{z_1 \\ C}}^{z_2} f(z)\, dz = \sum_{n=0}^{\infty} c_n \int_{z_1}^{z_2} (z - z_0)^n\, dz = \sum_{n=0}^{\infty} c_n \frac{(z - z_0)^{n+1}}{n + 1} \bigg|_{z_1}^{z_2},$$

or in terms of indefinite integrals,

$$\int f(z)\, dz = \sum_{n=0}^{\infty} c_n \frac{(z - z_0)^{n+1}}{n + 1} + \text{const}, \quad (|z - z_0| < r^*).$$

THEOREM 21. *A power series can be differentiated term by term; that is, if $r^* \neq 0$ and*

$$f(z) = \sum_{n=0}^{\infty} c_n(z - z_0)^n, \quad (|z - z_0| < r^*),$$

then

$$f'(z) = \sum_{n=1}^{\infty} n c_n(z - z_0)^{n-1}, \quad (|z - z_0| < r^*),$$

$$f''(z) = \sum_{n=2}^{\infty} n(n - 1)c_n(z - z_0)^{n-2}, \quad (|z - z_0| < r^*),$$

$\vdots$

Hence every power series with nonzero convergence radius defines an analytic function $f(z)$ within the circle of convergence, and the power series is the Taylor series of $f(z)$:

$$c_n = \frac{f^{(n)}(z_0)}{n!}.$$

THEOREM 22. *If two power series $\sum_{n=0}^{\infty} c_n(z - z_0)^n$, $\sum_{n=0}^{\infty} C_n(z - z_0)^n$ have nonzero convergence radii and have equal sums wherever both series converge, then the series are identical; that is,*

$$c_n = C_n \quad (n = 0, 1, 2, \ldots).$$

3–9 Power series expansion of general analytic function. In Section 3–8 it was shown that every power series with nonzero convergence radius represents an analytic function. We now proceed to show that all analytic functions are obtainable in this way. If a function $f(z)$ is analytic in a domain D of general shape, we cannot expect to represent $f(z)$ by one power series; for the power series converges only in a circular domain. However, we can show that for each circular domain D_0 in D, there is a power series converging in D_0 whose sum is $f(z)$. Thus several (perhaps infinitely many) power series are needed to represent $f(z)$ throughout all of D.

THEOREM 23. *Let $f(z)$ be analytic in the domain D. Let z_0 be in D and let R be the radius of the largest circle with center at z_0 and having its interior in D. Then there is a power series*

$$\sum_{n=0}^{\infty} c_n(z - z_0)^n$$

which converges to $f(z)$ for $|z - z_0| < R$. Furthermore,

$$c_n = \frac{f^{(n)}(z_0)}{n!} = \frac{1}{2\pi i} \oint_C \frac{f(z)}{(z - z_0)^{n+1}} \, dz, \qquad (3\text{–}90)$$

where C is a simple closed path in D enclosing z_0 and within which $f(z)$ is analytic.

Proof. For simplicity we take $z_0 = 0$. The general case can then be obtained by the substitution $z' = z - z_0$. Let the circle $|z| = R$ be the largest circle with center at z_0 and having its interior within D; the radius R is then positive or $+\infty$ (in which case D is the whole z-plane). Let z_1 be a point within this circle, so that $|z_1| < R$. Choose R_2 so that $|z_1| < R_2 < R$ (see Fig. 3–10). Then $f(z)$ is analytic in a domain including the circle C_2: $|z| = R_2$ plus interior. Hence by the Cauchy integral formula,

$$f(z_1) = \frac{1}{2\pi i} \oint_{C_2} \frac{f(z)}{z - z_1} \, dz.$$

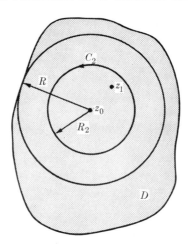

FIG. 3–10.   Taylor series of an analytic function.

Now the factor $1/(z - z_1)$ can be expanded in a geometric series:

$$\frac{1}{z - z_1} = \frac{1}{z\left(1 - \frac{z_1}{z}\right)} = \frac{1}{z}\left(1 + \frac{z_1}{z} + \cdots + \frac{z_1^n}{z^n} + \cdots\right).$$

The series can be considered as a power series in powers of $1/z$, for fixed z_1. It converges for $|z_1/z| < 1$ and converges uniformly for $|z_1/z| \leqq |z_1|/R_2 < 1$.

If we multiply by $f(z)$, we find

$$\frac{f(z)}{z - z_1} = \frac{f(z)}{z} + z_1\frac{f(z)}{z^2} + \cdots + z_1^n\frac{f(z)}{z^{n+1}} + \cdots;$$

since $f(z)$ is continuous for $|z| = R_2$, the series remains uniformly convergent on C_2. Hence we can integrate term by term on C_2:

$$\frac{1}{2\pi i}\oint_{C_2}\frac{f(z)}{z - z_1}\,dz = \frac{1}{2\pi i}\oint_{C_2}\frac{f(z)}{z}\,dz$$

$$+ \frac{z_1}{2\pi i}\oint_{C_2}\frac{f(z)}{z^2}\,dz + \cdots + \frac{z_1^n}{2\pi i}\oint_{C_2}\frac{f(z)}{z^{n+1}}\,dz + \cdots$$

The left-hand side is precisely $f(z_1)$, by the integral formula. Hence

$$f(z_1) = \sum_{n=0}^{\infty} c_n z_1^n, \qquad c_n = \frac{1}{2\pi i}\oint_{C_2}\frac{f(z)}{z^{n+1}}\,dz.$$

The path C_2 can be replaced by any path C as described in the theorem, since $f(z)/z^{n+1}$ is analytic in D except for $z = z_0 = 0$.

By Theorem 21, the series obtained is the Taylor series of f, so that

$$c_n = \frac{f^{(n)}(z_0)}{n!} \quad (z_0 = 0).$$

The theorem is now completely proved.

The consequences of this theorem are far-reaching. First of all, not only does it guarantee that every analytic function is representable by power series, but it ensures that the Taylor series converges to the function within each circular domain within the domain in which the function is given. Thus, *without further analysis*, we at once conclude that

$$e^z = 1 + z + \frac{z^2}{2!} + \cdots + \frac{z^n}{n!} + \cdots,$$

$$\sin z = z - \frac{z^3}{3!} + \frac{z^5}{5!} + \cdots + (-1)^n \frac{z^{2n+1}}{(2n+1)!} + \cdots,$$

$$\cos z = 1 - \frac{z^2}{2!} + \cdots + (-1)^n \frac{z^{2n}}{(2n)!} + \cdots$$

for all z. A variety of other familiar expansions can be obtained in the same way.

It should be recalled that a function $f(z)$ is defined to be analytic in a domain D if $f(z)$ has a continuous derivative $f'(z)$ in D (Section 3–4). By Theorem 23, $f(z)$ must have derivatives of all orders at every point of D. In particular, the derivative of an analytic function is itself analytic:

THEOREM 24. *If $f(z)$ is analytic in domain D, then $f'(z)$, $f''(z)$, ..., $f^{(n)}(z)$, ... exist and are analytic in D. Furthermore, for each n*

$$f^{(n)}(z_0) = \frac{n!}{2\pi i} \oint_C \frac{f(z)}{(z - z_0)^{n+1}} \, dz, \tag{3-91}$$

where C is any simple closed path in D enclosing z_0 and within which $f(z)$ is analytic.

Equation (3–91) is a restatement of (3–90).

Circle of convergence of the Taylor series. Theorem 23 guarantees convergence of the Taylor series of $f(z)$ about each z_0 in D in the largest circular domain $|z - z_0| < R$ in D, as shown in Fig. 3–10. However, this does not mean that R is the radius of convergence r^* of the series, for r^* can be larger than R, as suggested in Fig. 3–11. When this happens, the function $f(z)$ can be prolonged into a larger domain, while retaining

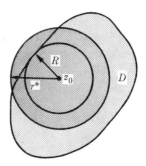

FIG. 3–11. Analytic continuation.

analyticity. For example, if $f(z) = \text{Log } z$ $(0 < \theta < \pi)$ is expanded in a Taylor series about the point $z = -1 + i$, the series has convergence radius $\sqrt{2}$, whereas $R = 1$ [Problem 4(c) below].

The process of prolonging the function suggested here is called *analytic continuation*.

PROBLEMS

1. Determine the radius of convergence of each of the following series:

(a) $\displaystyle\sum_{n=0}^{\infty} \frac{z^n}{n^2 + 1}$

(b) $\displaystyle\sum_{n=0}^{\infty} \frac{(z - 1)^n}{3^n}$

(c) $\displaystyle\sum_{n=0}^{\infty} n! z^n$

(d) $\displaystyle\sum_{n=0}^{\infty} \frac{z^n}{n!}$

2. Given the series $\sum_{n=1}^{\infty}(z^n/n)$, show that

(a) the series has radius of convergence 1;

(b) the series diverges for $z = 1$;

(c) the series converges for $z = i$ and for $z = -1$. It can be shown that the series converges for $|z| = 1$, except for $z = 1$.

3. By means of (3–91), evaluate each of the following:

(a) $\displaystyle\oint \frac{ze^z}{(z - 1)^4}\, dz$ on $|z| = 2$

(b) $\displaystyle\oint \frac{\sin z}{z^4}\, dz$ on $|z| = 1$

(c) $\displaystyle\oint \frac{dz}{z^3(z + 4)}$ on $|z| = 2$

4. Expand in a Taylor series about the point indicated, and determine the radius of convergence r^* and the radius R of the largest circle within which the

series converges to the function:

(a) $\sin z$ about $z = 0$
(b) $1/(z - 1)$ about $z = 2$
(c) $\text{Log } z \ (0 < \theta < \pi)$ about $z = -1 + i$

5. *Cauchy's inequalities.* Let $f(z)$ be analytic in a domain including the circle $C: |z - z_0| = R$ and interior, and let $|f(z)| \leqq M = \text{const on } C$. Prove that

$$|f^{(n)}(z_0)| \leqq \frac{Mn!}{R^n} \quad (n = 0, 1, 2, \ldots).$$

[*Hint:* Apply (3–91).]

6. A function $f(z)$ which is analytic in the whole z-plane is termed an *entire* function or an *integral* function. Examples are polynomials, e^z, $\sin z$, $\cos z$. Prove *Liouville's theorem:* If $f(z)$ is an entire function and $|f(z)| \leqq M$ for all z, where M is constant, then $f(z)$ reduces to a constant. [*Hint:* Take $n = 1$ in the Cauchy inequalities of Problem 5 to show that $f'(z_0) = 0$ for every z_0.]

ANSWERS

1. (a) 1　　(b) 3　　(c) 0　　(d) ∞

3. (a) $4\pi ei/3$　　(b) $-\pi i/3$　　(c) $\pi i/32$

4. (a) $\displaystyle\sum_{n=0}^{\infty} \frac{(-1)^n z^{2n+1}}{(2n + 1)!}$　$(r^* = R = \infty)$

(b) $\displaystyle\sum_{n=0}^{\infty} (-1)^n (z - 2)^n$　$(r^* = R = 1)$

(c) $\frac{1}{2} \log 2 + \frac{3}{4}\pi i - \displaystyle\sum_{n=1}^{\infty} \left(\frac{1+i}{2}\right)^n \frac{(z+1-i)^n}{n}$　$(r^* = \sqrt{2}, R = 1)$

3–10 Power series in positive and negative powers, Laurent expansion.

We have shown that every power series $\sum a_n(z - z_0)^n$ with nonzero convergence radius represents an analytic function and that every analytic function can be built up out of such series. It thus appears unnecessary to seek other explicit expressions for analytic functions. However, the power series represent functions only in circular domains and are hence awkward for representing a function in a more complicated type of domain. It is therefore worthwhile to consider other types of representations. A series of form

$$\sum_{n=1}^{\infty} \frac{b_n}{(z - z_0)^n} = \frac{b_1}{z - z_0} + \cdots + \frac{b_n}{(z - z_0)^n} + \cdots \quad (3\text{--}100)$$

will also represent an analytic function in a domain in which the series converges. For the substitution $z_1 = 1/(z - z_0)$ reduces the series to an ordinary power series,

$$\sum_{n=1}^{\infty} b_n z_1^n.$$

If this series converges for $|z_1| < r_1^*$, then its sum is an analytic function $F(z_1)$; hence the series (3–100) converges for

$$|z - z_0| > \frac{1}{r_1^*} = r_0^* \tag{3–101}$$

to the analytic function $g(z) = F\big(1/(z - z_0)\big)$. The value $z_1 = 0$ corresponds to $z = \infty$, in a limiting sense; accordingly we can also say that $g(z)$ is analytic at ∞ and $g(\infty) = 0$. This will be justified more fully in Section 3–12.

The domain of convergence of the series (3–100) is the region (3–101), which is the *exterior* of a circle. It can happen that $r_1^* = \infty$, in which case the series converges for all z except z_0; if $r_1^* = 0$, the series diverges for all z (except $z = \infty$, as above).

If we add to a series (3–100) a usual power series

$$\sum_{n=0}^{\infty} a_n(z - z_0)^n = a_0 + a_1(z - z_0) + \cdots,$$

converging for $|z - z_0| < r_2^*$, we obtain a sum

$$\sum_{n=1}^{\infty} \frac{b_n}{(z - z_0)^n} + \sum_{n=0}^{\infty} a_n(z - z_0)^n. \tag{3–102}$$

If $r_0^* < r_2^*$, the sum converges and represents an analytic function $f(z)$ in the *annular domain*: $r_0^* < |z - z_0| < r_2^*$; for each series has an analytic sum in this domain, so that the sum of the two series is analytic there. We can write this sum in the more compact form (after some relabeling)

$$f(z) = \sum_{n=-\infty}^{\infty} a_n(z - z_0)^n, \tag{3–103}$$

though this should be interpreted as the sum of two series, as in (3–102).

In this way we build up a new class of analytic functions, each defined in a ring-shaped domain. Every function analytic in such a domain can be obtained in this way:

THEOREM 25 (Laurent's theorem). *Let $f(z)$ be analytic in the ring:*

$$R_1 < |z - z_0| < R_2.$$

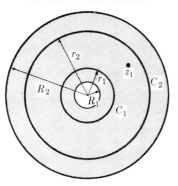

Fig. 3–12.   Laurent's theorem.

Then

$$f(z) = \sum_{n=-\infty}^{\infty} a_n(z - z_0)^n = [a_0 + a_1(z - z_0) + \cdots]$$

$$+ \left[\frac{a_{-1}}{z - z_0} + \frac{a_{-2}}{(z - z_0)^2} + \cdots\right],$$

where

$$a_n = \frac{1}{2\pi i} \oint_C \frac{f(z)}{(z - z_0)^{n+1}}\, dz \tag{3–104}$$

and C is any simple closed curve separating $|z| = R_1$ from $|z| = R_2$. The series converges uniformly for $R_1 < k_1 \leq |z - z_0| \leq k_2 < R_2$.

Proof. For simplicity we take $z_0 = 0$. Let z_1 be any point of the ring and choose r_1, r_2 so that $R_1 < r_1 < |z_1| < r_2 < R_2$, as in Fig. 3–12. We then apply the Cauchy integral formula in general form [Eq. (3–73)] to the region bounded by $C_1\colon |z| = r_1$ and $C_2\colon |z| = r_2$. Hence

$$f(z_1) = \frac{1}{2\pi i} \oint_{C_2} \frac{f(z)}{z - z_1}\, dz - \frac{1}{2\pi i} \oint_{C_1} \frac{f(z)}{z - z_1}\, dz.$$

The first term can be replaced by a power series

$$\sum_{n=0}^{\infty} a_n z_1^n, \qquad a_n = \frac{1}{2\pi i} \oint_{C_2} \frac{f(z)}{z^{n+1}}\, dz$$

as in the proof of Theorem 23 (Section 3–9). For the second term, the series expansion

$$\frac{1}{z - z_1} = -\frac{1}{z_1}\left(\frac{1}{1 - z/z_1}\right) = -\frac{1}{z_1} - \frac{z}{z_1^2} - \frac{z^2}{z_1^3} - \cdots,$$

valid for $|z_1| > |z| = r_1$, leads similarly to the series

$$\sum_{n=1}^{\infty} \frac{b_n}{z_1^n} = \sum_{n=-\infty}^{-1} a_n z_1^n, \qquad a_n = \frac{1}{2\pi i} \oint_{C_1} \frac{f(z)}{z^{n+1}} \, dz.$$

Hence

$$f(z_1) = \sum_{n=-\infty}^{\infty} a_n z_1^n, \qquad a_n = \frac{1}{2\pi i} \oint_{C_1} \frac{f(z)}{z^{n+1}} \, dz;$$

the path C_2 or C_1 can be replaced by any path C separating $|z| = R_1$ from $|z| = R_2$, since the function integrated is analytic throughout the annulus. The uniform convergence follows as for ordinary power series (Theorem 18). The theorem is now established.

Laurent's theorem continues to hold when $R_1 = 0$ or $R_2 = \infty$ or both. In the case $R_1 = 0$, the Laurent expansion represents a function $f(z)$ analytic in a *deleted neighborhood* of z_0, that is, in the circular domain $|z - z_0| < R_2$ minus its center z_0. If $R_2 = \infty$, we can say similarly that the series represents $f(z)$ in a *deleted neighborhood* of $z = \infty$.

3–11 Isolated singularities of an analytic function. Zeros and poles. Let $f(z)$ be defined and analytic in domain D. We say that $f(z)$ has an *isolated singularity* at the point z_0 if $f(z)$ is analytic throughout a neighborhood of z_0 except at z_0 itself; that is, to use the term mentioned at the end of the preceding section, $f(z)$ is analytic in a deleted neighborhood of z_0, but not at z_0. The point z_0 is then a boundary point of D and would be called an *isolated boundary point* (see Fig. 3–13).

A deleted neighborhood $0 < |z - z_0| < R_2$ forms a special case of the annular domain for which Laurent's theorem is applicable. Hence in this

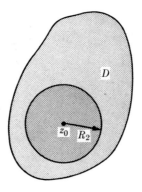

FIG. 3–13.   Isolated singularity.

deleted neighborhood $f(z)$ has a representation as a Laurent series:

$$f(z) = \sum_{n=-\infty}^{\infty} a_n(z - z_0)^n.$$

The form of this series leads to a classification of isolated singularities into three fundamental types:

Case I. *No terms in negative powers of $z - z_0$ appear.* In this case the series is a Taylor series and represents a function analytic in a neighborhood of z_0. Thus the singularity can be removed by setting $f(z_0) = a_0$. We call this a *removable singularity* of $f(z)$. It is illustrated by

$$\frac{\sin z}{z} = 1 - \frac{z^2}{3!} + \frac{z^4}{5!} - \cdots$$

at $z = 0$. In practice, we automatically remove the singularity by defining the function properly.

Case II. *A finite number of negative powers of $z - z_0$ appear.* Thus we have

$$f(z) = \frac{a_{-N}}{(z - z_0)^N} + \cdots + \frac{a_{-1}}{z - z_0} + a_0 + \cdots + a_n(z - z_0)^n + \cdots,$$

$$(N \geq 1, a_{-N} \neq 0). \quad (3\text{-}110)$$

Here $f(z)$ is said to have a *pole of order N* at z_0. We can write

$$f(z) = \frac{1}{(z - z_0)^N} g(z), \qquad g(z) = a_{-N} + a_{-N+1}(z - z_0) + \cdots,$$

$$(3\text{-}111)$$

so that $g(z)$ is analytic for $|z - z_0| < R_2$ and $g(z_0) \neq 0$. Conversely, every function $f(z)$ representable in the form (3-111) has a pole of order N at z_0. Poles are illustrated by rational functions of z, such as

$$f(z) = \frac{z - 2}{(z^2 + 1)(z - 1)^3}, \quad (3\text{-}112)$$

which has poles of order 1 at $\pm i$ and of order 3 at $z = 1$.

The rational function

$$\frac{a_{-N}}{(z - z_0)^N} + \cdots + \frac{a_{-1}}{z - z_0} = p(z) \quad (3\text{-}113)$$

is called the *principal part* of $f(z)$ at the pole z_0. Thus $f(z) - p(z)$ is analytic at z_0.

EXAMPLE 1.

$$f(z) = \frac{e^z \cos z}{z^3} \quad \text{at } z = 0.$$

To obtain the Laurent series, we expand the numerator in a Taylor series:

$$e^z \cos z = \left(1 + z + \frac{z^2}{2!} + \cdots\right)\left(1 - \frac{z^2}{2!} + \cdots\right) = 1 + z - \frac{z^3}{3} + \cdots$$

Hence

$$\frac{e^z \cos z}{z^3} = \frac{1}{z^3} + \frac{1}{z^2} - \frac{1}{3} + \cdots$$

Here the first two terms form the principal part; the pole is of order 3.

EXAMPLE 2.

$$f(z) = \frac{z}{(z + 1)^2(z^3 + 2)} \quad \text{at } z = -1.$$

We expand $z/(z^3 + 2)$ in a Taylor series about $z = -1$.

$$w = \frac{z}{z^3 + 2}, \qquad w' = \frac{2 - 2z^3}{(z^3 + 2)^2}, \qquad w'' = \frac{6(z^5 - 4z^2)}{(z^3 + 2)^3}, \cdots,$$

$$\frac{z}{z^3 + 2} = -1 + 4(z + 1) - 15(z + 1)^2 + \cdots,$$

$$\frac{z}{(z + 1)^2(z^3 + 2)} = \frac{-1}{(z + 1)^2} + \frac{4}{z + 1} - 15 + \cdots$$

The first two terms form the principal part; the pole is of order 2.

Case III. Infinitely many negative powers of $z - z_0$ appear. In this case $f(z)$ is said to have an *essential singularity* at z_0. This is illustrated by the function

$$f(z) = e^{1/z} = 1 + \frac{1}{z} + \frac{1}{2!}\frac{1}{z^2} + \frac{1}{3!}\frac{1}{z^3} + \cdots,$$

which has an essential singularity at $z = 0$.

In Case I, $f(z)$ has a finite limit at z_0 and accordingly $|f(z)|$ is bounded near z_0; that is, there is a real constant M such that $|f(z)| < M$ for z sufficiently close to z_0.

In Case II, $\lim_{z \to z_0} f(z) = \infty$, and it is customary to assign the value ∞ (complex) to $f(z)$ at a pole. At an essential singularity, $f(z)$ has a very complicated discontinuity. In fact, for every complex number c, we can find a sequence z_n converging to z_0 such that $\lim_{n \to \infty} f(z_n) = c$ (see Problem 8 below). It follows from this that if $|f(z)|$ is bounded near z_0, then z_0 must be a removable singularity, and if $\lim f(z) = \infty$ at z_0, then z_0 must be a pole.

If $f(z)$ is analytic at a point z_0, and $f(z_0) = 0$, then z_0 is termed a *root* or *zero* of $f(z)$. Thus the zeros of $\sin z$ are the numbers $n\pi$ ($n = 0, \pm 1, \pm 2, \ldots$). The Taylor series about z_0 has the form

$$f(z) = a_N(z - z_0)^N + a_{N+1}(z - z_0)^{N+1} + \cdots,$$

where $N \geq 1$ and $a_N \neq 0$, or else $f(z) \equiv 0$ in a neighborhood of z_0. It will be seen that the latter case can occur only if $f(z) \equiv 0$ throughout the domain in which it is given. If now $f(z)$ is not identically zero, then

$$f(z) = (z - z_0)^N \phi(z),$$

$$\phi(z) = a_N + a_{N+1}(z - z_0) + \cdots,$$

$$\phi(z_0) = \frac{f^{(N)}(z_0)}{N!} = a_N \neq 0.$$

We say that $f(z)$ has a zero of *order* N or *multiplicity* N at z_0. For example, $1 - \cos z$ has a zero of order 2 at $z = 0$, since

$$1 - \cos z = \frac{z^2}{2} - \frac{z^4}{24} + \cdots$$

If $f(z)$ has a zero of order N at z_0, then $F(z) = 1/f(z)$ has a pole of order N at z_0, and conversely. For if f has a zero of order N, then

$$f(z) = (z - z_0)^N \phi(z)$$

as above, with $\phi(z_0) \neq 0$. It follows from continuity that $\phi(z) \neq 0$ in a sufficiently small neighborhood of z_0. Hence $g(z) = 1/\phi(z)$ is analytic in the neighborhood and $g(z_0) \neq 0$. Now in this neighborhood, except for z_0,

$$F(z) = \frac{1}{f(z)} = \frac{1}{(z - z_0)^N \phi(z)} = \frac{g(z)}{(z - z_0)^N},$$

so that F has a pole at z_0. The converse is proved in the same way.

It remains to consider the case when $f \equiv 0$ in a neighborhood of z_0. This is covered by the following theorem.

THEOREM 26. *The zeros of an analytic function are isolated, unless the function is identically zero; that is, if $f(z)$ is analytic in domain D and $f(z)$ is not identically zero, then for each zero z_0 of $f(z)$ there is a deleted neighborhood of z_0 in which $f(z) \neq 0$.*

The proof is given on p. 557 of Reference 5.

3–12 The complex number ∞. The complex number ∞ has been introduced several times in connection with limiting processes; for example, in the discussion of poles in the preceding section. In each case ∞ has appeared in a natural way as the limiting position of a point receding indefinitely from the origin. We can incorporate this number into the complex number system with special algebraic rules:

$$\frac{z}{\infty} = 0 \quad (z \neq \infty), \qquad z \pm \infty = \infty \quad (z \neq \infty), \qquad \frac{z}{0} = \infty \quad (z \neq 0),$$

$$(3\text{–}120)$$

$$z \cdot \infty = \infty \quad (z \neq 0), \qquad \frac{\infty}{z} = \infty \quad (z \neq \infty).$$

Expressions such as $\infty + \infty$, $\infty - \infty$, and ∞/∞ are not defined.

A function $f(z)$ is said to be analytic in a deleted neighborhood of ∞ if $f(z)$ is analytic for $|z| > R_1$ for some R_1. In this case the Laurent expansion with $R_2 = \infty$ and $z_0 = 0$ is available, and we have

$$f(z) = \sum_{n=-\infty}^{\infty} a_n z^n \quad (|z| > R_1).$$

If there are no *positive* powers of z here, $f(z)$ is said to have a *removable singularity* at ∞ and we make f *analytic at* ∞ by defining $f(\infty) = a_0$:

$$f(z) = a_0 + \frac{a_{-1}}{z} + \cdots + \frac{a_{-n}}{z^n} + \cdots \quad (|z| > R_1). \quad (3\text{–}121)$$

This is clearly equivalent to the statement that if we set $z_1 = 1/z$, then $f(z)$ becomes a function of z_1 with removable singularity at $z_1 = 0$.

If a finite number of positive powers occurs, we have, with $N \geq 1$,

$$f(z) = a_N z^N + \cdots + a_1 z + a_0 + \frac{a_{-1}}{z} + \cdots$$

$$= z^N \phi(z), \quad (3\text{–}122)$$

$$\phi(z) = a_N + \frac{a_{N-1}}{z} + \cdots,$$

where $\phi(z)$ is analytic at ∞ and $\phi(\infty) = a_N \neq 0$. In this case $f(z)$ is said to have a *pole of order* N *at* ∞. The same holds for $f(1/z_1)$ at $z_1 = 0$. Furthermore,

$$\lim_{z \to \infty} f(z) = \infty. \quad (3\text{–}123)$$

If infinitely many positive powers appear, $f(z)$ is said to have an *essential singularity* at $z = \infty$.

If $f(z)$ is analytic at ∞ as in (3–121) and $f(\infty) = a_0 = 0$, then $f(z)$ is said to have a *zero* at $z = \infty$. If f is not identically zero, then necessarily

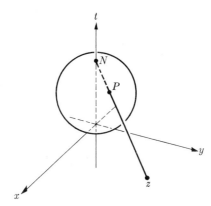

FIG. 3–14.   Stereographic projection.

some $a_{-N} \neq 0$ and

$$f(z) = \frac{a_{-N}}{z^N} + \frac{a_{-N-1}}{z^{N+1}} + \cdots \quad (|z| > R_1)$$

$$= \frac{1}{z^N} g(z), \tag{3–124}$$

$$g(z) = a_{-N} + \frac{a_{-N-1}}{z} + \cdots$$

Thus $g(z)$ is analytic at ∞ and $g(\infty) = a_{-N} \neq 0$. We say that $f(z)$ has a zero of order (or multiplicity) N at ∞. We can show that if $f(z)$ has a zero of order N at ∞, then $1/f(z)$ has a pole of order N at ∞, and conversely.

The significance of the complex number ∞ can be shown geometrically by the device of *stereographic projection*, i.e., a projection of the plane onto a sphere tangent to the plane at $z = 0$, as shown in Fig. 3–14. The sphere is given in xyt-space by the equation

$$x^2 + y^2 + (t - \tfrac{1}{2})^2 = \tfrac{1}{4}, \tag{3–125}$$

so that the radius is $\frac{1}{2}$. The letter N denotes the "north pole" of the sphere, the point $(0, 0, 1)$. If N is joined to an arbitrary point z in the xy-plane, the line segment Nz will meet the sphere at one other point P, which is the projection of z on the sphere. For example, the points of the circle $|z| = 1$ project on the "equator" of the sphere, i.e., the great circle $t = \frac{1}{2}$. As z recedes to infinite distance from the origin, P approaches N as limiting position. Thus N *corresponds to the complex number* ∞.

We refer to the z-plane plus the number ∞ as the *extended z-plane*. To emphasize that ∞ is *not* included, we refer to the *finite z-plane*.

PROBLEMS

1. For each of the following expand in a Laurent series at the isolated singularity given and state the type of singularity:

(a) $\dfrac{e^z - 1}{z}$ at $z = 0$

(b) $\dfrac{1}{z^2(z - 3)}$ at $z = 0$

(c) $\dfrac{z - \cos z}{z}$ at $z = 0$

(d) $\csc z$ at $z = 0$

[*Hint for* (d): Write

$$\csc z = \frac{1}{\sin z} = \frac{1}{z - (z^3/3!) + \cdots} = \frac{a_{-1}}{z} + a_0 + \cdots$$

and determine the coefficients $a_{-1}, a_0, a_1, \ldots$ so that

$$1 = (z - z^3/3! + \cdots) \cdot (a_{-1}z^{-1} + a_0 + a_1z + \cdots).]$$

2. For each of the following find the principal part at the pole given:

(a) $\dfrac{z^2 + 3z + 1}{z^4}$ $(z = 0)$

(b) $\dfrac{z^2 - 2}{z(z + 1)}$ $(z = 0)$

(c) $\dfrac{e^z \sin z}{(z - 1)^2}$ $(z = 1)$

(d) $\dfrac{1}{z^2(z^3 + z + 1)}$ $(z = 0)$

3. For each of the following expand in a Laurent series at $z = \infty$ and state the type of singularity:

(a) $\dfrac{1}{1 - z} = -\dfrac{1}{z} \dfrac{1}{1 - (1/z)}$

(b) $\dfrac{z^2}{z + 2}$

(c) $e^z + e^{1/z}$

4. Let $f(z)$ be a rational function in lowest terms:

$$f(z) = \frac{a_0z^n + a_1z^{n-1} + \cdots + a_n}{b_0z^m + b_1z^{m-1} + \cdots + b_m}.$$

The *degree* d of $f(z)$ is defined to be the larger of m and n. Assuming the fundamental theorem of algebra, show that $f(z)$ has precisely d zeros and d poles in the extended z-plane, a pole or zero of order N being counted as N poles or zeros.

5. For each of the following locate all zeros and poles in the extended plane (compare Problem 4):

(a) $\dfrac{z}{z - 1}$

(b) $\dfrac{z - 1}{z^2 + 3z + 2}$

(c) $\dfrac{z^3 + 3z^2 + 3z + 1}{z}$

6. Let $A(z)$ and $B(z)$ be analytic at $z = z_0$; let $A(z_0) \neq 0$ and let $B(z)$ have a zero of order N at z_0, so that

$$f(z) = \frac{A(z)}{B(z)} = \frac{a_0 + a_1(z - z_0) + \cdots}{b_N(z - z_0)^N + b_{N+1}(z - z_0)^{N+1} + \cdots}$$

has a pole of order N at z_0. Show that the principal part of $f(z)$ at z_0 is

$$\frac{a_0}{b_N}\frac{1}{(z-z_0)^N} + \frac{a_1 b_N - a_0 b_{N+1}}{b_N^2}\frac{1}{(z-z_0)^{N-1}} + \cdots$$

and obtain the next term explicitly. [*Hint:* Set

$$\frac{a_0 + a_1(z-z_0) + \cdots}{b_N(z-z_0)^N + b_{N+1}(z-z_0)^{N+1} + \cdots} = \frac{C_{-N}}{(z-z_0)^N} + \frac{C_{-N+1}}{(z-z_0)^{N-1}} + \cdots$$

Multiply across and solve for C_{-N}, C_{-N+1}, . . .]

7. Prove *Riemann's theorem: If $|f(z)|$ is bounded in a deleted neighborhood of an isolated singularity z_0, then z_0 is a removable singularity of $f(z)$.* [*Hint:* Proceed as in Problem 6 following Section 3–9 with the aid of (3–104).]

8. Prove the *Theorem of Weierstrass and Casorati: If z_0 is an essential singularity of $f(z)$, c is an arbitrary complex number, and $\epsilon > 0$, then $|f(z) - c| < \epsilon$ for some z in every neighborhood of z_0.* [*Hint:* If the property fails, then $1/[f(z) - c]$ is analytic and bounded in absolute value in a deleted neighborhood of z_0. Now apply Problem 7 and conclude that $f(z)$ has a pole or removable singularity at z_0.]

Answers

1. (a) $\displaystyle\sum_{n=1}^{\infty} \frac{z^{n-1}}{n!}$,   removable

   (b) $-\dfrac{1}{3z^2} - \dfrac{1}{9z} - \displaystyle\sum_{n=0}^{\infty} \frac{z^n}{3^{n+3}}$,   pole of order 2

   (c) $-\dfrac{1}{z} + 1 + \displaystyle\sum_{n=0}^{\infty} \frac{(-1)^n z^{2n+1}}{(2n+2)!}$,   pole of order 1

   (d) $\dfrac{1}{z} + \dfrac{z}{6} + \dfrac{7z^3}{360} + \cdots$,   pole of order 1

2. (a) $\dfrac{1}{z^4} + \dfrac{3}{z^3} + \dfrac{1}{z^2}$ (b) $-\dfrac{2}{z}$

   (c) $\dfrac{e \sin 1}{(z-1)^2} + \dfrac{e(\cos 1 + \sin 1)}{z-1}$ (d) $\dfrac{1}{z^2} - \dfrac{1}{z}$

3. (a) $\displaystyle\sum_{n=1}^{\infty} \frac{-1}{z^n}$,   removable,   zero of first order

   (b) $z - 2 + \displaystyle\sum_{n=1}^{\infty} \frac{(-2)^{n+1}}{z^n}$,   pole of order 1

   (c) $2 + \displaystyle\sum_{n=1}^{\infty} \frac{z^n}{n!} + \sum_{n=1}^{\infty} \frac{z^{-n}}{n!}$,   essential

5. (a) zero: 0,   pole: 1   (b) zeros: 1, ∞,   poles: -1, -2

   (c) zeros: -1, -1, -1,   poles: 0, ∞, ∞

6. $\dfrac{a_2 b_N^2 - a_1 b_N b_{N+1} - a_0 b_{N+2} b_N + a_0 b_{N+1}^2}{b_N^3 (z - z_0)^{N-2}}$.

3–13 Residues. Let $f(z)$ be analytic throughout a domain D except for an isolated singularity at a certain point z_0 of D. The integral

$$\oint f(z)\, dz$$

will not in general be zero on a simple closed path in D. However, the integral will have the same value on all curves C which enclose z_0 and no other singularity of f. This value, divided by $2\pi i$, is known as the *residue* of $f(z)$ at z_0 and is denoted by Res $[f(z), z_0]$. Thus

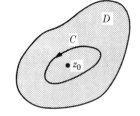

$$\text{Res}\,[f(z), z_0] = \frac{1}{2\pi i} \oint_C f(z)\, dz, \quad (3\text{–}130)$$

Fig. 3–15.   Residue.

where the integral is taken over any path C within which $f(z)$ is analytic except at z_0 (Fig. 3–15).

THEOREM 27.   *The residue of $f(z)$ at z_0 is given by the equation*

$$\text{Res}\,[f(z), z_0] = a_{-1}, \qquad (3\text{–}131)$$

where

$$f(z) = \cdots + \frac{a_{-N}}{(z - z_0)^N} + \cdots + \frac{a_{-1}}{z - z_0} + a_0 + a_1(z - z_0) + \cdots$$

$$(3\text{–}132)$$

is the Laurent expansion of $f(z)$ at z_0.

Proof. By (3–104)

$$a_{-1} = \frac{1}{2\pi i} \oint_C f(z)\, dz,$$

where C is chosen as in the definition of residue. Hence (3–131) follows at once.

If C is a simple closed path in D, within which $f(z)$ is analytic except for isolated singularities at $z_1, \ldots, z_k$, then by Theorem 16

$$\oint_C f(z)\, dz = \oint_{C_1} f(z)\, dz + \cdots + \oint_{C_k} f(z)\, dz,$$

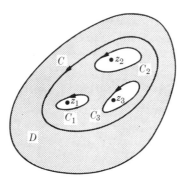

FIG. 3–16.   Cauchy residue theorem.

where C_1 encloses only the singularity at z_1, C_2 encloses only z_2, . . . , as in Fig. 3–16. We thus obtain the following basic theorem:

THEOREM 28 (Cauchy's residue theorem).   *If $f(z)$ is analytic in a domain D and C is a simple closed curve in D within which $f(z)$ is analytic except for isolated singularities at z_1, . . . , z_k, then*

$$\oint_C f(z)\,dz = 2\pi i\,\{\text{Res}\,[f(z),\,z_1] + \cdots + \text{Res}\,[f(z),\,z_k]\}. \quad (3\text{–}133)$$

This theorem permits rapid evaluation of integrals on closed paths, whenever it is possible to compute the coefficient a_{-1} of the Laurent expansion at each singularity inside the path. Various techniques for obtaining the Laurent expansion are illustrated in the problems preceding this section. However, if we wish only the term in $(z - z_0)^{-1}$ of the expansion, various simplifications are possible. We give several rules here:

Rule I.   At a simple pole z_0 (i.e., a pole of first order),

$$\text{Res}\,[f(z),\,z_0] = \lim_{z \to z_0}\,(z - z_0)f(z).$$

Rule II.   At a pole z_0 of order N $(N = 2, 3, \ldots)$,

$$\text{Res}\,[f(z),\,z_0] = \lim_{z \to z_0}\,\frac{g^{(N-1)}(z)}{(N - 1)!}\,,$$

where $g(z) = (z - z_0)^N f(z)$.

Rule III.   If $A(z)$ and $B(z)$ are analytic in a neighborhood of z_0, $A(z_0) \neq 0$, and $B(z)$ has a zero at z_0 of order 1, then

$$f(z) = \frac{A(z)}{B(z)}$$

has a pole of first order at z_0 and

$$\text{Res}\,[f(z),\,z_0] = \frac{A(z_0)}{B'(z_0)}.$$

Rule IV. *If $A(z)$ and $B(z)$ are as in Rule III, but $B(z)$ has a zero of second order at z_0, so that $f(z)$ has a pole of second order at z_0, then*

$$\text{Res}\,[f(z),\,z_0] = \frac{6A'B'' - 2AB'''}{3B''^2}, \qquad (3\text{--}134)$$

where A and the derivatives A', B'', B''' are evaluated at z_0.

Proofs of rules. Let $f(z)$ have a pole of order N:

$$f(z) = \frac{1}{(z - z_0)^N}[a_{-N} + a_{-N+1}(z - z_0) + \cdots] = \frac{1}{(z - z_0)^N}\,g(z),$$

where

$$g(z) = (z - z_0)^N f(z), \qquad g(z_0) = a_{-N}$$

and g is analytic at z_0. The coefficient of $(z - z_0)^{-1}$ in the Laurent series for $f(z)$ is the coefficient of $(z - z_0)^{N-1}$ in the Taylor series for $g(z)$. This coefficient, which is the residue sought, is

$$\frac{g^{(N-1)}(z_0)}{(N - 1)!} = \lim_{z \to z_0} \frac{g^{(N-1)}(z)}{(N - 1)!}.$$

For $N = 1$ this gives Rule I; for $N = 2$ or higher, we obtain Rule II. Rules III and IV follow from the identity of Problem 6 following Section 3–12:

$$\frac{A(z)}{B(z)} = \frac{a_0 + a_1(z - z_0) + \cdots}{b_N(z - z_0)^N + b_{N+1}(z - z_0)^{N+1} + \cdots}$$

$$= \frac{a_0}{b_N}\frac{1}{(z - z_0)^N} + \frac{a_1 b_N - a_0 b_{N+1}}{b_N^2}\frac{1}{(z - z_0)^{N-1}} + \cdots$$

For a first order pole, $N = 1$ and the residue is

$$\frac{a_0}{b_1} = \frac{A(z_0)}{B'(z_0)}.$$

For a second order pole, $N = 2$ and the residue is $[(a_1 b_2 - a_0 b_3)/b_2^2]$. Since

$$a_0 = A(z_0), \qquad a_1 = A'(z_0),$$

$$b_2 = \frac{B''(z_0)}{2!}, \qquad b_3 = \frac{B'''(z_0)}{3!},$$

this reduces to the expression (3–134).

EXAMPLE 1.

$$\oint_{|z|=2} \frac{ze^z}{z^2 - 1}\,dz = 2\pi i \,\{\text{Res}\,[f(z), 1] + \text{Res}\,[f(z), -1]\}.$$

Since $f(z)$ has first order poles at ± 1, we find by Rule I

$$\text{Res}\,[f(z), 1] = \lim_{z \to 1} (z - 1) \cdot \frac{ze^z}{z^2 - 1} = \lim_{z \to 1} \frac{ze^z}{z + 1} = \frac{e}{2},$$

$$\text{Res}\,[f(z), -1] = \lim_{z \to -1} (z + 1) \cdot \frac{ze^z}{z^2 - 1} = \lim_{z \to -1} \frac{ze^z}{z - 1} = \frac{-e^{-1}}{-2}.$$

Accordingly,

$$\oint_{|z|=2} \frac{ze^z}{z^2 - 1}\,dz = 2\pi i \left(\frac{e}{2} + \frac{e^{-1}}{2}\right) = 2\pi i \cosh 1.$$

Rule III could also have been used:

$$\text{Res}\,[f(z), 1] = \frac{ze^z}{2z}\bigg|_{z=1} = \frac{e}{2}, \qquad \text{Res}\,[f(z), -1] = \frac{ze^z}{2z}\bigg|_{z=-1} = \frac{-e^{-1}}{-2}.$$

This is simpler than Rule I, since the expression $A(z)/B'(z)$, once computed, serves for all poles of the prescribed type.

EXAMPLE 2.

$$\oint_{|z|=2} \frac{z}{z^4 - 1}\,dz = 2\pi i \,\{\text{Res}\,[f(z), 1] + \text{Res}\,[f(z), -1]$$
$$+ \text{Res}\,[f(z), i] + \text{Res}\,[f(z), -i]\}.$$

All poles are of first order. Rule III gives $A(z)/B'(z) = z/(4z^3) = 1/(4z^2)$ as the expression for the residue at any one of the four points. Moreover, $z^4 = 1$ at each pole, so that

$$\frac{1}{4z^2} = \frac{z^2}{4z^4} = \frac{z^2}{4}.$$

Hence

$$\oint_{|z|=2} \frac{z}{z^4 - 1}\,dz = \frac{2\pi i}{4}(1 + 1 - 1 - 1) = 0.$$

EXAMPLE 3.

$$\oint_{|z|=2} \frac{e^z}{z(z - 1)^2}\,dz = 2\pi i \,\{\text{Res}\,[f(z), 0] + \text{Res}\,[f(z), 1]\}.$$

At the first order pole $z = 0$, application of Rule I gives the residue 1.

At the second order pole $z = 1$, Rule II gives

$$\text{Res}\,[f(z), 1] = \frac{d}{dz}\left(\frac{e^z}{z}\right)\Big|_{z=1} = \frac{e^z(z-1)}{z^2}\Big|_{z=1} = 0.$$

Rule IV could also be used, with $A = e^z$, $B = z^3 - 2z^2 + z$:

$$\text{Res}\,[f(z), 1] = \frac{6e^z(6z-4) - 2e^z \cdot 6}{3(6z-4)^2}\Big|_{z=1} = 0.$$

Accordingly,

$$\oint_{|z|=2} \frac{e^z}{z(z-1)^2}\,dz = 2\pi i(1 + 0) = 2\pi i.$$

3–14 Residue at infinity. Let $f(z)$ be analytic for $|z| > R$. The *residue* of $f(z)$ *at* ∞ is defined as follows:

$$\text{Res}\,[f(z), \infty] = \frac{1}{2\pi i}\oint_C f(z)\,dz,$$

where the integral is taken in the *negative* direction on a simple closed path C, in the domain of analyticity of $f(z)$, and *outside* of which $f(z)$ has no singularity other than ∞. This is suggested in Fig. 3–17. Theorem 27 has an immediate extension to this case:

THEOREM 29. *The residue of $f(z)$ at ∞ is given by the equation*

$$\text{Res}\,[f(z), \infty] = -a_{-1}, \tag{3–140}$$

where a_{-1} is the coefficient of z^{-1} in the Laurent expansion of $f(z)$ at ∞:

$$f(z) = \cdots + \frac{a_{-n}}{z^n} + \cdots + \frac{a_{-1}}{z} + a_0 + a_1 z + \cdots \tag{3–141}$$

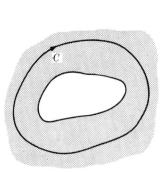

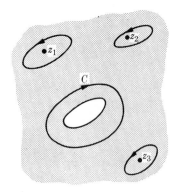

FIG. 3–17. Residue at infinity.   FIG. 3–18. Residue theorem for exterior.

The proof is the same as for Theorem 27. It should be stressed that presence of a nonzero residue at ∞ is not related to presence of a pole or essential singularity at ∞. That is, $f(z)$ can have a nonzero residue whether or not there is a pole or essential singularity, for the pole or essential singularity at ∞ is due to the *positive powers* of z, not to negative powers (Section 3–12 above). Thus the function $e^{1/z} = 1 + z^{-1} + (2!z^2)^{-1} + \cdots$ is analytic at ∞, but has the residue -1 there.

Cauchy's residue theorem has also an extension to include ∞:

THEOREM 30. *Let $f(z)$ be analytic in a domain D which includes a deleted neighborhood of ∞. Let C be a simple closed path in D outside of which $f(z)$ is analytic except for isolated singularities at $z_1, \ldots, z_k$. Then*

$$\oint_C f(z)\, dz = 2\pi i \left\{ \operatorname{Res}[f(z), z_1] + \cdots + \operatorname{Res}[f(z), z_k] + \operatorname{Res}[f(z), \infty] \right\}.$$
$$(3\text{–}142)$$

The proof, which is like that of Theorem 28, is left as an exercise (Problem 4, following Section 3–16). It is to be emphasized that the integral on C is taken in the *negative* direction (see Fig. 3–18) and that the *residue at ∞ must be included* on the right.

For a particular integral

$$\oint_C f(z)\, dz$$

on a simple closed path C, we have now two modes of evaluation: the integral equals $2\pi i$ times the sum of the residues inside the path (provided there are only a finite number of singularities there), and it also equals *minus* $2\pi i$ times the sum of the residues outside the path plus that at ∞ (provided there are only a finite number of singularities in the exterior domain). We can evaluate the integral both ways to check results. The principle involved here is summarized in the following theorem:

THEOREM 31. *If $f(z)$ is analytic in the extended z-plane except for a finite number of singularities, then the sum of all residues of $f(z)$ (including ∞) is zero.*

To evaluate the residues at ∞, we can formulate a set of rules like the ones above. However, the following two rules are adequate for most purposes:

Rule V. If $f(z)$ has a zero of first order at ∞, then

$$\operatorname{Res}[f(z), \infty] = -\lim_{z \to \infty} zf(z).$$

If $f(z)$ has a zero of second or higher order at ∞, the residue at ∞ is zero.

Rule VI.

$$\text{Res}\,[f(z),\,\infty] = -\,\text{Res}\left[\frac{1}{z^2}f(1/z),\,0\right].$$

The proof of Rule V is left as an exercise (Problem 8, following Section 3–16). To prove Rule VI, we write

$$f(z) = \cdots + a_n z^n + \cdots + a_1 z + a_0 + \frac{a_{-1}}{z} + \frac{a_{-2}}{z^2} + \cdots \quad (|z| > R).$$

Then for $0 < |z| < R^{-1}$,

$$f\left(\frac{1}{z}\right) = \cdots + \frac{a_n}{z^n} + \cdots + \frac{a_1}{z} + a_0 + a_{-1}z + a_{-2}z^2 + \cdots,$$

$$\frac{1}{z^2}f\left(\frac{1}{z}\right) = \cdots + \frac{a_0}{z^2} + \frac{a_{-1}}{z} + a_{-2} + \cdots$$

Hence

$$\text{Res}\left[\frac{1}{z^2}f\left(\frac{1}{z}\right),\,0\right] = a_{-1},$$

and the rule follows. This result reduces the problem to evaluation of a residue at zero, to which Rules I through IV are applicable.

EXAMPLE 1. We consider the integral

$$\oint_{|z|=2} \frac{z}{z^4 - 1}\,dz$$

of Example 2 in the preceding section. There is no singularity outside the path other than ∞, and at ∞ the function has a zero of order 3; hence the integral is zero.

EXAMPLE 2.

$$\oint_{|z|=2} \frac{1}{(z+1)^4(z^2-9)(z-4)}\,dz.$$

Here there is a fourth order pole inside the path, at which evaluation of the residue is tedious. Outside the path there are first order poles at ± 3 and 4 and a zero of order 7 at ∞. Hence by Rule I the integral equals

$$-2\pi i\left(\frac{1}{4^4 6(-1)} + \frac{1}{(-2)^4(-6)(-7)} + \frac{1}{5^4 \cdot 7}\right).$$

3–15 Logarithmic residues; argument principle. Let $f(z)$ be analytic in a domain D. Then $f'(z)/f(z)$ is analytic in D except at the zeros of $f(z)$. If an analytic branch of $\log f(z)$ is chosen in part of D [necessarily excluding the zeros of $f(z)$], then

$$\frac{d}{dz} \log f(z) = \frac{f'(z)}{f(z)}.$$ (3-150)

For this reason the expression f'/f is termed the *logarithmic deriva-tive* of $f(z)$. Its value is demonstrated by the following theorem:

THEOREM 32. *Let $f(z)$ be analytic in domain D. Let C be a simple closed path in D within which $f(z)$ is analytic except for a finite number of poles, and let $f(z) \neq 0$ on C. Then*

$$\frac{1}{2\pi i} \oint_C \frac{f'(z)}{f(z)} \, dz = N_0 - N_p,$$

where N_0 is the total number of zeros of f inside C and N_p is the total num-ber of poles of f inside C, zeros and poles being counted according to multiplicities.

Proof. The logarithmic derivative f'/f has isolated singularities precisely at the zeros and poles of f. At a zero z_0,

$$f(z) = (z - z_0)^N g(z) \quad [g(z_0) \neq 0],$$
$$f'(z) = (z - z_0)^N g'(z) + N(z - z_0)^{N-1} g(z),$$
$$\frac{f'(z)}{f(z)} = \frac{(z - z_0)^N g'(z) + N(z - z_0)^{N-1} g(z)}{(z - z_0)^N g(z)} = \frac{g'(z)}{g(z)} + \frac{N}{z - z_0}.$$

Hence the logarithmic derivative has a pole of first order, with residue N equal to the multiplicity of the zero. A similar analysis applies to each pole of f, with N replaced by $-N$. The theorem then follows from the Cauchy residue theorem (Theorem 28), provided we show that there are only a finite number of singularities. The poles of f are finite in number by assumption, and it is easily shown (Reference 5, p. 571) that there can be only a finite number of zeros. Thus the theorem is proved.

Remarks. Since

$$\frac{f'(z)}{f(z)} \, dz = d \log f(z) = d \log w = d \, (\log |w| + i \arg w),$$

we have

$$\frac{1}{2\pi i} \int_C \frac{f'(z)}{f(z)} \, dz = \frac{1}{2\pi i} \int_C d \log |w| + \frac{1}{2\pi} \int_C d \arg w.$$

Since $w \neq 0$ on C, $\log |w|$ is continuous on C and the first integral on the right is zero. The second measures the total charge in arg w, divided by

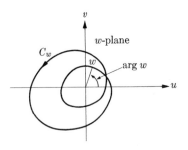

FIG. 3–19.   Argument principle.

2π, as w traces the path C_w, the image of C, in the w-plane. Hence it also measures the number of times C_w winds about the origin of the w-plane; in Fig. 3–19 the number is $+2$. The statement

$$\frac{1}{2\pi} [\text{increase in arg } f(z) \text{ on path}] = N_0 - N_p \qquad (3\text{–}151)$$

is known as the *argument principle*. This is of great value in finding zeros and poles of analytic functions.

3–16 Partial fraction expansion of rational functions.   The theory of analytic functions provides a simple proof for the familiar rules for partial fraction expansions.

Let

$$f(z) = \frac{a_0 z^n + \cdots + a_n}{b_0 z^m + \cdots + a_m} \quad (a_0 \neq 0, b_0 \neq 0) \qquad (3\text{–}160)$$

be given. We assume that $n < m$, so that f is a proper fraction. We also assume that the numerator and denominator have no common zeros. Let $z_1, z_2, \ldots, z_N$ be the *distinct* zeros of the *denominator* (no repetitions); these are the poles of $f(z)$. At z_1, $f(z)$ has a Laurent expansion:

$$f(z) = p_1(z) + g_1(z),$$

$$p_1(z) = \frac{A_{-k_1}}{(z - z_1)^{k_1}} + \frac{A_{-k_1+1}}{(z - z_1)^{k_1-1}} + \cdots + \frac{A_{-1}}{z - z_1}. \qquad (3\text{–}161)$$

Here $p_1(z)$ is the principal part of $f(z)$ at z_1 and $g_1(z)$ is analytic at z_1; k_1 is the order of the pole at z_1. Similar expressions hold at the other poles.

The partial fraction expansion of $f(z)$ is now simply the identity

$$f(z) = p_1(z) + p_2(z) + \cdots + p_N(z). \qquad (3\text{–}162)$$

To justify this, we let

$$F(z) = f(z) - [p_1(z) + p_2(z) + \cdots + p_N(z)].$$

Now $f(z) - p_1(z)$ has a removable singularity at z_1, while $p_2(z), \ldots, p_N(z)$ are analytic at z_1. Hence $F(z)$ has a removable singularity at z_1. In general, $F(z)$ has only removable singularities for finite z. At ∞, $f(z)$, $p_1(z)$, $\ldots$, $p_N(z)$ all have zeros; hence $F(z)$ has a zero at ∞. But $F(z)$ is a rational function with no poles. Hence $F(z)$ must be a polynomial. Thus $F(z)$ has a pole at ∞ unless F is constant; since we know $F = 0$ at ∞, F must be a constant, namely zero. This proves (3–162).

If $f(z)$ has only simple poles, the principal part at each pole z_j is simply $A_j/(z - z_j)$, where A_j is the residue of f at z_j, and hence in this case

$$f(z) = \frac{A_1}{z - z_1} + \cdots + \frac{A_m}{z - z_m} \quad (A_j = \text{Res}\,[f, z_j]) \qquad (3\text{--}163)$$

is the partial fraction expansion. If we write

$$f(z) = \frac{A(z)}{B(z)},$$

then Rule III can be applied:

$$A_j = \frac{A(z_j)}{B'(z_j)}, \qquad f(z) = \sum_{j=1}^{m} \frac{A(z_j)}{B'(z_j)} \frac{1}{z - z_j}. \qquad (3\text{--}164)$$

At a multiple pole z_j of order k, we can write

$$f(z) = \frac{1}{(z - z_j)^k}\, \phi(z).$$

The principal part $p_j(z)$ is then

$$p_j(z) = \frac{\phi(z_j)}{(z - z_j)^k} + \frac{\phi'(z_j)}{1!(z - z_j)^{k-1}} + \cdots + \frac{\phi^{(k-1)}(z_j)}{(k - 1)!(z - z_j)}. \qquad (3\text{--}165)$$

Hence $p_j(z)$ can be found without knowledge of the other poles.

EXAMPLE 1.

$$f(z) = \frac{z^2 + 1}{z^3 + 4z^2 + 3z} = \frac{z^2 + 1}{z(z + 1)(z + 3)}.$$

There are simple poles at 0, -1, -3. By (3–164) above,

$$f(z) = \frac{A(0)}{B'(0)} \frac{1}{z} + \frac{A(-1)}{B'(-1)} \frac{1}{z + 1} + \frac{A(-3)}{B'(-3)} \frac{1}{z + 3};$$

with $A = z^2 + 1$, $B = z^3 + 4z^2 + 3z$, $B' = 3z^2 + 8z + 3$, we find

$$f(z) = \frac{1}{3}\frac{1}{z} + \frac{2}{-2}\frac{1}{z+1} + \frac{10}{6}\frac{1}{z+3}.$$

EXAMPLE 2.

$$f(z) = \frac{z}{(z-1)^2(z^3+z+1)}.$$

At the pole $z = 1$, we write

$$f = \frac{1}{(z-1)^2}\phi(z) \quad \left(\phi = \frac{z}{z^3+z+1}\right).$$

Since $\phi(1) = \frac{1}{3}$, $\phi'(1) = -\frac{1}{9}$, the principal part at 1 is

$$\frac{1}{3}\frac{1}{(z-1)^2} - \frac{1}{9}\frac{1}{z-1}.$$

The cubic $z^3 + z + 1$ has one real root z_1 and two complex roots z_2, z_3. These are all simple. Hence we can write

$$f(z) = \frac{A(z_1)}{B'(z_1)}\frac{1}{z-z_1} + \frac{A(z_2)}{B'(z_2)}\frac{1}{z-z_2}$$

$$+ \frac{A(z_3)}{B'(z_3)}\frac{1}{z-z_3} + \frac{1}{3}\frac{1}{(z-1)^2} - \frac{1}{9}\frac{1}{z-1},$$

where $A(z) = z$, $B(z) = (z-1)^2(z^3+z+1)$. At the poles z_1, z_2, z_3, $B'(z)$ reduces to $z^2 + z + 7$, by the relation $z^3 + z + 1 = 0$.

PROBLEMS

1. Evaluate the following integrals on the paths given:

(a) $\oint \dfrac{z\,dz}{(z-1)(z-3)}$ $(|z| = 2)$ (b) $\oint \dfrac{e^{3z}\,dz}{z^2+4}$ $(|z| = 3)$

(c) $\oint \dfrac{\sin z}{z^4}\,dz$ $(|z| = 1)$ (d) $\oint \dfrac{z\,dz}{z^3+z+1}$ $(|z| = 4)$

(e) $\oint \dfrac{dz}{(z+1)^4(z+3)}$ $(|z| = 2)$ (f) $\oint \dfrac{dz}{(z+1)^5(z+3)}$ $(|z| = 4)$

(g) $\oint \dfrac{2z+2}{z^2+2z+2}\,dz$ $(|z| = 2)$ (h) $\oint \dfrac{3z^2-6z+1}{z^3-3z^2+z-3}\,dz$ $(|z| = 2)$

2. Expand each of the following in partial fractions:

(a) $\dfrac{1}{z^2 - 4}$ (b) $\dfrac{z+1}{(z-1)(z-2)(z-3)}$ (c) $\dfrac{z^2}{z^5 + 1}$ (d) $\dfrac{1}{z^n - 1}$

(e) $\dfrac{z}{(z-1)^2(z+1)^2}$ (f) $\dfrac{\phi(z)}{z^n(z+1)}$ $(\phi = c_0 + c_1 z + \cdots + c_n z^n)$

(g) $\dfrac{1}{(z-1)^3(z^4 + z + 1)}$ (h) $\dfrac{1}{(z^2+1)(z^2 + 2z + 2)}$

3. Prove the Fundamental Theorem of Algebra: *Every polynomial of degree at least 1 has a zero.* [*Hint:* Show that Res $[f'(z)/f(z), \infty]$ is not 0, but is in fact minus the degree n of the polynomial $f(z)$. Then use Theorem 32 to show that f has n zeros.]

4. Prove Theorem 30.

5. Formulate and prove Theorem 32 for integration around the boundary B of a region R in D, bounded by simple closed curves $C_1, \ldots, C_k$.

6. Prove that under the hypotheses of Theorem 32, if $g(z)$ is analytic in D and within C, then

$$\frac{1}{2\pi i} \oint_C \frac{g(z)f'(z)}{f(z)} \, dz = \sum_{k=1}^{n} g(z_k') - \sum_{l=1}^{m} g(z_l''),$$

where $z_1', \ldots, z_n'$ are the zeros of f, and $z_1'', \ldots, z_m''$ are the poles of f inside C, repeated according to multiplicity.

7. Extend Rule IV of Section 3–13 to the case in which $A(z)$ has a first order zero at z_0 and $B(z)$ has a second order zero.

8. Prove Rule V of Section 3–14.

ANSWERS

1. (a) $-\pi i$ (b) $\pi i \sin 6$ (c) $-\pi i/3$ (d) 0 (e) $-\pi i/8$ (f) 0 (g) $4\pi i$ (h) $4\pi i$

2. (a) $\dfrac{1}{4}\dfrac{1}{z-2} - \dfrac{1}{4}\dfrac{1}{z+2}$ (b) $\dfrac{1}{z-1} - \dfrac{3}{z-2} + \dfrac{2}{z-3}$

(c) $-\dfrac{1}{5}\left[\dfrac{z_2}{z - z_1} + \dfrac{z_5}{z - z_2} + \dfrac{z_3}{z - z_3} + \dfrac{z_1}{z - z_4} + \dfrac{z_4}{z - z_5}\right],$

$z_k = \exp[(2k - 1)\pi i/5], \ k = 1, \ldots, 5$

(d) $\dfrac{1}{n}\left[\dfrac{z_1}{z - z_1} + \cdots + \dfrac{z_n}{z - z_n}\right], \ z_k = \exp\left(\dfrac{2k\pi i}{n}\right), \ k = 1, \ldots, n$

(e) $\dfrac{1}{4}\dfrac{1}{(z-1)^2} - \dfrac{1}{4(z+1)^2}$

(f) $\dfrac{c_0}{z^n} + \dfrac{c_1 - c_0}{z^{n-1}} + \cdots + \dfrac{c_{n-1} - c_{n-2} + \cdots + (-1)^{n-1}c_0}{z}$

$+ (-1)^n \phi(-1) \dfrac{1}{z+1}$

(g) $\dfrac{1}{3}\dfrac{1}{(z-1)^3} - \dfrac{5}{9}\dfrac{1}{(z-1)^2} + \dfrac{7}{27}\dfrac{1}{z-1} + \displaystyle\sum_{j=1}^{4}\dfrac{g(z_j)}{z-z_j}\,,$

where $z_1, \ldots, z_4$ are the roots of $z^4 + z + 1 = 0$ and

$$g = (-7z^3 + 5z^2 + 3z - 13)^{-1}$$

(h) $\dfrac{1}{10}\left[\dfrac{-2-i}{z-i} + \dfrac{-2+i}{z+i} + \dfrac{2-i}{z+1-i} + \dfrac{2+i}{z+1+i}\right]$

7. $2A'/B''$

SUGGESTED REFERENCES

1. LARS V. AHLFORS, *Complex Analysis*. New York: McGraw-Hill, 1953.

2. RUEL V. CHURCHILL, *Introduction to Complex Variables and Applications*. 2nd ed. New York: McGraw-Hill, 1960.

3. ÉDOUARD GOURSAT, *A Course in Mathematical Analysis*, Vol. II, Part 1 (transl. by E. R. Hedrick and O. Dunkel). New York: Ginn, 1916.

4. EINAR HILLE, *Analytic Function Theory*, Vol. I. Boston: Ginn, 1959.

5. WILFRED KAPLAN, *Advanced Calculus*. Reading, Mass.: Addison-Wesley, 1952.

6. KONRAD KNOPP, *Theory of Functions*, 2 Vols. (transl. by F. Bagemihl). New York: Dover, 1945.

CHAPTER 4

FOURIER SERIES AND FINITE FOURIER TRANSFORM

4-1 Response to a sum of sinusoidal terms. Let a stable system be given, with transfer function $Y(s)$ for the output x relative to the input f. For example, let the system be described by a differential equation

$$(a_0 D^n + \cdots + a_n)x = f,$$

so that

$$Y(s) = \frac{1}{a_0 s^n + \cdots + a_n}.$$

We have seen in Section 2–9 that when $f = Ae^{i\omega t}$, a particular response is

$$x = Y(i\omega)Ae^{i\omega t}.$$

When $f = A \cos \omega t$, a particular response is obtained by superposition:

$$f = A \frac{e^{i\omega t} + e^{-i\omega t}}{2}, \qquad x = \frac{A}{2} [Y(i\omega)e^{i\omega t} + Y(-i\omega)e^{-i\omega t}].$$

When $Y(s)$ has real coefficients, $Y(-i\omega) = \overline{Y(i\omega)}$ and we can write, more simply,

$$f = \mathrm{Re} \, [Ae^{i\omega t}], \qquad x = \mathrm{Re} \, [A Y(i\omega)e^{i\omega t}].$$

Similarly, if $Y(s)$ has real coefficients and $f = A \sin \omega t = \mathrm{Im} \, [Ae^{i\omega t}]$, then we can choose

$$x = \mathrm{Im} \, [Y(i\omega)Ae^{i\omega t}].$$

If f is a sum of exponentials,

$$f = \sum_{n=1}^{N} A_n e^{i\omega_n t}, \tag{4–10}$$

we obtain the response by superposition:

$$x = \sum_{n=1}^{N} Y(i\omega_n) A_n e^{i\omega_n t}.$$

If f is a sum of sinusoidal terms:

$$f = \sum_{n=1}^{N} (a_n \cos \omega_n t + b_n \sin \omega_n t), \tag{4–11}$$

163

then f can be expressed in form (4–10):

$$f = \sum_{n=1}^{N} \left[\frac{a_n - ib_n}{2} e^{i\omega_n t} + \frac{a_n + ib_n}{2} e^{-i\omega_n t} \right], \qquad (4\text{–}12)$$

so that a particular response is

$$x = \sum_{n=1}^{N} \left[Y(i\omega_n) \frac{a_n - ib_n}{2} e^{i\omega_n t} + Y(-i\omega_n) \frac{a_n + ib_n}{2} e^{-i\omega_n t} \right].$$

In each case, the general solution for x is obtained by adding appropriate terms depending on the initial conditions. However, for a stable system these additional terms are transients and approach zero as $t \to +\infty$. Accordingly, each particular solution given above is the *steady-state solution*.

4–2 Periodic inputs. Fourier series. We consider an input which is a sum of sinusoidal terms whose frequencies are *multiples of a basic frequency*:

$$f(t) = a_1 \cos \omega t + b_1 \sin \omega t + a_2 \cos 2\omega t + b_2 \sin 2\omega t$$
$$+ \cdots + a_k \cos k\omega t + b_k \sin k\omega t \qquad (\omega > 0). \qquad (4\text{–}20)$$

Such a function has a property not in general shared by a function (4–11), namely, that of *periodicity*. We say that $f(t)$ has period τ ($\tau \neq 0$) if

$$f(t + \tau) = f(t)$$

for all t; that is, the function always returns to the same value after a time interval of length τ. For the function (4–20) we can choose $\tau = 2\pi/\omega$. For

$$\cos \omega \left(t + \frac{2\pi}{\omega} \right) = \cos (\omega t + 2\pi) = \cos \omega t,$$

$$\cos 2\omega \left(t + \frac{2\pi}{\omega} \right) = \cos (2\omega t + 4\pi) = \cos 2\omega t,$$

$$\vdots$$

Thus each term of (4–20) has period $2\pi/\omega$, and the same holds for the sum:

$$f \left(t + \frac{2\pi}{\omega} \right) = f(t).$$

Now a function $f(t)$ may be given in some form other than that of (4–20), and it may be known that $f(t)$ has period $2\pi/\omega$. If we could show that

$f(t)$ is expressible in form (4–20), then we could at once obtain the steady-state response x to f, as in the preceding section, by means of the frequency response function $Y(i\omega)$. Now it is not true that every periodic function of period $2\pi/\omega$ has the form (4–20). However, if we generalize this by adding a constant term (which is periodic, clearly) and allowing *infinitely many* terms, we do obtain an expression capable of representing essentially all periodic functions f of period $\tau = 2\pi/\omega$:

$$\frac{a_0}{2} + \sum_{n=1}^{\infty} (a_n \cos n\omega t + b_n \sin n\omega t). \tag{4–21}$$

A series such as (4–21) is called a *trigonometric series*. We shall show that a function $f(t)$ can, under very general conditions, be represented as the sum of such a series (4–21). The coefficients a_n, b_n are related to the function f by the equations

$$a_n = \frac{2}{\tau} \int_0^\tau f(t) \cos n\omega t \, dt,$$

$$b_n = \frac{2}{\tau} \int_0^\tau f(t) \sin n\omega t \, dt. \tag{4–22}$$

When (4–22) holds, the series (4–21) is called the *Fourier series* of $f(t)$.

If $f(t)$ is represented in the form (4–21), then, as in (4–12), we can express f in terms of exponentials:

$$f = \frac{a_0}{2} e^{0t} + \sum_{n=1}^{\infty} (c_n e^{in\omega t} + c'_n e^{-in\omega t}).$$

For the response $x(t)$ we would naturally expect the representation

$$x(t) = Y(0) \frac{a_0}{2} + \sum_{n=1}^{\infty} [Y(in\omega) c_n e^{in\omega t} + Y(-in\omega) c'_n e^{-in\omega t}].$$

This also turns out to be valid under very general conditions.

The fact that the Fourier series converges to the corresponding function is actually not of the greatest significance. To a quite general function f we can assign Fourier coefficients a_n, b_n, by (4–22). By means of the frequency response function we can compute the corresponding Fourier coefficients of $x(t)$. Knowledge of these coefficients alone completely determines $x(t)$, regardless of whether the series converges to $x(t)$. Indeed, it is in many cases possible to find $x(t)$ from the coefficients without employing the series.

4–3 Examples of Fourier series. EXAMPLE 1. Let $f(t) = t^2$ for $-\pi \leqq t < \pi$ and let $f(t)$ have period 2π, so that $f(t)$ has the graph of Fig. 4–1. Then $\omega = 1, \tau = 2\pi$, and formulas (4–22) give

$$a_n = \frac{1}{\pi} \int_{-\pi}^{\pi} t^2 \cos nt \, dt = \frac{4 \cos n\pi}{n^2} = \frac{4(-1)^n}{n^2} \quad (n \neq 0),$$

$$b_n = \frac{1}{\pi} \int_{-\pi}^{\pi} t^2 \sin nt \, dt = 0,$$

$$a_0 = \frac{1}{\pi} \int_{-\pi}^{\pi} t^2 \, dt = \frac{2\pi^2}{3}.$$

Hence we expect

$$f(t) = \frac{\pi^2}{3} + 4 \sum_{n=1}^{\infty} \frac{(-1)^n}{n^2} \cos nt.$$

In Fig. 4–1 the third partial sum, $(\pi^2/3) - 4 \cos t + \cos 2t$, is graphed; we can see that this is a good approximation to $f(t)$.

EXAMPLE 2. $f(t) = \pi$ for $-\pi \leqq t < 0$, $f(t) = \pi - t$ for $0 \leqq t < \pi$, and $f(t)$ has period 2π. The function is graphed in Fig. 4–2. We note that $f(t)$ has jump discontinuities at $\pm\pi, \pm3\pi, \ldots$ By (4–22),

$$a_n = \frac{1}{\pi} \left[\int_{-\pi}^{0} \pi \cos nt \, dt + \int_{0}^{\pi} (\pi - t) \cos nt \, dt \right] = \frac{1 - (-1)^n}{\pi n^2} \quad (n \neq 0),$$

$$b_n = \frac{1}{\pi} \left[\int_{-\pi}^{0} \pi \sin nt \, dt + \int_{0}^{\pi} (\pi - t) \sin nt \, dt \right] = \frac{\cos n\pi}{n},$$

while $a_0 = 3\pi/2$. Accordingly, we expect

$$f(t) = \frac{3\pi}{4} + \sum_{n=1}^{\infty} \left[\frac{1 - (-1)^n}{\pi n^2} \cos nt + \frac{(-1)^n}{n} \sin nt \right].$$

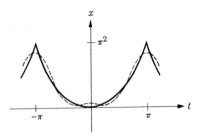

FIG. 4–1. Representation of continuous function by Fourier series.

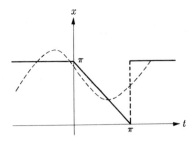

FIG. 4–2. Representation of discontinuous function by Fourier series.

The second partial sum, $(3\pi/4) + [(2 \cos t)/\pi] - \sin t$, is graphed in Fig. 4–2. There is good agreement except near the jumps. Here the sum is about 0.55π, which is close to the *average of left- and right-hand limits*. We shall show that the series converges to this average value at the jump points, and otherwise converges to $f(t)$.

EXAMPLE 3. $f(t) = \sin^5 t$. Rather than computing as above, we write

$$\sin^5 t = \left(\frac{e^{it} - e^{-it}}{2i} \right)^5$$

$$= \frac{1}{32i} \, (e^{5it} - 5e^{3it} + 10e^{it} - 10e^{-it} + 5e^{-3it} - e^{-5it})$$

$$= \tfrac{5}{8} \sin t - \tfrac{5}{16} \sin 3t + \tfrac{1}{16} \sin 5t.$$

Here the Fourier series has only three terms.

EXAMPLE 4. The power of complex numbers is further illustrated by the following example, which shows how known power series can be used to compute Fourier series. We start with the series expansion

$$\frac{1}{1 - z} = 1 + z + \cdots + z^n + \cdots, \tag{4–30}$$

which is valid for $|z| < 1$ (Section 3–8). We let $z = re^{it}$, where r is real and $|r| < 1$. Then (4–30) becomes

$$\frac{1}{1 - re^{it}} = 1 + re^{it} + \cdots + r^n e^{int} + \cdots$$

Taking real and imaginary parts, we find, for $|r| < 1$,

$$\frac{1 - r \cos t}{1 + r^2 - 2r \cos t} = 1 + r \cos t + \cdots + r^n \cos nt + \cdots, \tag{4–31}$$

$$\frac{r \sin t}{1 + r^2 - 2r \cos t} = r \sin t + r^2 \sin 2t + \cdots + r^n \sin nt + \cdots \tag{4–32}$$

PROBLEMS

1. Given the differential equation $(D^2 + 2D + 5)x = f(t)$, verify stability and obtain the steady-state response for the following choices of $f(t)$:

(a) $3 \sin 2t$ (b) $2 + 3 \cos 2t - 5 \sin \pi t$

(c) $\cos^5 t$ (compare Example 3 above)   (d) $\sin 2t \sin 5t$

(e) $\dfrac{\sin t}{5 - 4 \cos t}$ *[Hint: Use (4–32).]*

2. Expand each of the following in Fourier series:

(a) $\sin^3 t$ (b) $\cos^3 t$ (c) $\sin^2 t \cos^3 t$

(d) $\dfrac{r \sin t (1 - r \cos t)}{(1 + r^2 - 2r \cos t)^2}$, $|r| < 1$ [*Hint:* Use the power series for $(1 - z)^{-2}$.]

(e) $e^{r \cos t} \cos (r \sin t)$ [*Hint:* Use the power series for e^z.]

3. Expand in Fourier series. Graph $F(t)$ and the first few partial sums in each of the following cases:

(a) $F(t) = -t$ for $-\pi \leq t \leq 0$, $F(t) = t$ for $0 \leq t \leq \pi$, $F(t)$ of period 2π

(b) $F(t) = 1$ for $-\pi \leq t \leq 0$, $F(t) = 0$ for $0 < t < \pi$, $F(t)$ of period 2π

Answers

1. (a) $\dfrac{3}{2i}\left(\dfrac{e^{2it}}{1 + 4i} - \dfrac{e^{-2it}}{1 - 4i}\right)$

(b) $\dfrac{2}{5} + \dfrac{3}{2}\left(\dfrac{e^{2it}}{1 + 4i} + \dfrac{e^{-2it}}{1 - 4i}\right) - \dfrac{5}{2i}\left(\dfrac{e^{\pi it}}{5 - \pi^2 + 2\pi i} - \dfrac{e^{-\pi it}}{5 - \pi^2 - 2\pi i}\right)$

(c) $\dfrac{1}{32}\left(\dfrac{e^{5it}}{-20 + 10i} + \dfrac{5e^{3it}}{-4 + 6i} + \dfrac{10e^{it}}{4 + 2i} + \dfrac{10e^{-it}}{4 - 2i} + \dfrac{5e^{-3it}}{-4 - 6i}\right.$

$\left. + \dfrac{e^{-5it}}{-20 - 10i}\right)$

(d) $-\dfrac{1}{4}\left(\dfrac{e^{7it}}{-44 + 14i} - \dfrac{e^{3it}}{-4 + 6i} - \dfrac{e^{-3it}}{-4 - 6i} + \dfrac{e^{-7it}}{-44 - 14i}\right)$

(e) $\displaystyle\sum_{n=1}^{\infty} \dfrac{(5 - n^2)\sin nt - 2n \cos nt}{2^{n+1}[(5 - n^2)^2 + 4n^2]}$

2. (a) $\frac{1}{4}(3 \sin t - \sin 3t)$ \qquad (b) $\frac{1}{4}(3 \cos t + \cos 3t)$

(c) $\frac{1}{16}(2 \cos t - \cos 3t - \cos 5t)$

(d) $\frac{1}{2}[2r \sin t + 3r^2 \sin 2t + \cdots + (n + 1)r^n \sin nt + \cdots]$

(e) $1 + r \cos t + \cdots + \dfrac{r^n \cos nt}{n!} + \cdots$

3. (a) $\dfrac{\pi}{2} - \dfrac{2}{\pi}\left[2 \cos t + \dfrac{2 \cos 3t}{3^2} + \cdots + \dfrac{1 - (-1)^n}{n^2} \cos nt + \cdots\right]$

(b) $\dfrac{1}{2} - \dfrac{2}{\pi}\displaystyle\sum_{n=1}^{\infty} \dfrac{\sin (2n - 1)t}{2n - 1}$

4–4 Uniform convergence. A crucial notion for the theory is that of *uniform convergence*. A series of functions (real or complex)

$$\sum_{n=1}^{\infty} u_n(t) \quad (a \leqq t \leqq b) \tag{4–40}$$

is said to converge uniformly to $f(t)$ over the interval $a \leqq t \leqq b$ if, for every $\epsilon > 0$,

$$|u_1(t) + \cdots + u_n(t) - f(t)| < \epsilon \quad (a \leqq t \leqq b) \tag{4–41}$$

for all n greater than a suitable N which depends on ϵ but not on t. In most cases the functions $u_n(t)$ and $f(t)$ are continuous. If this is so, uniform convergence means that the *maximum error*,

$$E_n = \max |s_n(t) - f(t)| \quad (a \leqq t \leqq b),$$

where $s_n = u_1(t) + \cdots + u_n(t)$, approaches zero as $n \to \infty$. Thus for large n the partial sums approximate $f(t)$ uniformly over the whole interval.

We state the following theorems on uniform convergence without proof. For proofs and further discussion, see Reference 6, pp. 342–348, 498–499.

THEOREM 1. *If $\sum_{n=1}^{\infty} M_n$ is a convergent series of constants and*

$$|u_n(t)| \leqq M_n \quad (a \leqq t \leqq b),$$

then the series (4–40) converges uniformly for $a \leqq t \leqq b$.

This is the Weierstrass M-test.

THEOREM 2. *If each function $u_n(t)$ is continuous for $a \leqq t \leqq b$ and the series (4–40) converges uniformly for $a \leqq t \leqq b$ to $f(t)$, then $f(t)$ is continuous for $a \leqq t \leqq b$.*

THEOREM 3. *A uniformly convergent series of continuous functions can be integrated term by term; that is, if the conditions of Theorem 2 hold, then*

$$\int_a^b f(t)\, dt = \sum_{n=1}^{\infty} \int_a^b u_n(t)\, dt. \tag{4–42}$$

THEOREM 4. *If each $u_n(t)$ has a continuous derivative for $a \leqq t \leqq b$, if $\sum u_n(t) = f(t)$, and if $\sum u_n'(t)$ converges uniformly, then*

$$f'(t) = \sum_{n=1}^{\infty} u_n'(t) \quad (a \leqq t \leqq b). \tag{4–43}$$

THEOREM 5. *If $\sum u_n(t)$ and $\sum v_n(t)$ are uniformly convergent for $a \leqq t \leqq b$, and $h(t)$ is continuous for $a \leqq t \leqq b$, then the series*

$$\sum_{n=1}^{\infty} [u_n(t) + v_n(t)], \qquad \sum_{n=1}^{\infty} [u_n(t) - v_n(t)], \qquad \sum_{n=1}^{\infty} h(t)u_n(t)$$

are uniformly convergent for $a \leqq t \leqq b$.

4–5 Uniqueness theorem for Fourier series. We give the theory of Fourier series for real functions of period 2π. If $f(t)$ has period τ, then the substitution

$$t' = \frac{2\pi}{\tau} t$$

converts $f(t)$ into a function $g(t')$ of period 2π. The Fourier series for $g(t')$ becomes the Fourier series for $f(t)$ when we replace t' by $2\pi t/\tau$.

Let $f(t)$ have period 2π. Then the Fourier coefficients of $f(t)$ are the numbers $a_0, a_1, b_1, \ldots, a_n, b_n, \ldots$ defined by the formulas

$$a_n = \frac{1}{\pi} \int_{-\pi}^{\pi} f(t) \cos nt\, dt \quad (n = 0, 1, 2, \ldots),$$

$$b_n = \frac{1}{\pi} \int_{-\pi}^{\pi} f(t) \sin nt\, dt \quad (n = 1, 2, \ldots).$$

$$(4\text{–}50)$$

For (4–50) to have meaning, it is sufficient that $f(t)$ be continuous or, more generally, piecewise continuous, as in Fig. 4–3; that is, $f(t)$ is continuous except for a finite number of jump discontinuities for $-\pi \leqq t \leqq \pi$. If $f(t)$ is periodic of period 2π, then

$$\int_{\alpha}^{\alpha+2\pi} f(t)\, dt \qquad (4\text{–}51)$$

does not depend on α. We see this easily from a figure (see Problem 1, following Section 4–9). If we apply the result to $f(t) \cos nt$ and $f(t) \sin nt$,

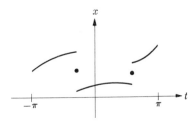

FIG. 4–3. Piecewise continuous function.

we conclude that in the definition (4–50) of a_n, b_n we can replace the limits of integration by α and $\alpha + 2\pi$. For example, the definitions

$$a_n = \frac{1}{\pi} \int_0^{2\pi} f(t) \cos nt \, dt, \qquad b_n = \frac{1}{\pi} \int_0^{2\pi} f(t) \sin nt \, dt \quad (4\text{–}52)$$

are commonly used.

If $f(t)$ is given only between $-\pi$ and π, we can repeat it outside this interval in such a manner as to obtain a periodic function of period 2π. However, even though f is continuous for $-\pi \leq t \leq \pi$, the extended function, called the *periodic extension of f*, may have jump discontinuities at $\pm\pi$, $\pm 3\pi$, ... If $f(\pi) \neq f(-\pi)$, the value at one of these points must be changed in any case, in forming the periodic extension. It should be remarked that the formulas (4–50) use only the values of f between $-\pi$ and π, and we can hence speak of the Fourier series of a function defined only in this interval. If the series converges to f inside the interval, it will converge to the periodic extension of f outside the interval.

The following theorem is of central importance in the application of Fourier series to systems analysis. It shows that a function is *uniquely determined* by its Fourier coefficients.

THEOREM 6. (Uniqueness theorem). *Let $f(t)$ and $f_1(t)$ be piecewise continuous in the interval $-\pi \leq t \leq \pi$ and let $f(t)$ and $f_1(t)$ have the same Fourier coefficients:*

$$\int_{-\pi}^{\pi} f(t) \cos nt \, dt = \int_{-\pi}^{\pi} f_1(t) \cos nt \, dt \quad (n = 0, 1, 2, \ldots),$$

$$\int_{-\pi}^{\pi} f(t) \sin nt \, dt = \int_{-\pi}^{\pi} f_1(t) \sin nt \, dt \quad (n = 1, 2, \ldots). \tag{4–53}$$

Then $f(t) = f_1(t)$ except perhaps at points of discontinuity.

Proof. Let $g(t) = f(t) - f_1(t)$. Then $g(t)$ is piecewise continuous, and by (4–53) all Fourier coefficients of $g(t)$ are 0. We shall show that $g(t) = 0$ except perhaps at discontinuity points.

Let us suppose $g(t_0) \neq 0$ at a point of continuity t_0; for example, $g(t_0) = 2c > 0$. Then by continuity, $g(t) > c$ for $|t - t_0| < \epsilon$ and ϵ sufficiently small. We can assume $-\pi < t_0 < \pi$, so that for ϵ sufficiently small $-\pi < t_0 - \epsilon < t_0 + \epsilon < \pi$.

We now achieve a contradiction by showing that there exists a "trigonometric polynomial"

$$P(t) = p_0 + p_1 \cos t + q_1 \sin t + \cdots + p_k \cos kt + q_k \sin kt \quad (4\text{–}54)$$

which represents a "pulse" at t_0 of arbitrarily large amplitude K and

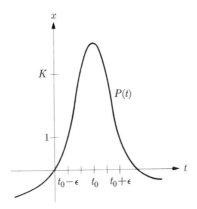

FIG. 4–4. Pulse function.

arbitrarily small width ϵ. This is suggested in Fig. 4–4. Specifically, the properties required are as follows:

(a) $P(t) \geqq K$ for $|t - t_0| \leqq \frac{1}{2}\epsilon$;
(b) $P(t) \geqq 1$ for $\frac{1}{2}\epsilon \leqq |t - t_0| \leqq \epsilon$;
(c) $|P(t)| < 1$ for $-\pi \leqq t < t_0 - \epsilon$ and for $t_0 + \epsilon < t \leqq \pi$.

Now if such a function P of the form (4–54) can be found, then we have a contradiction. For, on the one hand,

$$\int_{-\pi}^{\pi} P(t)g(t) \, dt = p_0 \int_{-\pi}^{\pi} g(t) \, dt + p_1 \int_{-\pi}^{\pi} g(t) \cos t \, dt + \cdots = 0, \quad (4\text{--}55)$$

since all Fourier coefficients of $g(t)$ are zero. On the other hand,

$$\int_{-\pi}^{\pi} P(t)g(t) \, dt = \int_{-\pi}^{t_0-(1/2)\epsilon} P(t)g(t) \, dt + \int_{t_0+(1/2)\epsilon}^{\pi} P(t)g(t) \, dt$$
$$+ \int_{t_0-(1/2)\epsilon}^{t_0+(1/2)\epsilon} P(t)g(t) \, dt.$$

By condition (a) the third term on the right can be made as large positive as desired by suitable choice of P; by virtue of (b) and (c), the first two terms remain above some negative constant. In particular, for proper choice of $P(t)$, the sum of all three terms is positive. This contradicts (4–55).

It remains to specify the pulse function $P(t)$. We set

$$P(t) = [1 + \cos (t - t_0) - \cos \epsilon]^N,$$

where N is an appropriately large positive integer. If $P(t)$ is expressed in terms of e^{it}, e^{-it}, we see that it is a trigonometric polynomial. The properties (a), (b), and (c) are readily verified. Hence the proof is complete.

4-6 Uniformly convergent Fourier series.

THEOREM 7. *If the trigonometric series*

$$\frac{a_0}{2} + \sum_{n=1}^{\infty} (a_n \cos nt + b_n \sin nt) \tag{4-60}$$

converges uniformly to $f(t)$ for all t, then $f(t)$ is a continuous periodic function of period 2π and (4-60) is the Fourier series of $f(t)$.

Proof. By Theorem 2, $f(t)$ is continuous in every interval of the t-axis, that is, continuous for all t. Since every term of the series has period 2π, $f(t)$ must have this period. From the equation

$$f(t) = \frac{a_0}{2} + \sum_{n=1}^{\infty} (a_n \cos nt + b_n \sin nt)$$

we obtain the relation

$$f(t) \cos mt = \frac{a_0}{2} \cos mt + \sum_{n=1}^{\infty} (a_n \cos mt \cos nt + b_n \cos mt \sin nt).$$

By Theorem 5, the series is still uniformly convergent. By Theorem 3,

$$\int_{-\pi}^{\pi} f(t) \cos mt \, dt = \frac{a_0}{2} \int_{-\pi}^{\pi} \cos mt \, dt$$

$$+ \sum_{n=1}^{\infty} \left(a_n \int_{-\pi}^{\pi} \cos mt \cos nt \, dt + b_n \int_{-\pi}^{\pi} \cos mt \sin nt \, dt \right).$$

Now when m and n are positive integers,

$$\int_{-\pi}^{\pi} \cos mt \cos nt \, dt = \begin{cases} 0 & (n \neq m) \\ \pi & (n = m) \end{cases}$$

and

$$\int_{-\pi}^{\pi} \cos mt \sin nt \, dt = 0.$$

Hence

$$\int_{-\pi}^{\pi} f(t) \cos mt \, dt = \pi a_m.$$

If $m = 0$, we find in the same way

$$\int_{-\pi}^{\pi} f(t) \, dt = \pi a_0$$

and, for $m = 1, 2, \ldots$,

$$\int_{-\pi}^{\pi} f(t) \sin mt \, dt = \pi b_m.$$

The last three equations show that (4–60) is indeed the Fourier series of $f(t)$.

Theorem 8. *If two trigonometric series converge uniformly for all t, and have the same sum for all t:*

$$\tfrac{1}{2}a_0 + \sum_{n=1}^{\infty} (a_n \cos nt + b_n \sin nt) \equiv \tfrac{1}{2}a_0' + \sum_{n=1}^{\infty} (a_n' \cos nt + b_n' \sin nt),$$

then the series are identical:

$$a_0 = a_0', \qquad a_n = a_n', \qquad b_n = b_n' \qquad for \quad n = 1, 2, \ldots$$

In particular, if a trigonometric series converges uniformly to 0 for all t, then all coefficients are zero.

Proof. Let $f(t)$ denote the sum for both series. Then by Theorem 7

$$a_n = a_n' = \frac{1}{\pi} \int_{-\pi}^{\pi} f(t) \cos nt \, dt \quad (n = 0, 1, 2, \ldots),$$

and similarly, $b_n = b_n'$ for all n. If $f(t) \equiv 0$, then all coefficients are zero.

This theorem is the basis of the method of undetermined coefficients for solution of problems (e.g., differential equations) by Fourier series. We replace an unknown periodic function by a Fourier series with coefficients to be determined. We can then compare coefficients on both sides of an equation, as in the theorem, and thereby determine the coefficients for the unknown function.

Theorem 9. *Let the function f(t) be continuous for $-\pi \leq t \leq \pi$ and let the Fourier series of f(t) converge uniformly in this interval. Then the series converges to f(t) for $-\pi \leq t \leq \pi$.*

Proof. Let the sum of the Fourier series of $f(t)$ be denoted by $f_1(t)$:

$$f_1(t) = \tfrac{1}{2}a_0 + \sum_{n=1}^{\infty} (a_n \cos nt + b_n \sin nt).$$

Since the series converges uniformly, it follows from Theorem 7 that $f_1(t)$ is continuous and that a_n, b_n are the Fourier coefficients of $f_1(t)$. But the series is given as the Fourier series of $f(t)$. Hence $f(t)$ and $f_1(t)$ have the same Fourier coefficients and, by Theorem 6, $f(t) = f_1(t)$; that is, $f(t)$ is the sum of its Fourier series for $-\pi \leq t \leq \pi$.

Theorem 10. *Let f(t) have period 2π and have continuous derivatives through the second order for all t. Then the Fourier series of f(t) converges uniformly to f(t) for all t.*

Proof. By integration by parts

$$b_n = \frac{1}{\pi} \int_{-\pi}^{\pi} f(t) \sin nt \, dt = \left. \frac{-f(t) \cos nt}{n\pi} \right|_{-\pi}^{\pi} + \frac{1}{n\pi} \int_{-\pi}^{\pi} f'(t) \cos nt \, dt.$$

The first term on the right is zero because of the periodicity of f. Similarly, a second integration by parts gives

$$b_n = - \frac{1}{n^2 \pi} \int_{-\pi}^{\pi} f''(t) \sin nt \, dt.$$

The function $f''(t)$ is continuous in the interval $-\pi \le t \le \pi$ and hence $|f''(t)| \le M$ for an appropriate constant M. We conclude that

$$|b_n| = \left| \frac{1}{n^2 \pi} \int_{-\pi}^{\pi} f''(t) \sin nt \, dt \right| \le \frac{2M}{n^2}.$$

In the same way we prove that $|a_n| \le 2M/n^2$ for $n = 1, 2, \ldots$ Hence each term of the Fourier series of $f(t)$ is in absolute value at most equal to the corresponding term of the convergent series of constants

$$\tfrac{1}{2}|a_0| + \left(\frac{2M}{1} + \frac{2M}{1} \right) + \left(\frac{2M}{2^2} + \frac{2M}{2^2} \right) + \cdots$$

Application of the Weierstrass M-test (Theorem 1 above) now establishes that the Fourier series converges uniformly for all t. By Theorem 9, the sum is $f(t)$.

Remark. The conclusion of the theorem can be established under much weaker assumptions. In particular it holds if $f(t)$ is continuous and only "piecewise smooth" (see the next section). If $f(t)$ is merely piecewise smooth, we can establish uniform convergence over each closed interval containing no discontinuity. (See References 3, 5, and 7.)

4-7 Convergence of Fourier series at a point. If $f(t)$ is periodic and merely continuous, we cannot conclude that its Fourier series converges for all t (or even, as far as is known, for some t). However, if, for a particular t_0, $f'(t_0)$ exists, then it can be shown that the series converges to $f(t_0)$ when $t = t_0$. Thus the behavior of the series for a particular value of t depends only on the behavior of $f(t)$ near that value. Away from that value, $f(t)$ can even have discontinuities. If $f(t)$ has a jump discontinuity at t_0, then the series converges to the average value

$$\tfrac{1}{2}[f(t_0+) + f(t_0-)] = \tfrac{1}{2}[\lim_{t \to t_0+} f(t) + \lim_{t \to t_0-} f(t)],$$

provided the limiting slopes to left and right exist:

$$f'_-(t_0) = \lim_{t \to t_0-} \frac{f(t_0-) - f(t)}{t_0 - t}, \qquad f'_+(t_0) = \lim_{t \to t_0+} \frac{f(t) - f(t_0+)}{t - t_0}.$$

If $f(t)$ is continuous at t_0, these limits can be interpreted as left- and right-side derivatives at t_0.

The proofs of these assertions will be obtained on the basis of several lemmas, which will also be of value in the theory of Fourier integrals.

LEMMA 1. *Let $g(t)$ be piecewise continuous for $a \leqq t \leqq b$. Then*

$$\lim_{u \to \infty} \int_a^b g(t) \sin ut \, dt = 0. \tag{4–70}$$

Proof. First we remark that we can assume that $g(t)$ is continuous for $a \leqq t \leqq b$. For if $g(t)$ is only piecewise continuous, the integral in (4–70) can be represented as a sum of a finite number of such integrals over subintervals of the interval $[a, b]$, with continuous integrands. If each term approaches zero as $u \to \infty$, then the same applies to the sum.

Next we note that (4–70) is valid when $g(t)$ is a linear function $At + B$, for the value of the integral is found to be

$$\frac{1}{u} [(Aa + B) \cos ua - (Ab + B) \cos ub] + \frac{A}{u^2} (\sin ub - \sin ua),$$

which does have limit zero as $u \to \infty$.

Now let $g(t)$ be continuous for $a \leqq t \leqq b$. By the result of the preceding paragraph, the limit in (4–70) is unaffected if we add a linear function to $g(t)$. By appropriate choice of this linear function, we can restrict our attention to the case of a function $g(t)$ for which $g(a) = g(b) = 0$. Furthermore we make $g(t)$ continuous for all t by setting $g(t) = 0$ for $t < a$ and for $t > b$. We assume $u > \pi$ and set $v = t - (\pi/u)$. Thus

$$\int_a^b g(t) \sin ut \, dt = \int_{a-(\pi/u)}^{b-(\pi/u)} g\left(v + \frac{\pi}{u}\right) \sin (uv + \pi) \, dv$$

$$= -\int_{a-1}^b g\left(v + \frac{\pi}{u}\right) \sin uv \, dv,$$

since $g[v + (\pi/u)] = 0$ for $v > b - (\pi/u)$ and for $v < a - 1$. Accordingly,

$$\int_a^b g(t) \sin ut \, dt = -\int_{a-1}^b g\left(t + \frac{\pi}{u}\right) \sin ut \, dt.$$

Since $g(t) = 0$ for $t < a$ and for $t > b$, we have

$$\int_a^b g(t) \sin ut \, dt = \int_{a-1}^b g(t) \sin ut \, dt.$$

If we add the last two equations and divide by 2, we find

$$\int_a^b g(t) \sin ut\, dt = \tfrac{1}{2}\int_{a-1}^b \left[g(t) - g\left(t + \frac{\pi}{u}\right)\right] \sin ut\, dt.$$

Accordingly,

$$\left|\int_a^b g(t) \sin ut\, dt\right| \leq \tfrac{1}{2}\int_{a-1}^b \left|g(t) - g\left(t + \frac{\pi}{u}\right)\right|\, dt$$

$$\leq \tfrac{1}{2}\int_{a-1}^b |g(t) - g(t + r)|\, dt, \qquad (4\text{--}71)$$

where $r = \pi/u$. The last integral is of the form $\int_{a-1}^b F(t, r)\, dt$, where $F(t, r)$ is continuous for all t, r. Accordingly, the integral is a continuous function of r (Chapter VI of Reference 4) and has limit $\int_{a-1}^b F(t, 0)\, dt$ as $r \to 0$. Hence

$$\lim_{u \to \infty} \tfrac{1}{2}\int_{a-1}^b \left|g(t) - g\left(t + \frac{\pi}{u}\right)\right|\, dt = \lim_{r \to 0} \tfrac{1}{2}\int_{a-1}^b |g(t) - g(t + r)|\, dt$$

$$= \int_{a-1}^b 0 \cdot dt = 0$$

and thus by (4-71)

$$\lim_{u \to \infty} \left|\int_a^b g(t) \sin ut\, dt\right| = 0.$$

Remark. The lemma is known as the theorem of Riemann-Lebesgue. It holds equally well with sin replaced by cos, and we can conclude that if $g(t)$ is piecewise continuous for $-\pi \leq t \leq \pi$, then as $n \to \infty$ *the Fourier coefficients* a_n, b_n *of* $g(t)$ *tend to zero.*

We now introduce the function

$$P_n(s) = \tfrac{1}{2} + \cos s + \cdots + \cos ns, \qquad (4\text{--}72)$$

whose importance will soon be seen. We can write

$$P_n(s) = \tfrac{1}{2}(1 + e^{is} + e^{-is} + \cdots + e^{ins} + e^{-ins}), \qquad (4\text{--}73)$$

from which it follows (Problem 10 below) that, for $s \neq 0, \pm 2\pi, \ldots$,

$$P_n(s) = \frac{\sin\,(n + \tfrac{1}{2})s}{2 \sin \tfrac{1}{2}s}. \qquad (4\text{--}74)$$

From (4-72) we deduce

$$\frac{1}{\pi} \int_0^\pi P_n(s)\, ds = \frac{1}{\pi} \int_{-\pi}^0 P_n(s)\, ds = \tfrac{1}{2}. \qquad (4\text{--}75)$$

LEMMA 2. *Let $f(t)$ be piecewise continuous for $-\pi \leqq t \leqq \pi$ and let $f(t)$ have period 2π. Let $S_n(t)$ denote the n-th partial sum of the Fourier series of $f(t)$:*

$$S_n(t) = \tfrac{1}{2}a_0 + a_1 \cos t + b_1 \sin t + \cdots + a_n \cos nt + b_n \sin nt.$$

Then

$$S_n(t) = \frac{1}{\pi} \int_{-\pi}^{\pi} f(t+s)P_n(s)\, ds. \qquad (4\text{--}76)$$

Proof. By the definition of the Fourier coefficients, we can write

$$S_n(t) = \frac{1}{2\pi} \int_{-\pi}^{\pi} f(u)\, du + \frac{\cos t}{\pi} \int_{-\pi}^{\pi} f(u) \cos u\, du$$

$$+ \frac{\sin t}{\pi} \int_{-\pi}^{\pi} f(u) \sin u\, du + \cdots$$

$$= \frac{1}{\pi} \int_{-\pi}^{\pi} f(u)[\tfrac{1}{2} + \cos(u - t) + \cos 2(u - t) + \cdots + \cos n(u - t)]\, du$$

$$= \frac{1}{\pi} \int_{-\pi}^{\pi} f(u)P_n(u - t)\, du = \frac{1}{\pi} \int_{-\pi}^{\pi} f(t+s)P_n(s)\, ds.$$

The last step follows from the substitution $s = u - t$ and the periodicity of $f(t)$.

DEFINITION. The function $f(t)$ is *piecewise smooth* over the interval $a \leqq t \leqq b$ if the interval can be divided into a finite number of sub-intervals $a \leqq t \leqq t_1, t_1 \leqq t \leqq t_2, \ldots, t_n \leqq t \leqq b$, in each of which $f(t)$ coincides, except perhaps at the endpoints, with a function having a continuous derivative in the whole subinterval.

For example, if $f(t) = t$ for $0 \leqq t < 1$, and $f(t) = -t$ for $1 \leqq t \leqq 2$, then $f(t)$ is piecewise smooth. There is a jump discontinuity at $t = 1$, but $f(t)$ coincides with the continuously differentiable function t for $0 \leqq t \leqq 1$, except at $t = 1$, and with the continuously differentiable function $-t$ for $1 \leqq t \leqq 2$.

If $f(t)$ is piecewise smooth, then $f'_+(t)$, $f'_-(t)$ exist at the subdivision points $t_1, t_2, \ldots, t_n$; they also exist and equal the ordinary derivative $f'(t)$ at the points interior to the subdivision intervals.

THEOREM 11. *Let $f(t)$ be piecewise smooth for $-\pi \leqq t \leqq \pi$ and let $f(t)$ have period 2π. Let $f(t)$ equal the average of left and right limits at each discontinuity. Then the Fourier series of $f(t)$ converges to $f(t)$ for all t.*

Proof. By assumption, for each discontinuity t,

$$f(t) = \tfrac{1}{2}[f(t+) + f(t-)].$$

This equation remains true if f is continuous at t, for then $f(t+) = f(t-) = f(t)$. From (4–75) and (4–76) we can write, for fixed t,

$$\tfrac{1}{2}f(t+) = \frac{1}{\pi} \int_0^\pi f(t+)P_n(s)\,ds,$$

$$\tfrac{1}{2}f(t-) = \frac{1}{\pi} \int_{-\pi}^0 f(t-)P_n(s)\,ds,$$

$$S_n(t) = \frac{1}{\pi} \int_0^\pi f(t+s)P_n(s)\,ds + \frac{1}{\pi} \int_{-\pi}^0 f(t+s)P_n(s)\,ds.$$

Hence

$$S_n(t) - f(t) = \frac{1}{\pi} \int_0^\pi [f(t+s) - f(t+)]P_n(s)\,ds$$

$$+ \frac{1}{\pi} \int_{-\pi}^0 [f(t+s) - f(t-)]P_n(s)\,ds. \qquad (4\text{–}77)$$

Consider the first integral on the right. It can be written as

$$\frac{1}{\pi} \int_0^\pi \frac{f(t+s) - f(t+)}{s}\, sP_n(s)\,ds.$$

Now the function

$$F(s) = \frac{f(t+s) - f(t+)}{s} \qquad (0 < s \leqq \pi)$$

is piecewise continuous for $s > 0$ and has a limit as $s \to 0+$, which is precisely $f'_+(t)$. If we assign this value to F at $s = 0$, then $F(s)$ becomes continuous at $s = 0$ and piecewise continuous for $s \geqq 0$. The integral is, by (4–74), now equal to

$$\frac{1}{\pi} \int_0^\pi F(s)sP_n(s)\,ds = \frac{1}{\pi} \int_0^\pi F(s) \frac{s}{2 \sin \tfrac{1}{2}s} \sin (n + \tfrac{1}{2})s\,ds.$$

The function $s/(2 \sin \tfrac{1}{2}s)$ is also continuous for $0 \leqq s \leqq \pi$ if we assign the value

$$1 = \lim_{s \to 0} \frac{s}{2 \sin \tfrac{1}{2}s}$$

at $s = 0$. Hence our integral is of the form

$$\int_0^\pi g(s) \sin (n + \tfrac{1}{2})s\,ds, \qquad (4\text{–}78)$$

where $g(s)$ satisfies the conditions of Lemma 1. Hence

$$\int_0^\pi g(s) \sin us \, ds \to 0 \qquad \text{as} \qquad u \to \infty,$$

and in particular the integral (4–78) tends to zero as $n \to \infty$. A similar reasoning applies to the second integral in (4–77). Therefore

$$\lim_{n \to \infty} [S_n(t) - f(t)] = 0$$

for all t, and the theorem is proved.

Remark. The proof remains valid if we assume only that $f(t)$ is piecewise continuous and that $f'_+(t)$, $f'_-(t)$ exist at the particular t considered; that is, the series converges to $f(t) = \frac{1}{2}[f(t+) + f(t-)]$ wherever $f'_+(t)$ and $f'_-(t)$ exist.

4–8 Convergence in the mean. The convergence of Fourier series can be considered from still another point of view, in some ways more natural than those discussed above.

Let $f_n(t)$ be a sequence of functions defined for $a \leqq t \leqq b$ and all piecewise continuous in this interval. We say that the sequence converges *in the mean* to a function $f(t)$, piecewise continuous for $a \leqq t \leqq b$, if

$$\lim_{n \to \infty} \int_a^b [f_n(t) - f(t)]^2 \, dt = 0. \qquad (4\text{–}80)$$

Thus the *mean-square error*

$$\frac{1}{b-a} \int_a^b [f_n(t) - f(t)]^2 \, dt$$

tends to zero as $n \to \infty$.

It can be shown that precisely this type of convergence holds for the Fourier series of a piecewise continuous function; that is, the sequence $S_n(t)$ converges in the mean to $f(t)$ on the interval $-\pi \leqq t \leqq \pi$. Furthermore, the Fourier coefficients are precisely those constants which make the expression

$$\frac{a_0}{2} + a_1 \cos t + \cdots + a_n \cos nt + b_n \sin nt$$

the best possible approximation in the mean (least-squares approximation) to $f(t)$ on the interval (Problem 9 below and Problem 11 following Section 4–18).

4–9 Differentiation and integration of Fourier series. We remark first of all that differentiation term by term of a trigonometric series

$$\frac{a_0}{2} + \Sigma(a_n \cos nt + b_n \sin nt)$$

multiplies the coefficients a_n, b_n by $\pm n$. Hence differentiation tends to slow down convergence and may even destroy it. On the other hand, integration has the effect of division of a_n, b_n by $\pm n$ and hence *improves* convergence. In fact, we have the following striking rule:

THEOREM 12. *Let $f(t)$ be piecewise continuous for $-\pi \leqq t \leqq \pi$, let $f(t)$ have period 2π, and let $f(t)$ have Fourier coefficients a_n, b_n. Then*

$$\int_{t_1}^{t_2} f(t)\, dt = \frac{a_0}{2} \int_{t_1}^{t_2} dt + a_1 \int_{t_1}^{t_2} \cos t\, dt + \cdots$$

$$= \frac{a_0}{2}(t_2 - t_1) + a_1(\sin t_2 - \sin t_1) + \cdots;$$

that is, the Fourier series of $f(t)$ can be integrated term by term.

Proof. Let

$$F(t) = \int_0^t f(u)\, du - \frac{a_0}{2}\, t.$$

Then $F(t)$ has period 2π (Problem 1 below). Furthermore $F(t)$ is continuous, and $F'(t)$ exists, equal to $f(t)$, except at the jumps of $f(t)$, where derivatives to the left and right exist. Hence $F(t)$ is piecewise smooth. By Theorem 11 the Fourier series of $F(t)$ converges to $F(t)$ for all t. Let A_n, B_n be the Fourier coefficients for $F(t)$. Then

$$A_n = -\frac{b_n}{n}, \qquad B_n = \frac{a_n}{n}, \qquad \text{for } n = 1, 2, \ldots$$

For

$$A_n = \frac{1}{\pi} \int_0^{2\pi} F(t) \cos nt\, dt = \frac{1}{\pi}\left[\frac{F(t) \sin nt}{n}\bigg|_0^{2\pi} - \frac{1}{n}\int_0^{2\pi} F'(t) \sin nt\, dt\right]$$

$$= -\frac{1}{n\pi}\int_0^{2\pi} f(t) \sin nt\, dt = -\frac{b_n}{n}.$$

The proof of $B_n = a_n/n$ is similar. Hence

$$F(t) = \frac{A_0}{2} + \sum_{n=1}^{\infty} (A_n \cos nt + B_n \sin nt)$$

$$= \frac{A_0}{2} + \sum_{n=1}^{\infty} \left(-b_n \frac{\cos nt}{n} + \frac{a_n \sin nt}{n}\right).$$

Now
$$F(t_2) - F(t_1) = \int_{t_1}^{t_2} f(u) \, du - \frac{a_0}{2} (t_2 - t_1).$$

Hence

$$\int_{t_1}^{t_2} f(t) \, dt = F(t_2) - F(t_1) + \frac{a_0}{2} (t_2 - t_1) = \frac{a_0}{2} (t_2 - t_1)$$
$$+ \sum_{n=1}^{\infty} \left\{ -\frac{b_n}{n} (\cos nt_2 - \cos nt_1) + \frac{a_n}{n} (\sin nt_2 - \sin nt_1) \right\}.$$

Thus the theorem is proved.

It should be remarked that it is not assumed (nor could it be proved from the hypotheses) that the Fourier series of $f(t)$ converges to $f(t)$.

PROBLEMS

1. Let $f(t)$ be piecewise continuous for $-\pi \leq t \leq \pi$ and let $f(t)$ have period 2π.

   (a) Prove that $\int_{\alpha}^{\alpha+2\pi} f(t) \, dt$ is independent of α.

   (b) Let $g(t) = \int_0^t f(u) \, du$. Prove that $g(t)$ has period 2π if and only if

$$\int_0^{2\pi} f(t) \, dt = 0.$$

   (c) Let $F(t) = \int_0^t f(u) \, du - (a_0/2)t$, where $a_0 = (1/\pi)\int_0^{2\pi} f(t) \, dt$. Prove that $F(t)$ has period 2π.

2. (a) Prove that if $f(t)$ is *even* $[f(-t) = f(t)]$ and has period 2π, then

$$a_n = \frac{2}{\pi} \int_0^{\pi} f(t) \cos nt \, dt, \qquad b_n = 0.$$

   (b) Prove that if $f(t)$ is *odd* $[f(-t) = -f(t)]$ and has period 2π, then $a_n = 0$ and

$$b_n = \frac{2}{\pi} \int_0^{\pi} f(t) \sin nt \, dt.$$

   (c) Expand $|\sin t|$ in a Fourier series.

   (d) Expand $\sin t \, |\sin t|$ in a Fourier series.

3. (a) Let $f(t)$ be defined for $0 \leq t \leq \pi$. The *Fourier cosine series* of $f(t)$ is defined as the series

$$\tfrac{1}{2}a_0 + \sum_{n=1}^{\infty} a_n \cos nt,$$

where

$$a_n = \frac{2}{\pi} \int_0^{\pi} f(t) \cos nt \, dt.$$

Show that if $f(t)$ is piecewise smooth and continuous, the Fourier cosine series of $f(t)$ converges to $f(t)$ for $0 \leq t \leq \pi$. [*Hint:* Extend the definition of f to all t to obtain an *even* function of period 2π. Expand in a Fourier series and use Problem 2(a).]

(b) Let $f(t)$ be defined for $0 \leq t \leq \pi$. The *Fourier sine* series of $f(t)$ is defined as the series $\sum_{n=1}^{\infty} b_n \sin nt$, where

$$b_n = \frac{2}{\pi} \int_0^\pi f(t) \sin nt \, dt.$$

Show that if $f(t)$ is piecewise smooth and continuous, the Fourier sine series of $f(t)$ converges to $f(t)$ for $0 \leq t \leq \pi$ except perhaps at $t = 0$ and $t = \pi$, where the series converges to zero.

4. Expand each of the following in a Fourier cosine series and graph the sum of the series:

(a) $f(t) = t$ $(0 \leq t \leq \pi)$ (b) $f(t) = e^t$ $(0 \leq t \leq \pi)$

5. Expand each of the following in a Fourier sine series and graph the sum of the series:

(a) $f(t) = \cos t$ $(0 \leq t \leq \pi)$ (b) $f(t) = 1 - t$ $(0 \leq t \leq \pi)$

6. For each of the following verify that the Fourier series converge uniformly for all t (with the aid of the M-test):

(a) series of Example 1 in Section 4–3;

(b) series of Problem 4(a).

7. A function $f(t)$ is called piecewise *very smooth* for $a \leq t \leq b$ if the interval can be divided into a finite number of subintervals in each of which $f(t)$ coincides, except perhaps at the ends, with a function having continuous first and second derivatives. Prove that if $f(t)$ is continuous and piecewise very smooth for $-\pi \leq t \leq \pi$ and $f(t)$ has period 2π, then the Fourier series of $f(t)$ converges uniformly to $f(t)$ for all t. [*Hint:* Modify the proof of Theorem 10.]

8. Verify that the following functions are piecewise smooth. Find $f(1+)$, $f(1-)$, $f'_+(1)$, $f'_-(1)$ for each of the following:

(a) $f(t) = t$ $(0 \leq t \leq 1)$, $f(t) = 2t$ $(1 < t \leq 2)$
(b) $f(t) = 1 - t^2$ $(0 \leq t < 1)$, $f(1) = 1$, $f(t) = 3 - t^2$ $(1 < t \leq 2)$

9. Let $f(t)$ be piecewise continuous for $-\pi \leq t \leq \pi$. Let $g_n(t) = \frac{1}{2}A_0 + A_1 \cos t + B_1 \sin t + \cdots + A_n \cos nt + B_n \sin nt$, where the A's and B's are constants. Let K_n be the square error

$$K_n = \int_{-\pi}^{\pi} [f(t) - g_n(t)]^2 \, dt.$$

(a) Show that

$$K_n = \int_{-\pi}^{\pi} [f(t)]^2 \, dt - \pi \left(\frac{a_0^2}{2} + a_1^2 + \cdots + b_n^2 \right)$$
$$+ \pi \left[2 \left(\frac{A_0}{2} - \frac{a_0}{2} \right)^2 + (A_1 - a_1)^2 + \cdots + (B_n - b_n)^2 \right],$$

where a_k, b_k are the Fourier coefficients of f.

(b) Show that K_n has its minimum value E_n when $A_0 = a_0$ and $A_k = a_k$, $B_k = b_k$ for $k = 1, \ldots, n$, and show that

$$E_n = \int_{-\pi}^{\pi} [f(t)]^2 \, dt - \pi \left(\frac{a_0^2}{2} + a_1^2 + \cdots + b_n^2 \right).$$

(c) Show that *Bessel's inequality* holds:

$$\tfrac{1}{2} a_0^2 + \sum_{k=1}^{n} (a_k^2 + b_k^2) \leq \frac{1}{\pi} \int_{-\pi}^{\pi} [f(t)]^2 \, dt.$$

(d) Show that $a_n \to 0$ and $b_n \to 0$ as $n \to \infty$. Relate this result to Lemma 1 above.

Remark. We can prove that $E_n \to 0$ as $n \to \infty$, and hence that

$$\int_{-\pi}^{\pi} [f(t)]^2 \, dt = \pi \frac{a_0^2}{2} + \pi \sum_{n=1}^{\infty} (a_n^2 + b_n^2).$$

This equation is known as *Parseval's relation* (see Problem 11 following Section 4–18).

10. Prove (4–74). [*Hint:* Let $r = e^{is}$ and treat (4–73) as the sum of a geometric progression

$$a + ar + \cdots + ar^N = a \frac{1 - r^{N+1}}{1 - r}.]$$

ANSWERS

2. (c) $\dfrac{2}{\pi} - \dfrac{4}{\pi} \displaystyle\sum_{n=1}^{\infty} \dfrac{\cos 2nt}{4n^2 - 1}$ (d) $\dfrac{8}{\pi} \displaystyle\sum_{n=1}^{\infty} \dfrac{\sin (2n - 1)t}{-8n^3 + 12n^2 + 2n - 3}$

4. (a) $\dfrac{\pi}{2} - \dfrac{4}{\pi} \left(\cos t + \dfrac{\cos 3t}{3^2} + \dfrac{\cos 5t}{5^2} + \cdots \right)$

   (b) $\dfrac{e^{\pi} - 1}{\pi} + \dfrac{2}{\pi} \displaystyle\sum_{n=1}^{\infty} \dfrac{e^{\pi}(-1)^n - 1}{n^2 + 1} \cos nt$

5. (a) $\dfrac{2}{\pi} \displaystyle\sum_{n=2}^{\infty} \dfrac{n(1 + (-1)^n)}{n^2 - 1} \sin nt$ (b) $\dfrac{2}{\pi} \displaystyle\sum_{n=1}^{\infty} \dfrac{1 + (-1)^n(\pi - 1)}{n} \sin nt$

8. (a) $f(1+) = 2$, $f(1-) = 1$, $f'_+(1) = 2$, $f'_-(1) = 1$
   (b) $f(1+) = 2$, $f(1-) = 0$, $f'_+(1) = -2$, $f'_-(1) = -2$

4–10 Change of period. As remarked earlier, it is sufficient to consider Fourier series for functions of period 2π, the general case being reducible to this one by a simple substitution. However it is convenient to state the main conclusions for a function of general period $\tau > 0$.

Let $f(t)$ have period τ. Then

$$g(u) = f\left(\frac{\tau}{2\pi} u\right)$$

has period 2π; for when u increases by 2π, $t = \tau u/(2\pi)$ increases by τ. Associated with $g(u)$ is its Fourier series

$$\frac{a_0}{2} + \sum_{n=1}^{\infty} (a_n \cos nu + b_n \sin nu),$$

which converges to $g(u)$ under appropriate hypotheses. The substitution $u = 2\pi t/\tau = \omega t$ yields the *Fourier series* of f:

$$\frac{a_0}{2} + \sum_{n=1}^{\infty} (a_n \cos n\omega t + b_n \sin n\omega t) \quad \left(\omega = \frac{2\pi}{\tau}\right). \qquad (4\text{-}100)$$

The coefficients can be defined by the formulas

$$a_n = \frac{1}{\pi} \int_0^{2\pi} g(u) \cos nu\, du, \qquad b_n = \frac{1}{\pi} \int_0^{2\pi} g(u) \sin nu\, du.$$

If we set $u = \omega t$, this becomes (Problem 1 following Section 4-12)

$$a_n = \frac{2}{\tau} \int_0^{\tau} f(t) \cos n\omega t\, dt, \qquad b_n = \frac{2}{\tau} \int_0^{\tau} f(t) \sin n\omega t\, dt. \qquad (4\text{-}101)$$

These are called the *Fourier coefficients* of $f(t)$ (for period τ). In (4-101) we can replace the interval of integration by $(-\tau/2, \tau/2)$ or any interval of length τ.

We can now restate Theorems 6 through 12 for the period τ: $f(t)$ is uniquely determined by its Fourier coefficients; a uniformly convergent series of the form (4-100) is the Fourier series of its sum; the coefficients of a uniformly convergent series of the form (4-100) are uniquely determined by its sum; if the Fourier series of $f(t)$ converges uniformly for all t, then it converges to $f(t)$; the Fourier series of $f(t)$ converges uniformly to $f(t)$ if $f''(t)$ is continuous and $f(t)$ has period τ; the Fourier series converges to $f(t)$ if $f(t)$ is piecewise smooth and has period τ, and if $f(t) = \frac{1}{2}[f(t+) + f(t-)]$ at every discontinuity; the Fourier series of a piecewise continuous function can be integrated term by term.

4-11 Complex form of Fourier series. As in Section 4-2, we use the substitutions

$$\cos n\omega t = \frac{e^{in\omega t} + e^{-in\omega t}}{2}, \qquad \sin n\omega t = \frac{e^{in\omega t} - e^{-in\omega t}}{2i}$$

to write the Fourier series (4–100) in complex form. The series becomes

$$\frac{a_0}{2} + \sum_{n=1}^{\infty} (c_n e^{in\omega t} + c'_n e^{-in\omega t}),$$

(4–110)

$$c_n = \frac{a_n - ib_n}{2}, \qquad c'_n = \frac{a_n + ib_n}{2} = \bar{c}_n.$$

If we write

$$c_0 = \frac{a_0}{2}, \qquad c_{-n} = c'_n \quad (n = 1, 2, \ldots),$$

(4–111)

then (4–110) becomes

$$c_0 + \sum_{n=1}^{\infty} (c_n e^{in\omega t} + c_{-n} e^{-in\omega t}) = \sum_{n=-\infty}^{\infty} c_n e^{in\omega t}.$$

(4–112)

We can give one general formula for the c_n:

$$c_n = \frac{1}{\tau} \int_0^\tau f(t) e^{-in\omega t} \, dt.$$

(4–113)

For the integral on the right is

$$\frac{1}{\tau} \int_0^\tau f(t) \, (\cos n\omega t - i \sin n\omega t) \, dt.$$

When $n = 0$ this gives $a_0/2$ [see (4–101)]; when $n > 0$ the expression equals $\frac{1}{2}(a_n - ib_n)$; when $n < 0$ it equals $\frac{1}{2}(a_{-n} + ib_{-n})$, in agreement with (4–111) and (4–110). Hence *the complex Fourier series and Fourier coefficients of $f(t)$ are defined as the expressions*

$$\sum_{-\infty}^{\infty} c_n e^{in\omega t}, \qquad c_n = \frac{1}{\tau} \int_0^\tau f(t) e^{-in\omega t} \, dt,$$

(4–114)

where the series is interpreted as in (4–112).

EXAMPLE. Let $f(t) = e^t$ for $0 \leq t < 1$ and let f have period 1. Here $\tau = 1, \omega = 2\pi/\tau = 2\pi$. Hence

$$c_n = \int_0^1 e^t e^{-2\pi i n t} \, dt = \frac{e^{t(1-2\pi i n)}}{1 - 2\pi i n} \Big|_0^1 = \frac{e^{1-2\pi i n} - 1}{1 - 2\pi i n} = \frac{e - 1}{1 - 2\pi i n},$$

$$f(t) = \sum_{n=-\infty}^{\infty} \frac{e - 1}{1 - 2\pi i n} e^{2\pi i n t} \quad (t \neq 0, \pm 1, \pm 2, \ldots).$$

At $0, \pm 1, \ldots, f(t)$ has jumps and the series converges to $\frac{1}{2}(1 + e)$.

Application to complex periodic functions. We have considered only the case when $f(t)$ is real-valued. However, essentially no change is required in the complex case. If $F(t) = f(t) + ig(t)$, where f and g are real and have period τ, then $F(t)$ also has period τ. The Fourier series of $F(t)$ is defined by the formulas (4–114), with f replaced by F. It follows that the complex Fourier coefficients of F are simply the numbers $C_n = c_n + id_n$, where c_n and d_n are the complex Fourier coefficients of $f(t)$, $g(t)$.

There is one significant difference between the complex case and the real case. For the real case, we have by (4–110) and (4–111), $c_{-n} = \bar{c}_n$. For the complex case,

$$C_{-n} = c_{-n} + id_{-n} = \bar{c}_n + i\bar{d}_n, \qquad \bar{C}_n = \bar{c}_n - i\bar{d}_n.$$

Hence $C_{-n} = \bar{C}_n$ only if $\bar{d}_n = -d_n$, that is, $d_n = 0$, so that $F(t)$ reduces to the real function $f(t)$.

The principal results stated for real Fourier series in the preceding section can be restated without change for the complex Fourier series of a complex function of t. In fact, by taking real and imaginary parts, the statements for the complex case reduce to those for the real case.

4–12 Orthogonal functions. The expansion in terms of the exponential functions $e^{in\omega t}$ in Section 4–11, as well as the Fourier sine series and Fourier cosine series of Problem 3 following Section 4–9, suggests that other sets of functions may be chosen for expansions like Fourier series. Confining our attention to real functions, we consider some illustrations here.

We call real functions $f(t)$, $g(t)$ *orthogonal* over an interval $a \leq t \leq b$ if

$$\int_a^b f(t)g(t)\, dt = 0. \tag{4–120}$$

For example, $\sin nt$ is orthogonal to $\sin mt$ for $m \neq n$ (m, n integers) over the interval $0 \leq t \leq 2\pi$. A sequence of real functions $\psi_n(t)$ ($n = 1, 2, \ldots$), $a \leq t \leq b$, is said to be an orthogonal sequence of functions if $\psi_n(t)$ and $\psi_m(t)$ are orthogonal whenever $n \neq m$. We assume all functions to be continuous and furthermore, to rule out trivial cases, that

$$k_n = \int_a^b \psi_n^2(t)\, dt > 0 \quad (n = 1, 2, \ldots) \tag{4–121}$$

If a series of form

$$\sum_{n=1}^{\infty} a_n \psi_n(t) \tag{4–122}$$

converges uniformly for $a \leq t \leq b$ to a function $f(t)$, then we can compute the coefficients just as for Fourier series. Thus

$$\int_a^b f(t)\psi_m(t)\, dt = \sum_{n=1}^\infty a_n \int_a^b \psi_n(t)\psi_m(t)\, dt$$

$$= a_m \int_a^b \psi_m^2(t)\, dt = k_m a_m$$

by orthogonality. Hence

$$a_n = \frac{1}{k_n} \int_a^b f(t)\psi_n(t)\, dt \quad (n = 1, 2, \ldots) \tag{4–123}$$

Given an arbitrary piecewise continuous function $f(t)$, $a \leqq t \leqq b$, and an orthogonal sequence $\psi_n(t)$ over this interval, we can compute the coefficients a_n by (4–123) and compute the series (4–122). This series may fail to converge, but even if it converges uniformly, we cannot be sure that the sum is $f(t)$ [see Problem 4(b) below]. The difficulty is that the sequence of ψ_n may not be *large* enough. An orthogonal sequence which is large enough so that, whenever the series (4–122) converges uniformly it converges to $f(t)$, is called *complete*. (For the theory of completeness, see Reference 6, pp. 417–423.)

A number of examples of complete orthogonal sequences are known. A few are mentioned here.

The *Legendre polynomials* $P_n(t)$:

$$P_n(t) = \frac{1}{2^n n!} D^n[(t^2 - 1)^n] \quad (n = 0, 1, 2, \ldots).$$

The interval is $-1 \leqq t \leqq 1$ and $k_n = 2/(2n + 1)$.

The functions $\sqrt{t}\, J_0(\lambda_n t)$ $(n = 1, 2, \ldots)$, where $J_0(x)$ is the *Bessel function* of order zero:

$$J_0(x) = \sum_{n=0}^\infty \frac{(-1)^n x^{2n}}{2^{2n}(n!)^2},$$

and where $\lambda_1, \lambda_2, \ldots$ are the positive zeros of $J_0(x)$ arranged in increasing order. The interval is $0 \leqq t \leqq 1$ and $k_n = \frac{1}{2}[J_0'(\lambda_n)]^2$.

The functions $e^{-t/2} L_n(t)$, where $L_n(t)$ is the nth *Laguerre polynomial*:

$$L_n(t) = \frac{1}{n!} e^t D^n(e^{-t} t^n) \quad (n = 0, 1, 2, \ldots).$$

The interval is $0 \leqq t < \infty$ and $k_n = 1$. (The infiniteness of the interval causes certain complications in the theory.)

These and many other orthogonal sequences arise naturally in analysis of the partial differential equations of mathematical physics. For further discussion refer to References 1, 3, 5, and 7.

PROBLEMS

1. In Section 4–10 the Fourier coefficients of a function $f(t)$ of period τ are defined to be the same as the Fourier coefficients of $g(u) = f(\tau u/[2\pi]) = f(u/\omega)$. Show that this leads to the formulas (4–101).

2. Let $f(t) = 1 - t^2$. Expand $f(t)$ in a Fourier series over each of the following intervals and graph the sum of the series in each case:

(a) $0 \leq t \leq 1$ (period 1)
(b) $-1 \leq t \leq 1$ (period 2)
(c) $-\frac{1}{2} \leq t \leq \frac{1}{2}$ (period 1)

3. For each of the following obtain the Fourier series of $f(t)$ in complex form:

(a) $f(t) = t,\ -\pi \leq t \leq \pi$ (period 2π)
(b) $f(t) = \sin^4 t,\ 0 \leq t \leq \pi$ (period π)
(c) $f(t) = e^{(1+2i)t},\ 0 \leq t \leq 1$ (period 1)

4. (a) Show that the functions $\psi_n = \sin nt\ (n = 1, 2, \ldots)$ form an orthogonal sequence for the interval $0 \leq t \leq 2\pi$ and find the corresponding k_n.

(b) Show that the sequence of part (a) is not complete; in particular, find a continuous function $f(t)$, not identically zero, orthogonal to all ψ_n. [The series $\sum a_n \psi_n(t)$ then converges uniformly to zero, not to $f(t)$.]

(c) Show generally that completeness, as defined in Section 4–12, is equivalent to the following condition: there is no continuous function $f(t)$, not identically zero, orthogonal to all ψ_n.

ANSWERS

2. (a) $\dfrac{2}{3} + \sum\limits_{n=1}^{\infty}\left(-\dfrac{1}{\pi^2 n^2}\cos 2\pi nt + \dfrac{1}{\pi n}\sin 2\pi nt\right)$

(b) $\dfrac{2}{3} - \dfrac{4}{\pi^2}\sum\limits_{n=1}^{\infty}(-1)^n\dfrac{\cos \pi nt}{n^2}$ (c) $\dfrac{11}{12} - \dfrac{1}{\pi^2}\sum\limits_{n=1}^{\infty}(-1)^n\dfrac{\cos 2\pi nt}{n^2}$

3. (a) $\sum\limits_{n=-\infty}^{\infty}{}'\dfrac{i(-1)^n}{n}e^{int}$ ($\sum'$ here denotes the sum over all n except 0)

(b) $\dfrac{1}{16}(e^{4it} - 4e^{2it} + 6 - 4e^{-2it} + e^{-4it})$

(c) $(e^{1+2i} - 1)\sum\limits_{n=-\infty}^{\infty}\dfrac{e^{2\pi int}}{1 + 2i(1 - \pi n)}$

4. (a) $k_n = \pi$

4–13 Solution of differential equations by Fourier series. Let $f(t)$ have period τ and be representable by its Fourier series $\sum c_n e^{in\omega t}$, $(\omega = 2\pi/\tau)$. Then, as remarked in Section 4–2, we can obtain a particular solution of a differential equation

$$(a_0 D^n + \cdots + a_n)x = f(t) = \sum_{n=-\infty}^{\infty} c_n e^{in\omega t} \qquad (4\text{--}130)$$

by means of the frequency response function:

$$x = \sum_{n=-\infty}^{\infty} c_n Y(in\omega)e^{in\omega t}. \tag{4–131}$$

This procedure is merely formal. However, if f is piecewise continuous, then we shall see that the series (4–131) does converge and define a periodic solution of (4–130).

EXAMPLE. Let $f(t)$ have period 2π and equal 1 for $0 \leq t < \pi$, and 0 for $\pi \leq t < 2\pi$. Thus f is a square wave. We find $c_0 = \frac{1}{2}$, $c_n = i[(-1)^n - 1]/(2\pi n)$ for $n \neq 0$. Hence the differential equation $(3D^2 + 2D + 4)x = f(t)$ has a particular solution

$$x = \sum_{n=-\infty}^{\infty} \frac{c_n e^{int}}{4 - 3n^2 + 2in} = \frac{1}{8} + \frac{i}{2\pi} {\sum_{n=-\infty}^{\infty}}' \frac{(-1)^n - 1}{n(4 - 3n^2 + 2in)} e^{int}.$$

Here $\sum'$ denotes a summation over n, omitting the value $n = 0$.

4–14 The finite Fourier transform. We can consider the problem of the preceding section in another way. The numbers c_n are the Fourier coefficients of $f(t)$:

$$c_n = \frac{1}{\tau} \int_0^\tau f(t)e^{-in\omega t}\, dt \quad (n = 0, \pm 1, \dots). \tag{4–140}$$

If $x(t)$ is the periodic solution sought, then $x(t)$ has Fourier coefficients k_n:

$$k_n = \frac{1}{\tau} \int_0^\tau x(t)e^{-in\omega t}\, dt \quad (n = 0, \pm 1, \dots). \tag{4–141}$$

Our procedure suggests the relationship

$$k_n = Y(in\omega)c_n. \tag{4–142}$$

Thus the Fourier coefficients of $x(t)$ are determined once the Fourier coefficients of $f(t)$ and the frequency response function are known. In Section 4–5 we saw that a function is uniquely determined except at jump points by its Fourier coefficients. Thus $x(t)$ is determined by $f(t)$ through

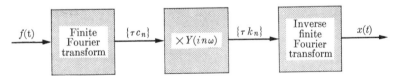

FIG. 4–5.   Input and output via finite Fourier transform.

the medium of Fourier coefficients but without reference to the corresponding Fourier series. The relationship is suggested schematically in Fig. 4–5.
For convenience, the sequences $\{c_n\}$, $\{k_n\}$ are replaced by the sequences
$\{\tau c_n\}$, $\{\tau k_n\}$. These sequences are the *finite Fourier transforms* of $f(t)$ and
$x(t)$, respectively.

The finite Fourier transform of a function $f(t)$ is thus a sequence $\{\tau c_n\}$;
that is, it is a *function of* n $(n = 0, \pm 1, \ldots)$, which we can denote by
$\phi(n)$:

$$\phi(n) = \tau c_n = \int_0^\tau f(t)e^{-in\omega t}\, dt. \tag{4–143}$$

We can describe the relationship symbolically:

$$\phi = \Phi_\tau[f], \tag{4–144}$$

where Φ_τ is an operator assigning a function ϕ (of n) to each function
f (of t). The operator Φ_τ is called the *finite Fourier transformation*, while
ϕ is the *finite Fourier transform* of f.

Given f, we find its finite Fourier transform ϕ by (4–143). Given ϕ, we
know f is uniquely determined (except at jump points); that is, there is a
transformation assigning f to ϕ. We call this transformation the *inverse
finite Fourier transformation* and denote it by Φ_τ^{-1}:

$$\Phi_\tau^{-1}[\phi] = f. \tag{4–145}$$

It may be possible to recover f from ϕ by means of Fourier series; other
methods are given below.

The transformation Φ_τ is *linear* in the linear space of all complex functions f piecewise continuous for $0 \leqq t \leqq \tau$. This follows at once from
(4–143). Accordingly, the inverse transformation Φ_τ^{-1} is linear in the
corresponding class of functions ϕ (see Section 2–2).

Examples of finite Fourier transforms. EXAMPLE 1. Let $f(t) = t$ for
$0 \leqq t < \tau$ and have period τ. Then for $n \neq 0$

$$\phi(n) = \int_0^\tau te^{-in\omega t}\, dt = \frac{1 - e^{-in\omega\tau} - in\omega\tau e^{-in\omega\tau}}{(in\omega)^2} = \frac{2\pi i}{n\omega^2},$$

while $\phi(0) = \tau^2/2$. (We have used the relations $\omega\tau = 2\pi$ and $e^{2\pi ik} = 1$,
which are valid when k is an integer.)

EXAMPLE 2. Let $f(t) = e^{at}$, where $a - in\omega \neq 0$ for $n = 0, \pm 1, \ldots$
Then

$$\Phi_\tau[f] = \phi, \qquad \phi(n) = \int_0^\tau e^{at}e^{-in\omega t}\, dt = \frac{e^{a\tau} - 1}{a - in\omega}.$$

EXAMPLE 3. $f(t) = 3t - 5e^{2t}$ The result can be computed from that of Examples 1 and 2 by linearity.

$$\Phi_\tau[f] = 3\Phi_\tau[t] - 3\Phi_\tau[e^{2t}] = \frac{6\pi i}{n\omega^2} - 5\frac{e^{2\tau} - 1}{2 - in\omega} \quad (n \neq 0),$$

$$\Phi_\tau[f] = \frac{3\tau^2}{2} - \frac{5}{2}(e^{2\tau} - 1) \quad (n = 0).$$

PROBLEMS

1. Obtain a periodic solution of each of the following differential equations, with the aid of the series expansion of the previous problem suggested:

(a) $(2D^2 + D + 2)x = f(t)$, $f(t)$ as in Problem 3(a) following Section 4–12.
(b) $(D^3 + 3D^2 + D + 1)x = f(t)$, $f(t)$ as in Problem 3(b) following Section 4–12.

2. Find the finite Fourier transforms, for $\tau = 2\pi$, of the following functions:

(a) $f = 1$, $0 \leq t \leq \pi$; $f = 0$, $\pi < t < 2\pi$
(b) $f = t$, $0 \leq t \leq \pi$; $f = 0$, $\pi < t < 2\pi$
(c) $f = 3 + 5t$, $0 \leq t \leq \pi$; $f = 0$, $\pi < t < 2\pi$

3. For each of the following, find the inverse finite Fourier transforms, for $\tau = 2\pi$, with the aid of the results of Problem 2:

(a) 2π for $n = 0$, $2[1 - (-1)^n]/(in)$ for $n \neq 0$
(b) $\pi^2/2$ for $n = 0$, $[1 - (-1)^n - \pi in]/n^2$ for $n \neq 0$

ANSWERS

1. (a) $\displaystyle\sum_{n=-\infty}^{\infty}{}' i(-1)^n e^{int}[n(2 - 2n^2 + in)]^{-1}$

(b) $\dfrac{1}{16}\left[\dfrac{e^{4it}}{-47 - 60i} + \dfrac{4e^{2it}}{11 + 6i} + 6 + \dfrac{4e^{-2it}}{11 - 6i} + \dfrac{e^{-4it}}{-47 + 60i}\right]$

2. (a) π for $n = 0$, $[1 - (-1)^n]/(in)$ for $n \neq 0$
(b) $\pi^2/2$ for $n = 0$, $[(-1)^n(1 + \pi in) - 1]/n^2$ for $n \neq 0$
(c) $3\pi + (5\pi^2/2)$ for $n = 0$, $[(-1)^n\{5 + in(5\pi + 3)\} - 5 - 3in]/n^2$ for $n \neq 0$

3. (a) 2 for $0 < t < \pi$, 0 for $\pi < t < 2\pi$
(b) $\pi - t$ for $0 < t < \pi$, 0 for $\pi < t < 2\pi$

4–15 The truncated Laplace transform as an aid to computation of finite Fourier transforms. From the examples of Section 4–14 it is clear that we can simplify the computation of finite Fourier transforms if we first compute

$$\int_0^\tau f(t)e^{-st}\,dt \tag{4–150}$$

and then replace s by $in\omega$. The expression (4–150) is called the *truncated Laplace transform* of f; it defines a function of (the complex variable) s, depending on the parameter τ. We write

$$F_\tau(s) = \int_0^\tau f(t)e^{-st}\, dt. \qquad (4\text{--}151)$$

Again, F is obtained from f by a linear transformation, the truncated Laplace transformation. We write $\mathcal{L}_\tau$ for this transformation:

$$\mathcal{L}_\tau[f] = F_\tau, \qquad F_\tau(s) = \int_0^\tau f(t)e^{-st}\, dt. \qquad (4\text{--}152)$$

The ordinary Laplace transformation corresponds to the limiting case $\tau = \infty$:

$$\mathcal{L}[f] = \mathcal{L}_\infty[f] = F, \qquad F(s) = \int_0^\infty f(t)e^{-st}\, dt.$$

This will be studied in detail in Chapter 6.

To obtain $\Phi_\tau[f]$ from $\mathcal{L}_\tau[f]$ we simply replace s by $in\omega$:

$$\text{if} \qquad \mathcal{L}_\tau[f] = F_\tau(s), \qquad \text{then} \qquad \Phi_\tau[f] = \phi(n) = F_\tau(in\omega).$$

Thus the operation is similar to that for passage from $Y(s)$ to $Y(in\omega)$. Here we start with a function of the complex variable s, $F_\tau(s)$, and restrict attention to its values $\phi(n)$ at the points $in\omega(n = 0, \pm 1 \ldots)$ on the imaginary axis.

In Table 4–1 are presented several truncated Laplace transforms.

Some points of interest arise from the table. The first entry is for $f(t) = h(t)$. The transform appears to have a discontinuity for $s = 0$. However, e^{-st} is a continuous function of s, and hence (4–151) *must define a continuous function of s for all s.* In fact we shall show in Chapter 6 that $F_\tau(s)$ can be represented by a convergent power series in s for all complex s. Hence the discontinuity must somehow be illusory. First, direct computation shows that when $s = 0$, $F = \tau$. Second,

$$\lim_{s\to 0} \frac{1 - e^{-s\tau}}{s} = \lim_{s\to 0} \frac{1 - [1 - st + (s^2\tau^2/2!) - \cdots]}{s} = \tau.$$

(By Section 3–1, the operations are valid for complex s.) Hence if we assign the proper limiting value τ at $s = 0$, the transform is continuous. We are dealing here with a *removable discontinuity* (Section 3–11). We could eliminate the difficulty by giving separately the value for $s = 0$. However, it is far simpler to *write all the transforms as given with removable discontinuities.* When $F_\tau(s)$ has to be evaluated at such a value of s, the discontinuity will be immediately evident, and there will be little trouble in finding the proper value, either from the definition of F or from a limit process.

TABLE 4–1

TRUNCATED LAPLACE TRANSFORMS

No.	$f(t)$	$\mathcal{L}_\tau[f] = \displaystyle\int_0^\tau f(t)e^{-st}\,dt$ $= F_\tau(s) \quad (\tau > 0)$
1	$h(t)$	$\dfrac{1 - e^{-s\tau}}{s}$
2	t	$\dfrac{1 - e^{-s\tau}(1 + s\tau)}{s^2}$
3	$t^k \quad (k = 1, 2, \ldots)$	$\dfrac{k! - e^{-s\tau}g_k(s\tau)}{s^{k+1}}$ (See Note)
4	e^{at}	$\dfrac{1 - e^{(a-s)\tau}}{s - a}$
5	$t^k e^{at} \quad (k = 1, 2, \ldots)$	$\dfrac{k! - e^{(a-s)\tau}g_k((s - a)\tau)}{(s - a)^{k+1}}$
6	$h(t - c) \quad (c > 0)$	$\dfrac{e^{-sc} - e^{-s\tau}}{s} h(\tau - c)$

Note: $g_k(x) = x^k + kx^{k-1} + k(k - 1)x^{k-2} + \cdots + k!$

Entry 2 of Table 4–1 presents the same difficulty at $s = 0$. By power series,

$$\frac{1 - e^{-s\tau} - s\tau e^{-s\tau}}{s^2} = \frac{1 - [1 - s\tau + (s^2\tau^2/2) - \cdots] - s\tau(1 - s\tau + \cdots)}{s^2}$$
$$= \frac{(s^2\tau^2/2) + \cdots}{s^2}.$$

Hence at $s = 0$, we assign the value $\tau^2/2$. Entries 4 and 5 present removable discontinuities at $s = a$.

In Entry 3, we use the abbreviation

$$g_k(x) = x^k + kx^{k-1} + k(k - 1)x^{k-2} + \cdots + k! \qquad (4\text{–}153)$$

Thus $g_k(x)$ is a polynomial of degree k, with simple properties (Problem 5 below). Entry 3 expanded states

$$\mathcal{L}_\tau[t^k] = \frac{k! - e^{-s\tau}[(s\tau)^k + k(s\tau)^{k-1} + \cdots + k!]}{s^{k+1}}. \qquad (4\text{–}154)$$

In Entry 6 the function f is $h(t - c)$, equal to 1 for $t \geqq c$, and equal to 0 for $t < c$. The transform involves $h(\tau - c)$; hence it is identically 0 if $c > \tau$.

With the aid of Table 4–1 we deduce a corresponding list of finite Fourier transforms shown in Table 4–2. In the third column of Table 4–2 the entries are obtained by simply replacing s by $in\omega$ in the corresponding expression in Table 4–1; in the fourth column, the values are spelled out in detail. In practice we can use the form in the third column, recalling that when zero appears in the denominator the value can be obtained by a limit process (with n treated as a continuous variable).

PROBLEMS

1. Verify the following entries of Table 4–1:

(a) No. 1   (b) No. 2   (c) No. 3   (d) No. 4   (e) No. 5   (f) No. 6

2. With the aid of Table 4–1, compute the truncated Laplace transforms of the following functions:

(a) $1 + 2t - t^2$ (b) $e^{-t}(2 - t)$ (c) $\sin bt, \cos bt$
(d) $f(t) = 0$ for $t < 0$, $f(t) = 1$ for $0 \leqq t < 1$, $f(t) = 2$ for $1 \leqq t < 2$, $f(t) = 0$ for $t \geqq 2$

3. With the aid of Table 4–2, compute the finite Fourier transforms of the following functions:

(a) $e^{-t} + 2e^{-2t}$, $\tau = 1$ (b) te^t, $\tau = 1$
(c) $f(t) = 0$ for $0 \leqq t < 1$, $f(t) = 1$ for $1 \leqq t < 2$, $f(t) = 0$ for $2 \leqq t < 3$, $\tau = 3$

4. Sum the following Fourier series:

(a) $\displaystyle\sum_{n=-\infty}^{\infty} \frac{e^{int}}{1 - in}$ (b) $\displaystyle\sum_{n=-\infty}^{\infty} \left(\frac{2}{1 + in} - \frac{3}{1 - in} \right) e^{int}$

[*Hint:* For (a) we have $\tau = 2\pi$, $\omega = 1$. If $f(t)$ is the sum, then $\Phi_{2\pi}[f] = 2\pi c_n = 2\pi/(1 - in)$.]

5. Prove the following properties of $g_k(x)$:

(a) $g'_k(x) = k g_{k-1}(x)$ (b) $g_k(x) = x^k + k g_{k-1}(x)$

ANSWERS

2. (a) $\dfrac{1}{s} + \dfrac{2}{s^2} - \dfrac{2}{s^3} + e^{-s\tau} \dfrac{2 + 2s(\tau - 1) + s^2(\tau^2 - 2\tau - 1)}{s^3}$

(b) $\dfrac{2s + 1 + e^{(-1-s)\tau}[s(\tau - 2) + \tau - 1]}{(s + 1)^2}$

TABLE 4-2

FINITE FOURIER TRANSFORMS

No.	$f(t)$	$\phi(n) = \Phi_\tau[f] = \int_0^\tau f(t)e^{-in\omega t}\,dt$	Remarks
1	1	$\dfrac{1 - e^{-2\pi i n}}{in\omega}$	$\phi(0) = \tau$ $\phi(n) = 0$ for $n \neq 0$
2	t	$\dfrac{1 - e^{-2\pi i n}(1 + 2\pi i n)}{(in\omega)^2}$	$\phi(0) = \tau^2/2$ $\phi(n) = 2\pi i/(n\omega^2)$ for $n \neq 0$
3	$t^k,$ $k = 1, 2, \ldots$	$\dfrac{k! - e^{-2\pi i n} g_k(2\pi i n)}{(in\omega)^{k+1}}$	$\phi(0) = \tau^{k+1}/(k+1)$ $\phi(n) = [k! - g_k(2\pi i n)]/(in\omega)^{k+1}$ for $n \neq 0$
4	$e^{at},$ $a - in\omega \neq 0$	$\dfrac{e^{(a-in\omega)\tau} - 1}{a - in\omega}$	$\phi(n) = \dfrac{e^{a\tau} - 1}{a - in\omega}$
5	$t^k e^{at},$ $a - in\omega \neq 0,$ $k = 1, 2, \ldots$	$\dfrac{k! - e^{(a-in\omega)\tau} g_k[(in\omega - a)\tau]}{(in\omega - a)^{k+1}}$	$\phi(n) = \dfrac{k! - e^{a\tau} g_k(2\pi i n - a\tau)}{(in\omega - a)^{k+1}}$
6	$h(t - c),$ $0 < c < \tau$	$\dfrac{e^{-in\omega c} - e^{-2\pi i n}}{in\omega}$	$\phi(0) = \tau - c$ $\phi(n) = \dfrac{e^{-in\omega c} - 1}{in\omega}$ for $n \neq 0$

7	$(at+b)[h(t-\alpha)-h(t-\beta)]$, $0 \le \alpha < \beta \le \tau$	$(n^2\omega^2)^{-1}\{e^{-\beta in\omega}[(a\beta+b)in\omega+a]$ $-e^{-\alpha in\omega}[(a\alpha+b)in\omega+a]\}$	$\phi(0) = \frac{1}{2}a(\beta^2-\alpha^2)+b(\beta-\alpha)$
8	$t[h(t)-h(t-\frac{1}{2}\tau)]$ $+(\tau-t)[h(t-\frac{1}{2}\tau)-h(t-\tau)]$	$\left(\dfrac{1-e^{-\pi in}}{in\omega}\right)^2$	$\phi(0) = \tau^2/4$ $\phi(n) = \dfrac{2[(-1)^n-1]}{n^2\omega^2}$ for $n \ne 0$
9	$t[1-h(t-\alpha)]+\alpha h(t-\alpha)$, $0 \le \alpha \le \tau$	$\dfrac{e^{-\alpha in\omega}+\alpha in\omega e^{-2\pi in}-1}{n^2\omega^2}$	$\phi(0) = \alpha\tau-\frac{1}{2}\alpha^2$ $\phi(n) = \dfrac{e^{-\alpha in\omega}+\alpha in\omega-1}{n^2\omega^2}$ for $n \ne 0$
10	$\tau-\alpha+(\alpha-t)h(t-\alpha)$, $0 \le \alpha \le \tau$	$\dfrac{e^{-in\omega\alpha}-e^{-2\pi in}-in\omega(\tau-\alpha)}{n^2\omega^2}$	$\phi(0) = \dfrac{\tau^2-\alpha^2}{2}$ $\phi(n) = \dfrac{e^{-in\omega\alpha}-1-2\pi in+in\omega\alpha}{n^2\omega^2}$ for $n \ne 0$
11	$[h(t-\alpha)-h(t-\beta)]\cdot$ $\sin(at+b)$, $0 \le \alpha < \beta \le \tau$, a, b real, $a-n\omega \ne 0$	$(a^2-n^2\omega^2)^{-1}[e^{-in\omega\alpha}(a\cos\theta+in\omega\sin\theta)$ $-e^{-in\omega\beta}(a\cos\mu+in\omega\sin\mu)]$, $\theta = a\alpha+b, \quad \mu = a\beta+b$	
12	$\left[h(t)-h\left(t-\dfrac{\pi}{a}\right)\right]\sin at$, $a \ne n\omega, \quad a > \pi/\tau$	$\dfrac{a}{a^2-n^2\omega^2}(1+e^{-in\omega\pi/a})$	

(c) $\dfrac{b - e^{-s\tau}(s \sin b\tau + b \cos b\tau)}{s^2 + b^2}$, $\dfrac{s + e^{-s\tau}(b \sin b\tau - s \cos b\tau)}{s^2 + b^2}$

(d) $\dfrac{1}{s}[(1 - e^{-s\tau}) + (e^{-s} - e^{-s\tau})h(\tau - 1) - 2(e^{-2s} - e^{-s\tau})h(\tau - 2)]$

3. (a) $\dfrac{1 - e^{-1}}{1 + 2\pi in} + 2\dfrac{1 - e^{-2}}{2 + 2\pi in}$ (b) $\dfrac{1 - 2e\pi in}{(2\pi in - 1)^2}$

(c) $\dfrac{3}{2\pi in}(e^{-2\pi in/3} - e^{-4\pi in/3})$, value is 1 for $n = 0$

4. (a) $2\pi e^t/(e^{2\pi} - 1), 0 < t < 2\pi$ (b) $2\pi\left(\dfrac{2e^{-t}}{1 - e^{-2\pi}} - \dfrac{3e^t}{e^{2\pi} - 1}\right)$

4-16 Properties of the finite Fourier transform. The finite Fourier transform has a number of special properties. Linearity has been pointed out above. The following two properties are basic for differential equations.

THEOREM 13. *Let $f(t)$ be continuous and piecewise smooth for $0 \leqq t \leqq \tau$. Then*

$$\Phi_\tau[f'] = f(\tau) - f(0) + in\omega\Phi_\tau[f]. \tag{4-160}$$

In particular, if $f(0) = f(\tau)$, then

$$\Phi_\tau[f'] = in\omega\Phi_\tau[f]. \tag{4-161}$$

Remark. Under the hypotheses made, f' may fail to exist at a finite number of points. However, if we assign arbitrary values to f' at these points, then f' is piecewise continuous and $\Phi_\tau[f']$ is well defined; the value of $\Phi_\tau[f']$ does not depend on the value assigned to f' at the discontinuity points.

Proof of the theorem. If f' is continuous for $0 \leqq t \leqq \tau$, then we have by integration by parts

$$\Phi_\tau[f'] = \int_0^\tau f'(t)e^{-in\omega t}\, dt = f(t)e^{-in\omega t}\Big|_0^\tau + in\omega \int_0^\tau f(t)e^{-in\omega t}\, dt$$

$$= f(\tau) - f(0) + in\omega\Phi_\tau[f].$$

If f' is only piecewise continuous, we obtain a similar result for each sub-interval. Addition of the partial results and application of continuity of f give the desired formula (4-160).

COROLLARY TO THEOREM 13. *If $D^{k-1}f$ is continuous and piecewise smooth for $0 \leqq t \leqq \tau$, then for $k = 1, 2, 3, \ldots$,*

$$\Phi_\tau[D^k f] = f^{(k-1)}(\tau) - f^{(k-1)}(0) + in\omega\Phi_\tau[D^{k-1}f].$$

In particular, if f has period τ, then

$$\Phi_\tau[D^k f] = (in\omega)^k \Phi_\tau[f].$$

THEOREM 14. *Let f be piecewise continuous for $0 \leqq t \leqq \tau$ and let*

$$g(t) = D_0^{-1} f = \int_0^t f(u)\, du.$$

Then for $n \neq 0$,

$$\Phi_\tau[g] = \Phi_\tau[D_0^{-1} f] = \frac{\Phi_\tau[f] - g(\tau)}{in\omega}, \tag{4-162}$$

and for $n = 0$,

$$\Phi_\tau[g] = \int_0^\tau g(t)\, dt.$$

Proof. Since $g(t)$ is piecewise smooth, (4–160) is applicable:

$$\Phi_\tau[g'] = g(\tau) - g(0) + in\omega\Phi_\tau[g].$$

But $g' = f$ (except at discontinuities) and $g(0) = 0$. Hence

$$\Phi_\tau[f] = g(\tau) + in\omega\Phi_\tau[g].$$

If we solve for $\Phi_\tau[g]$ (for $n \neq 0$), we obtain (4–162). The value for $n = 0$ is simply the definition of that value.

If f has period τ, then $g = D_0^{-1} f$ will not in general have period τ, for

$$g(t + \tau) - g(t) = \int_t^{t+\tau} f(u)\, du = \int_0^\tau f(u)\, du.$$

Hence g is periodic if and only if $\Phi_\tau[f]$ is 0 when $n = 0$; that is, when the constant term of the Fourier series of f is zero. When this condition holds, $g(t)$ will be periodic, but $D_0^{-1} g = D_0^{-2} f$ will not necessarily be periodic; for the constant term in the Fourier series for g need not be zero. However, $g(t) - c$ will be periodic and have constant term zero, provided that

$$\int_0^\tau [g(t) - c]\, dt = 0;$$

that is,

$$c = \frac{1}{\tau}\int_0^\tau g(t)\, dt.$$

Accordingly, when $\int_0^\tau f(u)\, du = 0$, then an indefinite integral of f, namely, $g(t) - c$, can be found which is periodic and has constant term zero. For this choice of $D^{-1} f$, $D^{-2} f$ will be periodic (for every choice of its arbitrary constant); for proper choice of the arbitrary constant in $D^{-2}f$, $D^{-3}f$ is periodic, etc. In fact, if $\int_0^\tau f(u)\,du = 0$, then the operator D^{-1},

$$D^{-1}f = D_0^{-1}f - \frac{1}{\tau}\int_0^\tau D_0^{-1}f\,dt,$$

is a linear operator, transforming f onto a function f_1 with the property $\int_0^\tau f_1(u)\,du = 0$; also D^{-1} is a choice of the indefinite integral of f. The powers of this operator, which are choices of D^{-2}, D^{-3}, ..., also transform f onto functions with average value zero over a period. By means of this operator, we can state a corollary analogous to that for Theorem 13.

COROLLARY TO THEOREM 14. *Let f be piecewise continuous for* $0 \leqq t \leqq \tau$ *and let f have period* τ. *Let* $\int_0^\tau f(u)\,du = 0$. *For an arbitrary piecewise continuous function* $F(t)$ *defined for all t, define the operator* D^{-k} *by the equations*

$$D^{-1}F = D_0^{-1}F - \frac{1}{\tau}\int_0^\tau D_0^{-1}F\,dt,$$

$$\tag{4-163}$$

$$D^{-k}F = D^{-1}[D^{-(k-1)}F] \quad (k = 2, 3, \ldots).$$

Then $D^{-1}f$, $D^{-2}f$, ... *are all functions of period* τ, *and for* $k = 1, 2, \ldots$

$$\Phi_\tau[D^{-k}f] = \frac{\Phi_\tau[f]}{(in\omega)^k} \quad (n \neq 0), \tag{4-164}$$

while $\Phi_\tau[D^{-k}f] = 0$ *for* $n = 0$.

THEOREM 15. *Let* $f(t)$ *be piecewise continuous for* $0 \leqq t \leqq \tau$ *and let* $f(t)$ *have period* τ. *If* $\Phi_\tau[f] = \phi(n)$, *then*

$$\Phi_\tau[f(t - c)] = e^{-in\omega c}\phi(n). \tag{4-165}$$

Proof. We have

$$\Phi_\tau[f(t - c)] = \int_0^\tau f(t - c)e^{-in\omega t}\,dt = \int_{-c}^{\tau - c} f(u)e^{-in\omega(u+c)}\,du$$

$$= e^{-in\omega c}\int_0^\tau f(u)e^{-in\omega u}\,du,$$

by the periodicity of $f(t)$. In the last integral u is a dummy variable, and the integral is simply $\Phi_\tau[f]$.

The translation process can also be applied to nonperiodic functions:

THEOREM 16. *Let* $f(t)$ *be identically zero for* $t < 0$ *and piecewise continuous for* $0 \leqq t \leqq \infty$. *Then for* $0 < c < \tau$,

$$\mathcal{L}_\tau[f(t - c)] = e^{-sc}\mathcal{L}_{\tau - c}[f]. \tag{4-166}$$

The formula is stated for $\mathcal{L}_\tau$, rather than Φ_τ, since it is easier to describe the result in terms of $\mathcal{L}_\tau$; also in practice, we can most easily use (4–166) to compute $\mathcal{L}_\tau[f(t - c)]$, and then replace s by $in\omega$ to obtain $\Phi_\tau[f(t - c)]$. For an example refer to Entries 1 and 6 of Tables 4–1 and 4–2. By (4–166), for $0 < c < \tau$,

$$\mathcal{L}_\tau[h(t - c)] = e^{-sc}\mathcal{L}_{\tau-c}[h] = e^{-sc}\frac{1 - e^{-s(\tau-c)}}{s} = \frac{e^{-sc} - e^{-s\tau}}{s}.$$

Hence, as in Entry 6 of Table 4–2,

$$\Phi_\tau[h(t - c)] = \frac{e^{-in\omega c} - e^{-in2\pi}}{in\omega}.$$

The proof of Theorem 16 is left as an exercise [Problem 3 following Section 4–18].

THEOREM 17. *Let $f(t)$ be piecewise continuous for $0 \leqq t \leqq \tau$ and let $\mathcal{L}_\tau[f] = F(s)$. Then*

$$\mathcal{L}_\tau[e^{at}f] = F(s - a). \tag{4–167}$$

Proof. We have

$$\mathcal{L}_\tau[e^{at}f] = \int_0^\tau e^{at}f(t)e^{-st}\,dt = \int_0^\tau f(t)e^{-(s-a)t}\,dt = F(s - a).$$

Again the formula can be used to obtain $\Phi_\tau[e^{at}f]$. In particular, it can be used to obtain Entry 5 of Table 4–2 from Entry 3.

4–17 Convolution. The convolution is defined in Section 2–8. Here we define a different convolution, the *f-convolution* (*f* standing for "finite Fourier"). It will turn out to have the same properties for finite Fourier transforms as the previous convolution has for Fourier transforms.

DEFINITION. Let $f(t)$, $g(t)$ be piecewise continuous and have period τ. Then the *f-convolution* of $f(t)$, $g(t)$ (for period τ) is the function

$$p(t) = f(t) \, \Delta \, g(t) = \int_0^\tau f(u)g(t - u)\,du. \tag{4–170}$$

THEOREM 18. *The f-convolution of two functions of period τ is a function of period τ. Furthermore,*

$$f \, \Delta \, g = g \, \Delta \, f, \tag{4–171}$$

$$f \, \Delta \, (g_1 + g_2) = f \, \Delta \, g_1 + f \, \Delta \, g_2, \tag{4–172}$$

$$f \, \Delta \, (cg) = (cf) \, \Delta \, g = c(f \, \Delta \, g), \tag{4–173}$$

$$e^{at}(f \, \Delta \, g) = (e^{at}f) \, \Delta \, (e^{at}g), \ a = ki\omega, \ k = 0, \pm 1, \ldots, \tag{4–174}$$

$$D(f \, \Delta \, g) = (Df) \, \Delta \, g; \tag{4–175}$$

for (4–175), f is assumed to be continuous and piecewise smooth.

The proofs are left as exercises. Because of (4–172) and (4–173), for fixed f, $f \Delta g$ is a linear operator on g; because of (4–171), the same holds with f and g interchanged. Accordingly, $f \Delta g$ can be considered as a *bilinear* operator.

In general, evaluation of $p(t)$ in (4–170) requires values of g outside the interval $0 \leqq t \leqq \tau$. It is convenient to have a formula using only the values of g inside the interval. Such a formula is the following:

$$f \Delta g = \int_0^t f(u)g(t - u)\, du + \int_t^\tau f(u)g(t + \tau - u)\, du. \quad (4\text{--}176)$$

When $0 \leqq t \leqq \tau$, then $t - u$ lies between zero and t in the first integral, while $t + \tau - u$ lies between t and τ in the second. To justify (4–176), we remark that $g(t + \tau - u)$ can be replaced by $g(t - u)$ in the second integral; combining the two integrals gives (4–170) again.

THEOREM 19. *Let $f(t)$, $g(t)$ be piecewise continuous for $0 \leqq t \leqq \tau$ and have period τ. Then $p(t) = f(t) \Delta g(t)$ is a continuous function of period τ, and*

$$\Phi_\tau[p] = \Phi_\tau[f \Delta g] = \Phi_\tau[f] \cdot \Phi_\tau[g]. \quad (4\text{--}177)$$

In words, the finite Fourier transform of the f-convolution of two functions equals the product of the transforms of the functions.

Proof. The continuity of $p = f \Delta g$ is proved as in Section 2–8. To prove (4–177) we write

$$\Phi_\tau[p] = \int_0^\tau p(t)e^{-in\omega t}\, dt$$

$$= \int_0^\tau \int_0^\tau f(u)g(t - u)e^{-in\omega t}\, du\, dt$$

$$= \int_0^\tau \int_0^\tau f(u)g(t - u)e^{-in\omega t}\, dt\, du$$

$$= \int_0^\tau f(u)e^{-in\omega u} \int_0^\tau g(t - u)e^{-in\omega(t-u)}\, dt\, du$$

$$= \int_0^\tau f(u)e^{-in\omega u} \left(\int_0^\tau g(v)e^{-in\omega v}\, dv \right) du,$$

by the substitution $v = t - u$. The inner integral is simply $\Phi_\tau[g]$; it is independent of u. Hence

$$\Phi_\tau[p] = \Phi_\tau[g]\int_0^\tau f(u)e^{-in\omega u}\, du = \Phi_\tau[g]\Phi_\tau[f].$$

Thus the theorem is proved.

COROLLARY. *The f-convolution obeys the associative law:*

$$f \Delta (g \Delta q) = (f \Delta g) \Delta q. \qquad (4\text{--}178)$$

For by Theorem 19

$$\Phi_\tau[f \Delta (g \Delta q)] = \Phi_\tau[f]\Phi_\tau[g \Delta q] = \Phi_\tau[f]\Phi_\tau[g]\Phi_\tau[q]$$
$$= \Phi_\tau[(f \Delta g) \Delta q].$$

Since both sides of (4–178) have the same transform and both are continuous, they must coincide for all t by the uniqueness theorem of Section 4–5.

4–18 Special convolutions. Of much importance for applications are convolutions in which one of the two factors is $g = e^{at}/(1 - e^{a\tau})$, $0 \leqq t < \tau$. (The function is always assumed extended outside this interval to have period τ.) We abbreviate the numerical factor

$$\gamma_a = (1 - e^{a\tau})^{-1} \qquad (4\text{--}180)$$

(τ being considered fixed) and write the convolution as $\gamma_a e^{at} \Delta f$.

THEOREM 20. *Let $f(t)$ be piecewise continuous for $0 \leqq t \leqq \tau$ and let f have period τ. Let*

$$\mathcal{L}_t[f] = \int_0^t e^{-su} f(u)\, du = F_t(s) \qquad (4\text{--}181)$$

for $0 \leqq t \leqq \tau$. Then for these values of t,

$$\gamma_a e^{at} \Delta f = e^{at}[F_t(a) + \gamma_a e^{a\tau} F_\tau(a)]. \qquad (4\text{--}182)$$

Proof. By (4–171) and (4–176),

$$e^{at} \Delta f = f \Delta e^{at} = \int_0^t f(u)e^{a(t-u)}\, du + \int_t^\tau f(u)e^{a(t+\tau-u)}\, du$$

$$= e^{at}\int_0^t f(u)e^{-au}\, du + e^{a(t+\tau)}\int_t^\tau f(u)e^{-au}\, du$$

$$= e^{at}\int_0^t f(u)e^{-au}\, du + e^{a(t+\tau)}\left[\int_0^\tau f(u)e^{-au}\, du - \int_0^t f(u)e^{-au}\, du\right]$$

$$= (e^{at} - e^{a(t+\tau)})\int_0^t f(u)e^{-au}\, du + e^{a(t+\tau)}\int_0^\tau f(u)e^{-au}\, du.$$

By virtue of (4–180) and (4–181), this is the same as (4–182).

With the aid of Table 4–1, we can now tabulate $\gamma_a e^{at} \Delta f$ for various choices of f, as shown in Table 4–3. For Entry 6 of Table 4–3, we use

TABLE 4-3

f-CONVOLUTIONS

No.	$f(t),\ 0 \leqq t < \tau$	$\gamma_a e^{at}\, \Delta f,\ 0 \leqq t < \tau$
1	$h(t)$	$-a^{-1}h(t)$
2	t	$a^{-2}(-1 - at - \gamma_a a\tau e^{at})$
3	$t^k \quad (k = 1, 2, \ldots)$	$a^{-k-1}[k!e^{at} - g_k(at) + \gamma_a e^{at}\{k!e^{a\tau} - g_k(a\tau)\}]$
4	$e^{bt} \quad (b \neq a)$	$(b - a)^{-1}[e^{bt} - e^{at} + \gamma_a e^{at}(e^{b\tau} - e^{a\tau})]$
5	$t^k e^{bt} \quad (k = 1, 2, \ldots)$	$(a - b)^{-k-1}(k!e^{at} - e^{bt}g_k[(a - b)t] + \gamma_a e^{at}\{k!e^{a\tau} - e^{b\tau}g_k[(a - b)\tau]\})$
6	$h(t - c),\ 0 \leqq c \leqq \tau$	$a^{-1}[(e^{a(t-c)} - 1)h(t - c) - \gamma_a e^{at}(1 - e^{a(\tau-c)})]$
7	$(At + B)[h(t - \alpha) - h(t - \beta)],$ $0 \leqq \alpha < \beta \leqq \tau$	$Aa^{-2}\{e^{a(t-\alpha)}(1 + a\alpha)h(t - \alpha) - e^{a(t-\beta)}(1 + a\beta)h(t - \beta)$ $- (1 + at)[h(t - \alpha) - h(t - \beta)]\} + Ba^{-1}\{[e^{a(t-\alpha)} - 1]h(t - \alpha)$ $- [e^{a(t-\beta)} - 1]h(t - \beta) + \gamma_a e^{at}[e^{a(\tau-\alpha)} - e^{a(\tau-\beta)}]\}$

8	$t[h(t) - h(t - \frac{1}{2}\tau)]$ $+ (\tau - t)h(t - \frac{1}{2}\tau)$	$a^{-2}\{e^{at} - 1 - at + [2 + 2at - a\tau - 2e^{a(2t-\tau)/2}]h(t - \frac{1}{2}\tau)$ $+ \gamma_a e^{at}[e^{a\tau} - 2e^{a\tau/2} + 1]\}$
9	$t[1 - h(t - \alpha)] + \alpha h(t - \alpha)$, $0 \leqq \alpha \leqq \tau$	$a^{-2}\{(1 + at)[h(t - \alpha) - 1] - [e^{a(t-\alpha)} + a\alpha]h(t - \alpha)$ $- \gamma_a e^{at}[e^{a(\tau-\alpha)} - 1 + a\alpha]\}$
10	$\tau - \alpha - (t - \alpha)h(t - \alpha)$, $0 \leqq \alpha \leqq \tau$	$-a^{-2}\{[e^{a(t-\alpha)} - 1 - at + a\alpha]h(t - \alpha) + \gamma_a e^{at}[e^{a(\tau-\alpha)} - 1]$ $+ (a\tau - a\alpha)(1 - \gamma_a e^{at})\}$
11	$[h(t - \alpha) - h(t - \beta)]\sin(At + B)$, A, B real, $0 \leqq \alpha < \beta \leqq \tau$	$(a^2 + A^2)^{-1}e^{at}\{[e^{-a\alpha}\{a \sin \eta + A \cos \eta\} - e^{-at}\{a \sin(At + B)$ $+ A \cos(At + B)\}][h(t - \alpha) - h(t - \beta)] + [e^{-a\alpha}\{a \sin \eta + A \cos \eta\}$ $- e^{-a\beta}\{a \sin \zeta + A \cos \zeta\}][\gamma_a e^{a\tau} + h(t - \beta)]\}$, $\eta = A\alpha + B, \qquad \zeta = A\beta + B$
12	$[h(t) - h(t - \frac{\pi}{A})]\sin At$, $A > \pi/\tau$	$(a^2 + A^2)^{-1}\{[ke^{at} - a \sin At - A \cos At][h(t) - h(t - \frac{\pi}{A})]$ $+ e^{at}[Ae^{-a\pi/A} + k][h(t - \frac{\pi}{A})]\}, \quad k = A\gamma_a(1 + e^{a\tau}e^{-a\pi/A})$

Entry 6 of Table 4–1 to find

$$F_t(s) = \frac{e^{-sc} - e^{-st}}{s} h(t - c);$$

hence

$$F_t(a) = \frac{e^{-ac} - e^{-at}}{a} h(t - c),$$

$$F_\tau(a) = \frac{e^{-ac} - e^{-a\tau}}{a} h(\tau - c) = \frac{e^{-ac} - e^{-a\tau}}{a}.$$

By substitution in (4–182) we have Entry 6. Entries 1 through 5 are obtained in the same way (Problem 8 below). Entry 8 can be deduced from Entries 2 and 7, for the function of Entry 8 (triangular wave) can be written as $t - 2(t - c)h(t - c)$, where $c = \tau/2$.

The following theorem will prove to be fundamental in the application of finite Fourier transforms to differential equations.

THEOREM 21. *Under the hypotheses of Theorem 20, the function* $g = \gamma_a e^{at}\Delta f$ *is a continuous and piecewise smooth function of period* τ. *Furthermore,*

$$(D - a)g = f(t)$$

wherever f is continuous.

Proof. As an f-convolution, g is necessarily periodic. By (4–181), $F_t(a)$ is a continuous function of t and

$$DF_t(a) = \frac{d}{dt} \int_0^t e^{-au} f(u) \, du = e^{-at} f(t)$$

wherever f is continuous. Hence $F_t(a)$ is a piecewise smooth and continuous function. By (4–182) the same holds for $g(t)$. Finally, by (4–182)

$$Dg = D(\gamma_a e^{at} \, \Delta \, f) = e^{at} e^{-at} f(t) + ae^{at}[F_t(a) + \gamma_a e^{a\tau} F_\tau(a)] = f + ag.$$

Hence the theorem is proved.

PROBLEMS

1. Verify that (4–161) is satisfied for the following choices of f (values given for $0 \leqq t < \tau$):

(a) $t - t^2,\ \tau = 1$ (b) $1 - e^t + (e - 1)t,\ \tau = 1$

2. With the aid of Theorem 16, compute $\mathcal{L}_\tau[f]$ for the following functions:

(a) $(t - 1)^2 h(t - 1),\ \tau > 1$ (b) $e^{t-2} h(t - 2),\ \tau > 2$

3. Prove Theorem 16.

4. With the aid of Theorem 17, verify the following entries of Table 4–1:

(a) Entry 4 (use Entry 1)   (b) Entry 5

5. Compute $f \triangle g$ for the following functions, all of period 2π (values given for $0 \leqq t < 2\pi$):

(a) $f = \sin t, g = \cos t$ (b) $f = t, g = t^2$
(c) $f = e^{it}, g = \sin^2 t$ (d) $f = 1 + t, g = 1 - t$

6. Find $\Phi_{2\pi}[f]$, $\Phi_{2\pi}[g]$ for each of the pairs of functions of Problem 5 and then, with the aid of Theorem 19, find $\Phi_{2\pi}[f \triangle g]$.

7. Prove the rules (4–171) through (4–175).

8. Verify Entries 1 through 7 of Table 4–3.

9. For each of the following functions f of period 3 graph the function and find $\gamma_a e^{at} \triangle f$ (values given for $0 \leqq t < 3$):

(a) $h(t - 1) - h(t - 2)$ (b) $h(t) + h(t - 1) - 2h(t - 2)$
(c) $t - (t - 1)h(t - 1)$

10. Let $g(t)$ be piecewise continuous and have period τ; let $\Phi_\tau[g] = \psi(n)$. Let $G(t) = \overline{g(-t)}$. Prove that $\Phi_\tau[G] = \overline{\psi(n)}$.

11. Let $f(t)$ and $g(t)$ have period τ and let $\Phi_\tau[f] = \phi(n)$, $\Phi_\tau[g] = \psi(n)$. Prove each of the following under appropriate hypotheses:

(a) $\displaystyle \int_0^\tau f(u)g(t - u)\, du = \frac{1}{\tau}\sum_{-\infty}^{\infty} \phi(n)\psi(n)e^{in\omega t}$

(b) $\displaystyle \int_0^\tau f(u)g(-u)\, du = \frac{1}{\tau}\sum_{-\infty}^{\infty} \phi(n)\psi(n)$

(c) $\displaystyle \int_0^\tau f(u)\overline{g(u)}\, du = \frac{1}{\tau}\sum_{-\infty}^{\infty} \phi(n)\overline{\psi(n)}$

(d) $\displaystyle \int_0^\tau |f(u)|^2\, du = \frac{1}{\tau}\sum_{-\infty}^{\infty} |\phi(n)|^2$

[*Hint for* (c): Apply the result of Problem 10.]

The equations (c) and (d) are known as *Parseval relations*. (See Problem 8 following Section 5–12.)

Answers

2. (a) $\dfrac{2e^{-s} - e^{-s\tau}g_2(s(\tau - 1))}{s^3}$ (b) $\dfrac{e^{\tau - s\tau - 2} - e^{-2s}}{1 - s}$

5. (a) $\pi \sin t$ (b) $\dfrac{8\pi^3}{3}t + \dfrac{4\pi^4}{3} - \dfrac{2\pi}{3}t^3$

(c) 0   (d) $2\pi - \dfrac{4\pi^3}{3} - 2\pi^2 t + \pi t^2$

6. (a) $\Phi_{2\pi}[f] = -\pi i \ (n = 1), \ = \pi i \ (n = -1), \ = 0$ otherwise

   $\Phi_{2\pi}[g] = \pi \ (n = \pm 1), \ = 0$ otherwise

   $\Phi_{2\pi}[f \Delta g] = -\pi^2 i \ (n = 1), \ = \pi^2 i \ (n = -1), \ = 0$ otherwise

   (b) $\Phi_{2\pi}[f] = 2\pi^2 \ (n = 0), \ = 2\pi i/n \ (n \neq 0)$

   $\Phi_{2\pi}[g] = 8\pi^3/3 \ (n = 0), \ = 4\pi(1 + \pi n i)/n^2 \ (n \neq 0)$

   $\Phi_{2\pi}[f \Delta g] = 16\pi^5/3 \ (n = 0), \ = 8\pi^2 i(1 + \pi n i)/n^3 \ (n \neq 0)$

   (c) $\Phi_{2\pi}[f] = 2\pi \ (n = 1), \ = 0 \ (n \neq 1)$

   $\Phi_{2\pi}[g] = -\pi/2 \ (n = \pm 2), \ = \pi \ (n = 0), \ = 0$ otherwise

   $\Phi_{2\pi}[f \Delta g] \equiv 0$

   (d) $\Phi_{2\pi}[f] = 2\pi + 2\pi^2 \ (n = 0), \ = 2\pi i/n \ (n \neq 0)$

   $\Phi_{2\pi}[g] = 2\pi - 2\pi^2 \ (n = 0), \ = -2\pi i/n \ (n \neq 0)$

   $\Phi_{2\pi}[f \Delta g] = 4\pi^2 - 4\pi^4 \ (n = 0), \ = 4\pi^2/n^2 \ (n \neq 0)$

9. (a) $a^{-1}\{[(e^{at-a} - 1)h(t - 1) - (e^{at-2a} - 1)h(t - 2)]$

   $+ e^{at}(e^{2a} - e^a)(1 - e^{3a})^{-1}\}$

   (b) $a^{-1}\{[(e^{at-a} - 1)h(t - 1) - 2(e^{at-2a} - 1)h(t - 2)]$

   $+ [e^{at}(1 + e^{2a} - 2e^a) + e^{3a} - 1](1 - e^{3a})^{-1}\}$

   (c) $a^{-2}[(1 + at - a - e^{a(t-1)})h(t - 1)$

   $+ (1 - e^{3a})^{-1}e^{at}(1 - a - e^{2a}) - 1 - at]$

4–19 The inverse finite Fourier transform. If $\Phi_\tau[f] = \phi(n)$, then we write

$$f = \Phi_\tau^{-1}[\phi]$$

and call Φ_τ^{-1} the *inverse finite Fourier transform*. Given ϕ, there is at most one f such that $\Phi_\tau[f] = \phi(n)$, by the uniqueness theorem of Section 4–5; that is, Φ_τ^{-1} is uniquely defined. The class of functions $\phi(n)$ for which Φ_τ^{-1} is defined forms a *linear space* and Φ_τ^{-1} is a *linear operator* in this space (Section 4–14).

Given a function $\phi(n)$, it is not easy to determine whether $\Phi_\tau^{-1}[\phi]$ is defined. By the theorem of Riemann-Lebesgue (Section 4–7), if $\phi(n) = \Phi_\tau[f]$, then $\phi(n)$ must approach zero as $n \to \pm\infty$; however, this alone is not sufficient. In many cases, the following condition is adequate:

THEOREM 22. *Let $\phi(n)$ be defined for $n = 0, \pm 1, \pm 2, \ldots,$ and let the series*

$$\frac{1}{\tau}\sum_{n=-\infty}^{\infty} \phi(n)e^{in\omega t} = \frac{1}{\tau}\left[\phi(0) + \sum_{n=1}^{\infty} \{\phi(n)e^{in\omega t} + \phi(-n)e^{-in\omega t}\}\right]$$

converge uniformly to $f(t)$ for $0 \leq t \leq \tau$. Then $\Phi_\tau^{-1}[\phi] = f$.

This is a restatement of the uniform convergence theorem of Section 4–6. To prove the uniform convergence, we can often apply the Weierstrass M-test (Section 4–4). In particular if the series

$$\sum_{n=-\infty}^{\infty} |\phi(n)| = |\phi(0)| + \sum_{n=1}^{\infty} \{|\phi(n)| + |\phi(-n)|\}$$

converges, then we can apply the M-test with $M_n = |\phi(n)| + |\phi(-n)|$. For

$$|\phi(n)e^{in\omega t} + \phi(-n)e^{-in\omega t}| \leq |\phi(n)| + |\phi(-n)|.$$

For example, if $\phi(n) = 1/n^2$ for $n \neq 0$ and $\phi(0) = 0$, then the series $\sum |\phi(n)|$ converges, and $\Phi_{2\pi}^{-1}[\phi]$ is defined and is equal to

$$f(t) = \frac{1}{2\pi} \sum_{n=-\infty}^{\infty}{}' \frac{e^{int}}{n^2} = \frac{1}{2\pi} \sum_{n=1}^{\infty} \frac{e^{int} + e^{-int}}{n^2} = \frac{1}{\pi} \sum_{n=1}^{\infty} \frac{\cos nt}{n^2}.$$

The function $f(t)$ is perfectly well defined by this expression and can be computed to any desired accuracy for each t. However, it is of interest to note that in this case

$$f(t) = \frac{\pi}{6} - \frac{t}{2} + \frac{t^2}{4\pi} \qquad (0 \leq t \leq 2\pi),$$

as can be verified directly.

It is natural to ask how we can recognize the inverse transform as an elementary function. One way to achieve this goal is to set up a large table of transforms $\Phi_\tau[f]$; this can be read backwards to give inverse transforms. By linearity, a vast number of additional transforms can be found. Theorems 13 through 19 above (especially Theorem 19) expand the possibilities. States in terms of inverse transforms Theorem 19 reads: Given $\phi_1(n)$ and $\phi_2(n)$, if

$$\Phi_\tau^{-1}[\phi_1] = f_1, \qquad \Phi_\tau^{-1}[\phi_2] = f_2, \tag{4--190}$$

then

$$\Phi_\tau^{-1}[\phi_1 \cdot \phi_2] = f_1 \, \Delta \, f_2. \tag{4--191}$$

Thus the *product of two transforms is a transform*. Accordingly, the class of transforms $\phi(n)$ forms a *linear algebra* (Problem 2 following Section 2-2).

From Entries 1 and 2 of Table 4–2, we find easily that

$$\Phi_\tau \left[t - \frac{\tau}{2} \right] = \frac{2\pi i}{n\omega^2} \qquad (n \neq 0)$$

and has value 0 for $n = 0$. Hence if $\phi(n) = 1/n$ for $n \neq 0$ and $\phi(0) = 0$, then

$$\Phi_\tau^{-1}[\phi] = \frac{\omega^2}{2\pi i} \left(t - \frac{\tau}{2} \right) = f,$$

so that $f(t)$ is of the form $at + b$ for $0 \leq t \leq \tau$. We conclude that

$$\Phi_\tau^{-1}[\phi^2] = f \, \Delta \, f, \quad \Phi_\tau^{-1}[\phi^3] = (f \, \Delta \, f) \, \Delta \, f, \ldots,$$

where $\phi^2(n) = 1/n^2$, $\phi^3(n) = 1/n^3, \ldots$, except for the value 0 when

TABLE 4–4

INVERSE FINITE FOURIER TRANSFORMS

No.	$\phi(n)$	$\Phi_\tau^{-1}[\phi] = f = f(t), \quad 0 \leqq t < \tau$
1	$\begin{array}{l}1, \quad n = 0 \\ 0, \quad n \neq 0\end{array}$	$\dfrac{1}{\tau} h(t) = \dfrac{1}{\tau}$
2	$0, \quad n = 0; \quad \dfrac{1}{n}, \quad n \neq 0$	$\dfrac{2\pi i}{\tau}\left(\dfrac{1}{2} - \dfrac{t}{\tau}\right)$
3	$0, \quad n = 0; \quad \dfrac{1}{n^2}, \quad n \neq 0$	$\dfrac{(2\pi i)^2}{\tau}\left(-\dfrac{1}{12} + \dfrac{1}{2}\dfrac{t}{\tau} - \dfrac{1}{2}\dfrac{t^2}{\tau^2}\right)$
4	$0, \quad n = 0; \quad \dfrac{1}{n^3}, \quad n \neq 0$	$\dfrac{(2\pi i)^3}{\tau}\left(-\dfrac{1}{12}\dfrac{t}{\tau} + \dfrac{1}{4}\dfrac{t^2}{\tau^2} - \dfrac{1}{6}\dfrac{t^3}{\tau^3}\right)$
5	$0, \quad n = k; \quad \dfrac{1}{n - k}, \quad n \neq k$	$e^{ik\omega t}\dfrac{2\pi i}{\tau}\left(\dfrac{1}{2} - \dfrac{t}{\tau}\right)$
6	$\dfrac{1}{n - b}, \quad b \neq 0, \pm 1, \ldots$	$\dfrac{i\omega e^{bi\omega t}}{1 - q}, \quad q = e^{2\pi bi}$
7	$\dfrac{1}{(n - b)^2}, \quad b \neq 0, \pm 1, \ldots$	$\dfrac{e^{bi\omega t}[(q - 1)\omega^2 t - 2\pi\omega q]}{(q - 1)^2},$ $q = e^{2\pi bi}$
8	$\dfrac{1}{in\omega - a}, \quad \dfrac{ai}{\omega} \neq 0, \pm 1, \ldots$	$\gamma_a e^{at}, \quad \gamma_a = (1 - e^{a\tau})^{-1}$
9	$\dfrac{1}{(in\omega - a)^2}, \quad \dfrac{ai}{\omega} \neq 0, \pm 1, \ldots$	$\gamma_a e^{at}[t + \tau(\gamma_a - 1)]$
10	$\dfrac{1}{(in\omega - a)^k}, \quad \dfrac{ai}{\omega} \neq 0, \pm 1, \ldots$	$\dfrac{1}{(k - 1)!}\dfrac{\partial^{k-1}}{\partial a^{k-1}}(\gamma_a e^{at}), \quad k = 1, 2, \ldots$
11	$\dfrac{1}{(in\omega - \alpha)^2 + \beta^2},$ α, β real, $\beta \neq 0,$ $\dfrac{(\alpha i - \beta)}{\omega} \neq 0, \pm 1, \ldots,$	$\dfrac{e^{\alpha t}[\sin \beta t - e^{\alpha\tau}\sin \beta(t - \tau)]}{\beta(1 + e^{2\alpha\tau} - 2e^{\alpha\tau}\cos \beta\tau)}$

$n = 0$. The results are shown in Table 4–4. By linearity, we can find the inverse transform of a function $\phi(n)$ of the form

$$\frac{a_1}{n} + \frac{a_2}{n^2} + \cdots + \frac{a_k}{n^k},$$

with value zero at $n = 0$. The value at $n = 0$ can be adjusted by adding an appropriate constant times $h(t)$. Another way of finding $\Phi_\tau^{-1}[\phi^2]$, $\Phi_\tau^{-1}[\phi^3]$, ... is indicated in Problem 1 below.

The other entries in the table, except the last two, are simple consequences of Table 4–2. Entries 9 and 10 are special cases of the rule

$$\Phi_\tau\left[\frac{\partial}{\partial a} f(t, a)\right] = \frac{\partial}{\partial a} \Phi_\tau[f(t, a)]; \tag{4–192}$$

this is valid if $\partial f/\partial a$ is continuous in a, t for $0 \leqq t \leqq \tau$, and for some interval of a. For (4–192) is simply a case of Leibnitz's rule

$$\int_0^\tau \frac{\partial}{\partial a} f(t, a)e^{-in\omega t}\, dt = \frac{\partial}{\partial a} \int_0^\tau f(t, a)e^{-in\omega t}\, dt. \tag{4–193}$$

Entry 11 is reducible to Entry 8 by partial fractions.

4–20 Finite Fourier transforms of generalized functions. By Section 1–13, especially Eq. (1–134), we have

$$\int_0^\tau \delta(t - c)e^{-in\omega t}\, dt = e^{-in\omega c} \quad (0 < c < \tau),$$

$$\int_0^\tau \delta^{(m)}(t - c)e^{-in\omega t}\, dt = (-1)^m(-in\omega)^m e^{-in\omega c}$$

$$= (in\omega)^m e^{-in\omega c} \quad (0 < c < \tau).$$

As $c \to 0+$, the function $e^{-in\omega c}$ approaches 1. For this reason and other reasons given below, we define $\Phi_\tau[\delta(t)]$ to be $\phi(n) \equiv 1$ and in general

$$\Phi_\tau[\delta^{(m)}(t - c)] = (in\omega)^m e^{-in\omega c} \quad (0 \leqq c < \tau,\ m = 0, 1, 2, \ldots). \tag{4–200}$$

By (4–200) and by linearity, the finite Fourier transform of any periodic generalized function can be found. For example, for $\tau = 2$, $\omega = \pi$,

$$\Phi_\tau[e^t + 2\, \delta(t) - 3\, \delta'(t - 1)] = \frac{e^2 - 1}{1 - in\pi} + 2 - 3\pi ine^{-in\pi}.$$

The very concept of a periodic generalized function requires an extension of the definition of generalized function. For example, the periodic function equal to $\delta(t)$ for $0 \leqq t < \tau$ is a new generalized function which we denote

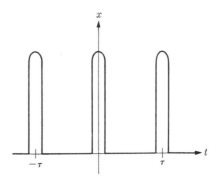

FIG. 4-6. Periodic impulse $\delta_\tau(t)$.

by $\delta_\tau(t)$. We can identify this function with the sum of the series

$$\delta(t) + \delta(t - \tau) + \cdots + \delta(t - n\tau) + \cdots$$
$$+ \delta(t + \tau) + \cdots + \delta(t + n\tau) + \cdots,$$

which can be interpreted as a periodic "pulse" as in Fig. 4-6. For such a periodic function the finite Fourier transform could be obtained by integrating over an arbitrary interval of length τ. For example,

$$\Phi_\tau\left[\delta_\tau(t)\right] = \int_{-\tau/2}^{\tau/2} \delta(t)e^{-in\omega t}\, dt = e^{-0\, in\omega} = 1,$$

in agreement with (4–200).

We note that equations (4–200) imply that $\Phi_\tau[f'] = in\omega\Phi_\tau[f]$ for $f = \delta_\tau(t - c)$, $f = \delta_\tau'(t - c), \ldots$ Thus the rule (4–161) holds, rather than an analogue of the rule (4–160). Indeed, when periodic generalized functions are used, *the rule $\Phi_\tau[f'] = in\omega\Phi_\tau[f]$ becomes universal.* To justify this statement we first consider a periodic ordinary function $f(t)$, coinciding for $0 < t < \tau$ with $g(t)$, where $g(t)$ is continuous and piecewise smooth for $0 \leq t \leq \tau$. At $t = 0$, $\pm\tau$, $\pm 2\tau, \ldots$, $f(t)$ has jump discontinuities; the jump, equal to the right-hand limit minus the left-hand limit, equals $q(0) - g(\tau)$. Accordingly, $f'(t)$ must be considered as a periodic generalized function equal to $g'(t) + [g(0) - g(\tau)]\,\delta_\tau(t)$. Therefore,

$$\Phi_\tau[f'(t)] = \Phi_\tau[g'(t)] + [g(0) - g(\tau)] \cdot 1.$$

But $\Phi_\tau[g'(t)]$ can be evaluated by (4–160); hence

$$\Phi_\tau[f'(t)] = in\omega\Phi_\tau[g] + [g(\tau) - g(0)] \cdot 1$$
$$+ [g(0) - g(\tau)] \cdot 1 = in\omega\Phi_\tau[f].$$

A similar discussion applies if f has jump discontinuities inside the interval, and, by virtue of (4-200), the rule is valid for each periodic generalized function which is a linear combination of functions $\delta_\tau^{(m)}(t - c)$ and a piecewise smooth function $g(t)$.

The rules for transforms of integrals can also be restated. In particular, the corollary to Theorem 14 is valid for periodic generalized functions, provided we interpret $D_0^{-1} \delta_\tau(t - c)$ as $h(t - c)$ and $D_0^{-1} \delta_\tau^{(m)}(t - c)$ as $\delta_\tau^{(m-1)}(t - c)$.

The convolution of generalized functions is treated most easily by employing (4-177) as a *definition*. Thus for each ordinary function f, $f \, \Delta \, \delta_\tau = f$, since

$$\Phi_\tau[f \, \Delta \, \delta_\tau] = \Phi_\tau[f] \cdot \Phi_\tau[\delta_\tau]$$

and $\Phi_\tau[\delta_\tau] = 1$. Similarly, $f \, \Delta \, \delta_\tau^{(m)}(t) = f^{(m)}(t)$, provided $f^{(m)}(t)$ exists as a generalized function, since $\Phi_\tau[\delta_\tau^{(m)}(t)] = (in\omega)^m$. By the same reasoning,

$$\delta_\tau^{(m)}(t - a) \, \Delta \, \delta_\tau^{(p)}(t - b) = \delta_\tau^{(m+p)}(t - a - b).$$

Finally, we remark that the important Theorem 21 remains valid when f is a periodic generalized function. For example, if $f = \delta_\tau(t)$, then as above $\gamma_a e^{at} \, \Delta \, f = \gamma_a e^{at}$, a periodic function which jumps by 1 at 0, $\pm\tau$, $\pm 2\tau, \ldots$; for

$$e^{at} \Big|_\tau^0 = 1 - e^{a\tau} = \gamma_a^{-1}.$$

Hence $D(\gamma_a e^{at}) = a\gamma_a e^{at} + \delta_\tau(t)$, so that $(D - a)(\gamma_a e^{at}) = \delta_\tau(t)$.

PROBLEMS

1. (a) Let $\Phi_\tau[f] = \phi(n)$, where $\phi(0) = 0$. Let

$$g(t) = D_0^{-1} f = \int_0^t f(u) \, du, \qquad c = \frac{1}{\tau} \int_0^\tau g(t) \, dt,$$

$$f_1(t) = \frac{2\pi i}{\tau} [g(t) - c], \qquad \Phi_\tau[f_1] = \phi_1(n).$$

Show that $\phi_1(0) = 0$ and $\phi_1(n) = \phi(n)/n$ for $n \neq 0$. [*Hint:* Apply the corollary to Theorem 14.]

(b) Let

$$f = \frac{2\pi i}{\tau} \left(\frac{1}{2} - \frac{t}{\tau} \right)$$

as in Entry 2 of Table 4-4. Verify that $\Phi_\tau[f] = \phi(n)$, where $\phi(n) = 1/n$ for $n \neq 0$, $\phi(0) = 0$. Compute $f_1(t)$ as in part (a), so that $\phi_1(n) = [\phi(n)]^2$; compare with Entry 3 of Table 4-4.

(c) Proceed as in part (b) to obtain Entry 4 of Table 4-4.

2. With the aid of Table 4–4, find $\Phi_T^{-1}[\phi]$ if $\phi(0) = 0$ and if $\phi(n)$ has the value given for $n \neq 0$:

(a) $\dfrac{2}{n} - \dfrac{3}{n^2}$ (b) $\dfrac{1}{n} + \dfrac{2}{n^2} + \dfrac{1}{n^3}$ (c) $\dfrac{1}{n} - \dfrac{1}{n^3}$

3. Find $\Phi_T^{-1}[\phi]$ if $\phi(0) = 2$ and $\phi(n) = \dfrac{5}{n^3}$ for $n \neq 0$.

4. Sum each of the following series ($\sum'$ denotes summation excluding $n = 0$):

(a) $\displaystyle\sum_{n=-\infty}^{\infty}{}' \left(\dfrac{2}{n} - \dfrac{3}{n^2}\right) e^{2\pi i n t}$ (b) $\displaystyle\sum_{n=-\infty}^{\infty}{}' \left(\dfrac{1}{n} + \dfrac{4}{n^3}\right) e^{int}$

5. For each of the following, find the inverse transforms (i) as Fourier series and (ii) in terms of elementary functions:

(a) $\Phi_{2\pi}^{-1}[\phi]$, where $\phi(n) = \dfrac{1}{n^2 + 1}$ (b) $\Phi_{2\pi}^{-1}[\phi]$, where $\phi(n) = \dfrac{1}{n^2 - 3}$

6. (a) Let $f(t) = \sum_{n=-\infty}^{\infty} c_n e^{in\omega t}$, $g(t) = \sum_{n=-\infty}^{\infty} d_n e^{in\omega t}$. Granting conditions such that the series can be multiplied term by term and terms of the product can be arranged as desired, show that

$$f(t) \cdot g(t) = \sum_{n=-\infty}^{\infty} C_n e^{in\omega t}, \qquad C_n = \sum_{m=-\infty}^{\infty} c_{n-m}\, d_m.$$

We can call

$$X(n) = \sum_{m=-\infty}^{\infty} \phi(n-m)\psi(m)$$

the inverse convolution of $\phi(n) = \Phi_T[f]$ and $\psi(n) = \Phi_T[g]$, and write $X = \phi \nabla \psi$. Show that

$$\Phi_T^{-1}[\phi \nabla \psi] = T\Phi_T^{-1}[\phi] \cdot \Phi_T^{-1}[\psi].$$

(b) Show that if, in part (a), $c_n = d_n = 0$ for $n < 0$, then C_n becomes the finite sum $c_0 d_n + c_1 d_{n-1} + \cdots + c_n d_0$. Show that in this case $f(t)$ and $g(t)$ can be interpreted as $F(e^{i\omega t})$ and $G(e^{i\omega t})$, where $F(z)$ and $G(z)$ are analytic for $|z| < 1$.

7. Find the finite Fourier transforms of each of the following generalized functions f of period 2π:

(a) $f = e^{2t} + 3\,\delta_{2\pi}(t) - 2\,\delta_{2\pi}'(t - \pi)$ $(0 \leq t < 2\pi)$
(b) $f = 1 + 2\,\delta_{2\pi}''(t) + 3\,\delta_{2\pi}'''(t)$ $(0 \leq t < 2\pi)$

8. For each of the following functions f of period 2π find f' as a generalized function of period 2π and verify that $\Phi_T[f'] = in\omega\Phi_T[f]$:

(a) $f = t[h(t) - h(t - \pi)]$ $(0 \leq t < 2\pi)$
(b) $f = h(t - \tfrac{1}{2}\pi) - h(t - \pi)$ $(0 \leq t < 2\pi)$

9. Evaluate the convolution $f \triangle g$ for period 2 if

(a) $f = h(t - 1)$ $(0 \leq t < 2)$, $g = \delta_2(t)$
(b) $f = \delta_2'(t - 1)$, $g = \delta_2(t - 1)$

Answers

2. (a) $\dfrac{2\pi i - \pi^2}{\tau} + \dfrac{6\pi^2 - 4\pi i}{\tau^2} t - \dfrac{6\pi^2}{\tau^3} t^2$

   (b) $\dfrac{3\pi i + 2\pi^2}{3\tau} + \dfrac{-6\pi i - 12\pi^2 + 2\pi^3 i}{3\tau^2} t + \dfrac{4\pi^2 - 2\pi^3 i}{\tau^3} t^2 + \dfrac{4\pi^3 i}{3\tau^4} t^3$

   (c) $\dfrac{\pi i}{\tau} - \dfrac{6\pi i + 2\pi^3 i}{3\tau^2} t + \dfrac{2\pi^3 i}{\tau^3} t^2 - \dfrac{4\pi^3 i}{3\tau^4} t^3$

3. $\dfrac{2}{\tau} + \dfrac{10\pi^3 i}{3\tau^2} t - \dfrac{10\pi^3 i}{\tau^3} t^2 + \dfrac{20\pi^3 i}{3\tau^4} t^3$

4. (a) $2\pi i - \pi^2 + (6\pi^2 - 4\pi i)t - 6\pi^2 t^2$ $(0 < t < 1,\ \text{period } 1)$

   (b) $\pi i + t\left(\dfrac{4\pi^3 i}{3} - i\right) - 2\pi i t^2 + \tfrac{2}{3} i t^3$ $(0 < t < 2\pi,\ \text{period } 2\pi)$

5. (a) $\dfrac{1}{2\pi} \displaystyle\sum_{n=-\infty}^{\infty} \dfrac{e^{int}}{n^2 + 1},\quad \dfrac{1}{2}\left(\dfrac{e^{-t}}{1 - e^{-2\pi}} - \dfrac{e^{t}}{1 - e^{2\pi}}\right)$

   (b) $\dfrac{1}{2\pi} \displaystyle\sum_{n=-\infty}^{\infty} \dfrac{e^{int}}{n^2 - 3},\quad \dfrac{1}{2c}\left(\dfrac{e^{-ct}}{1 - e^{-2\pi c}} - \dfrac{e^{ct}}{1 - e^{2\pi c}}\right)$ $(c = \sqrt{3}i)$

7. (a) $\dfrac{e^{4\pi} - 1}{2 - in} + 3 - 2in(-1)^n$ (b) $\dfrac{1 - e^{-2\pi in}}{in} - 2n^2 - 3in^3$

8. (a) $f' = g(t) - \delta_{2\pi}(t - \pi),\quad g(t) = h(t) - h(t - \pi)$ $(0 \leqq t < 2\pi)$

   (b) $f' = \delta_{2\pi}(t - \tfrac{1}{2}\pi) - \delta_{2\pi}(t - \pi)$ $(0 \leqq t < 2\pi)$

9. (a) $h(t - 1)$ $(0 \leqq t < 2)$ (b) $\delta_2'(t)$

4–21 Application of finite Fourier transforms to linear differential equations. We consider a differential equation

$$(a_0 D^m + \cdots + a_m)x = f(t)   (a_0 \neq 0)   (4\text{–}210)$$

with constant coefficients. Let $f(t)$ be periodic, with period τ, and be piecewise continuous for $0 \leq t \leq \tau$. We then seek a periodic solution of (4–210). If $x(t)$ is such a periodic solution, then, by Theorem 13 and its corollary,

$$\Phi_\tau[Dx] = in\omega\Phi_\tau[x],\quad \Phi_\tau[D^2 x] = (in\omega)^2\Phi_\tau[x],\ \ldots$$

if we assume appropriate continuity. Hence

$$\Phi_\tau[(a_0 D^m + \cdots + a_m)x] = [a_0(in\omega)^m + a_1(in\omega)^{m-1} + \cdots + a_m]\Phi_\tau[x]$$
$$= V(in\omega)\Phi_\tau[x].$$

Therefore, by (4–210)

$$V(in\omega)\Phi_\tau[x] = \Phi_\tau[f],$$
$$\Phi_\tau[x] = Y(in\omega)\Phi_\tau[f]. \tag{4–211}$$

Thus the Fourier coefficients of $x(t)$ can be computed from those of f, so that we can write $x(t)$ as the sum of a Fourier series:

$$x = \frac{1}{\tau}\sum_{n=-\infty}^{\infty} Y(in\omega)\phi(n)e^{in\omega t}, \tag{4–212}$$

where $\Phi_\tau[f] = \phi(n)$. This is the same as the solution obtained in Section 4–2.

We can also write

$$x = \Phi_\tau^{-1}[Y(in\omega)\phi(n)], \tag{4–213}$$

and attempt to find $x(t)$ explicitly from a table of inverse transforms.

Finally, we can write

$$W_\tau(t) = \Phi_\tau^{-1}[Y(in\omega)]. \tag{4–214}$$

Then (4–211) reads

$$\Phi_\tau[x] = \Phi_\tau[W_\tau]\cdot\Phi_\tau[f] = \Phi_\tau[W_\tau \Delta f].$$

Accordingly,

$$x = W_\tau \Delta f = \int_0^\tau W_\tau(u)f(t-u)\,du. \tag{4–215}$$

We have proceeded formally. However, all the above steps can be justified.

THEOREM 23. *Let $a_0, \ldots, a_m$ be constants, with $a_0 \neq 0$. Let $V(s) = a_0 s^m + \cdots + a_m$ and let $V(in\omega) \neq 0$ for $n = 0, \pm1, \pm2, \ldots$ Let $f(t)$ have period τ and let $f(t)$ be piecewise continuous for $0 \leqq t \leqq \tau$. Then the differential equation (4–210) has a unique solution $x(t)$ of period τ. The function $x(t)$ has continuous derivatives through the $(m-1)$st order; $D^m x$ is continuous except for jump discontinuities where $f(t)$ is discontinuous. The Fourier series of $x(t)$ is given by (4–212); it converges to $x(t)$ for all t. The finite Fourier transform of $x(t)$ is defined by (4–211). The function*

$$Y(in\omega) = \frac{1}{V(in\omega)}$$

has an inverse finite Fourier transform $W_\tau(t)$ which is periodic and piecewise continuous, and $x(t)$ is equal to the f-convolution of W_τ with f.

The proof of the theorem is given in the following section. Here we consider some examples.

EXAMPLE 1. $(D + 2)x = f(t)$, where $f(t) = t$ for $0 \leq t < 1$ and f has period 1. Accordingly,

$$\omega = 2\pi \quad \text{and} \quad V(in\omega) = 2\pi in + 2, \quad Y(in\omega) = \frac{1}{2\pi in + 2},$$

$$\Phi_\tau[f] = \tfrac{1}{2} \quad \text{for } n = 0,$$

$$\Phi_\tau[f] = \frac{i}{2\pi n} \quad \text{for } n \neq 0.$$

Therefore by (4–211) and (4–212)

$$\Phi_\tau[x] = \tfrac{1}{4} \quad \text{for } n = 0,$$

$$\Phi_\tau[x] = \frac{i/(2\pi)}{n(2\pi in + 2)} \quad \text{for } n \neq 0,$$

$$x = \frac{1}{4} + \frac{i}{2\pi} \sum_{n=-\infty}^{\infty}{}' \frac{1}{n(2\pi in + 2)} e^{2\pi int}$$

$$= \frac{1}{4} + \frac{1}{2\pi} \sum_{n=1}^{\infty} \frac{\pi n \cos 2\pi nt - \sin 2\pi nt}{n(\pi^2 n^2 + 1)}.$$

By Entry 8 of Table 4–4, for $0 \leq t < 1$,

$$W_\tau(t) = \Phi_\tau^{-1}[Y(in\omega)] = \Phi_\tau^{-1}\left[\frac{1}{2\pi in + 2}\right] = \gamma_{-2}e^{-2t} = \frac{e^{-2t}}{1 - e^{-2}}.$$

Therefore by (4–215)

$$x = W_\tau \Delta f = \frac{e^{-2t} \Delta f}{1 - e^{-2}} = \gamma_{-2}e^{-2t} \Delta f, \quad \gamma_{-2} = (1 - e^{-2})^{-1} = 1.16.$$

To evaluate the convolution, we use Entry 2 of Table 4–3, with $a = -2$. Hence for $0 \leq t < 1$,

$$x = \tfrac{1}{4}(-1 + 2t + \gamma_{-2}2e^{-2t}) = 0.58e^{-2t} + \tfrac{1}{2}t - \tfrac{1}{4}.$$

The solution x and input f are shown in Fig. 4–7.

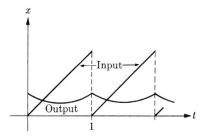

FIG. 4–7.   Periodic response to sawtooth wave.

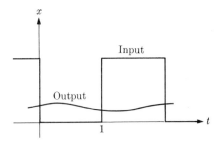

FIG. 4–8.   Periodic response to square wave.

EXAMPLE 2. $(D^2 + 3D + 2)x = f(t)$,   where $f(t) = h(t - 1)$ for $0 \leq t < 2$ and f has period 2. Here

$$Y(in\omega) = \frac{1}{(in\omega)^2 + 3in\omega + 2} \quad (\omega = \pi).$$

To find W_τ, we note that

$$Y(s) = \frac{1}{(s + 1)(s + 2)} = \frac{1}{s + 1} - \frac{1}{s + 2},$$

by partial fractions. Hence

$$Y(in\omega) = \frac{1}{in\omega + 1} - \frac{1}{in\omega + 2}.$$

Therefore by Entry 8 of Table 4–4,

$$W_\tau = \Phi_\tau^{-1}\left[\frac{1}{in\omega + 1}\right] - \Phi_\tau^{-1}\left[\frac{1}{in\omega + 2}\right] = \gamma_{-1}e^{-t} - \gamma_{-2}e^{-2t}.$$

Now by (4–215) and by Entry 6 of Table 4–3,

$$\begin{aligned}
x &= \gamma_{-1}e^{-t}\,\Delta f - \gamma_{-2}e^{-2t}\,\Delta f \\
&= -[(e^{-(t-1)} - 1)h(t - 1) - \gamma_{-1}e^{-t}(1 - e^{-1})] \\
&\quad + \tfrac{1}{2}[(e^{-2(t-1)} - 1)h(t - 1) - \gamma_{-2}e^{-2t}(1 - e^{-2})] \\
&= (\tfrac{1}{2} - 2.72e^{-t} + 3.69e^{-2t})h(t - 1) + 0.73e^{-t} - 0.44e^{-2t}.
\end{aligned}$$

Again x and f are shown in Fig. 4–8.

The procedures described here lend themselves easily to generalization.

THEOREM 24. *Under the conditions of Theorem 23, let $V(s)$ have simple roots $s_1, \ldots, s_m$, so that*

$$Y(s) = \sum_{j=1}^{m} \frac{A_j}{s - s_j}.$$

Then the periodic solution $x(t)$ is given by $x = \sum_{j=1}^{m} A_j G(t; s_j)$, where
$G(t; s) = \gamma_s e^{st} \Delta f.$

For

$$Y(in\omega) = \sum_{j=1}^{m} \frac{A_j}{in\omega - s_j},$$

so that

$$W_\tau = \sum_{j=1}^{m} \frac{A_j e^{s_j t}}{1 - e^{s_j \tau}} = \sum_{j=1}^{m} A_j \gamma_{s_j} e^{s_j t}.$$

If $V(s)$ has a double root, the partial fraction expansion of $Y(s)$ will have a term of form $A/(s - s_j)^2$. By Entry 9 of Table 4-4, the corresponding term in W_τ is

$$A\gamma_a e^{at}[t + \tau(\gamma_a - 1)] = A \frac{\partial}{\partial a}(\gamma_a e^{at}) \quad (a = s_j).$$

Roots of higher multiplicity are covered by Entry 10 of Table 4-4. Computation of $W_\tau \Delta f$ then requires the f-convolutions

$$\left[\frac{\partial^k}{\partial a^k}(\gamma_a e^{at})\right]\Delta f, \quad k = 1, 2, \ldots \tag{4-216}$$

Here we have a rule analogous to (4-192):

$$\left[\frac{\partial}{\partial a} g(t, a)\right]\Delta f = \frac{\partial}{\partial a}[g(t, a) \Delta f], \tag{4-217}$$

whose proof is left as a problem (Problem 3 below). From (4-217) we conclude that the convolutions (4-216) are equal to

$$\frac{\partial^k}{\partial a^k}[(\gamma_a e^{at}) \Delta f].$$

Thus with the aid of Table 4-3, the necessary f-convolutions can be found.

4-22 Proof of Theorem 23. We first verify that $x = W_\tau \Delta f$ is well defined and satisfies all conditions stated. Let $Y(s)$ be expanded in partial fractions; each term will be of form $A/(s - s_j)^k$, where s_j is a root of $V(s)$. Since $V(s)$ has no root of form $in\omega$, $s_j \neq in\omega$ for any n. Accordingly, each term of $Y(in\omega)$ has an inverse transform, obtained from Entry 10 of Table 4-4. Thus W_τ is well-defined and is in fact a sum of terms of the form $bt^\alpha e^{at}$.

We can also write $Y(s)$ in factored form:

$$Y(s) = \frac{1}{a_0(s - s_1)(s - s_2) \cdots (s - s_m)}.$$

Then, if $\Phi_\tau[f] = \phi(n)$,

$$Y(in\omega)\Phi_\tau[f] = \frac{\phi(n)}{a_0(in\omega - s_1)(in\omega - s_2)\cdots(in\omega - s_m)}.$$

The right-hand side can be regarded as a product of transforms:

$$\frac{1}{in\omega - s_1} \cdot \frac{1}{in\omega - s_2} \cdots \frac{1}{in\omega - s_m} \cdot \frac{\phi(n)}{a_0}.$$

Hence its inverse transform is

$$x = W_\tau(t) \,\Delta\, f = \gamma_{s_1} e^{s_1 t} \,\Delta\, \gamma_{s_2} e^{s_2 t} \,\Delta\, \cdots \,\Delta\, \gamma_{s_m} e^{s_m t} \,\Delta\, \frac{f}{a_0},$$

where no parentheses are needed, by the associative law for the convolution (Eq. (4–178)).

Now we can regard x as

$$\gamma_{s_1} e^{s_1 t} \,\Delta\, p(t),$$

where p is continuous. Hence by Theorem 21

$$(D - s_1)x = p = \gamma_{s_2} e^{s_2 t} \,\Delta\, \gamma_{s_3} e^{s_3 t} \,\Delta\, \cdots \,\Delta\, \frac{f}{a_0}.$$

Similarly,

$$(D - s_2)(D - s_1)x = \gamma_{s_3} e^{s_3 t} \,\Delta\, \gamma_{s_4} e^{s_4 t} \,\Delta\, \cdots \,\Delta\, \frac{f}{a_0},$$

until finally,

$$(D - s_{m-1})(D - s_{m-2}) \cdots (D - s_1)x = \gamma_{s_m} e^{s_m t} \,\Delta\, \frac{f}{a_0}.$$

Accordingly, by Theorem 21, wherever f is continuous,

$$(D - s_m)(D - s_{m-1}) \cdots (D - s_1)x = \frac{f}{a_0}.$$

This is equivalent to the differential equation

$$(a_0 D^m + a_1 D^{m-1} + \cdots + a_m)x = f.$$

The reasoning also shows that $D^{m-1}x$ is continuous, and $D^m x$ is continuous except for jumps at the jumps of f. Finally, $x(t)$ is periodic since it is an f-convolution. Hence $x(t)$ has all the properties asserted.

Next we remark that x is continuous and piecewise smooth; hence by Section 4–7 it is equal to the sum of its Fourier series. By definition of x and by Theorem 19, $\Phi_\tau[x] = Y(in\omega)\phi(n)$ and the series is given by (4–212).

It remains to show that x is unique. But if $x(t)$ is a periodic solution with the properties described, then the reasoning of Section 4–21 shows that (4–211) must hold, so that $\Phi_\tau[x] = Y(in\omega)\phi(n)$. Hence $x(t)$ must be the function $W_\tau \,\Delta\, f$.

PROBLEMS

1. For each of the following find the periodic solution of period τ:

(a) $(D^2 + 5D + 4)x = f(t); f = 1 + t^2, 0 \leq t < 1, \tau = 1$

(b) $(D^2 + 5D + 4)x = f(t); f = e^t, 0 \leq t < 1, \tau = 1$

(c) $(D + 2)x = f(t); f = t, 0 \leq t \leq 1; f = 2 - t, 1 \leq t \leq 2; \tau = 2$

(d) $(D^2 + 5D + 4)x = f(t); f = 0, 0 \leq t \leq 1; f = t - 1, 1 \leq t < 2; \tau = 2$

(e) $(D^2 + 2D + 2)x = f(t), f(t)$ as in part (b), $\tau = 1$

(f) $(D + 3)x = f(t), f(t) = h(t - 1) - h(t - 2), 0 \leq t \leq 3, \tau = 3$

2. (a) Show that $(D^2 + D)x = 1 + \sin t$ has no periodic solution.

(b) Let $f(t)$ have period τ and let $\Phi_\tau[f] = \phi(n)$; let $V(s)$ have simple roots $\pm ik\omega$, where k is an integer and let $V(in\omega) \neq 0$ for $n^2 \neq k^2$. Show that Eq. (4–210) has a solution of period τ if and only if $\phi(\pm k) = 0$. Is the solution unique? [*Hint:* Show that $\psi(n) = \Phi_\tau[x]$ can be chosen to satisfy $V(in\omega)\psi(n) = \phi(n)$ if and only if $\phi(\pm k) = 0$. To find x, let $Y(in\omega) = 1/V(in\omega)$ for $n^2 \neq k^2$, $Y(in\omega) = 0$ for $n^2 = k^2$. With the aid of Entry 5 of Table 4–4, show that $Y(in\omega)$ has inverse transform W_τ and that x can be chosen as $W_\tau \Delta f$.]

3. Prove (4–217).

ANSWERS

1. (a) $\frac{1}{32}(8t^2 - 20t + 29) - \dfrac{e^{-t}}{3(1 - e^{-1})} - \dfrac{e^{-4t}}{24(1 - e^{-4})}$

(b) $\dfrac{e^t}{10} + \dfrac{e}{6}e^{-t} + \dfrac{1 - e}{15(1 - e^{-4})}e^{-4t}$

(c) $\dfrac{e^{-2t}}{2(1 + e^{-2})} + \dfrac{(2t - 1)}{4} + [3 - 2t - e^{-2(t-1)}]\dfrac{h(t - 1)}{2}$

(d) $\frac{1}{48}h(t - 1)(16e^{1-t} - e^{4-4t} + 12t - 27)$

$$+ \frac{1}{3}\left(\frac{e^{-1-t}}{1 - e^{-2}} - \frac{3 + e^{-4}}{16(1 - e^{-8})}e^{-4t}\right)$$

(e) $\dfrac{e^t}{5} + \dfrac{1}{2i}\left[\dfrac{(e - 1)e^{(-1+i)t}}{(2 - i)(1 - e^{-1+i})} - \dfrac{(e - 1)e^{(-1-i)t}}{(2 + i)(1 - e^{-1-i})}\right]$

(f) $\frac{1}{3}\left\{[(1 - e^{-3t+3})h(t - 1) - (1 - e^{-3t+6})h(t - 2)] + \dfrac{e^{-3} - e^{-6}}{1 - e^{-9}}e^{-3t}\right\}$

2. (b) Solution is not unique since $\psi(\pm k)$ is arbitrary.

4–23 The weighting function. As in Section 2–8, the convolution integral can be interpreted as a weighted average. However, for the f-convolution the weighted average is over an interval of length τ, instead of over the whole "past".

Consider a differential equation

$$(a_0 D^m + \cdots + a_m)x = f(t), \tag{4-230}$$

with transfer function $Y(s)$, satisfying the conditions of Theorem 23. Then the unique periodic solution $x(t)$ is given by

$$x(t) = W_\tau(t) \,\Delta f = \int_0^\tau W_\tau(u)f(t - u) \, du, \tag{4-231}$$

where $W_\tau = \Phi_\tau^{-1}[Y(in\omega)]$. Thus the value of x at time t is obtained by averaging the values of $f(t)$ from time $t - \tau$ to time t, the value at time $t - u$ receiving weight $W_\tau(u)$. We call $W_\tau(t)$ the *weighting function*.

Accordingly, $x(t)$ can be computed graphically for each t as suggested in Fig. 4–9. We graph W_τ against u, with positive direction to the *left*. This graph is placed above that of $f(t)$, the origin lying above the point $(t, f(t))$. For each u, $0 \leq u \leq \tau$, the value of f at $t - u$ is multiplied by $W_\tau(u)$, which is simply the ordinate on the W-graph above $t - u$. Thus the product $f(t - u) \cdot W_\tau(u)$ is graphed over the interval from $t - \tau$ to t. The area under this curve, that is, the integral (4–231), is $x(t)$.

The total weight here is

$$\int_0^\tau W_\tau(u) \, du = Y(0) = \frac{1}{a_m}. \tag{4-232}$$

For

$$Y(in\omega) = \Phi_\tau[W_\tau] = \int_0^\tau W_\tau(u)e^{-in\omega u} \, du.$$

When $n = 0$, this equation reduces to (4–232). Accordingly, if $a_m = 1$, the total weight is 1 and (4–231) is a true weighted average. Otherwise (4–231) can be considered as a weighted average with a scale factor.

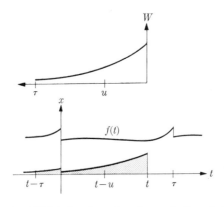

FIG. 4–9. Weighting function for periodic solutions.

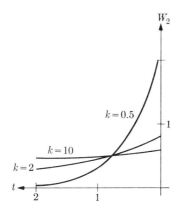

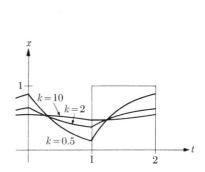

FIG. 4–10. Weighting function W_2 for $(kD + 1)x = f$.

FIG. 4–11. Periodic response to square wave for $(kD + 1)x = f$.

When $a_m = 1$, then $x(t)$ must be, on the average, of the same order of magnitude as $f(t)$, that is,

$$input \sim output. \tag{4–233}$$

In the extreme case when all coefficients in (4–230) are zero except $a_m = 1$, then $x = f(t)$. When $a_m = 1$, but all other coefficients are negligible compared to a_m, the approximation (4–233) is a good one. As the other coefficients become larger, the output will be a more and more distorted picture of the input.

EXAMPLE. $(kD + 1)x = f(t)$, where $f(t) = h(t - 1)$ for $0 \leqq t < 2$ and f has period 2. Thus the input is a square wave. Here $a_m = 1$, so that we have a true weighted average. The weighting function is

$$W_2 = \Phi_2^{-1}\left[\frac{1}{kin\pi + 1}\right] = \frac{1}{k}\frac{e^{-t/k}}{1 - e^{-2/k}}.$$

The graph of W_2 for various choices of k is shown in Fig. 4–10. As $k \to +\infty$, W_2 approaches the constant $\frac{1}{2}$; thus for very large k, $x(t)$ is simply the average of $f(t)$ from $t - \tau$ to t. This average is always $\frac{1}{2}$. Accordingly, $x(t) = \frac{1}{2}$ and $x(t)$ has no resemblance to $f(t)$. As $k \to 0$, $W_2(t)$ approaches a delta function; that is, W_2 approaches zero for all t except zero, where W_2 approaches $+\infty$, and $\int_0^2 W_2(t)\,dt$ is always 1. Accordingly, for very small k, $x(t)$ is obtained by averaging $f(t)$ from time $t - \tau$ to time t, with times very close to t receiving almost all the weight.

In Fig. 4–11, $x(t)$ is graphed for $k = 0.5$, $k = 2$, and $k = 10$. The gradual distortion of $f(t)$ is evident. The graphs can be obtained as described above, or by formula (Entry 6 of Table 4–3):

$$x = W_2 \,\Delta f = \frac{1}{k} \frac{e^{-t/k} \,\Delta f}{1 - e^{-2/k}} = \frac{1}{k} \gamma_a e^{at} \,\Delta f \quad (a = -1/k)$$

$$= (1 - e^{-(t-1)/k}) h(t-1) + \frac{1 - e^{-1/k}}{1 - e^{-2/k}} e^{-t/k},$$

for $0 \leqq t \leqq 2$. Hence

$$x = \frac{1 - e^{-1/k}}{1 - e^{-2/k}} e^{-t/k} \quad (0 \leqq t \leqq 1),$$

$$x = 1 + \frac{1 - e^{1/k}}{1 - e^{-2/k}} e^{-t/k} \quad (1 \leqq t \leqq 2).$$

4–24 Response to periodic generalized functions. We first consider the periodic response to the periodic input $\delta_\tau(t) = \delta(t) + \delta(t-1) + \delta(t+1) + \cdots$ as in Fig. 4–6.

THEOREM 25. *Under the hypotheses of Theorem 23, we have*

$$(a_0 D^m + \cdots + a_m) W_\tau(t) = \delta_\tau(t); \tag{4–241}$$

that is, the weighting function is the periodic response to a periodic impulse input.

Proof. Let $g(t)$ have period τ and equal $-t/\tau$ for $0 \leqq t < \tau$. We then verify that $g(t)$ has derivative $\delta_\tau(t) - (1/\tau)$. Let $y(t)$ be the solution of period τ of the differential equation

$$(a_0 D^m + \cdots + a_m) y = g(t). \tag{4–242}$$

Then Dy is a periodic solution of the equation

$$(a_0 D^m + \cdots + a_m) Dy = g'(t) = \delta_\tau(t) - \frac{1}{\tau}.$$

Furthermore, if $(\tau a_m)^{-1}$ denotes the constant function always equal to $1/(\tau a_m)$, we have

$$(a_0 D^m + \cdots + a_m)(\tau a_m)^{-1} = \frac{1}{\tau}.$$

Accordingly, $x = Dy + (\tau a_m)^{-1}$ has period τ and satisfies the equation

$$(a_0 D^m + \cdots + a_m) x = \delta_\tau(t). \tag{4–243}$$

It remains to show that $x = W_\tau(t)$. Now

$$\Phi_\tau[g(t)] = \Phi_\tau\left[-\frac{t}{\tau}\right] = \begin{cases} -\dfrac{\tau}{2} & (n = 0), \\[2mm] \dfrac{1}{in\omega} & (n \neq 0). \end{cases}$$

Hence from (4–242)

$$\Phi_\tau[y] = Y(in\omega)\Phi_\tau[g] = \begin{cases} -\dfrac{\tau}{2}\,Y(0) & (n = 0), \\[2mm] \dfrac{Y(in\omega)}{in\omega} & (n \neq 0), \end{cases}$$

and thus

$$\Phi_\tau[Dy] = in\omega\Phi_\tau[y] = \begin{cases} 0 & (n = 0), \\[1mm] Y(in\omega) & (n \neq 0). \end{cases}$$

Therefore

$$\Phi_\tau[x] = \Phi_\tau[Dy] + \Phi_\tau[(\tau a_m)^{-1}] = \begin{cases} 1/a_m & (n = 0), \\[1mm] Y(in\omega) & (n \neq 0). \end{cases}$$

But $Y(0) = 1/a_m$. Thus $\Phi_\tau[x] = Y(in\omega)$ and $x = W_\tau(t)$.

We can also obtain the conclusion by taking transforms in Eq. (4–243). By Eq. (4–200) $\Phi_\tau[\delta_\tau(t)] = 1$, so that we find

$$V(in\omega)\Phi_\tau[x] = 1$$

and have $\Phi_\tau[x] = Y(in\omega)$, $x = W_\tau(t)$. Finally we can write, from Eq. (4–243),

$$x = W_\tau \,\Delta\, \delta_\tau = W_\tau,$$

as in Section 4–20.

Theorem 25 allows us to conclude, as in Section 1–14, that $W_\tau(t)$, $W'_\tau(t), \ldots, W_\tau^{(m-2)}(t)$ are continuous, while $W_\tau^{(m-1)}(t)$ has jump discontinuities and $W_\tau^{(m)}(t)$, $W_\tau^{(m+1)}(t), \ldots$ are generalized functions.

By differentiation and translation we deduce from (4–241) the general rule

$$(a_0 D^m + \cdots + a_m)W_\tau^{(k)}(t - c) = \delta_\tau^{(k)}(t - c). \qquad (4\text{–}244)$$

By superposition, the periodic response to an arbitrary generalized function can now be found.

4–25 Systems of differential equations. We illustrate the method for a system of two equations

$$(\alpha_0 D^M + \cdots + \alpha_M)x_1 + (\beta_0 D^N + \cdots + \beta_N)x_2 = f_1,$$
$$(4\text{–}250)$$
$$(\gamma_0 D^M + \cdots + \gamma_M)x_1 + (\eta_0 D^N + \cdots + \eta_N)x_2 = f_2.$$

Here $\alpha_0, \alpha_1, \ldots, \beta_0, \ldots, \gamma_0, \ldots, \eta_0, \ldots, \eta_N$ are constants; M and N

are > 0, and we assume that

$$\begin{vmatrix} \alpha_0 & \beta_0 \\ \gamma_0 & \eta_0 \end{vmatrix} \neq 0, \tag{4-251}$$

so that (4–250) can be solved for the highest derivatives $D^M x_1$, $D^N x_2$. The functions f_1, f_2 are assumed to have period τ and to be piecewise continuous. We seek a solution pair $x_1(t)$, $x_2(t)$ of period τ.

We introduce the characteristic matrix (Section 2–9)

$$\begin{bmatrix} V_{11}(s) = \alpha_0 s^M + \cdots + \alpha_M & V_{12}(s) = \beta_0 s^N + \cdots + \beta_N \\ V_{21}(s) = \gamma_0 s^M + \cdots + \gamma_M & V_{22}(s) = \eta_0 s^N + \cdots + \eta_N \end{bmatrix}$$

with determinant

$$V(s) = \begin{vmatrix} V_{11}(s) & V_{12}(s) \\ V_{21}(s) & V_{22}(s) \end{vmatrix}.$$

By (4–251), $V(s)$ is a polynomial of degree $N + M$. We assume that $V(s) \neq 0$ for $s = in\omega$ ($n = 0, \pm 1, \pm 2, \ldots$).

It will be convenient below to introduce the corresponding differential operators $V_{11}(D) = \alpha_0 D^M + \cdots + \alpha_M, \ldots, V_{22}(D) = \eta_0 D^N + \cdots$, and

$$V(D) = \begin{vmatrix} V_{11}(D) & V_{12}(D) \\ V_{21}(D) & V_{22}(D) \end{vmatrix}.$$

We also employ the transfer matrix (Section 2–9)

$$\begin{bmatrix} Y_{11}(s) & Y_{12}(s) \\ Y_{21}(s) & Y_{22}(s) \end{bmatrix}.$$

This is obtained by solving (4–250) formally for x_1 and x_2 while replacing D by s:

$$x_1 = Y_{11}(s)f_1 + Y_{12}(s)f_2,$$

$$x_2 = Y_{21}(s)f_1 + Y_{22}(s)f_2.$$

Thus

$$Y_{11}(s) = \frac{V_{22}(s)}{V(s)}, \qquad Y_{12}(s) = \frac{-V_{12}(s)}{V(s)},$$

$$Y_{21}(s) = \frac{-V_{21}(s)}{V(s)}, \qquad Y_{22}(s) = \frac{V_{11}(s)}{V(s)}. \tag{4-252}$$

(The transfer matrix is the inverse of the characteristic matrix.)

Now let $x_1(t)$, $x_2(t)$ be the periodic solution sought. We then apply the finite Fourier transform to equations (4–250). Proceeding formally, we obtain the equations

$$V_{11}(in\omega)\psi_1 + V_{12}(in\omega)\psi_2 = \phi_1,$$
$$V_{21}(in\omega)\psi_1 + V_{22}(in\omega)\psi_2 = \phi_2; \tag{4-253}$$

where

$$\psi_1 = \Phi_\tau[x_1], \quad \psi_2 = \Phi_\tau[x_2], \quad \phi_1 = \Phi_\tau[f_1], \quad \phi_2 = \Phi_\tau[f_2].$$

We can solve (4–253) for ψ_1, ψ_2 since $V(in\omega) \neq 0$:

$$\psi_1 = Y_{11}(in\omega)\phi_1 + Y_{12}(in\omega)\phi_2,$$
$$\psi_2 = Y_{21}(in\omega)\phi_1 + Y_{22}(in\omega)\phi_2. \tag{4-254}$$

From its definition (4–252), $Y_{11}(s)$ is a ratio of a polynomial of degree N to one of degree $N + M$; hence $Y_{11}(s)$ can be expanded in partial fractions, and we conclude that $Y_{11}(in\omega)$ has an inverse transform $W_{11}(t)$. Similarly $W_{12}(t)$, $W_{21}(t)$, $W_{22}(t)$ are well defined. Equations (4–254) can now be written

$$\Phi_\tau[x_1] = \Phi_\tau[W_{11}\,\Delta f_1 + W_{12}\,\Delta f_2],$$
$$\Phi_\tau[x_2] = \Phi_\tau[W_{21}\,\Delta f_1 + W_{22}\,\Delta f_2].$$

Accordingly, we are led to the formulas

$$x_1 = W_{11}\,\Delta f_1 + W_{12}\,\Delta f_2,$$
$$x_2 = W_{21}\,\Delta f_1 + W_{22}\,\Delta f_2. \tag{4-255}$$

We can verify that these define a periodic solution of (4–250). For let $Y(s) = 1/V(s)$ and let $W(t) = \Phi_\tau^{-1}[Y(in\omega)]$, so that $W(t)$ is the weighting function for the differential equation $V(D)x = f$; hence

$$V(D)(W\,\Delta f) = f. \tag{4-256}$$

We know that W is periodic and has derivatives through order $N + M - 1$ (the derivative of order $N + M - 1$ being piecewise continuous), so that

$$\Phi_\tau[D^k W] = (in\omega)^k \Phi_\tau[W] = \frac{(in\omega)^k}{V(in\omega)}$$

for $k = 1, \ldots, N + M - 1$. Hence by superposition

$$\Phi_\tau[V_{22}(D)W] = V_{22}(in\omega)\Phi_\tau[W] = \frac{V_{22}(in\omega)}{V(in\omega)},$$

so that

$$W_{11}(t) = \Phi_\tau^{-1}\left[\frac{V_{22}(in\omega)}{V(in\omega)}\right] = V_{22}(D)W(t).$$

Now by the rule (4–175) we have $D^k(f \Delta g) = (D^k f \Delta g)$, provided $D^{k-1}f$ is continuous and piecewise smooth. Hence again by superposition

$$W_{11} \Delta f_1 = [V_{22}(D)W] \Delta f_1 = V_{22}(D)(W \Delta f_1).$$

Similarly,

$$W_{12} \Delta f_2 = -V_{12}(D)(W \Delta f_2), \quad W_{21} \Delta f_1 = -V_{21}(D)(W \Delta f_1),$$

$$W_{22} \Delta f_2 = V_{11}(D)(W \Delta f_2).$$

Accordingly,

$$
\begin{aligned}
V_{11}(D)x_1 + V_{12}(D)x_2 &= V_{11}(D)(W_{11} \Delta f_1 + W_{12} \Delta f_2) \\
&\quad + V_{12}(D)(W_{21} \Delta f_1 + W_{22} \Delta f_2) \\
&= V_{11}(D)[V_{22}(D)(W \Delta f_1) - V_{12}(D)(W \Delta f_2)] \\
&\quad + V_{12}(D)[-V_{21}(D)(W \Delta f_1) + V_{11}(D)(W \Delta f_2)] \\
&= V(D)(W \Delta f_1) = f_1
\end{aligned}
$$

by (4–256). Thus the first equation (4–250) is satisfied, and the second can be verified in the same way.

Now $W \Delta f_1$ and $W \Delta f_2$ have continuous derivatives through order $N + M - 1$, the derivatives of order $N + M$ being piecewise continuous. Hence

$$
\begin{aligned}
x_1 &= W_{11} \Delta f_1 + W_{12} \Delta f_2 = V_{22}(D)(W \Delta f_1) - V_{12}(D)(W \Delta f_2) \\
&= (\eta_0 D^N + \cdots)(W \Delta f_1) - (\beta_0 D^N + \cdots)(W \Delta f_2)
\end{aligned}
$$

has a piecewise smooth derivative $D^{M-1}x_1$; if $M = 1$, we can state only that x_1 is continuous and piecewise smooth. We conclude that x_1 can be represented by its Fourier series; similar remarks apply to x_2. The Fourier coefficients are given by (4–254). Hence

$$x_1(t) = \frac{1}{\tau} \sum_{n=-\infty}^{\infty} [Y_{11}(in\omega)\phi_1(n) + Y_{12}(in\omega)\phi_2(n)]e^{in\omega t},$$

$$(4\text{–}257)$$

$$x_2(t) = \frac{1}{\tau} \sum_{n=-\infty}^{\infty} [Y_{21}(in\omega)\phi_1(n) + Y_{22}(in\omega)\phi_2(n)]e^{in\omega t}.$$

Finally $x_1(t)$, $x_2(t)$ is the only periodic solution pair of (4–250); for if x_1, x_2 satisfy (4–250) and have period τ, then (4–254) follows, so that x_1, x_2 must be given by (4–255).

We summarize the conclusions as follows:

THEOREM 26. *Let the system of equations (4–250) be given, where the coefficients are constant, $M > 0$, $N > 0$, and (4–251) holds. Let $f_1(t)$,*

$f_2(t)$ be piecewise continuous and have period τ. Let $V_{11}(s), \ldots, V_{22}(s)$, $V(s)$, $Y_{11}(s), \ldots, Y_{22}(s)$ be defined as above and let $V(in\omega) \neq 0$ for $n = 0, \pm 1, \pm 2, \ldots,$ where $\omega = 2\pi/\tau$. Then (4–250) has a unique solution pair $x_1(t)$, $x_2(t)$ of period τ. The solution is given by (4–255), where $W_{kj} = \Phi_\tau^{-1}[Y_{kj}(in\omega)]$ or, equivalently, by the series (4–257), where $\phi_1(n) = \Phi_\tau[f_1]$, $\phi_2(n) = \Phi_\tau[f_2]$.

The theorem clearly generalizes to a general system of k equations in k unknowns.

EXAMPLE. $(D - 1)x - 2y = f,$ $12x - (D + 1)y = 0,$ where $f = h(t - 1)$ for $0 \leq t < 3$ and f has period 3. The condition (4–251) is satisfied:

$$\begin{vmatrix} 1 & 0 \\ 0 & -1 \end{vmatrix} \neq 0.$$

We also have

$$V(s) = \begin{vmatrix} s - 1 & -2 \\ 12 & -(s+1) \end{vmatrix} = 25 - s^2.$$

We now apply the finite Fourier transform. With the notation $s_n = in\omega = in2\pi/3$, we obtain the equations

$$(s_n - 1)\psi_1 - 2\psi_2 = \phi, \qquad 12\psi_1 - (s_n + 1)\psi_2 = 0,$$

where $\psi_1 = \Phi_3[x]$, $\psi_2 = \Phi_3[y]$, $\phi = \Phi_3[f]$. We solve for ψ_1, ψ_2:

$$\psi_1 = \frac{\begin{vmatrix} \phi & -2 \\ 0 & -(s_n + 1) \end{vmatrix}}{V(s_n)} = \frac{-(s_n + 1)}{25 - s_n^2}\phi,$$

$$\psi_2 = \frac{\begin{vmatrix} s_n - 1 & \phi \\ 12 & 0 \end{vmatrix}}{V(s_n)} = \frac{-12\phi}{25 - s_n^2}.$$

If we replace s_n by s, we obtain the transfer functions

$$Y_{11}(s) = \frac{-(s + 1)}{25 - s^2}, \qquad Y_{21}(s) = \frac{-12}{25 - s^2}.$$

The other two transfer functions are not needed, since the second differential equation is homogeneous. We have the partial fraction expansions

$$Y_{11} = \frac{1}{5}\left(\frac{3}{s - 5} + \frac{2}{s + 5}\right), \qquad Y_{21}(s) = \frac{6}{5}\left(\frac{1}{s - 5} - \frac{1}{s + 5}\right).$$

Hence

$$W_{11} = \Phi_3^{-1}[Y_{11}] = \tfrac{1}{5}(3\gamma_5 e^{5t} + 2\gamma_{-5}e^{-5t}),$$
$$W_{21} = \tfrac{6}{5}(\gamma_5 e^{5t} - \gamma_{-5}e^{-5t}),$$

and

$$x = W_{11}\,\Delta f = \tfrac{1}{5}(3\gamma_5 e^{5t}\,\Delta f + 2\gamma_{-5}e^{-5t}\,\Delta f), \qquad y = W_{21}\,\Delta f.$$

By Entry 6 of Table 4–3, for $\tau = 3$,

$$\gamma_a e^{at}\,\Delta f = a^{-1}[(e^{a(t-1)} - 1)h(t - 1) - \gamma_a e^{at}(1 - e^{2a})],$$

where $\gamma_a = (1 - e^{3a})^{-1}$. Hence

$$x = \tfrac{1}{25}[(3e^{5(t-1)} - 2e^{-5(t-1)} - 1)h(t - 1) - 3\gamma_5 e^{5t}(1 - e^{10})$$
$$+ 2\gamma_{-5}e^{-5t}(1 - e^{-10})],$$
$$\gamma_5 = (1 - e^{15})^{-1} = -3.06 \times 10^{-7}, \qquad \gamma_{-5} = (1 - e^{-15})^{-10} = 1.00.$$

A similar expression is obtained for y.

4–26 Integrodifferential equations. Since integrodifferential equations are reducible to simultaneous differential equations, the results of the preceding section are applicable. For example, the equation

$$(D + 3 + 2D^{-1})x = f(t) \tag{4–260}$$

is equivalent to the two equations

$$(D + 3)x + 2y = f(t), \qquad Dy = x. \tag{4–261}$$

We assume f is piecewise continuous and has period τ. We can apply the transform method to equations (4–261). Equivalent results are obtained by taking transforms in the integrodifferential equation (4–260) and applying the corollary to Theorem 14:

$$\Phi_\tau[D^{-k}x] = (in\omega)^{-k}\Phi_\tau[x] \quad (n \neq 0). \tag{4–262}$$

Here $D^{-k}x$ is always chosen to yield a periodic function; as in the corollary, such a choice is possible if $\int_0^\tau x\,dt = 0$, that is, if $\Phi_\tau[x] = 0$ for $n = 0$. From Eq. (4–260) we then obtain, for $n \neq 0$,

$$\left(in\omega + 3 + \frac{2}{in\omega}\right)\Phi_\tau[x] = \Phi_\tau[f] = \phi(n),$$
$$\Phi_\tau[x] = \frac{in\omega\phi(n)}{(in\omega)^2 + 3in\omega + 2}. \tag{4–263}$$

The same expression is obtained from equations (4–261) by the method of Section 4–25. We note that Eq. (4–263) forces $\Phi_\tau[x]$ to be 0 for $n = 0$.

For a general equation

$$(a_0 D^m + \cdots + a_m + a_{m+1}D^{-1} + \cdots + a_{m+p}D^{-p})x = f(t) \quad (4\text{–}264)$$

the periodic solution is expressible in terms of the transfer function

$$Y(s) = \frac{s^p}{a_0 s^{m+p} + \cdots + a_{m+p}} \quad (4\text{–}265)$$

as follows:

$$x = W_\tau \, \Delta \, f, \qquad W_\tau = \Phi_\tau^{-1}[Y(in\omega)]_1,$$

$$x = \frac{1}{\tau} \sum_{n=-\infty}^{\infty} \phi(n)\, Y(in\omega) e^{in\omega t}, \qquad \phi(n) = \Phi_\tau[f]. \quad (4\text{–}266)$$

Since $Y(0) = 0$, $\Phi_\tau[x] = 0$ for $n = 0$, so that $D^{-1}x, \ldots, D^{-p}x$ can be chosen to be periodic in (4–264); the arbitrary constant in $D^{-1}x, \ldots,$ $D^{-(p-1)}x$ is thereby fixed; that in $D^{-p}x$ is then determined by Eq. (4–264). For validity of (4–266) we must assume that $Y(in\omega)$ is defined for all n; that is, that no characteristic root (Section 2–9) has form $in\omega$.

Similar remarks apply to simultaneous integrodifferential equations. The formal procedures of Section 4–25 can be followed, with the aid of the rule (4–262). It is again necessary to assume that no characteristic root has form $in\omega$.

EXAMPLE.　$(D - D^{-1})x + (D + D^{-1})y = 1$,　$(2D -- D^{-1})x + (D + D^{-1})y = 0$. By the procedure of Section 2–9, the characteristic equation is

$$s^2 \begin{vmatrix} s - s^{-1} & s + s^{-1} \\ 2s - s^{-1} & s + s^{-1} \end{vmatrix} = s^2(-s^2 - 1) = 0.$$

Hence zero is a characteristic root. We therefore cannot apply the transform method. Subtraction of the two integrodifferential equations gives the equation $Dx = -1$. Hence $x = -t + c$ for some constant c, and x cannot be periodic, even though the input has every period.

Remark.　Integrodifferential equations in which the operators D_0^{-1}, $D_0^{-2}, \ldots$ appear cannot be analyzed as above, and such equations do not in general have periodic responses to periodic inputs. The difficulty arises from the *initial conditions* fixed by the operators D_0^{-k}; if $y = D_0^{-k}x$, then $y(0) = 0$, $y'(0) = 0, \ldots, y^{(k-1)}(0) = 0$. A periodic solution will in general not satisfy such conditions.

4–27 Systems analysis for periodic inputs.   The conclusions of the previous sections lead to the following picture of the performance of a stable linear system subject to inputs of period τ.

Let the system be described by a nondegenerate system of integrodifferential equations with constant coefficients. Let the inputs be $f_1(t)$, $f_2(t)$, ..., all of period τ and piecewise continuous. Let the outputs be $x_1(t)$, $x_2(t)$, ... Then after transients have died out, the outputs will be periodic, of period τ. For each output x_k there is a transfer function $Y_{kj}(s)$ of x_k relative to the input f_j. The function $Y_{kj}(in\omega)$ has an inverse Fourier transform $W_{kj}(t)$, the weighting function of x_k relative to f_j; $W_{kj}(t)$ is the output x_k due to a periodic delta function input f_j. In general, the output x_k due to input f_j is $W_{kj} \Delta f_j$, and the total output x_k is

$$x_k = W_{k1} \Delta f_1 + W_{k2} \Delta f_2 + \cdots + W_{kN} \Delta f_N \qquad (4\text{–}270)$$

if there are N inputs. Equation (4–270) is of the form

$$x_k = T_{k1}[f_1] + T_{k2}[f_2] + \cdots + T_{kN}[f_N], \qquad (4\text{–}271)$$

where the transformations T_{kj} are *linear*, and are defined for the linear space of all piecewise continuous functions of period τ. The outputs x_k are continuous and piecewise smooth functions of period τ and are represented by their Fourier series

$$x_k = \frac{1}{\tau} \sum_{n=-\infty}^{n=\infty} [Y_{k1}(in\omega)\phi_1(n) + \cdots + Y_{kN}(in\omega)\phi_N(n)]e^{in\omega t},$$

$$\phi_j(n) = \Phi_\tau [f_j]. \qquad (4\text{–}272)$$

PROBLEMS

1. Find the weighting function for period 1 for the following equation and graph for $a = 0.1$, $a = 1$, $a = 10$:

$$(3a^2 D^2 + 4aD + 1)x = f$$

2. Given the equation $(D^2 + 2D + 2)x = f$, find the weighting function for period 5, and graph.

3. For the equation of Problem 2, find the periodic response of period 5 if
(a) $f = 2[\delta(t - 1) + \delta(t - 6) + \delta(t + 4) + \cdots] = 2\,\delta_5(t - 1)$
(b) $f = \delta'(t) + \delta'(t - 5) + \delta'(t + 5) + \cdots = \delta'_5(t)$
(c) $f = \delta''(t) + \delta''(t - 5) + \delta''(t + 5) + \cdots = \delta''_5(t)$

4. Let a differential equation have weighting function $W_\tau(t)$. Let $f_\epsilon(t)$ be the periodic function of period τ, equal to $[h(t) - h(t - \epsilon)]/\epsilon$ for $0 \leq t < \tau$. Show that as $\epsilon \to 0$ the periodic response $x_\epsilon(t)$ to $f_\epsilon(t)$ approaches $W_\tau(t)$. [We can consider the periodic pulse $\delta_\tau(t)$ as $\lim f_\epsilon(t)$, as $\epsilon \to 0$.]

5. Given the integrodifferential equation $(2D + 5 + 3D^{-1})x = f$,
(a) find the transfer function and weighting function for period 2;

(b) find the output of period 2, if $f = h(t) - h(t - 1)$, $0 \le t < 2$, and f has period 2.

6. For each of the following integrodifferential equations show that the periodic output of period 1 is identically 0. Discuss the reason in each case.

(a) $(D + D^{-1})x = 1$

(b) $(D + 1 + D^{-1} + D^{-2})x = 1$

7. Show that for the system described by $(D + D^{-1})x = f$, the periodic output is the same for input f as for input $f + c$, $c = $ const (cf. Prob. 6(a)).

8. For each of the following find the periodic solution of period 2:

(a) $(2D + 2)x + (D + 1)y = f$, $Dx + (D + 1)y = 0$, $f = t$, $0 \le t < 2$

(b) $(2D + 4)x + (D + 2)y = f$, $2Dx + (2D + 1)y = g$, $f = g = h(t - 1) - h(t - 2)$, $0 \le t < 2$

(c) $(D + D^{-1})x + (D + 26D^{-1})y = f$, $(D - D^{-1})x + (2D + 10D^{-1})y = g$, $f = h(t) - h(t - 1)$, $g = h(t - 1) - h(t - 2)$, $0 \le t < 2$

9. In a certain system, the transfer function of output x relative to input f is given by the function $Y(s) = (s + 2)/(s^2 + 4s + 3)$.

(a) Find the weighting function for period 1.

(b) Write the output, for period 1, as a convolution and as a Fourier series.

ANSWERS

1. $W = \dfrac{1}{2a}\left[\dfrac{e^{-t/(3a)}}{1 - e^{-1/(3a)}} - \dfrac{e^{-t/a}}{1 - e^{-1/a}}\right]$

2. $e^{-t}\,[\sin t - e^{-5}\sin(t - 5)]/(1 + e^{-10} - 2e^{-5}\cos 5)$

3. (a) $2Ae^{-(t+4)}\,[\sin(t + 4) - e^{-5}\sin(t - 1)]$, $0 \le t \le 1$,

   $2Ae^{-(t-1)}\,[\sin(t - 1) - e^{-5}\sin(t - 6)]$, $1 \le t \le 5$, where

   $A = (1 + e^{-10} - 2e^{-5}\cos 5)^{-1}$

   (b) $Ae^{-t}\,[\cos t - \sin t + e^{-5}\{\sin(t - 5) - \cos(t - 5)\}]$

   (c) $Ae^{-t}[-2\cos t + 2e^{-5}\cos(t - 5)] + \delta_5(t)$

5. (a) $\dfrac{3/2}{s + (3/2)} - \dfrac{1}{s + 1}$, $\dfrac{(3/2)e^{-(3/2)t}}{1 - e^{-3}} - \dfrac{e^{-t}}{1 - e^{-2}}$

   (b) $(e^{-(3/2)(t-1)} - e^{-(t-1)})h(t - 1) + [e^{-t}(1 - e^{-1})/(1 - e^{-2})]$

   $- [e^{-3t/2}(1 - e^{-3/2})/(1 - e^{-3})]$

8. (a) $x = \tfrac{1}{2}t - \tfrac{1}{4} + e^{-2t}(1 - e^{-4})^{-1}$

   $y = -\tfrac{1}{2} + 2e^{-t}(1 - e^{-2})^{-1} - 2e^{-2}(1 - e^{-4})^{-1}$

   (b) $x = -[(e^{-t}\Delta f)/(1 - e^{-2})] + \tfrac{3}{2}[(e^{-2t}\Delta f)/(1 - e^{-4})]$

   $y = 2[(e^{-t}\Delta f)/(1 - e^{-2})] - 2[(e^{-2t}\Delta f)/(1 - e^{-4})]$

(c) $x = \left(\dfrac{14u - 9v}{5}\right)\Delta f - \left(\dfrac{7u - 6v}{2}\right)\Delta g,$

$y = \left(\dfrac{-8u + 3v}{10}\right)\Delta f + \left(\dfrac{2u - v}{2}\right)\Delta g,$ where

$u = [e^{-3t}/(1 - e^{-6})] + [e^{3t}/(1 - e^{6})],$

$v = [e^{-2t}/(1 - e^{-4})] + [e^{2t}/(1 - e^{4})]$

9. (a) $\frac{1}{2}[\{e^{-3t}/(1 - e^{-3})\} + \{e^{-t}/(1 - e^{-1})\}]$

(b) $\frac{1}{2}[\{(e^{-3t}\Delta f)/(1 - e^{-3})\} + \{(e^{-t}\Delta f)/(1 - e^{-1})\}],$

$$\sum_{n=-\infty}^{\infty} \frac{1 + 2\pi i n}{3 - 4\pi^2 n^2 + 8\pi i n}\, \phi(n)e^{2\pi i n t}, \quad \phi(n) = \Phi_1[f]$$

Suggested References

1. Ruel V. Churchill, *Fourier Series and Boundary Value Problems*. New York: McGraw-Hill, 1941.

2. Ruel V. Churchill, *Operational Mathematics*, 2nd ed. New York: McGraw-Hill, 1958.

3. R. Courant and D. Hilbert, *Methoden der Mathematischen Physik*. Vol. 1, 2nd ed., Berlin: Springer, 1931. Vol. 2, Berlin: Springer, 1937.

4. Édouard Goursat, *A Course in Mathematical Analysis*, Vol. 1 (transl. by E. R. Hedrick). New York: Ginn, 1904.

5. Dunham Jackson, *Fourier Series and Orthogonal Polynomials* (Carus Mathematical Monographs, No. 6). Menasha, Wisconsin: Mathematical Association of America, 1941.

6. Wilfred Kaplan, *Advanced Calculus*. Reading, Mass.: Addison-Wesley, 1952.

7. A. Zygmund, *Trigonometric Series*. 2 Vols. London: Cambridge University Press, 1959.

CHAPTER 5

THE FOURIER INTEGRAL AND FOURIER TRANSFORM

5-1 Introduction of the Fourier integral. We consider a stable linear system with transfer function $Y(s)$ of output x relative to input f. When f is of the form $Ae^{i\omega t}$, the steady-state solution for x is $AY(i\omega)e^{i\omega t}$. When f is of the form

$$f = \sum_{j=1}^{n} A_j e^{i\omega_j t}, \tag{5-10}$$

then it would be natural to call the output

$$x = \sum_{j=1}^{n} A_j Y(i\omega_j)e^{i\omega_j t} \tag{5-11}$$

the steady state corresponding to the given input. However, unless the ω_j are all integral multiples of some fundamental frequency ω_0, neither f nor x will be periodic (they are "almost periodic"); hence the meaning of "steady state" is not as clear as in the periodic case. We can clarify the term by remarking that f has the property of being bounded in absolute value for all t and so has x, as defined by (5–11). For

$$|f| \leqq \sum |A_j|, \qquad |x| \leqq \sum |A_j|\,|Y(i\omega_j)|.$$

Furthermore, x is the only output which is bounded for $-\infty < t < \infty$. For the general output will be formed of (5–11) plus terms of the complementary function. By the assumption of stability, each characteristic root has negative real part. Hence the exponential functions $e^{\lambda t}$ in the complementary function are all unbounded for *negative* t (they go to zero for positive t). Thus (5–11) represents the only output which is bounded for both positive and negative t.

We could generalize the input by permitting infinitely many terms in (5–10). This does not result in a Fourier series, unless all ω_j have a common submultiple. We shall disregard this case of an infinite series of the form (5–10) and immediately make one further passage to the limit, replacing (5–10) by an *integral:*

$$f(t) = \int_{-\infty}^{\infty} A(\omega)e^{i\omega t}\,d\omega. \tag{5-12}$$

Such an integral is called a *Fourier integral.* It can be regarded as a sum of infinitely many sinusoidal oscillations of frequencies ω, varying from

$-\infty$ to ∞, and associated amplitudes $A(\omega)$ [more precisely, $A(\omega)\,d\omega$].

If $f(t)$ has the form (5–12), then (proceeding formally) we expect a possible output to be

$$x(t) = \int_{-\infty}^{\infty} Y(i\omega)A(\omega)e^{i\omega t}\,d\omega. \qquad (5\text{–}13)$$

This will be shown to be valid, provided $f(t)$ remains reasonably small for large t (positive and negative) or more precisely, if

$$\int_{-\infty}^{\infty} |f(t)|\,dt$$

is finite. It will then be shown that $x(t)$ satisfies a similar condition:

$$\int_{-\infty}^{\infty} |x(t)|\,dt < \infty.$$

Again, it is natural to term $x(t)$ the *steady-state solution*. It is the only output x of the system satisfying such a smallness condition.

In this chapter we shall develop the theory of the Fourier integral and the associated *Fourier transform*.

5–2 Basic properties of the Fourier integral. We now ask the same questions as for Fourier series: What functions $f(t)$ can be represented in the form (5–12), that is, as a Fourier integral? How are the "coefficients" $A(\omega)$ determined? Are they uniquely determined?

We obtain clues to the answers by an intuitive passage to the limit from a Fourier series. We write the series (see Section 4–11) in the form

$$f(t) = \sum_{n=-\infty}^{\infty} c_n e^{in\lambda t}, \qquad c_n = \frac{\lambda}{2\pi}\int_{-\tau/2}^{\tau/2} f(t)e^{-in\lambda t}\,dt, \qquad (5\text{–}20)$$

where $\lambda = 2\pi/\tau$. We then write $A_n = c_n/\lambda$, so that

$$f(t) = \sum_{n=-\infty}^{\infty} A_n \lambda e^{in\lambda t}, \qquad A_n = \frac{1}{2\pi}\int_{-\tau/2}^{\tau/2} f(t)e^{-in\lambda t}\,dt.$$

Now we write $\omega_n = n\lambda$, so that λ can be considered as $\Delta\omega$:

$$\lambda = \Delta\omega = \omega_{n+1} - \omega_n.$$

We can also think of A_n as a function of ω evaluated at ω_n:

$$f(t) = \sum_{n=-\infty}^{\infty} A(\omega_n)e^{i\omega_n t}\,\Delta\omega, \qquad A(\omega_n) = \frac{1}{2\pi}\int_{-\tau/2}^{\tau/2} f(t)e^{-i\omega_n t}\,dt.$$

If we now let $\lambda = \Delta\omega$ approach zero, so that $\tau = 2\pi/\lambda \to \infty$, then we obtain the limiting expressions

$$f(t) = \int_{-\infty}^{\infty} A(\omega)e^{i\omega t}\, d\omega, \qquad A(\omega) = \frac{1}{2\pi}\int_{-\infty}^{\infty} f(t)e^{-i\omega t}\, dt. \qquad (5\text{-}21)$$

These are the fundamental relations for Fourier integrals. They are the counterpart of (5–20) for Fourier series. The two formulas in (5–21) can be combined into one:

$$f(t) = \frac{1}{2\pi}\int_{-\infty}^{\infty}\int_{-\infty}^{\infty} f(u)e^{i\omega(t-u)}\, du\, d\omega; \qquad (5\text{-}22)$$

this relation is known as *Fourier's identity*.

The preceding discussion does not assume $f(t)$ to be real. If $f(t)$ is real, we can write (5–21) in purely real form. We set $A(\omega) = \alpha(\omega) - i\beta(\omega)$. Then since f is real, the first formula of (5–21) becomes

$$f(t) = \int_{-\infty}^{\infty} \alpha(\omega)\cos\omega t\, d\omega + \int_{-\infty}^{\infty} \beta(\omega)\sin\omega t\, d\omega, \qquad (5\text{-}23)$$

and the second gives (by comparing real and imaginary parts)

$$\alpha(\omega) = \frac{1}{2\pi}\int_{-\infty}^{\infty} f(t)\cos\omega t\, dt, \qquad \beta(\omega) = \frac{1}{2\pi}\int_{-\infty}^{\infty} f(t)\sin\omega t\, dt. \qquad (5\text{-}24)$$

From these we see that

$$\alpha(-\omega) = \alpha(\omega),$$
$$\beta(-\omega) = -\beta(\omega);$$

thus both integrands in (5–23) are even, and if we set

$$a(\omega) = 2\alpha(\omega),$$
$$b(\omega) = 2\beta(\omega),$$

we obtain finally

$$f(t) = \int_{0}^{\infty} a(\omega)\cos\omega t\, d\omega + \int_{0}^{\infty} b(\omega)\sin\omega t\, d\omega,$$

$$(5\text{-}25)$$

$$a(\omega) = \frac{1}{\pi}\int_{-\infty}^{\infty} f(t)\cos\omega t\, dt, \qquad b(\omega) = \frac{1}{\pi}\int_{-\infty}^{\infty} f(t)\sin\omega t\, dt.$$

These are in direct analogy with the formulas for Fourier series in real form (Section 4–2).

If $f(t)$ is an even function $[f(-t) = f(t)]$, then $f(t) \sin \omega t$ is odd, so that $b(\omega) = 0$ and

$$\left. \begin{array}{l} f(t) = \displaystyle\int_0^\infty a(\omega) \cos \omega t \, d\omega \\[1.5em] a(\omega) = \dfrac{2}{\pi} \displaystyle\int_0^\infty f(t) \cos \omega t \, dt \end{array} \right\} \quad (f \text{ even}). \qquad (5\text{-}26)$$

Similarly, if $f(t)$ is an odd function $[f(-t) = -f(t)]$, then

$$\left. \begin{array}{l} f(t) = \displaystyle\int_0^\infty b(\omega) \sin \omega t \, d\omega \\[1.5em] b(\omega) = \dfrac{2}{\pi} \displaystyle\int_0^\infty f(t) \sin \omega t \, dt \end{array} \right\} \quad (f \text{ odd}). \qquad (5\text{-}27)$$

If $f(t)$ is real, we can take real parts in the Fourier identity (5–22), which becomes

$$f(t) = \frac{1}{2\pi} \int_{-\infty}^\infty \int_{-\infty}^\infty f(u) \cos \omega(t - u) \, du \, d\omega$$

or, since $\cos \omega(t - u)$ is *even* with respect to ω,

$$f(t) = \frac{1}{\pi} \int_0^\infty \int_{-\infty}^\infty f(u) \cos \omega(t - u) \, du \, d\omega. \qquad (5\text{-}28)$$

5–3 Fourier transforms. In the formulas (5–21), both $f(t)$ and $A(\omega)$ are functions of a real variable, defined from $-\infty$ to $+\infty$. We emphasize this similarity further by writing

$$f(t) = \frac{1}{\sqrt{2\pi}} \int_{-\infty}^\infty f_1(\omega) e^{i\omega t} \, d\omega,$$

$$f_1(\omega) = \frac{1}{\sqrt{2\pi}} \int_{-\infty}^\infty f(t) e^{-i\omega t} \, dt. \qquad (5\text{-}30)$$

Given f, the second equation defines a function f_1, which is often called the Fourier transform of f. The first formula shows that if $f_1(\omega)$ is the Fourier transform of $f(t)$, then $f(-t)$ is the Fourier transform of $f_1(\omega)$.

Despite the symmetry of these relations, we shall omit the numerical factor and define the *Fourier transform* of $f(t)$ to be

$$\Phi[f] = \phi(\omega) = \int_{-\infty}^\infty f(t) e^{-i\omega t} \, dt;$$

this is in direct analogy with the definition of the finite Fourier transform (Section 4–14). Indeed, for the finite transform of a periodic function,

$$\Phi_\tau[f] = \int_{-\tau/2}^{\tau/2} f(t) e^{-in\omega t}\, dt = \phi(n).$$

If we let $\tau \to \infty$, and replace $n\omega$ by ω_n or, in the limit, by ω, we can write

$$\Phi_\infty[f] = \int_{-\infty}^{\infty} f(t) e^{-i\omega t}\, dt = \phi(\omega).$$

Thus $\Phi[f]$ can be regarded as the limiting case of $\Phi_\tau[f]$.

With this definition of $\phi(\omega)$ the formulas (5–21) become

$$f(t) = \frac{1}{2\pi} \int_{-\infty}^{\infty} \phi(\omega) e^{i\omega t}\, d\omega,$$

$$\phi(\omega) = \int_{-\infty}^{\infty} f(t) e^{-i\omega t}\, dt. \tag{5–31}$$

These will be used consistently in the following discussion. With the new definition, $\phi(\omega)$ is the Fourier transform of $f(t)$, and $f(-t)$ is the Fourier transform of $\phi(\omega)/2\pi$.

If $f(t)$ is given only for $0 \leq t < \infty$, we can define $f(t)$ for negative t by the equation $f(-t) = f(t)$, so that the resulting function is *even*. Under appropriate conditions, (5–26) is then applicable. These give a representation of $f(t)$, defined for $0 \leq t < \infty$, as a *Fourier cosine integral*. As for (5–30), we can make the formulas symmetrical by writing

$$f(t) = \sqrt{2/\pi} \int_0^\infty f_1(\omega) \cos \omega t\, d\omega,$$

$$f_1(\omega) = \sqrt{2/\pi} \int_0^\infty f(t) \cos \omega t\, dt. \tag{5–32}$$

We call $f_1(\omega)$ the *Fourier cosine transform* of $f(t)$; hence $f(t)$ is also the Fourier cosine transform of $f_1(\omega)$. (Again the definition can be varied by dropping the numerical factor.)

In the same way (5–27) can be used for a function given for $0 \leq t < \infty$ and leads to the symmetrical formulas

$$f(t) = \sqrt{2/\pi} \int_0^\infty f_1(\omega) \sin \omega t\, d\omega,$$

$$f_1(\omega) = \sqrt{2/\pi} \int_0^\infty f(t) \sin \omega t\, dt. \tag{5–33}$$

Here $f(t)$ and $f_1(\omega)$ are *Fourier sine transforms* of each other.

5–4 Validity of the formulas. The preceding manipulations are purely formal and require justification. Above all, the integrals with infinite limits are *improper* and will have meaning only under special assumptions. From Section 1–8 we recall that

$$\int_a^\infty f(t)\,dt = \lim_{b\to\infty}\int_a^b f(t)\,dt,$$

if the limit exists. The integral $\int_{-\infty}^a f(t)\,dt$ is defined similarly, and

$$\int_{-\infty}^\infty f(t)\,dt = \int_0^\infty f(t)\,dt + \int_{-\infty}^0 f(t)\,dt;$$

both integrals on the right must exist in order that the integral on the left exist. However, when these conditions fail,

$$(P)\int_{-\infty}^\infty f(t)\,dt = \lim_{b\to\infty}\int_{-b}^b f(t)\,dt, \tag{5-40}$$

the *principal value* of the integral, may exist.

The existence of the integral can often be established by the *comparison theorem: Let $f(t)$ and $g(t)$ be piecewise continuous for $a \leqq t < \infty$; if $|f(t)| \leqq |g(t)|$ and $\int_a^\infty |g(t)|\,dt$ exists, then $\int_a^\infty f(t)\,dt$ exists.*

The integral $\int_a^\infty f(t)\,dt$ is called *absolutely convergent* if $\int_a^\infty |f(t)|\,dt$ exists. As a special case of the comparison theorem we note that if the integral is absolutely convergent, then it exists (or *converges*). Also, if the integral is absolutely convergent, then $\int_a^\infty f(t)e^{i\omega t}\,dt$ exists; for $|f(t)e^{i\omega t}| = |f(t)|$.

A similar discussion applies to integrals from $-\infty$ to a and from $-\infty$ to ∞, and all functions may be complex-valued. On the basis of these remarks, we can state the following theorem:

THEOREM 1. *Let $f(t)$ be piecewise continuous for $-\infty < t < \infty$, and let $\int_{-\infty}^\infty f(t)\,dt$ be absolutely convergent. Then the Fourier transform of f,*

$$\Phi[f] = \phi(\omega) = \int_{-\infty}^\infty f(t)e^{-i\omega t}\,dt, \tag{5-41}$$

is defined for $-\infty < \omega < \infty$.

This theorem gives conditions under which $\phi(\omega)$ has meaning. We can then attempt to form the Fourier integral for $f(t)$:

$$\frac{1}{2\pi}\int_{-\infty}^\infty \phi(\omega)e^{i\omega t}\,d\omega. \tag{5-42}$$

We state the following conditions for convergence of this integral to $f(t)$. We call $f(t)$ piecewise smooth for $-\infty < t < \infty$ if $f(t)$ is piecewise smooth on each finite interval (Section 4–7).

THEOREM 2. *Let $f(t)$ be piecewise smooth for $-\infty < t < \infty$ and let $\int_{-\infty}^{\infty} f(t)\,dt$ be absolutely convergent. Then the function $\phi(\omega)$ defined by Eq. (5–41) is continuous for all ω and the integral (5–42) converges for all t in the sense that*

$$\lim_{b\to\infty} \frac{1}{2\pi} \int_{-b}^{b} \phi(\omega)e^{i\omega t}\,d\omega = \frac{1}{2\pi}(P)\int_{-\infty}^{\infty} \phi(\omega)e^{i\omega t}\,d\omega \qquad (5\text{–}43)$$

exists. The value of the integral is $f(t)$, wherever $f(t)$ is continuous, and is equal to the average of left and right limits wherever $f(t)$ has jumps.

A proof of the theorem is given later in the chapter (Section 5–8). As a consequence, all the above formulas are justified if $f(t)$ satisfies the conditions described. From Theorem 2 it follows that $f(t)$ is uniquely determined by its Fourier transform, provided $f(t)$ is piecewise smooth.

5–5 Examples of Fourier integrals. EXAMPLE 1. Let $f(t) = 1$ for $-1 < t < 1, f(-1) = f(1) = \frac{1}{2}, f(t) = 0$ otherwise. Then

$$\phi(\omega) = \int_{-1}^{1} e^{-i\omega t}\,dt = \frac{e^{-i\omega} - e^{i\omega}}{-i\omega} = \frac{2\sin\omega}{\omega};$$

for $\omega = 0$, $\phi = 2$ [removable singularity of $(2/\omega)\sin\omega$]. Hence by Theorem 2,

$$\frac{1}{2\pi}(P)\int_{-\infty}^{\infty} \frac{2\sin\omega}{\omega}e^{i\omega t}\,d\omega = f(t) = \begin{cases} 1, & -1 < t < 1, \\ 0, & |t| > 1, \\ \frac{1}{2}, & t = \pm 1. \end{cases}$$

We remark that the imaginary part of the integral is

$$(P)\int_{-\infty}^{\infty} \frac{2\sin\omega\sin\omega t}{\omega}\,d\omega;$$

it can be verified that this integral exists, so that the principal value is not needed, except for $t = \pm 1$, when the integral reduces to

$$\pm\int_{-\infty}^{\infty} \frac{2\sin^2\omega}{\omega}\,d\omega.$$

However, the integrand is *odd*, so that in all cases the principal value is *zero*. Accordingly,

$$\frac{1}{\pi}(P)\int_{-\infty}^{\infty} \frac{\sin\omega\cos\omega t}{\omega}\,d\omega = f(t);$$

here the principal value is not needed since we can verify existence of the integral; in fact, the integrand is even, so that we can write

$$\frac{2}{\pi} \int_0^\infty \frac{\sin \omega \cos \omega t}{\omega} \, d\omega = f(t).$$

For $t = 0$, this gives the valuable formula

$$\int_0^\infty \frac{\sin \omega}{\omega} \, d\omega = \frac{\pi}{2}. \tag{5-50}$$

EXAMPLE 2.   Let $f(t) = a \cos at$, $-\pi/(2a) \leqq t \leqq \pi/(2a)$, $f(t) = 0$ otherwise.   Then

$$\phi(\omega) = \int_{-\pi/(2a)}^{\pi/(2a)} a \cos at \, e^{-i\omega t} \, dt = \frac{2a^2}{a^2 - \omega^2} \cos \frac{\pi \omega}{2a}$$

(removable singularity at $\omega = \pm a$), so that

$$f(t) = \frac{1}{2\pi} (P) \int_{-\infty}^{\infty} \frac{2a^2}{a^2 - \omega^2} \cos \frac{\pi \omega}{2a} e^{i\omega t} \, d\omega.$$

Here the integral is absolutely convergent since, for $\omega > a$,

$$\left| \frac{2a^2}{a^2 - \omega^2} \cos \frac{\pi \omega}{2a} e^{i\omega t} \right| \leqq \frac{2a^2}{\omega^2 - a^2}.$$

Taking real parts and taking advantage of evenness of $\phi(\omega)$, we find

$$f(t) = \frac{2a^2}{\pi} \int_0^\infty \frac{\cos \omega t}{a^2 - \omega^2} \cos \frac{\pi \omega}{2a} \, d\omega. \tag{5-51}$$

EXAMPLE 3.   Let $f(t) = e^{-at}$ for $t > 0$, $f(0) = \frac{1}{2}$, $f(t) = 0$ for $t < 0$. Then for $a > 0$ Theorem 2 is applicable:

$$\phi(\omega) = \int_0^\infty e^{-at} e^{-i\omega t} \, dt = \frac{1}{a + i\omega} ;$$

$$f(t) = \frac{1}{2\pi} (P) \int_{-\infty}^{\infty} \frac{e^{i\omega t}}{a + i\omega} \, d\omega.$$

Comparing real and imaginary parts, we find

$$f(t) = \frac{1}{2\pi} (P) \int_{-\infty}^{\infty} \frac{a \cos \omega t + \omega \sin \omega t}{a^2 + \omega^2} \, d\omega,$$

$$0 = \frac{1}{2\pi} (P) \int_{-\infty}^{\infty} \frac{a \sin \omega t - \omega \cos \omega t}{a^2 + \omega^2} \, d\omega.$$

It should be remarked that in the second formula the integrand is odd, so that the principal value is zero; the first integrand is even, so that we can write

$$f(t) = \frac{1}{\pi} \int_0^\infty \frac{a \cos \omega t + \omega \sin \omega t}{a^2 + \omega^2} \, d\omega. \tag{5-52}$$

It will be seen below that the *imaginary* part of the Fourier integral representation of a real function is always the integral of an odd function, and hence is zero; furthermore, the real part always converges, so that we do not need the principal value; it is needed only in some cases to evaluate the imaginary part to zero.

From (5–52) for $t = 0$, we find

$$\int_0^\infty \frac{a \, d\omega}{a^2 + \omega^2} = \pi f(0) = \frac{\pi}{2}. \tag{5-53}$$

This equation can be verified directly. For values of t other than zero the integral is difficult.

EXAMPLE 4. $F(t) = \sin t$. Here $\int_{-\infty}^\infty |f(t)| \, dt = \int_{-\infty}^\infty |\sin t| \, dt$ does not exist. The theory is *inapplicable*. Similar remarks apply to the following functions: $f(t) = 1$, $f(t) = t$, $f(t) = $ polynomial in t, and $f(t) = e^{at}$. (Fourier transforms of many such functions can be defined with the aid of generalized functions; see Section 5–16.)

PROBLEMS

1. For each of the following express $f(t)$ as a Fourier integral, wherever possible (express the answer in both complex and real forms):

(a) $f(t) = t, 0 \leqq t < c; f(t) = 0$ for $t < 0$ and for $t > c; f(c) = c/2$
(b) $f(t) = te^{-t}, -\infty < t < \infty$
(c) $f(t) = \sin t, 0 \leqq t \leqq \pi; f(t) = 0$ otherwise
(d) $f(t) = e^{-t} \sin 2t, t \geqq 0; f(t) = 0$ otherwise
(e) $f(t) = (\cos t)/(\pi^2 - 4t^2), -\infty < t < \infty$ [*Hint:* Interchange ω and t in Example 2 above.]
(f) $f(t) = e^{-|t|}, -\infty < t < \infty$
(g) $f(t) = 1/(1 + t^2), -\infty < t < \infty$ [*Hint:* Interchange t and ω in the relations of part (f).]

2. Express each of the following as Fourier cosine integrals:

(a) $f(t) = e^{-t}, t \geqq 0$ (b) $f(t) = 1 - t, 0 \leqq t \leqq 1$
                                          $f(t) = 0, t \geqq 1$

3. Represent the functions of Problem 2 as Fourier sine integrals.
4. Show that the substitutions

$$g(t) = f(t)e^{\sigma t}, \quad s = \sigma + i\omega \ (\sigma = \text{real const})$$

convert the formulas (5–31) into the formulas

$$g(t) = \frac{1}{2\pi i} \int_{-\infty}^{\infty} \psi(\sigma + i\omega)e^{st}i\, d\omega,$$

$$\psi(s) = \psi(\sigma + i\omega) = \int_{-\infty}^{\infty} g(t)e^{-st}\, dt.$$

[The first integral can be interpreted as an integral with respect to the complex variable s along the line $\sigma = \text{const}$, $-\infty < \omega < \infty$, with $ds = id\omega$. The second is the *two-sided Laplace transform* of $g(t)$. If $g(t) = 0$ for $t < 0$, this reduces to the ordinary Laplace transform $\psi(s) = \int_0^\infty g(t)e^{-st}\, dt$.]

5. Show that the Mellin transform

$$F(s) = \int_0^\infty f(x)x^{s-1}\, dx$$

can be written as a Laplace transform (Problem 4) after an appropriate substitution.

ANSWERS

1. (a) $\dfrac{1}{2\pi}(P)\displaystyle\int_{-\infty}^{\infty} \dfrac{(1 + i\omega c)e^{i\omega(t-c)} - e^{i\omega t}}{\omega^2}\, d\omega,$

$\dfrac{1}{\pi}\displaystyle\int_0^\infty \dfrac{\cos\omega(t - c) - \omega c\sin\omega(t - c) - \cos\omega t}{\omega^2}\, d\omega$

(b) impossible

(c) $\dfrac{1}{2\pi}\displaystyle\int_{-\infty}^{\infty} \dfrac{e^{i\omega t} + e^{i\omega(t-\pi)}}{1 - \omega^2}\, d\omega,$ $\dfrac{1}{\pi}\displaystyle\int_0^\infty \dfrac{\cos\omega t + \cos\omega(t - \pi)}{1 - \omega^2}\, d\omega$

(d) $\dfrac{1}{\pi}\displaystyle\int_{-\infty}^{\infty} \dfrac{e^{i\omega t}}{5 - \omega^2 + 2i\omega}\, d\omega,$ $\dfrac{2}{\pi}\displaystyle\int_0^\infty \dfrac{(5 - \omega^2)\cos\omega t + 2\omega\sin\omega t}{25 - 6\omega^2 + \omega^4}\, d\omega$

(e) $\dfrac{1}{4\pi}\displaystyle\int_{-1}^{1} \cos\dfrac{\pi\omega}{2}e^{i\omega t}\, d\omega,$ $\dfrac{1}{2\pi}\displaystyle\int_0^1 \cos\dfrac{\pi\omega}{2}\cos\omega t\, d\omega$

(f) $\dfrac{1}{\pi}\displaystyle\int_{-\infty}^{\infty} \dfrac{e^{i\omega t}}{1 + \omega^2}\, d\omega,$ $\dfrac{2}{\pi}\displaystyle\int_0^\infty \dfrac{\cos\omega t}{1 + \omega^2}\, d\omega$

(g) $\dfrac{1}{2}\displaystyle\int_{-\infty}^{\infty} e^{-|\omega|}e^{i\omega t}\, d\omega,$ $\dfrac{1}{2}\displaystyle\int_{-\infty}^{\infty} e^{-|\omega|}\cos\omega t\, d\omega$

2. (a) $\dfrac{2}{\pi}\displaystyle\int_0^\infty \dfrac{\cos\omega t}{1 + \omega^2}\, d\omega$ (b) $\dfrac{2}{\pi}\displaystyle\int_0^\infty \dfrac{1 - \cos\omega}{\omega^2}\cos\omega t\, d\omega$

3. (a) $\dfrac{2}{\pi}\displaystyle\int_0^\infty \dfrac{\omega\sin\omega t}{1 + \omega^2}\, d\omega$ (b) $\dfrac{2}{\pi}\displaystyle\int_0^\infty \dfrac{\omega - \sin\omega}{\omega^2}\sin\omega t\, d\omega$

5–6 Uniform convergence for improper integrals. An improper integral such as $\int_0^\infty f(t)\,dt$ is in many ways analogous to an infinite series. An integral $\int_0^\infty g(\omega, t)\,dt$ is analogous to a *series of functions*. We review the properties of such integrals here; they parallel those of Section 4–4. For proofs refer to Reference 2, pp. 342–348 and pp. 378–380. Integrals from $-\infty$ to 0 and from $-\infty$ to ∞ can be analyzed in the same way.

An integral $\int_0^\infty g(\omega, t)\,dt$ is said to *converge uniformly* to $F(\omega)$ for ω in a given interval (perhaps infinite) if $\int_0^\infty g(\omega, t)\,dt = F(\omega)$ and given $\epsilon > 0$, we can choose c so large that for $b > c$ the "partial integral" $\int_0^b g(\omega, t)\,dt$ differs from $F(\omega)$ by less than ϵ for all ω of the chosen interval:

$$\left| \int_0^b g(\omega, t)\,dt - F(\omega) \right| < \epsilon \quad (b > c).$$

THEOREM 3 (*M*-test). *Let $M(t)$ be defined for $t \geqq 0$, $M(t) \geqq 0$ and let $\int_0^\infty M(t)\,dt$ exist. If $g(\omega, t)$ is piecewise continuous in t for $0 \leqq t < \infty$ for each ω of a certain interval and $|g(\omega, t)| \leqq M(t)$ for all these values of ω and t, then $\int_0^\infty g(\omega, t)\,dt$ converges uniformly to a function $F(\omega)$ over the given interval of ω.*

THEOREM 4. *For $0 \leqq t < \infty$ and for a given interval of ω, let $g(\omega, t)$ be continuous in ω for each t and piecewise continuous in t for each ω. If $\int_0^\infty g(\omega, t)\,dt$ converges uniformly to $F(\omega)$ for ω in the given interval, then $F(\omega)$ is continuous in ω over this interval.*

THEOREM 5 (Integration of integrals). *Under the hypotheses of Theorem 4, if the finite interval $a \leqq \omega \leqq b$ is included in the given interval, then*

$$\int_a^b F(\omega)\,d\omega = \int_a^b \int_0^\infty g(\omega, t)\,dt\,d\omega = \int_0^\infty \int_a^b g(\omega, t)\,d\omega\,dt.$$

THEOREM 6. *For a certain interval of ω and for $0 \leqq t < \infty$, let $g_1(\omega, t)$, $g_2(\omega, t)$, $g(t)$ be piecewise continuous in t, let $|g(t)|$ be bounded, and let $\int_0^\infty g_1(\omega, t)\,dt$, $\int_0^\infty g_2(\omega, t)\,dt$ converge uniformly. Then over this interval of ω the following integrals converge uniformly:*

$$\int_0^\infty [g_1(\omega, t) \pm g_2(\omega, t)]\,dt, \qquad \int_0^\infty g(t)g_1(\omega, t)\,dt.$$

5–7 Preliminary lemmas. The main convergence theorem for Fourier integrals is Theorem 2 (Section 5–4). First we verify that $\phi(\omega)$, the Fourier transform of $f(t)$, is continuous for all ω. By assumption, $\int_{-\infty}^\infty f(t)\,dt$ is absolutely convergent. It follows that $\int_{-\infty}^\infty f(t)e^{-i\omega t}\,dt$ converges uniformly for all ω. For, with $M(t) = |f(t)|$, we know that $\int_{-\infty}^\infty M(t)\,dt$ exists, and

$$|f(t)e^{-i\omega t}| = |f(t)| = M(t).$$

Hence by Theorem 3 above (applied to the interval $-\infty < t < \infty$), $\int_{-\infty}^{\infty} f(t)e^{-i\omega t}\,dt$ converges uniformly to $\phi(\omega)$. By Theorem 4, $\phi(\omega)$ is continuous.

It remains to show that

$$\frac{1}{2\pi}\,(P)\int_{-\infty}^{\infty} \phi(\omega)e^{i\omega t}\,d\omega = f(t). \tag{5-70}$$

We shall assume $f(t)$ equals the average of left and right limits at its jump points, and then prove (5–70) for all t. The proof parallels that of Section 4–7.

LEMMA 1. *Let $g(t)$ be piecewise continuous for $a \leqq t \leqq b$. Then*

$$\lim_{u\to\infty}\int_a^b g(t)\sin ut\,dt = 0. \tag{5-71}$$

If we take real and imaginary parts, this reduces to Lemma 1 of Section 4–7.

LEMMA 2. *Let $g(t)$ be piecewise continuous for $a \leqq t < \infty$ and let $\int_a^\infty |g(t)|\,dt$ exist. Then*

$$\lim_{u\to\infty}\int_a^\infty g(t)\sin ut\,dt = 0. \tag{5-72}$$

Proof. Given $\epsilon > 0$, we can choose b so large that

$$\int_b^\infty |g(t)|\,dt < \frac{\epsilon}{2}.$$

This follows from convergence of $\int_a^\infty |g(t)|\,dt$. Next

$$\left|\int_a^b g(t)\sin ut\,dt\right| < \frac{\epsilon}{2}$$

for u sufficiently large, by Lemma 1. Hence

$$\left|\int_a^\infty g(t)\sin ut\,dt\right| \leqq \left|\int_a^b g(t)\sin ut\,dt\right| + \left|\int_b^\infty g(t)\sin ut\,dt\right|$$

$$< \frac{\epsilon}{2} + \int_b^\infty |g(t)|\,dt \quad < \frac{\epsilon}{2} + \frac{\epsilon}{2} = \epsilon$$

for u sufficiently large. Thus (5–72) is proved.

We now introduce the function

$$P_b(t) = \tfrac{1}{2}\int_{-b}^b e^{-i\omega t}\,d\omega \quad (b > 0), \tag{5-73}$$

which is the analogue of the function $P_n(s)$ of Section 4–7. From (5–73) we find at once:

$$P_b(t) = \frac{\sin bt}{t}. \tag{5–74}$$

Furthermore,

$$\frac{1}{\pi} \int_0^\infty P_b(t) \, dt = \frac{1}{\pi} \int_0^\infty \frac{\sin bt}{t} \, dt = k$$

exists (Reference 2, p. 375). The number k is independent of b, for

$$\int_0^\infty \frac{\sin bt}{t} \, dt = \int_0^\infty \frac{\sin u}{u} \, du \quad (b > 0),$$

by the substitution $u = bt$. Later we shall show that

$$\frac{1}{\pi} \int_0^\infty \frac{\sin bt}{t} \, dt = k = \tfrac{1}{2} \quad (b > 0). \tag{5–75}$$

We note that

$$\frac{1}{\pi} \int_0^\infty P_b(t) \, dt = \frac{1}{\pi} \int_{-\infty}^0 P_b(t) \, dt = \frac{1}{\pi} \int_{-\infty}^0 \frac{\sin bt}{t} \, dt$$

$$= \frac{1}{\pi} \int_0^\infty \frac{\sin u}{u} \, du = k. \tag{5–76}$$

LEMMA 3.   *Let $f(t)$ be piecewise continuous for $-\infty < t < \infty$ and let $\int_{-\infty}^\infty |f(t)| \, dt$ exist. Let*

$$S_b(t) = \frac{1}{2\pi} \int_{-b}^b \phi(\omega) e^{i\omega t} \, d\omega, \tag{5–77}$$

where $\phi(\omega) = \Phi[f]$. Then

$$S_b(t) = \frac{1}{\pi} \int_{-\infty}^\infty f(t + v) P_b(v) \, dv. \tag{5–78}$$

Proof. We have

$$S_b(t) = \frac{1}{2\pi} \int_{-b}^b \phi(\omega) e^{i\omega t} \, d\omega = \frac{1}{2\pi} \int_{-b}^b \left(\int_{-\infty}^\infty f(u) e^{-i\omega u} \, du \right) e^{i\omega t} \, d\omega$$

$$= \frac{1}{2\pi} \int_{-b}^b \int_{-\infty}^\infty f(u) e^{-i\omega(u-t)} \, du \, d\omega$$

$$= \frac{1}{2\pi} \int_{-b}^b \int_{-\infty}^\infty f(t + v) e^{-i\omega v} \, dv \, d\omega,$$

by the substitution $v = u - t$. Now the inner integral converges uniformly, by the absolute convergence of $\int_{-\infty}^{\infty} f(t + v)\, dv$. Hence by Theorem 5,

$$S_b(t) = \frac{1}{2\pi} \int_{-\infty}^{\infty} \int_{-b}^{b} f(t + v)e^{-i\omega v}\, d\omega\, dv$$

$$= \frac{1}{\pi} \int_{-\infty}^{\infty} f(t + v) \left(\frac{1}{2} \int_{-b}^{b} e^{-i\omega v}\, d\omega \right) dv$$

$$= \frac{1}{\pi} \int_{-\infty}^{\infty} f(t + v)P_b(v)\, dv.$$

5–8 Proof of Theorem 2. We now complete the proof of (5–70). By assumption, for each discontinuity t,

$$f(t) = \tfrac{1}{2}[f(t+) + f(t-)].$$

This equation remains true if f is continuous at t; for then $f(t+) = f(t-) = f(t)$. From (5–76) and (5–78) we can write

$$kf(t+) = \frac{1}{\pi} \int_{0}^{\infty} f(t+)P_b(v)\, dv,$$

$$kf(t-) = \frac{1}{\pi} \int_{-\infty}^{0} f(t-)P_b(v)\, dv,$$

$$S_b(t) = \frac{1}{\pi} \int_{0}^{\infty} f(t + v)P_b(v)\, dv + \frac{1}{\pi} \int_{-\infty}^{0} f(t + v)P_b(v)\, dv.$$

Hence

$$S_b(t) - k[f(t+) + f(t-)]$$

$$= \frac{1}{\pi} \int_{0}^{\infty} [f(t + v) - f(t+)]P_b(v)\, dv + \frac{1}{\pi} \int_{-\infty}^{0} [f(t + v) - f(t-)]P_b(v)\, dv.$$

We consider the first integral on the right. It can be written as

$$\frac{1}{\pi} \int_{0}^{\infty} \frac{f(t + v) - f(t+)}{v} \sin bv\, dv = \frac{1}{\pi} \int_{0}^{1} \frac{f(t + v) - f(t+)}{v} \sin bv\, dv$$

$$+ \frac{1}{\pi} \int_{1}^{\infty} \frac{f(t + v)}{v} \sin bv\, dv - \frac{f(t+)}{\pi} \int_{1}^{\infty} \frac{\sin bv}{v}\, dv.$$

Now the function

$$g(v) = \frac{f(t + v) - f(t+)}{v} \quad (v > 0)$$

has a limit as $v \to 0$, namely, $f'_+(t)$. If we assign this value to $g(v)$ at $v = 0$, then $g(v)$ becomes continuous at $v = 0$ and piecewise continuous for $v \geqq 0$. Thus the first term is of the form

$$\frac{1}{\pi} \int_0^1 g(v) \sin bv \, dv,$$

where $g(v)$ satisfies the hypotheses of Lemma 1, with b replacing u. Hence this term approaches 0 as $b \to \infty$. Next we observe that Lemma 2 is applicable to the second term, since $|f(t + v)/v| \leqq |f(t + v)|$, so that $f(t + v)/v$ is absolutely integrable from 1 to ∞; accordingly, this term also approaches 0 as $b \to \infty$. Finally, the third term can be written

$$-\frac{f(t+)}{\pi} \int_b^\infty \frac{\sin u}{u} \, du = -\frac{f(t+)}{\pi} \left[\int_0^\infty \frac{\sin u}{u} \, du - \int_0^b \frac{\sin u}{u} \, du \right],$$

by the substitution $u = bv$. Accordingly, this term has limit zero as $b \to \infty$. We conclude that

$$\frac{1}{\pi} \int_0^\infty [f(t + v) - f(t+)] P_b(v) \, dv \to 0 \qquad \text{as } b \to \infty.$$

Similarly the analogous integral from $-\infty$ to 0 approaches 0 as $b \to \infty$. Hence by (5–77)

$$\lim_{b \to \infty} S_b(t) = \frac{1}{2\pi} (P) \int_{-\infty}^\infty \phi(\omega) e^{i\omega t} \, d\omega = k[f(t+) + f(t-)]. \qquad (5\text{–}80)$$

If we knew that $k = \frac{1}{2}$, we would have proved our theorem:

$$\frac{1}{2\pi} (P) \int_{-\infty}^\infty \phi(\omega) e^{i\omega t} \, d\omega = \frac{f(t+) + f(t-)}{2} = f(t). \qquad (5\text{–}81)$$

To verify that $k = \frac{1}{2}$, we apply (5–80) to a particular function, e.g., to the function of Example 3 of Section 5–5. Here $f(t) = e^{-at}$ for $t > 0$, $f(t) = 0$ for $t < 0$, and $f(0) = \frac{1}{2}$; $\phi(\omega) = 1/(a + i\omega)$. Hence by (5–80)

$$\frac{1}{2\pi} (P) \int_{-\infty}^\infty \frac{e^{i\omega t}}{a + i\omega} \, d\omega = k[f(t+) + f(t-)].$$

If we take real parts and set $t = 0$, we obtain the equation [see (5–52) and (5–53)]

$$\frac{1}{\pi} \int_0^\infty \frac{a \, d\omega}{a^2 + \omega^2} = k[f(0+) + f(0-)] = 2kf(0) = k.$$

The integral on the left is an elementary one, equal to $\frac{1}{2}$. Hence

$$k = \frac{1}{\pi} \int_0^\infty \frac{\sin v}{v}\, dv = \frac{1}{2}.$$

Thus (5–81) is proved.

Remarks. The above proof is valid for $f(t)$ real or complex. If $f(t)$ is real, then

$$\phi(-\omega) = \overline{\phi(\omega)}. \tag{5–82}$$

For since f is real, then

$$\phi(-\omega) = \int_{-\infty}^\infty f(t)e^{i\omega t}\, dt = \int_{-\infty}^\infty \overline{f(t)e^{-i\omega t}}\, dt = \overline{\phi(\omega)}.$$

Thus if we write $\phi(\omega) = \alpha(\omega) + i\beta(\omega)$, then (5–82) reads

$$\alpha(-\omega) + i\beta(-\omega) = \alpha(\omega) - i\beta(\omega).$$

Hence

$$\alpha(-\omega) = \alpha(\omega), \qquad \beta(-\omega) = -\beta(\omega); \tag{5–83}$$

that is, $\alpha(\omega) = \text{Re}\,[\phi]$ is *even* and $\beta(\omega) = \text{Im}\,[\phi]$ is *odd*. From (5–77)

$$S_b(t) = \frac{1}{2\pi} \int_{-b}^b [\alpha(\omega)\cos\omega t - \beta(\omega)\sin\omega t]\, d\omega$$

$$+ \frac{i}{2\pi} \int_{-b}^b [\alpha(\omega)\sin\omega t + \beta(\omega)\cos\omega t]\, d\omega.$$

By (5–83), the imaginary part of S_b is the integral of an odd function and hence is zero for every b; therefore as $b \to \infty$, its limit is zero. The real part is the integral of an even function, hence equals twice the integral from 0 to b. Since we know the limit of S_b exists as $b \to \infty$, we conclude that for the real part the integrals from 0 to ∞ and from $-\infty$ to 0 exist separately (and are equal). Thus the principal value is needed in (5–80) and (5–81) only to take care of the imaginary part, which is an integral of an odd function.

From (5–78) and (5–81) we obtain an interesting corollary:

COROLLARY TO THEOREM 2. *Let $f(t)$ be piecewise smooth for $-\infty < t < \infty$, let $\int_{-\infty}^\infty |f(t)|\, dt$ exist and let $f(t)$ equal the average of left and right limits at jump points. Then*

$$f(t) = \lim_{b\to\infty} \frac{1}{\pi} \int_{-\infty}^\infty f(t+v) \frac{\sin bv}{v}\, dv. \tag{5–84}$$

Equation (5–84) is known as *Fourier's single integral representation of* $f(t)$.

Remark 1. As in Section 4–7, in Theorem 2 and in the corollary $f(t)$ need only be piecewise continuous; Eqs. (5–70) and (5–84) are then proved valid only at those points at which $f'_+(t)$ and $f'_-(t)$ exist.

Remark 2. A convergence theorem for Fourier transforms can be established for functions f for which $|f|^2$, rather than $|f|$, is integrable from $-\infty$ to ∞. The function $f = t/(1 + t^2)$ is of this type. It is convenient to allow functions with complicated discontinuities which are "measurable" and for which $|f|^2$ is integrable "in the Lebesgue sense" from $-\infty$ to ∞. (For theory of Lebesgue integrals see References 4 and 6.) These functions (complex-valued) form a linear space denoted by $L_2(-\infty, \infty)$. If $f(t)$ and $f_\alpha(t)$ are in $L_2(-\infty, \infty)$, for all $\alpha > 0$, then we write

$$f(t) = \underset{\alpha \to \infty}{\text{l.i.m.}}\, f_\alpha(t)$$

(l.i.m. meaning "limit in the mean") if

$$\lim_{\alpha \to \infty} \int_{-\infty}^{\infty} |f_\alpha(t) - f(t)|^2\, dt = 0.$$

The Fourier transform of a function $f(t)$ in $L_2(-\infty, \infty)$ is then defined as

$$\phi(\omega) = \underset{\alpha \to \infty}{\text{l.i.m.}}\, \frac{1}{\sqrt{2\pi}} \int_{-\alpha}^{\alpha} f(t) e^{-i\omega t}\, dt \tag{5–85}$$

(the factor $1/\sqrt{2\pi}$ being inserted, as in Section 5–3, to gain symmetry). It can be shown that $\phi(\omega)$ is well defined (except for a set of "measure zero") as a member of $L_2(-\infty, \infty)$ and that

$$f(t) = \underset{\alpha \to \infty}{\text{l.i.m.}}\, \frac{1}{\sqrt{2\pi}} \int_{-\alpha}^{\alpha} \phi(\omega) e^{i\omega t}\, d\omega. \tag{5–86}$$

Thus the Fourier transform defines a one-to-one transformation of the linear space $L_2(-\infty, \infty)$ onto itself. We have furthermore

$$\int_{-\infty}^{\infty} |\phi(\omega)|^2\, d\omega = \int_{-\infty}^{\infty} |f(t)|^2\, dt. \tag{5–87}$$

The results described form *Plancherel's theorem*. For a proof, refer to Chapter 1 of Reference 6. Equation (5–87) is *Parseval's relation*, which has important physical applications. [See Problem 8 following Section 5–12 for a proof of (5–87) under other assumptions.]

5–9 Uniqueness theorem.

THEOREM 7. *Let $f_1(t)$ and $f_2(t)$ be piecewise smooth for $-\infty < t < \infty$ and let $\int_{-\infty}^{\infty} |f_1(t)|\, dt$ and $\int_{-\infty}^{\infty} |f_2(t)|\, dt$ exist. Let $\Phi[f_1] = \phi_1(\omega)$ and $\Phi[f_2] = \phi_2(\omega)$. If $\phi_1(\omega) \equiv \phi_2(\omega)$, $-\infty < \omega < \infty$, then $f_1(t) = f_2(t)$, except perhaps at jump points.*

Proof. Since the hypotheses of Theorem 2 are satisfied,

$$f_1(t) = \frac{1}{2\pi}\,(P)\int_{-\infty}^{\infty} \phi_1(\omega)e^{i\omega t}\, d\omega = \frac{1}{2\pi}\,(P)\int_{-\infty}^{\infty} \phi_2(\omega)e^{i\omega t}\, d\omega = f_2(t),$$

except perhaps at the jump points.

As for Fourier series, the uniqueness theorem holds more generally, e.g., when "piecewise smooth" is replaced by "piecewise continuous."

Theorem 7 amounts to the assertion that *a function is uniquely determined by its Fourier transform.*

5–10 Properties of the Fourier transform.
The Fourier transform is defined for all f in the linear space of functions which are piecewise continuous for $-\infty < t < \infty$ and for which $\int_{-\infty}^{\infty} |f(t)|\, dt$ exists. Within this class it is a *linear operator.* For

$$\Phi[c_1 f_1 + c_2 f_2] = \int_{-\infty}^{\infty} [c_1 f_1(t) + c_2 f_2(t)]e^{-i\omega t}\, dt$$

$$= c_1 \int_{-\infty}^{\infty} f_1(t)e^{-i\omega t}\, dt + c_2 \int_{-\infty}^{\infty} f_2(t)e^{-i\omega t}\, dt$$

$$= c_1 \Phi[f_1] + c_2 \Phi[f_2].$$

Additional properties are given in Theorems 8 through 12.

THEOREM 8. *Let $f(t)$ be continuous and piecewise smooth for $-\infty < t < \infty$; let both $f(t)$ and $f'(t)$ have absolutely convergent integrals for $-\infty < t < \infty$. Then*

$$\Phi[f'] = i\omega\Phi[f]. \tag{5–100}$$

Proof. We first remark that $f(t)$ must $\to 0$ as $t \to \pm\infty$. For

$$f(t) = f(0) + \int_0^t f'(u)\, du.$$

Since $\int_0^\infty f'(u)\, du$ exists, $f(t)$ has a limit as $t \to +\infty$; a similar argument proves existence of a limit as $t \to -\infty$. Now if $f(t) \to c \neq 0$, as $t \to +\infty$, then $\int_0^\infty f(t)\, dt$ would diverge; hence $f(t) \to 0$ as $t \to +\infty$, and similarly $f(t) \to 0$ as $t \to -\infty$.

Now by integration by parts,

$$\Phi[f'] = \int_{-\infty}^{\infty} f'(t)e^{-i\omega t}\, dt = f(t)e^{-i\omega t}\Big|_{-\infty}^{\infty} + i\omega \int_{-\infty}^{\infty} f(t)e^{-i\omega t}\, dt.$$

Since $f(t)$ has limits zero as $t \to \pm\infty$, the first term is zero and $\Phi[f'] = i\omega\Phi[f]$.

By repeated application of (5–100), we obtain the rule

$$\Phi[D^k f] = (i\omega)^k \Phi[f] \quad (k = 1, 2, \ldots). \tag{5–101}$$

It is valid if $f, Df, \ldots, D^{k-1}f$ are continuous and $D^{k-1}f$ is piecewise smooth, and if $f, Df, \ldots, D^k f$ all have absolutely convergent integrals from $-\infty$ to $+\infty$.

Remark 1. Theorem 8 has a converse: if $\Phi[g] = i\omega\Phi[f]$, then $g = f'$. See the corollary to Theorem 16 in Section 5–12.

Remark 2. The rule (5–100) can be generalized to a function $f(t)$ which is piecewise smooth but has a finite number of jumps at $t = t_k$ $(k = 1, \ldots, N)$. We find

$$\Phi[f'] = i\omega\Phi[f] - \sum_{k=1}^{N} \epsilon_k e^{-i\omega t_k}, \tag{5–100'}$$

where

$$\epsilon_k = f(t_k+) - f(t_k-)$$

is the jump at t_k. In (5–100') $f'(t)$ is considered as a piecewise continuous function (absolutely integrable) equal to the derivative of f wherever the derivative exists; if, instead, $f'(t)$ were considered as a generalized function, then (5–100') would have to be changed—in fact, would have to become (5–100) again (see Section 5–16).

We prove (5–100') for $N = 1$. By integrating f' from t_1 to t and letting $t \to \pm\infty$, we again conclude that $f \to 0$ as $t \to \pm\infty$. We then find

$$\Phi[f'] = \int_{-\infty}^{t_1} f'(t)e^{-i\omega t}\, dt + \int_{t_1}^{\infty} f'(t)e^{-i\omega t}\, dt$$

$$= f(t)e^{-i\omega t}\Big|_{-\infty}^{t_1-} + f(t)e^{-i\omega t}\Big|_{t_1+}^{\infty} + i\omega \int_{-\infty}^{\infty} f(t)e^{-i\omega t}\, dt,$$

after integration by parts. If we take into account the different limits of f at $t_1\pm$ and the fact that $f(t)$ approaches zero as $t \to \pm\infty$, we find that

$$\Phi[f'] = e^{-i\omega t_1}[f(t_1-) - f(t_1+)] + i\omega\Phi[f],$$

in agreement with (5–100').

The rule (5–101) can be generalized similarly. For example, if f, Df, $\ldots$, $D^{k-1}f$ have jumps ϵ_0, ϵ_0', $\ldots$, $\epsilon_0^{(k-1)}$ only at $t = 0$, then

$$\Phi[D^k f] = (i\omega)^k \Phi[f] - [\epsilon_0(i\omega)^{k-1} + \epsilon_0'(i\omega)^{k-2} + \cdots + \epsilon_0^{(k-1)}]. \quad (5\text{--}101')$$

THEOREM 9. *Let $f(t)$ be piecewise continuous for $-\infty < t < \infty$; let $g(t) = \int_0^t f(u)\,du + c$, where c is a constant. Let both $f(t)$ and $g(t)$ have absolutely convergent integrals for $-\infty < t < \infty$. Then*

$$\Phi[g(t)] = \frac{\Phi[f]}{i\omega} \quad \text{for } \omega \neq 0, \quad\quad (5\text{--}102)$$

and for $\omega = 0$,

$$\Phi[g(t)] = \int_{-\infty}^{\infty} g(t)\,dt.$$

Proof. Since $g'(t) = f(t)$ (except at the jumps of f), the theorem is a consequence of Theorem 8. It should be remarked that only one choice of c is permitted here. For if $g(t)$ has an absolutely convergent integral from $-\infty$ to ∞, then $g(t) + c_1$ will not if c_1 is a constant other than zero.

We can write the rule (5–102) in the form

$$\Phi[D^{-1}f] = \frac{\Phi[f]}{i\omega} \quad (\omega \neq 0). \quad\quad (5\text{--}102')$$

Thus for *one* choice of the indefinite integral of f, (5–102′) is valid. Repetition of (5–102′) gives the rule

$$\Phi[D^{-k}f] = \frac{\Phi[f]}{(i\omega)^k} \quad (\omega \neq 0, \quad k = 1, 2, \ldots); \quad\quad (5\text{--}102'')$$

this is valid if $D^{-1}f$, $D^{-2}f$, $\ldots$, $D^{-k}f$ can be chosen to have absolutely convergent integrals for $-\infty < t < \infty$.

THEOREM 10. *Let $f(t)$ be piecewise continuous for $-\infty < t < \infty$ and let $\int_{-\infty}^{\infty} |f(t)|\,dt$ exist. Then for each constant c, $f(t - c)$ also has an absolutely convergent integral, and*

$$\Phi[f(t - c)] = e^{-i\omega c}\Phi[f]. \quad\quad (5\text{--}103)$$

THEOREM 11. *Let $i\alpha$ be a pure imaginary constant. Then under the hypotheses of Theorem 10, if $\Phi[f] = \phi(\omega)$,*

$$\Phi[e^{i\alpha t}f(t)] = \phi(\omega - \alpha). \quad\quad (5\text{--}104)$$

THEOREM 12. *For every function $f(t)$, $-\infty < t < \infty$, let Θf denote the reflection of f, that is, the function $f(-t)$. Let $\mathcal{C}f$ denote the conjugate of f,* *that is, the function $\bar{f}(t)$. Then Θ is a linear operator, but $\mathfrak{C}$ is not a linear operator. Furthermore,*

$$\Theta^2 f = \Theta[\Theta f] = f, \tag{5–105}$$

$$\Phi[\Theta f] = \Theta \Phi[f], \tag{5–106}$$

$$\Phi[\Phi[f]] = 2\pi \Theta f, \tag{5–107}$$

$$(\Theta f) \cdot (\Theta g) = \Theta(f \cdot g), \tag{5–108}$$

$$\Phi[\mathfrak{C} f] = \mathfrak{C}\Theta \Phi[f]. \tag{5–109}$$

For (5–106) and (5–109) it is assumed that $\Phi[f]$ exists; for (5–107) it is assumed that f and $\Phi[f]$ are piecewise smooth, and absolutely integrable from $-\infty$ to ∞.

The proofs of Theorems 10 through 12 are left as exercises (Problems 2 through 8 below).

On the basis of the properties listed, we can easily develop a table of Fourier transforms (Table 5–1, see next page). Various exercises below refer to the entries of this table.

PROBLEMS

1. Obtain the following entries in Table 1 as suggested:

(a) No. 2 from No. 1
(b) No. 5 from No. 4
(c) No. 6 from No. 4
(d) No. 7 from Nos. 4 and 6
(e) No. 8 from Nos. 4 and 6
(f) No. 9 by direct evaluation
(g) No. 9 from No. 4, by differentiating both f and $\Phi[f]$ with respect to the parameter a
(h) No. 10 from No. 2
(i) No. 11 from No. 7
(j) No. 12 from No. 11, by differentiation
(k) No. 13 from No. 11
(l) Nos. 14 and 15 from No. 13 (Why can we not simply take real and imaginary parts?)
(m) No. 17 by direct evaluation (Show that the discontinuity at $\omega = 0$ is removable and find the value $\phi(0)$.]
(n) No. 18 from Nos. 16 and 1, by writing the function as

$$2t[h(t) - h(t - c)] - t[h(t) - h(t - 2c)] + 2c[h(t - c) - h(t - 2c)]$$

(o) No. 18 from No. 1 by differentiating the function, finding $\Phi[f']$ and applying Theorem 9 to obtain $\Phi[f]$

2. Prove Theorem 10.   3. Prove Theorem 11.   4. Prove (5–105).
5. Prove (5–106). 6. Prove (5–107). 7. Prove (5–108). 8. Prove (5–109).

TABLE 5–1

FOURIER TRANSFORMS

No.	$f(t)$	$\Phi[f] = \int_{-\infty}^{\infty} f(t)e^{-i\omega t}\,dt = \phi(\omega)$
1	$h(t - c_1) - h(t - c_2),$ $c_1 < c_2$	$\dfrac{e^{-i\omega c_1} - e^{-i\omega c_2}}{i\omega}$
2	$h(t + c) - h(t - c),$ $c > 0$	$\dfrac{2 \sin c\omega}{\omega}$
3	$e^{at}[h(t - c_1) - h(t - c_2)],$ $c_1 < c_2$	$\dfrac{e^{(a-i\omega)c_2} - e^{(a-i\omega)c_1}}{a - i\omega}$
4	$e^{at}h(t),\quad \mathrm{Re}\,(a) < 0$	$\dfrac{1}{i\omega - a}$
5	$e^{at}h(t - c),\quad \mathrm{Re}\,(a) < 0$	$\dfrac{e^{(a-i\omega)c}}{i\omega - a}$
6	$e^{-at}h(-t),\quad \mathrm{Re}\,(a) < 0$	$\dfrac{-1}{i\omega + a}$
7	$e^{a|t|},\quad \mathrm{Re}\,(a) < 0$	$\dfrac{-2a}{\omega^2 + a^2}$
8	$e^{at}h(t) - e^{-at}h(-t),$ $\mathrm{Re}\,(a) < 0$	$\dfrac{-2i\omega}{\omega^2 + a^2}$
9	$t^k e^{at}h(t),\quad k = 1, 2, \ldots,$ $\mathrm{Re}\,(a) < 0$	$\dfrac{k!}{(i\omega - a)^{k+1}}$
10	$e^{ibt}[h(t + c) - h(t - c)],$ $c > 0$	$\dfrac{2 \sin c(\omega - b)}{\omega - b}$
11	$\dfrac{1}{a^2 + t^2},\quad \mathrm{Re}\,(a) < 0$	$-\dfrac{\pi}{a}\,e^{a|\omega|}$
12	$\dfrac{t}{(a^2 + t^2)^2},\quad \mathrm{Re}\,(a) < 0$	$\dfrac{i\omega\pi}{2a}\,e^{a|\omega|}$
13	$\dfrac{e^{ibt}}{a^2 + t^2},\quad \mathrm{Re}\,(a) < 0,\quad b\ \text{real}$	$-\dfrac{\pi}{a}\,e^{a|\omega - b|}$

Table 5-1 Continued

No.	$f(t)$	$\Phi[f] = \int_{-\infty}^{\infty} f(t) e^{-i\omega t}\, dt = \phi(\omega)$
14	$\dfrac{\cos bt}{a^2 + t^2},\quad \mathrm{Re}\,(a) < 0,\quad b\ \text{real}$	$-\dfrac{\pi}{2a}[e^{a|\omega-b|} + e^{a|\omega+b|}]$
15	$\dfrac{\sin bt}{a^2 + t^2},\quad \mathrm{Re}\,(a) < 0,\quad b\ \text{real}$	$-\dfrac{\pi}{2ai}[e^{a|\omega-b|} - e^{a|\omega+b|}]$
16	$t[h(t) - h(t - c)]$	$\dfrac{1 - e^{-i\omega c}(1 + i\omega c)}{-\omega^2}$
17	$t^k[h(t) - h(t - c)],$ $k = 1, 2, \ldots$	$\dfrac{k! - e^{-i\omega c}g_k(i\omega c)}{(i\omega)^{k+1}},$ $g_k(x) = x^k + kx^{k-1} + \cdots + k!$
18	$t[h(t) - h(t - c)] + (2c - t)$ $\times\, [h(t - c) - h(t - 2c)]$	$\dfrac{1 - 2e^{-i\omega c} + e^{-2i\omega c}}{-\omega^2}$
19	$e^{-at^2},\quad a > 0$	$\sqrt{\dfrac{\pi}{a}}\, e^{-\omega^2/(4a)}$

9. Establish Entry 19 of Table 5-1. [*Hint:* Show by integration by parts that the transform $\phi(\omega)$ satisfies the differential equation $\phi'(\omega) = -\omega\phi/(2a)$, so that $\phi = \text{const} \cdot e^{-\omega^2/(4a)}$. To evaluate the constant, evaluate ϕ for $\omega = 0$ (see Reference 2, p. 218).]

10. Prove that if $\Phi[f(t)] = \phi(\omega)$ and $a > 0$, then $\Phi[f(at)] = a^{-1}\phi(\omega/a)$.

5-11 Convolution. Let $f(t)$ and $g(t)$ be piecewise continuous for $-\infty < t < \infty$. The convolution of f and g is defined to be the function

$$F(t) = f * g = \int_{-\infty}^{\infty} f(u)g(t - u)\, du. \qquad (5\text{-}110)$$

It will be seen that, under appropriate conditions, $F(t)$ is well defined and continuous for $-\infty < t < \infty$. An important special case is that in which $f(t) \equiv 0$ for $t < 0$ and $g(t) \equiv 0$ for $t < 0$. Then (5-110) becomes

$$f * g = \int_0^t f(u)g(t - u)\, du. \qquad (5\text{-}110')$$

In the theory of the Laplace transform, we consider only functions equal to zero for $t < 0$; hence the definition (5-110') is the correct one for that case.

The convolution has the basic property

$$\Phi[f * g] = \Phi[f] \cdot \Phi[g]. \tag{5-111}$$

This will be proved below (under appropriate hypotheses). It is instructive to derive the rule formally:

$$\Phi[f * g] = \int_{-\infty}^{\infty} \int_{-\infty}^{\infty} f(u)g(t - u)e^{-i\omega t} \, du \, dt$$

$$= \int_{-\infty}^{\infty} \int_{-\infty}^{\infty} f(u)g(t - u)e^{-i\omega t} \, dt \, du$$

$$= \int_{-\infty}^{\infty} f(u) \left(\int_{-\infty}^{\infty} g(t - u)e^{-i\omega(t-u)} \, dt \right) e^{-i\omega u} \, du$$

$$= \int_{-\infty}^{\infty} f(u) \left(\int_{-\infty}^{\infty} g(v)e^{-i\omega v} \, dv \right) e^{-i\omega u} \, du$$

$$= \int_{-\infty}^{\infty} f(u)e^{-i\omega u} \, du \int_{-\infty}^{\infty} g(v)e^{-i\omega v} \, dv$$

$$= \Phi[f] \cdot \Phi[g].$$

THEOREM 13. *Let $f(t)$ and $g(t)$ be piecewise continuous for $-\infty < t < \infty$, let both functions have absolutely convergent integrals for $-\infty < t < \infty$, and let one of the two functions be bounded; for example, $|f(t)| \leq M$ for all t. Then $F(t) = f * g$ is defined for all t, is continuous for all t, and has an absolutely convergent integral for $-\infty < t < \infty$. Furthermore, the identity (5–111) is valid, and*

$$f * g = g * f. \tag{5-112}$$

If g_1 and g_2 satisfy the same conditions as g, then

$$f * (g_1 + g_2) = f * g_1 + f * g_2. \tag{5-113}$$

Proof. If $|f(t)| \leq M$, then

$$\left| \int_0^b |f(u)g(t - u)| \, du \right| \leq M \left| \int_0^b |g(t - u)| \, du \right| \cdot$$

By the assumptions on $g(t)$, the integral on the right has a finite limit as $b \to \pm\infty$. Hence the integral in (5–110) exists for all t:

$$F(t) = \lim_{b \to \infty} \int_0^b f(u)g(t - u) \, du + \lim_{c \to -\infty} \int_c^0 f(u)g(t - u) \, du.$$

Now

$$g * f = \int_{-\infty}^{\infty} g(u)f(t - u) \, du = \int_{-\infty}^{\infty} f(v)g(t - v) \, dv,$$

by the substitution $v = t - u$. Hence $g * f = f * g$. The proof of (5–113) is left as an exercise (Problem 4 below).

Now we can write

$$F(t) = f * g = \int_{-\infty}^{\infty} g(u)f(t - u)\, du$$

$$= \lim_{b \to \infty} \int_0^b g(u)f(t - u)\, du + \lim_{c \to -\infty} \int_c^0 g(u)f(t - u)\, du.$$

The convergence in both cases is uniform for all t; for the error is

$$\int_b^{\infty} g(u)f(t - u)\, du$$

or

$$\int_{-\infty}^c g(u)f(t - u)\, du;$$

since $|f| \leqq M$, the absolute error is, respectively, at most

$$M \int_b^{\infty} |g(u)|\, du$$

or

$$M \int_{-\infty}^c |g(u)|\, du.$$

Since $g(t)$ is absolutely integrable from $-\infty$ to ∞, each can be made as small as desired by choosing b sufficiently large positive or c sufficiently large negative. Thus the convergence is uniform for all t. Since g and f are piecewise continuous, $\int_0^b g(u)f(t - u)\, du$ is continuous in t; see Section 2–8. Thus $F(t) = f * g$ is a uniform limit of continuous functions [for example, $F(t) = \lim_{n \to \infty} \int_{-n}^n g(u)f(t - u)\, du$]; hence $F(t)$ is continuous for all t (see pp. 345–346 of Reference 2).

We proceed to prove (5–111). To this end we first remark that it is sufficient to prove the formula for two cases: (i) $f \equiv 0$ for $t < 0$, $g \equiv 0$ for $t < 0$; (ii) $f \equiv 0$ for $t < 0$, $g \equiv 0$ for $t > 0$. For we can write

$$f = f(t)h(t) + f(t)[1 - h(t)] = f_1 + f_2,$$

$$g = g(t)h(t) + g(t)[1 - h(t)] = g_1 + g_2.$$

Then by (5–113),

$$f * g = (f_1 + f_2) * (g_1 + g_2) = f_1 * g_1 + f_2 * g_1 + f_1 * g_2 + f_2 * g_2.$$

Now $f_1 = f(t)h(t) \equiv 0$ for $t < 0$, $g_1 = g(t)h(t) \equiv 0$ for $t < 0$, etc. Hence each of the products $f_1 * g_1, f_2 * g_1, \ldots$ is of one of the types, (i) or (ii) (or one reducible to such by replacing t by $-t$). If we have proved (5–111) and absolute integrability of the convolution for these cases, then

$$\Phi[f * g] = \Phi[f_1 * g_1 + f_2 * g_1 + f_1 * g_2 + f_2 * g_2]$$
$$= \Phi[f_1 * g_1] + \Phi[f_2 * g_1] + \cdots$$
$$= \Phi[f_1]\Phi[g_1] + \Phi[f_2]\Phi[g_1] + \cdots$$
$$= \Phi[f_1 + f_2]\Phi[g_1 + g_2] = \Phi[f]\Phi[g].$$

Thus (5–111) and absolute integrability of $f * g$ will follow in general. We consider (i). In this case

$$\Phi[f]\Phi[g] = \int_0^\infty f(u)e^{-i\omega u}\, du \int_0^\infty g(v)e^{-i\omega v}\, dv.$$

We remark that this is the same as the double integral

$$\iint_R f(u)g(v)e^{-i\omega u}e^{-i\omega v}\, du\, dv = \iint_R \psi(u, v)\, du\, dv$$

over the first quadrant $R: 0 \leq u < \infty, 0 \leq v < \infty$. The double integral is an *improper one* and is said to converge and have value k if for every $\epsilon > 0$ we can choose a bounded region R_0 in R such that, for every bounded region R_1 containing R_0 and in R,

$$\left| \iint_{R_1} \psi(u, v)\, du\, dv - k \right| < \epsilon.$$

The improper integral is called absolutely convergent when

$$\iint_R |\psi(u, v)|\, du\, dv$$

exists; absolute convergence implies convergence. To test for absolute convergence it is sufficient to show that for an expanding sequence of regions R_n which approach R as limit, the integral over R_n approaches a limit; for example, here we need only consider the integrals over squares $0 \leq u \leq b, 0 \leq v \leq b$, with side b tending to ∞ (see p. 217 of Reference 2). Applying the test, we consider

$$\int_0^b \int_0^b |f(u)g(v)e^{-i\omega u}e^{-i\omega v}|\, du\, dv = \int_0^b \int_0^b |f(u)|\, |g(v)|\, du\, dv$$
$$= \int_0^b |f(u)|\, du \int_0^b |g(v)|\, dv.$$

As $b \to \infty$, this has a limit, by the assumptions on f and g.

We now consider $\Phi[F] = \Phi[f * g]$. For (i),

$$F(t) = \int_0^t f(u)g(t-u)\,du \quad (t \geqq 0),$$

while $F(t) = 0$ for $t \leqq 0$. Accordingly,

$$
\begin{aligned}
\Phi[F] &= \int_0^\infty F(t)e^{-i\omega t}\,dt = \lim_{b\to\infty} \int_0^b F(t)e^{-i\omega t}\,dt \\
&= \lim_{b\to\infty} \int_0^b \int_0^t f(u)g(t-u)e^{-i\omega t}\,du\,dt \\
&= \lim_{b\to\infty} \int_0^b \int_u^b g(t-u)e^{-i\omega(t-u)}\,dt\, f(u)e^{-i\omega u}\,du \\
&= \lim_{b\to\infty} \int_0^b \int_0^{b-u} g(v)e^{-i\omega v}\,dv\, f(u)e^{-i\omega u}\,du.
\end{aligned}
$$

The integral is equal to a double integral over the triangle R_1: $0 \leqq u \leqq b$, $0 \leqq v \leqq b - u$. As $b \to \infty$, R_1 approaches the whole first quadrant R. Hence the limit exists and equals the double integral over the quadrant, which was seen above to equal $\Phi[f]\Phi[g]$. Hence for (i)

$$\Phi[f * g] = \Phi[f]\Phi[g].$$

A similar argument shows that $f * g$ is absolutely integrable from $-\infty$ to ∞.

In case (ii) we have

$$F(t) = \int_0^\infty f(u)g(t-u)\,du = \lim_{c\to\infty} \int_0^c f(u)g(t-u)\,du.$$

Hence

$$\Phi[F] = \int_{-\infty}^\infty F(t)e^{-i\omega t}\,dt$$

$$= \lim_{b\to\infty} \int_0^b F(t)e^{-i\omega t}\,dt + \lim_{B\to-\infty} \int_B^0 F(t)e^{-i\omega t}\,dt.$$

Now

$$\int_0^b F(t)e^{-i\omega t}\,dt = \int_0^b \int_0^\infty f(u)g(t-u)\,du\, e^{-i\omega t}\,dt$$

$$= \int_0^\infty \int_0^b g(t-u)e^{-i\omega(t-u)}\,dt\, f(u)e^{-i\omega u}\,du,$$

the interchange of order being justified by uniform convergence. Next

$$\int_0^b F(t)e^{-i\omega t}\,dt = \int_0^\infty \int_{-u}^{b-u} g(v)e^{-i\omega v}\,dv\, f(u)e^{-i\omega u}\,du.$$

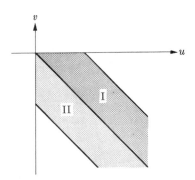

FIG. 5–1.   Proof of convolution theorem.

Since $g(v) = 0$ for $v > 0$, this is equal to the double integral over the region I of Fig. 5–1, which exists for the same reason as above. As $b \to \infty$, this tends to the double integral over the half-quadrant $0 \leqq u < \infty$, $-u \leqq v \leqq 0$. Similarly the integral from B to 0 represents the double integral over region II of Fig. 5–1. As $B \to -\infty$, this has as limit the integral over the other half-quadrant. Hence the sum of the two limits, which is $\Phi[F]$, is equal to the absolutely convergent double integral

$$\iint_R f(u)g(v)e^{-i\omega u}e^{-i\omega v}\, du\, dv,$$

where R is the fourth quadrant. But, as above, this integral is equal to

$$\int_0^\infty f(u)e^{-i\omega u}\, du \int_{-\infty}^0 g(v)e^{-i\omega v}\, dv = \Phi[f] \cdot \Phi[g].$$

This proves (5–111) for (ii), and a similar argument shows integrability of $|F(t)|$. Thus the proof is complete.

COROLLARY. *Let $f(t)$ and $g(t)$ be piecewise smooth and absolutely integrable for $-\infty < t < \infty$; let $\Phi[f] = \phi$ and $\Phi[g] = \psi$ also be absolutely integrable for $-\infty < \omega < \infty$. Then the convolution of ϕ and ψ is well defined:*

$$\phi * \psi = \int_{-\infty}^\infty \phi(u)\psi(\omega - u)\, du; \tag{5–114}$$

$\phi * \psi$ is continuous for $-\infty < \omega < \infty$, and

$$\Phi[f \cdot g] = \frac{\phi * \psi}{2\pi} = \frac{\Phi[f] * \Phi[g]}{2\pi}. \tag{5–115}$$

Proof. Both ϕ and ψ must be bounded for $-\infty < \omega < \infty$. For example,

$$|\phi(\omega)| = \left| \int_{-\infty}^\infty f(t)e^{-i\omega t}\, dt \right| \leqq \int_{-\infty}^\infty |f(t)|\, dt.$$

Hence ϕ, ψ satisfy the hypotheses of Theorem 13, with t replaced by ω; therefore $\phi * \psi$ is well defined and continuous for all ω and

$$\Phi[\phi * \psi] = \Phi[\phi]\Phi[\psi].$$

We now apply Theorem 12: $\Phi[\phi] = \Phi[\Phi[f]] = 2\pi\Theta f$, and similarly $\Phi[\psi] = 2\pi\Theta g$, so that

$$\Phi[\phi * \psi] = (2\pi\Theta f) \cdot (2\pi\Theta g) = 4\pi^2\Theta(f \cdot g).$$

Similarly,

$$\Phi[\Phi[\phi * \psi]] = 2\pi\Theta(\phi * \psi) = 4\pi^2\Phi[\Theta(f \cdot g)] = 4\pi^2\Theta\Phi[f \cdot g],$$

so that

$$\Phi[f \cdot g] = \frac{\phi * \psi}{2\pi}.$$

THEOREM 14. *The convolution $f * g$ obeys the rules*

$$f * (cg) = (cf) * g = c(f * g) \quad (c = \text{complex const}), \quad (5\text{–}116)$$

$$(f * g) * q = f * (g * q), \quad (5\text{–}117)$$

$$e^{at}(f * g) = (e^{at}f) * (e^{at}g) \quad (a = \text{complex const}). \quad (5\text{–}118)$$

The rules are valid if all functions are piecewise continuous, bounded, and absolutely integrable from $-\infty$ to ∞.

The proofs are left as exercises (Problems 5 through 7 below).

5–12 Special convolutions. Of much importance for applications are convolutions in which one of the two "factors" is $e^{st}h(t)$ with Re $(s) < 0$.

THEOREM 15. *Let $f(t)$ be piecewise continuous for $-\infty < t < \infty$. Let*

$$\mathfrak{L}_t[f] = \int_0^t e^{-su}f(u)\, du = F_t(s) \quad (\text{Re }(s) < 0) \quad (5\text{–}120)$$

for $-\infty < t < \infty$. Then $F_t(s)$ is defined for all t. Let $F_t(s)$ have a finite limit $F_{-\infty}(s)$ as $t \to -\infty$. Then $[e^{st}h(t)] * f(t)$ exists and

$$[e^{st}h(t)] * f(t) = e^{st}[F_t(s) - F_{-\infty}(s)]. \quad (5\text{–}121)$$

Proof. We first remark that $\mathfrak{L}_t[f]$ is the truncated Laplace transform of f, as introduced in Section 4–15 and tabulated for $t > 0$ in Table 4–1. Since $e^{-st}f(t)$ is piecewise continuous, the integral $(5\text{–}120)$ exists. Since $F_{-\infty}(s)$ exists, we can write

$$F_{-\infty}(s) = -\int_{-\infty}^0 e^{-su}f(u)\, du = \lim_{t \to -\infty} \int_0^t e^{-su}f(u)\, du.$$

We verify that $[e^{st}h(t)] * f(t) = f(t) * [e^{st}h(t)]$, if either exists. Thus

$$[e^{st}h(t)] * f(t) = \int_{-\infty}^{\infty} f(u)e^{s(t-u)}h(t-u)\,du = \int_{-\infty}^{t} f(u)e^{s(t-u)}\,du$$

$$= e^{st}\int_{-\infty}^{t} f(u)e^{-su}\,du$$

$$= e^{st}\left[\int_{0}^{t} f(u)e^{-su}\,du + \int_{-\infty}^{0} f(u)e^{-su}\,du\right]$$

$$= e^{st}[F_t(s) - F_{-\infty}(s)].$$

Thus $[e^{st}h(t)] * f(t)$ is well defined, and (5–121) is satisfied.

Remark. The integral $F_{-\infty}(s)$ will exist if $|f(t)| <$ const for negative t, for then $|e^{-su}f(u)| <$ const $\cdot\, e^{-su}$, and $\int_{-\infty}^{0} e^{-su}\,du$ exists. More generally, it will exist if $|f(t)| <$ const $\cdot\, |t|^k$ for some number k, for $t < 0$. It will not exist for all s such that Re $(s) < 0$ if $f(t) = e^{-t}$. One broad class for which the integral exists consists of those functions of *less than exponential growth*, as $t \to -\infty$; that is, of those f for which

$$|f(t)| < e^{at} \quad (t < c),$$

for every $a < 0$, where c depends on a and f.

THEOREM 16. *Under the hypotheses of Theorem 15, the function $g = [e^{st}h(t)] * f$ is a continuous and piecewise smooth function for all t. Furthermore, wherever f is continuous,*

$$(D - s)g = f(t). \tag{5–122}$$

If f is of less than exponential growth as $t \to -\infty$, then so also is $g(t)$.

Proof. By its definition, $F_t(s)$ is continuous and

$$DF_t(s) = \frac{d}{dt}\int_{0}^{t} e^{-su}f(u)\,du = e^{-st}f(t),$$

wherever f is continuous. Hence by (5–121),

$$D[\{e^{st}h(t)\} * f(t)] = D[e^{st}\{F_t(s) - F_{-\infty}(s)\}]$$
$$= e^{st}[e^{-st}f(t)] + se^{st}[F_t(s) - F_{-\infty}(s)]$$
$$= f(t) + s\{[e^{st}h(t)] * f(t)\}.$$

Thus (5–122) is proved, and g is continuous and piecewise smooth.

If f is of less than exponential growth, then given $a < 0$, we choose b so that $0 > b > \sigma = \mathrm{Re}\,(s)$, $0 > b > a$; then we choose c so that $|f(t)| < e^{bt}$ for $t < c$. Then for $t < c$, $s = \sigma + i\omega$,

$$|g(t)| = \left| e^{st} \int_{-\infty}^{t} f(u)e^{-su}\,du \right|$$

$$\leqq e^{\sigma t} \int_{-\infty}^{t} |f(u)|e^{-\sigma u}\,du$$

$$\leqq e^{\sigma t} \int_{-\infty}^{t} e^{bu}e^{-\sigma u}\,du = e^{\sigma t} \left. \frac{e^{(b-\sigma)u}}{b-\sigma} \right|_{-\infty}^{t}$$

$$\leqq \frac{e^{bt}}{b-\sigma}.$$

But $e^{bt} < (b-\sigma)e^{at}$ for t sufficiently large negative, $t < c_1$. Hence $|g(t)| < e^{at}$ for $t < c_2 = \min\,(c, c_1)$. Thus g is of less than exponential growth as $t \to -\infty$.

COROLLARY. *Let $p(t)$, $q(t)$ be piecewise continuous for $-\infty < t < \infty$. Let p and q be absolutely integrable from $-\infty$ to ∞. Let $\Phi[p] = \phi(\omega)$, $\Phi[q] = i\omega\phi(\omega)$. Then if $p(t)$ is properly defined at discontinuities, $p(t)$ becomes a continuous function, $p'(t)$ exists wherever $q(t)$ is continuous, and $p'(t) = q(t)$.*

Proof. By Theorem 13, the function $e^{-t}h(t) * (p + q)$ is well defined and continuous for all t and has Fourier transform

$$\Phi[e^{-t}h(t) * (p + q)] = \frac{1}{1 + i\omega}\,[\phi(\omega) + i\omega\phi(\omega)] = \phi(\omega) = \Phi[p].$$

Hence by Theorem 7, if p is properly defined at discontinuities, then

$$e^{-t}h(t) * (p + q) = p,$$

so that p is continuous. Since the convolution exists, the conditions of Theorem 15 hold with $f = p + q$; by Theorem 16 with $s = -1$, wherever q is continuous,

$$(D + 1)p = p + q, \quad Dp + p = p + q,$$

so that $Dp = q$.

THEOREM 17. *Let $f(t)$ be piecewise continuous for $-\infty < t < \infty$ and let $f(t)$ be of less than exponential growth as $t \to -\infty$. Then for $\mathrm{Re}\,(a) < 0$, $\mathrm{Re}\,(b) < 0$,*

$$e^{at}h(t) * [e^{bt}h(t) * f] = [e^{at}h(t) * e^{bt}h(t)] * f. \tag{5-123}$$

TABLE 5–2

CONVOLUTIONS

No.	$f(t), \quad -\infty < t < \infty$	$[e^{st}h(t)] * f(t), \quad \text{Re }(s) < 0$
1	1	$-\dfrac{1}{s}$
2	t	$\dfrac{-1 - st}{s^2},$
3	$t^k, \quad k = 1, 2, \ldots$	$\dfrac{-g_k(st)}{s^{k+1}},$ $g_k(x) = x^k + kx^{k-1} + \cdots$
4	$e^{at}, \quad \text{Re }(a) \geqq 0$	$\dfrac{e^{at}}{a - s}$
5	$t^k e^{at}, \quad \text{Re }(a) \geqq 0$	$\dfrac{-e^{at}g_k[(s - a)t]}{(s - a)^{k+1}}$
6	$\cos bt, \; b$ real	$\dfrac{b \sin bt - s \cos bt}{s^2 + b^2}$
7	$\sin bt, \; b$ real	$\dfrac{-b \cos bt - s \sin bt}{s^2 + b^2}$
8	$h(t)$	$\dfrac{e^{st} - 1}{s} h(t)$
9	$h(t - c)$	$\dfrac{e^{s(t-c)} - 1}{s} h(t - c)$
10	$h(t) - h(t - c)$	$\dfrac{e^{st} - 1}{s} [h(t) - h(t - c)]$ $- \dfrac{e^{s(t-c)} - e^{st}}{s} h(t - c)$
11	$th(t)$	$\dfrac{e^{st} - 1 - st}{s^2} h(t)$
12	$(t - c)h(t - c)$	$\dfrac{e^{s(t-c)} - 1 - s(t - c)}{s^2} h(t - c)$

TABLE 5-2 *Continued*

No.	$f(t), \quad -\infty < t < \infty$	$[e^{st}h(t)] * f(t), \quad \text{Re}\,(s) < 0$
13	$t[h(t) - h(t - c)]$	$\dfrac{e^{st} - 1 - st}{s^2}[h(t) - h(t - c)]$ $+ \dfrac{e^{st}[1 - (1 + cs)e^{-sc}]}{s^2}\,h(t - c)$
14	$t[h(t) - h(t - c)] + (2c - t)$ $\times [h(t - c) - h(t - 2c)]$	$\dfrac{e^{st} - 1 - st}{s^2}[h(t) - h(t - c)]$ $+ \dfrac{e^{st} - 2e^{s(t-c)} + 1 + st - 2cs}{s^2}$ $\times [h(t - c) - h(t - 2c)]$ $+ \dfrac{e^{st}(1 - 2e^{-cs} + e^{-2cs})}{s^2}$ $\times h(t - 2c)$
15	$e^{at}h(t)$	$\dfrac{e^{at} - e^{st}}{a - s}\,h(t), \quad s \neq a$ $te^{st}h(t) \text{ for } s = a$
16	$t^k e^{at}h(t), \quad k = 1, 2, \ldots$	$\dfrac{k!e^{st} - e^{at}g_k[(s - a)t]}{(s - a)^{k+1}}\,h(t),$ $s \neq a,$ $\dfrac{e^{st}t^{k+1}}{k + 1} \text{ for } s = a$
17	$e^{at}h(-t), \quad \text{Re}\,(a) \geq 0$	$\dfrac{e^{at}[1 - h(t)] + e^{st}h(t)}{a - s}$

Proof. We remark that the rule is a special case of the associative law (5-117); however, the assumptions are more general here.

Now let $f_1(t) = f(t)$ for $t \leq t_1$, $f_1(t) = 0$ for $t > t_1$. Choose c so that $\text{Re}\,(a) < c < 0$, $\text{Re}\,(b) < c < 0$. Since $f(t)$ is of less than exponential growth as $t \to -\infty$, we conclude that $f_1(t)e^{-ct} \to 0$ as $t \to -\infty$; in fact, since $f_1(t) \equiv 0$ for $t > t_1$, $f_1(t)e^{-ct}$ is bounded and absolutely integrable from $-\infty$ to $+\infty$. A similar statement applies to the functions $e^{(a-c)t}h(t)$, $e^{(b-c)t}h(t)$, both of which approach zero as $t \to +\infty$. Therefore, the associative law (5–117) is applicable:

$$e^{(a-c)t}h(t) * [e^{(b-c)t}h(t) * e^{-ct}f_1(t)] = [e^{(a-c)t}h(t) * e^{(b-c)t}h(t)] * e^{-ct}f_1(t).$$

By (5–118), e^{-ct} can be factored out of all the convolutions, and we conclude that

$$e^{at}h(t) * [e^{bt}h(t) * f_1(t)] = [e^{at}h(t) * e^{bt}h(t)] * f_1(t).$$

But for $t \leq t_1$, $e^{bt}h(t) * f_1(t) = e^{bt}h(t) * f(t)$, since $f(t) = f_1(t)$ for $t \leq t_1$, so that

$$e^{bt}h(t) * f_1(t) = \int_{-\infty}^{t} f_1(u)e^{b(t-u)} \, du$$

$$= \int_{-\infty}^{t} f(u)e^{b(t-u)} \, du = e^{bt}h(t) * f(t).$$

By similar reasoning we conclude that the left-hand side of (5–123) coincides with the right-hand side for $t \leq t_1$. But t_1 is arbitrary. Therefore, (5–123) is valid for all t.

Remark. By similar reasoning, the result can be extended to more factors; for example,

$$e^{at}h(t) * [e^{bt}h(t) * \{e^{ct}h(t) * f\}] = \{e^{at}h(t) * [e^{bt}h(t) * e^{ct}h(t)]\} * f$$
$$= \{[e^{at}h(t) * e^{bt}h(t)] * e^{ct}h(t)\} * f.$$

With the aid of Theorem 15, we can easily tabulate $[e^{st}h(t)] * f(t)$ for a variety of choices of $f(t)$; see Table 5–2. It should be remarked that the only restrictions on f are piecewise continuity and existence of $\int_{-\infty}^{0} f(u)e^{-su} \, du$; f may grow as rapidly as desired for positive t.

A rule for translation is useful. Let $g(t) = [e^{st}h(t)] * f(t)$. Then

$$[e^{st}h(t)] * f(t - c) = g(t - c). \tag{5–124}$$

For

$$[e^{st}h(t)] * f(t - c) = \int_{-\infty}^{t} f(u - c)e^{s(t-u)} \, du = \int_{-\infty}^{t-c} f(v)e^{s(t-v-c)} \, dv$$

$$= e^{s(t-c)} \int_{-\infty}^{t-c} f(v)e^{-sv} \, dv = e^{s(t-c)}[F_{t-c}(s) - F_{-\infty}(s)]$$

$$= g(t - c).$$

Thus replacement of t by $t - c$ in f corresponds to replacement of t by $t - c$ in $e^{st}h(t) * f = g(t)$.

PROBLEMS

1. Find the function f such that

$$\Phi[f] = \frac{\sin \omega}{\omega(i\omega + 1)}.$$

[*Hint:* Form the convolution of Entries 2 and 4 of Table 5–1. Use Table 5–2 to evaluate the convolution.]

2. Find the function f such that

$$\Phi[f] = \frac{e^{-|\omega|}(1 - e^{-i\omega}(1 + i\omega))}{\omega^2}.$$

[*Hint:* Use Entries 11 and 16 of Table 5–1.]

3. Find the Fourier transform of $(t^2 + 1)^{-1}(t^2 + 4)^{-1}$ with the aid of (5–113).

4. Prove (5–113).

5. Prove (5–116).

6. Prove (5–117).

7. Prove (5–118).

8. Let $\Phi[f] = \phi$, $\Phi[g] = \psi$. Prove each of the following under appropriate hypotheses:

(a) $\displaystyle \int_{-\infty}^{\infty} \phi(\omega)\psi(\omega)e^{i\omega t}\, d\omega = 2\pi \int_{-\infty}^{\infty} f(u)g(t-u)\, du$

(b) $\displaystyle \int_{-\infty}^{\infty} \phi(\omega)\psi(\omega)\, d\omega = 2\pi \int_{-\infty}^{\infty} f(t)g(-t)\, dt$

(c) $\displaystyle \int_{-\infty}^{\infty} \phi(\omega)\overline{\psi(\omega)}\, d\omega = 2\pi \int_{-\infty}^{\infty} f(t)\overline{g(t)}\, dt$

(d) $\displaystyle \int_{-\infty}^{\infty} |\phi(\omega)|^2\, d\omega = 2\pi \int_{-\infty}^{\infty} |f(t)|^2\, dt$

[*Hints:* For (a) use (5–111) and Theorem 2. Deduce (b) from (a). Deduce (c) from (b) with the aid of (5–106) and (5–109). Deduce (d) from (c).]

9. Verify the following entries of Table 5–2:

(a) No. 3   (b) No. 4
(c) No. 5 [use (5–118)]   (d) Nos. 6 and 7 from No. 4
(e) No. 9 from No. 8 [use (5–124)]   (f) No. 12 from No. 11
(g) No. 14 from Nos. 12 and 9   (h) No. 15 from No. 8 [use (5–118)]
(i) No. 17 from Nos. 4 and 15

ANSWERS

1. $\frac{1}{2}[(1 - e^{-t-1})h(t + 1) + (e^{-t+1} - 1)h(t - 1)]$

2. $\dfrac{1}{2\pi} \log \dfrac{1 + t^2}{2 - 2t + t^2} - \dfrac{t}{\pi}\, \text{arc tan}\, \dfrac{1}{1 - t + t^2}$

3. $\dfrac{\pi}{3}\, (e^{-|\omega|} - \tfrac{1}{2}e^{-2|\omega|})$

5–13 The inverse Fourier transform. If $\Phi[f] = \phi(\omega)$, then we write

$$f = \Phi^{-1}[\phi]$$

and call Φ^{-1} the inverse Fourier transformation. By the uniqueness theorem of Section 5–9, there is at most one f such that $\Phi[f] = \phi$; that is, the inverse transformation is single-valued. Since Φ is a linear transformation, it follows that Φ^{-1} is defined in a linear space of functions and is a linear operator in that class (Section 2–2).

Because of the symmetry of the Fourier transform, we have the special relation

$$\Phi^{-1}[\phi] = \frac{1}{2\pi}\, \Phi[\phi(-\omega)] \tag{5–130}$$

under appropriate hypotheses. Indeed, by (5–107), if $f = (2\pi)^{-1}\Phi[\phi(-\omega)]$, then

$$\Phi[f] = \frac{1}{2\pi}\, \Phi[\Phi\{\phi(-\omega)\}] = \Theta\{\phi(-\omega)\} = \phi(\omega).$$

The identity (5–130) will be valid if both $\phi(\omega)$ and $\Phi[\phi]$ are piecewise smooth and have absolutely convergent integrals over the whole real axis. Equation (5–130) is the same as the equation

$$f(t) = \frac{1}{2\pi} \int_{-\infty}^{\infty} \phi(\omega) e^{i\omega t}\, d\omega, \tag{5–130'}$$

expressing f as a Fourier integral; (5–130) is obtained from (5–130') by replacing ω by $-\omega$. Under the hypotheses stated, $f(t)$ is well defined and is represented by its Fourier integral; that is,

$$f(t) = \Phi^{-1}[\phi] = \frac{1}{2\pi} \int_{-\infty}^{\infty} \phi(\omega) e^{i\omega t}\, d\omega$$

$$= \frac{1}{2\pi} \int_{-\infty}^{\infty} \phi(-\omega) e^{-i\omega t}\, d\omega = \frac{1}{2\pi}\, \Phi[\phi(-\omega)]. \tag{5–130''}$$

It is convenient to allow principal values in the preceding integrals. Thus the Fourier transform of $f(t)$ is defined as

$$\Phi[f] = (P) \int_{-\infty}^{\infty} f(t) e^{-i\omega t}\, dt, \tag{5–131}$$

whenever the principal value exists for all ω, $-\infty < \omega < \infty$. With this convention, (5–130) and (5–130'') are valid much more generally, for example, when f is piecewise smooth and absolutely integrable from $-\infty$ to ∞ and $\phi(\omega) = \Phi[f]$.

It follows that a table of Fourier transforms can be used in two ways as a table of inverse Fourier transforms; whenever $\Phi[f] = g$, then $\Phi^{-1}[g] = f$ and $\Phi^{-1}[f] = (2\pi)^{-1}g(-t)$. Similarly, the convolution formula can be used in two ways, as in (5–111) and (5–115), to obtain new transforms and inverse transforms.

Of great importance for applications are the inverse transforms of $(i\omega - a)^{-k}$ for $k = 1, 2, \ldots$ From Entries 4 and 9 of Table 5–1, we read off:

$$\Phi^{-1}\left[\frac{1}{i\omega - a}\right] = e^{at}h(t),$$

$$\Phi^{-1}\left[\frac{1}{(i\omega - a)^2}\right] = te^{at}h(t), \tag{5-132}$$

$$\Phi^{-1}\left[\frac{1}{(i\omega - a)^k}\right] = \frac{t^{k-1}e^{at}h(t)}{(k-1)!}.$$

Here Re $(a) < 0$, and we shall be interested mainly in this case. For Re $(a) > 0$, Entry 6 provides the rule

$$\Phi^{-1}\left[\frac{1}{i\omega - a}\right] = -e^{at}h(-t). \tag{5-133}$$

5–14 Evaluation of inverse Fourier transforms by residues. Given $f(t)$, piecewise smooth for all t and having an absolutely convergent integral for $-\infty < t < \infty$, the Fourier transform $\Phi[f] = \phi(\omega)$ is well defined, and we know that (for proper choice of $f(t)$ at jumps)

$$f(t) = \frac{1}{2\pi}(P)\int_{-\infty}^{\infty}\phi(\omega)e^{i\omega t}\,d\omega.$$

It may happen that there is a function $F(s)$ analytic in a portion of the complex s-plane, $s = \sigma + i\omega$, which includes the imaginary axis and such that $F(i\omega) = \phi(\omega)$. In that case we can consider the above integral as a line integral of $F(s)$ along the imaginary axis. For $ds = id\omega$ on the axis, and hence

$$f(t) = \frac{1}{2\pi i}(P)\int_{C}F(s)e^{st}\,ds, \tag{5-140}$$

where $s = i\omega$ and C is the imaginary axis, with positive direction upward. The symbol (P) denotes principal value, that is,

$$\lim_{R\to\infty}\int_{C_R}F(s)e^{st}\,ds,$$

where C_R is the interval from $-iR$ to iR on the imaginary axis (Fig. 5–2).

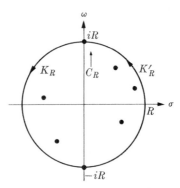

FIG. 5–2.   Evaluation of $\Phi^{-1}[\phi]$ by residues.

Now let us suppose that $F(s)$ is analytic for $\sigma < 0$ except for a finite number of singularities $s_1, \ldots, s_k$. We choose R so large that all these singularities lie inside the circle $|s| = R$. Then

$$\int_{C_R} F(s)e^{st}\, ds + \int_{K_R} F(s)e^{st}\, ds = 2\pi i \sum_{j=1}^{k} \text{Res}\,[e^{st}F(s),\, s_j],$$

by the Cauchy residue theorem (Section 3–13), where K_R is the semicircle from iR to $-iR$ (Fig. 5–2). If we let R increase further we get the limiting relation

$$(P)\int_{C} F(s)e^{st}\, ds + \lim_{R\to\infty} \int_{K_R} F(s)e^{st}\, ds = 2\pi i \sum_{j=1}^{k} \text{Res}\,[e^{st}F(s),\, s_j].$$

For the first term approaches a limit, namely, $2\pi i f(t)$, by the theory of Fourier transforms; the right-hand side does not change as R increases. Hence we conclude from (5–140):

$$f(t) = \sum_{j=1}^{k} \text{Res}\,[e^{st}F(s),\, s_j] - \frac{1}{2\pi i} \lim_{R\to\infty} \int_{K_R} F(s)e^{st}\, ds. \qquad (5\text{–}141)$$

In many cases it is easy to evaluate the limit of the integral on K_R, so that (5–141) becomes a valuable tool for finding the inverse transform. In particular, we shall see that in many cases the limit is zero. If the limit is difficult, we may be able to proceed similarly in the right half-plane. The same reasoning leads to the formula

$$f(t) = - \sum_{j=1}^{k'} \text{Res}\,[e^{st}F(s),\, s_j'] + \frac{1}{2\pi i} \lim_{R\to\infty} \int_{K_R'} F(s)e^{st}\, ds, \qquad (5\text{–}142)$$

where $s_1', \ldots, s_{k'}'$ are the singularities in the right half-plane, and K_R' is the semicircle from $-iR$ to iR. It is also possible that for some values of t, (5–141) is easier, while for others, (5–142) is easier.

We now give a theorem which provides information on the crucial limits:

THEOREM 18. *If $F(s)$ is analytic for $|s|$ sufficiently large and has a zero at ∞, then*

$$\lim_{R \to \infty} \int_{K_R} F(s)e^{st}\, ds = 0 \quad \text{for } t > 0, \tag{5-143}$$

$$\lim_{R \to \infty} \int_{K_R'} F(s)e^{st}\, ds = 0 \quad \text{for } t < 0, \tag{5-144}$$

while (for $t = 0$)

$$\lim_{R \to \infty} \int_{K_R} F(s)\, ds = \lim_{R \to \infty} \int_{K_R'} F(s)\, ds = -\pi i \operatorname{Res}[F(s), \infty]. \tag{5-145}$$

The proof is given in the following section. Here we state a corollary and give some examples.

COROLLARY. *Let $F(s)$ be analytic, except for singularities at the points s_j, s_j', and let $F(s)$ have a zero at ∞. If $\Phi[f] = \phi(\omega) = F(i\omega)$, then*

$$f(t) = \sum_{j=1}^{k} \operatorname{Res}[e^{st}F(s), s_j] \quad \text{for } t > 0, \tag{5-146}$$

$$f(t) = -\sum_{j=1}^{k'} \operatorname{Res}[e^{st}F(s), s_j'] \quad \text{for } t < 0, \tag{5-147}$$

$$f(0) = \sum_{j=1}^{k} \operatorname{Res}[F(s), s_j] + \tfrac{1}{2}\operatorname{Res}[F(s), \infty]$$

$$= -\sum_{j=1}^{k'} \operatorname{Res}[F(s), s_j'] - \tfrac{1}{2}\operatorname{Res}[F(s), \infty]. \tag{5-148}$$

EXAMPLE 1. $\Phi[f] = 1/(i\omega - 2)$. $F(s) = 1/(s - 2)$. Then $F(s)$ has a pole of first order at $s = 2$ and a zero at ∞. Hence there is no point s_j for $\sigma < 0$, and $f(t) = 0$ for $t > 0$. For $t < 0$,

$$f(t) = -\operatorname{Res}\left[\frac{e^{st}}{s - 2}, 2\right] = -e^{2t}.$$

Hence $f(t)$ has a jump for $t = 0$. The corollary gives the average value at $t = 0$ as $\tfrac{1}{2}\operatorname{Res}[1/(s - 2), \infty]$. Now

$$\frac{1}{s - 2} = \frac{1}{s\left(1 - \dfrac{2}{s}\right)} = \frac{1}{s} + \frac{2}{s^2} + \cdots$$

Hence the residue is -1 and $f(0) = -\tfrac{1}{2}$.

EXAMPLE 2. $\Phi[f] = \dfrac{1}{-\omega^2 - 4} = \dfrac{1}{(i\omega)^2 - 4}.$

We choose

$$F(s) = \frac{1}{s^2 - 4}.$$

Thus $F(s)$ has poles of first order at $s = \pm 2$ and a zero of *second* order at ∞. Hence there is zero residue at ∞. By Rule III of Section 3–13,

$$\text{for } t \geqq 0, \quad f(t) = \text{Res}\left[\frac{e^{st}}{s^2 - 4}, -2\right] = \frac{e^{st}}{2s}\bigg|_{-2} = \frac{e^{-2t}}{-4},$$

$$\text{for } t \leqq 0, \quad f(t) = -\text{Res}\left[\frac{e^{st}}{s^2 - 4}, +2\right] = -\frac{e^{st}}{2s}\bigg|_{2} = -\frac{e^{2t}}{4}.$$

Thus $f(t)$ is continuous at $t = 0$, with value $-\frac{1}{4}$. The continuity at $t = 0$ will clearly hold whenever $F(s)$ is rational with a zero of second or higher order at ∞.

EXAMPLE 3. $\Phi[f] = e^{1/(i\omega - 1)} - 1$; $F(s) = e^{1/(s-1)} - 1$. The function $1/(s - 1)$ is analytic at $s = \infty$, with value zero there; hence $e^{1/(s-1)}$ is also analytic at $s = \infty$, with value 1, and $F(s)$ has a zero at ∞. $F(s)$ has a singularity only at $s = 1$:

$$F(s) = \frac{1}{s - 1} + \frac{1}{2!}\frac{1}{(s - 1)^2} + \cdots + \frac{1}{n!(s - 1)^n} + \cdots$$

Hence $f(t) = 0$ for $t > 0$. For $t < 0$,

$$f(t) = -\text{Res}\,[e^{st}F(s), 1].$$

Now

$$e^{st}F(s)$$

$$= e^t\left(1 + \frac{t(s - 1)}{1!} + \frac{t^2(s - 1)^2}{2!} + \cdots\right)\left(\frac{1}{s - 1} + \frac{1}{2!}\frac{1}{(s - 1)^2} + \cdots\right)$$

Since both series converge absolutely for $s \neq 1$, we can multiply and arrange terms in any order. If we collect the terms in $(s - 1)^{-1}$, we find

$$\frac{e^t}{s - 1}\left(1 + \frac{t}{2!} + \frac{t^2}{2!3!} + \cdots \frac{t^n}{n!(n + 1)!} + \cdots\right).$$

Hence for $t < 0$,

$$f(t) = -\text{Res}\,[e^{st}F(s), 1]$$

$$= -e^t\left(1 + \frac{t}{2!} + \frac{t^2}{2!3!} + \cdots + \frac{t^n}{n!(n + 1)!} + \cdots\right). \tag{5–149}$$

This gives $f(t)$ as an infinite series. We can also write $f(t)$ as a contour integral:

$$f(t) = - \frac{1}{2\pi i} \oint_{C_0} e^{st}(e^{1/(s-1)} - 1)\, ds,$$

where C_0 is a simple closed path (e.g., a circle) enclosing the singularity $s = 1$. If C_0 is the circle $|s - 1| = \frac{1}{2}$, then Re $(s) \geq \frac{1}{2}$ on C_0. Hence as $t \to -\infty$, $e^{st} \to 0$ uniformly on C_0, so that $f(t) \to 0$ as $t \to -\infty$. The series (5–149) gives $f(t)$ easily for small negative t, while for large negative t, $f(t) \to 0$. Accordingly we obtain the graph of Fig. 5–3.

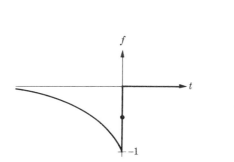

Fɪɢ. 5–3.   Inverse Fourier transform of $e^{1/(i\omega - 1)} - 1$.

Fɪɢ. 5–4.   Paths for evaluation of inverse Fourier transform.

For $t = 0$, we can compute the value by residues, but we can predict the result:

$$f(0) = -\tfrac{1}{2}.$$

This example suggests various new techniques which become available. In particular, it may be of value to replace the residue expressions in the corollary by the corresponding contour integrals. Thus (5–146) can also be written

$$f(t) = \frac{1}{2\pi i} \oint_{C_1} e^{st} F(s)\, ds \quad (t > 0), \qquad (5\text{–}146')$$

where C_1 is a path enclosing all the points s_j (Fig. 5–4). Similarly, (5–147) becomes

$$f(t) = \frac{1}{2\pi i} \oint_{C_1'} e^{st} F(s)\, ds \quad (t < 0). \qquad (5\text{–}147')$$

Thus the path of integration along the imaginary axis has been deformed into the two types of closed paths.

5-15 Proof of Theorem 18. We first prove (5–143). Since $F(s)$ has a zero at ∞,

$$F(s) = \frac{g(s)}{s},$$

where $g(s)$ is analytic at ∞. Hence, in particular, $|g(s)| \leqq M$ for some constant M, for $|s|$ sufficiently large, and

$$|F(s)| \leqq \frac{M}{|s|} \quad \text{for } |s| > R_0. \tag{5–150}$$

Next we note that

$$|e^{st}| = |e^{(\sigma+i\omega)t}| = e^{\sigma t} = e^{-Rt\sin\phi}, \tag{5–151}$$

where ϕ is the angle shown in Fig. 5–5. Along the path K_R,

$$s = Re^{i[\phi+(\pi/2)]}, \qquad ds = iRe^{i[\phi+(\pi/2)]}\,d\phi.$$

Accordingly, for $R > R_0$,

$$\left| \int_{K_R} F(s)e^{st}\,ds \right| = \left| \int_0^\pi F(s)e^{st} \cdot Re^{i[\phi+(\pi/2)]}\,d\phi \right|$$

$$\leqq \int_0^\pi |F(s)|\,|e^{st}|\,R\,d\phi \leqq \int_0^\pi \frac{M}{R} e^{-tR\sin\phi}R\,d\phi \leqq M\int_0^\pi e^{-tR\sin\phi}\,d\phi.$$

Hence our assertion will be proved if we show that for $t > 0$,

$$\lim_{R\to\infty} \int_0^\pi e^{-tR\sin\phi}\,d\phi = 0. \tag{5–152}$$

The function being integrated is $G_R(\phi) = e^{-tR\sin\phi}$. Its graph is shown in Fig. 5–6. As $R \to \infty$, $G_R(\phi) \to 0$, except for $\phi = 0$, π. From this we can deduce the conclusion (5–152). First, given $\epsilon > 0$, we choose δ so that $\delta < \epsilon/3$ and $0 < \delta < \pi/2$. Then $0 < G_R(\phi) \leqq e^{-tR\sin\delta}$ for $\delta \leqq \phi \leqq \pi - \delta$. For R sufficiently large, $e^{-tR\sin\delta} < \epsilon/(3\pi)$, since $t > 0$. Hence for such R

$$0 < \int_0^\pi G_R(\phi)\,d\phi = \int_0^\delta G_R(\phi)\,d\phi + \int_{\pi-\delta}^\pi G_R(\phi)\,d\phi + \int_\delta^{\pi-\delta} G_R(\phi)\,d\phi$$

$$\leqq 2\delta + e^{-tR\sin\delta}(\pi - 2\delta)$$

$$< \frac{2\epsilon}{3} + \frac{\epsilon}{3\pi}\cdot\pi = \epsilon.$$

Therefore (5–152) and its consequence (5–143) are proved.

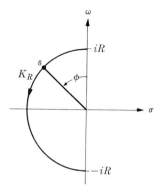

FIG. 5–5. Estimation of $\int_{K_R} F(s)e^{st}\,ds$.

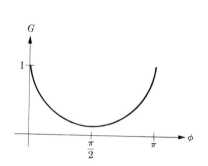

FIG. 5–6. The function $\exp(-tR\sin\phi)$.

The relation (5–144) is proved in the same way. We now prove (5–145). Since $F(s)$ has a zero at ∞,

$$F(s) = \frac{a_{-1}}{s} + \frac{a_{-2}}{s^2} + \cdots = \frac{a_{-1}}{s} + \frac{P(s)}{s^2},$$

where $P(s)$ is analytic at ∞ and hence $|P(s)| \le M$ for $|s| \ge R_0$, for appropriate choice of M and R_0. Thus

$$\int_{K_R} F(s)\,ds = \int_{K_R} \left(\frac{a_{-1}}{s} + \frac{P(s)}{s^2}\right) ds = a_{-1}\cdot\pi i + \int_{K_R} \frac{P(s)}{s^2}\,ds.$$

Since $|P(s)| \le M$ for $|s| \ge R_0$, we conclude (Problem 4 below) that

$$\lim_{R\to\infty}\int_{K_R} \frac{P(s)}{s^2}\,ds = 0. \tag{5–153}$$

Therefore

$$\lim_{R\to\infty}\int_{K_R} F(s)\,ds = \pi i a_{-1} = -\pi i \operatorname{Res}[F(s),\infty].$$

The second limit in (5–145) is evaluated in the same way. Hence Theorem 18 is completely proved.

PROBLEMS

1. With the aid of Table 5–1 (Section 5–10), find $\Phi^{-1}[\phi]$ for the following choices of $\phi(\omega)$:

(a) $\dfrac{1}{\omega^2 + 4}$　　　(b) $\dfrac{1}{(\omega + 2i)(\omega + 3 + i)}$　　　(c) $e^{-|\omega|}(3\omega - 2)$

2. Evaluate the inverse Fourier transforms by residues for the following functions of ω:

(a) $\dfrac{1}{(i\omega - 2)(i\omega + 1)}$

(b) $\dfrac{1}{\omega^4 + 1}$

(c) $\dfrac{1}{(i\omega - a)^n}$ (Re $(a) \neq 0$, $n = 1, 2, \ldots$)

3. Represent the inverse Fourier transforms of the following functions by means of contour integrals:

(a) $\exp\left(\dfrac{1}{1 + \omega^2}\right) - 1$

(b) $\sin\dfrac{1}{i\omega - 2}$

4. Prove (5–153) under the hypotheses of the text.

Answers

1. (a) $\frac{1}{4}e^{-2|t|}$

   (b) $\frac{1}{10}(3i - 1)[e^{(1-3i)t} - e^{2t}]h(-t)$

   (c) $(-2t^2 + 6it - 2)/[\pi(1 + t^2)^2]$

2. (a) $-\frac{1}{3}e^{2t}[1 - h(t)] - \frac{1}{3}e^{-t}h(t)$

   (b) $\frac{1}{2}$ Re $(ce^{ct})[1 - h(t)] + \frac{1}{2}$ Re $(ce^{-ct})h(t)$ [$c = \exp(i\pi/4)$]

   (c) for Re $(a) > 0$, $-t^{n-1}e^{at}h(-t)/(n - 1)!$; for

Re $(a) < 0$, $t^{n-1}e^{at}h(t)/(n - 1)!$

3. (a) $\dfrac{1}{2\pi i} \oint_{C_1} \exp\dfrac{1}{1 - s^2} e^{st}\, ds$ for $t > 0$,

   $-\dfrac{1}{2\pi i} \oint_{C_1'} \exp\dfrac{1}{1 - s^2} e^{st}\, ds$ for $t < 0$,

where C_1 encloses -1 but not 1, and C_1' encloses 1 but not -1.

(b) $f = 0$ for $t > 0$,

$$f = -\dfrac{1}{2\pi i} \oint_C \sin\dfrac{1}{s - 2} e^{st}\, ds \text{ for } t < 0,$$

where C encloses 2.

5–16 Fourier transforms of generalized functions.

Our object is to define Fourier transforms of generalized functions and of certain ordinary functions for which the usual definition is meaningless. We wish to do this in a manner which is consistent with the properties of generalized functions and with the basic properties of the Fourier transform. The ones considered basic are the following:

(a) $\Phi[c_1 f_1 + c_2 f_2] = c_1 \Phi[f_1] + c_2 \Phi[f_2]$,
(b) $\Phi[f'] = i\omega \Phi[f]$,
(c) $\Phi[\Theta f] = \Theta \Phi[f]$,
(d) $\Phi[\Phi[f]] = 2\pi \Theta f$,
(e) $\Phi[f(t - c)] = e^{-i\omega c} \Phi[f]$,
(f) $\Phi[e^{i\alpha t} f] = \phi(\omega - \alpha)$, where $\phi = \Phi[f]$.

It will be important in the following discussion to write a transform relation either as $\Phi[f(t)] = \phi(\omega)$ or as $\Phi[f(\omega)] = \phi(t)$; that is, the notation chosen for independent variables is of no significance.

From the theory of Section 1–13, we are led to the definitions

$$\Phi[\delta(t)] = \int_{-\infty}^{\infty} \delta(t) e^{-i\omega t}\, dt = 1,$$

$$\Phi[\delta'(t)] = \int_{-\infty}^{\infty} \delta'(t) e^{-i\omega t}\, dt = i\omega,$$

and in general [see Eq. (1–134)], for $n = 0, 1, 2, \ldots$,

$$\Phi[\delta^{(n)}(t)] = (i\omega)^n. \tag{5-160}$$

Similarly,

$$\Phi[\delta^{(n)}(t - c)] = \int_{-\infty}^{\infty} \delta^{(n)}(t - c) e^{-i\omega t}\, dt = (i\omega)^n e^{-i\omega c}. \tag{5-161}$$

The transform of a linear combination of such functions and an ordinary function (having a Fourier transform) is defined by linearity, property (a). For example,

$$\Phi[e^{-t} h(t) + 3\, \delta(t) + 2\, \delta'(t - 1)] = \frac{1}{i\omega + 1} + 3 + 2i\omega e^{-i\omega}.$$

The definitions given thus far are in agreement with properties (a), (c), (e), and (f), if we recall that $\delta^{(n)}(t)$ is even or odd according as n is even or odd and the definition of multiplication of an ordinary function by a generalized function in Section 1–13. Property (b) certainly holds for the functions of form $\delta^{(n)}(t - c)$; we shall see that it holds quite generally.

To satisfy property (d), we are forced to make additional definitions. Thus since $\Phi[\delta] = 1$, we must define

$$\Phi[1] = \Phi[\Phi[\delta]] = 2\pi\Theta\, \delta = 2\pi\, \delta,$$

for δ is even. Similarly, since $\Phi[\delta'(t)] = i\omega$ or $\Phi[\delta'(\omega)] = it$, we define

$$\Phi[it] = \Phi[\Phi[\delta']] = 2\pi\Theta\, \delta' = -2\pi\, \delta'(\omega).$$

Hence $\Phi[t] = 2\pi i \, \delta'(\omega)$, and in general

$$\Phi[t^n] = 2\pi i^n \, \delta^{(n)}(\omega) \quad (n = 0, 1, 2, \ldots), \qquad (5\text{-}162)$$

$$\Phi[t^n e^{-ict}] = 2\pi i^n \, \delta^{(n)}(\omega + c) \quad (n = 0, 1, 2, \ldots). \qquad (5\text{-}163)$$

The functions $1, t, \ldots, t^n, \ldots, t^n e^{-ict}$ do not have Fourier transforms by the usual definition. Hence the new formulas introduce no inconsistencies. We can extend the definition to linear combinations of all functions thus far considered and can now verify properties (a), (c), (d), (e), and (f) for all these functions.

Property (b) holds for all the new transforms, as can be easily verified. For example, if $f = t^2$, then $f' = 2t$ and

$$\Phi[f'] = 2\Phi[t] = 4\pi i \, \delta'(\omega),$$

$$i\omega\Phi[f] = i\omega\Phi[t^2] = -2\pi i\omega \, \delta''(\omega)$$

$$= -2\pi i[0 \, \delta''(\omega) - 2 \, \delta'(\omega)]$$

$$= 4\pi i \, \delta'(\omega) = \Phi[f'],$$

in accordance with the rule (1-132). The property also holds for a linear combination of the new functions with an ordinary function $g(t)$ for which $g(t)$ and $g'(t)$ are continuous and absolutely integrable from $-\infty$ to ∞. We can now relax the continuity conditions by allowing $g(t)$ to have a finite number of jumps, so that $g'(t)$ becomes a generalized function. For example, let $g(t) = G(t)h(t)$, where $G(t)$ has a continuous derivative for all t. Then

$$g'(t) = G'(t)h(t) + G(0) \, \delta(t).$$

As in Section 5-10 (see Remark 2 following Theorem 8),

$$\Phi[G'(t)h(t)] = \int_0^\infty G'(t)e^{-i\omega t} \, dt$$

$$= G(t)e^{-i\omega t}\Big|_0^\infty + i\omega \int_0^\infty G(t)e^{-i\omega t} \, dt$$

$$= -G(0) + i\omega\Phi[G(t)h(t)].$$

Hence

$$\Phi[g'(t)] = -G(0) + i\omega\Phi[g] + G(0)\Phi[\delta] = i\omega\Phi[g],$$

as asserted.

We can enlarge the class of functions having Fourier transforms by defining $\Phi[h(t - c)]$ appropriately. First we let $c = 0$ and consider $\Phi[h(t)] = \phi(\omega)$. If property (c) is to hold, then $\Phi[h(-t)] = \phi(-\omega)$. But $h(t) + h(-t) \equiv 1$ (except at $t = 0$). Hence by property (a),

$$\phi(\omega) + \phi(-\omega) = \Phi[1] = 2\pi \, \delta(\omega).$$

We now assume that $\phi(\omega)$ is of the form $\psi(\omega) + a\,\delta(\omega)$, where $\psi(\omega)$ is an ordinary function. Then since $\delta(\omega)$ is even,

$$\phi(\omega) + \phi(-\omega) = \psi(\omega) + \psi(-\omega) + 2a\,\delta(\omega) = 2\pi\,\delta(\omega).$$

We thus conclude that $a = \pi$ and $\psi(\omega)$ is odd. To find $\psi(\omega)$, we reason that if property (b) is to hold, then

$$\Phi[h'(t)] = \Phi\,[\delta(t)] = 1 = i\omega\phi(\omega) = i\omega[\psi(\omega) + \pi\,\delta(\omega)].$$

By the multiplication rule of Section 1–13, $\omega\,\delta(\omega) = 0$, so that $i\omega\psi(\omega) = 1$ and

$$\psi(\omega) = \frac{1}{i\omega},$$

$$\phi(\omega) = \Phi[h(t)] = \frac{1}{i\omega} + \pi\,\delta(\omega). \tag{5-164}$$

The result is to be regarded as a definition; it is peculiar in that the ordinary function $1/(i\omega)$ is discontinuous at $\omega = 0$. If we postulate the validity of property (d), then we obtain a transform for this function:

$$\Phi\left[\frac{1}{i\omega} + \pi\,\delta(\omega)\right] = \Phi[\Phi[h(t)]] = 2\pi h(-t).$$

Since $\Phi[\pi\,\delta(\omega)] = \pi \cdot 1$, we conclude that

$$\Phi\left[\frac{1}{i\omega}\right] = 2\pi h(-t) - \pi = \pi - 2\pi h(t),$$

the last equality holding except at $t = 0$. We can also write

$$\Phi\left[\frac{1}{t}\right] = \pi i - 2\pi i h(\omega) \tag{5-165}$$

(again regarded as a definition). By translation we deduce the definitions

$$\Phi[h(t - c)] = \frac{e^{-i\omega c}}{i\omega} + \pi\,\delta(\omega), \tag{5-164'}$$

$$\Phi\left[\frac{1}{t - c}\right] = e^{-i\omega c}[\pi i - 2\pi i h(\omega)]. \tag{5-165'}$$

We ensure the validity of properties (d) and (f) by also defining

$$\Phi[e^{i\alpha t}h(t - c)] = \frac{e^{-ic(\omega-\alpha)}}{i(\omega - \alpha)} + \pi\,\delta(\omega - \alpha), \tag{5-164''}$$

$$\Phi\left[\frac{e^{i\alpha t}}{t - c}\right] = e^{-ic(\omega-\alpha)}[\pi i - 2\pi i h(\omega - \alpha)]. \tag{5-165''}$$

TABLE 5–3

FOURIER TRANSFORMS OF GENERALIZED FUNCTIONS*

No.	$f(t)$	$\phi(\omega) = \Phi[f]$
1	$\delta(t)$	1
2	$\delta(t - c)$	$e^{-ic\omega}$
3	$\delta^{(n)}(t)$	$(i\omega)^n$
4	$\delta^{(n)}(t - c)$	$e^{-ic\omega}(i\omega)^n$
5	1	$2\pi\,\delta(\omega)$
6	t	$2\pi i\,\delta'(\omega)$
7	t^n	$2\pi i^n\,\delta^{(n)}(\omega)$
8	$e^{i\alpha t}$	$2\pi\,\delta(\omega - \alpha)$
9	$t^n e^{i\alpha t}$	$2\pi i^n\,\delta^{(n)}(\omega - \alpha)$
10	$h(t)$	$\dfrac{1}{i\omega} + \pi\,\delta(\omega)$
11	$t^n h(t)$	$\dfrac{n!}{(i\omega)^{n+1}} + \pi i^n\,\delta^{(n)}(\omega)$
12	$h(t - c)$	$\dfrac{e^{-i\omega c}}{i\omega} + \pi\,\delta(\omega)$
13	$(t - c)^n h(t - c)$	$\dfrac{n!\,e^{-ic\omega}}{(i\omega)^{n+1}} + \pi i^n \sum_{r=0}^{n} \binom{n}{r}(ic)^{n-r}\,\delta^{(r)}(\omega)$
14	$e^{i\alpha t} h(t - c)$	$\dfrac{e^{-ic(\omega-\alpha)}}{i(\omega - \alpha)} + \pi\,\delta(\omega - \alpha)$
15	$e^{i\alpha t} t^n h(t)$	$\dfrac{n!}{[i(\omega - \alpha)]^{n+1}} + \pi i^n\,\delta^{(n)}(\omega - \alpha)$
16	$e^{i\alpha t}(t - c)^n h(t - c)$	$\dfrac{n!\,e^{-ic(\omega-\alpha)}}{[i(\omega - \alpha)]^{n+1}} + \pi i^n \sum_{r=0}^{n} \binom{n}{r}$ $\times (ic)^{n-r}\,\delta^{(r)}(\omega - \alpha)$
17	$\dfrac{1}{t}$	$\pi i - 2\pi i h(\omega)$

* *Note:* Throughout, $n = 1, 2, 3, \ldots$, and c and α are real constants.

TABLE 5–3 *Continued*

No.	$f(t)$	$\phi(\omega) = \Phi[f]$
18	$\dfrac{1}{t^n}$	$\dfrac{(-i\omega)^{n-1}}{(n-1)!}\,[\pi i - 2\pi i h(\omega)]$
19	$\dfrac{1}{t-c}$	$e^{-ic\omega}[\pi i - 2\pi i h(\omega)]$
20	$\dfrac{1}{(t-c)^n}$	$\dfrac{e^{-ic\omega}(-i\omega)^{n-1}}{(n-1)!}\,[\pi i - 2\pi i h(\omega)]$
21	$\dfrac{e^{i\alpha t}}{t-c}$	$e^{-ic(\omega-\alpha)}[\pi i - 2\pi i h(\omega-\alpha)]$
22	$\dfrac{e^{i\alpha t}}{(t-c)^n}$	$\dfrac{e^{-ic(\omega-\alpha)}[-i(\omega-\alpha)]^{n-1}}{(n-1)!}\,[\pi i - 2\pi i h(\omega-\alpha)]$

To maintain property (b), we are led from (5–165) to the definitions

$$\Phi\!\left[\frac{-1}{t^2}\right] = i\omega[\pi i - 2ih(\omega)],$$

$$\Phi\!\left[\frac{2}{t^3}\right] = (i\omega)^2[\pi i - 2\pi i h(\omega)],$$

$$\vdots$$

and in general, for $n = 1, 2, \ldots,$

$$\Phi\!\left[\frac{1}{t^n}\right] = \frac{(-i\omega)^{n-1}}{(n-1)!}\,[\pi i - 2ih(\omega)]. \qquad (5\text{–}166)$$

To maintain properties (d), (e), and (f), we adjoin the definitions 11, 13, 15, 16, 20, and 22 of Table 5–3. For a more thorough discussion of this topic, refer to Reference 3. Table I on p. 43 of Reference 3 corresponds to our Table 5–3, with x replacing t and y replacing $\omega/(2\pi)$.

Convolutions involving generalized functions are not in general meaningful, particularly when they lead to multiplication of two generalized functions. However, the convolution of $\delta^{(n)}$ with an ordinary function can be evaluated easily (as in Section 1–13) to yield the rule

$$\delta^{(n)}(t) * f(t) = f^{(n)}(t), \qquad (5\text{–}167)$$

provided only that $f^{(n)}(t)$ is continuous.

PROBLEMS

1. Evaluate the Fourier transforms of the following functions:

(a) $1 - 2\,\delta(t) + 3\,\delta'(t-1)$

(b) $\cos 5t$ (c) $\sin^3 t$

2. Verify that $\Phi[f'] = i\omega\Phi[f]$ for the following choices of f:

(a) $e^{-t}h(t)$ (b) $t[h(t) - h(t-1)]$

3. Certain linear combinations of the functions $f(t)$ in Table 5–3 are functions having a Fourier transform in the usual sense. Verify that the definitions of Table 5–3 give results in agreement with the usual definitions for each of the following cases:

(a) $h(t) - h(t-c)$

(b) $t[h(t) - h(t-c)] = th(t) - (t-c)h(t-c) - ch(t-c)$

(c) $e^{i\alpha t}[h(t) - h(t-c)]$

4. Verify that Property (d) holds for the following entries in Table 5–3:

(a) No. 14   (b) No. 19   (c) No. 21

5. Verify that property (e) holds for the following entries in Table 5–3:

(a) No. 8   (b) No. 7

ANSWERS

1. (a) $2\pi\,\delta(\omega) - 2 + 3i\omega e^{-i\omega}$ (b) $\pi\,[\delta(\omega+5) + \delta(\omega-5)]$

(c) $(\pi i/4)\,[\delta(\omega-3) - 3\,\delta(\omega-1) + 3\,\delta(\omega+1) - \delta(\omega+3)]$

5–17 Application of Fourier transforms to linear differential equations. We consider a differential equation

$$(a_0 D^n + \cdots + a_n)x = f(t) \quad (a_0 \neq 0), \tag{5–170}$$

where $f(t)$ is piecewise continuous for $-\infty < t < \infty$ and the coefficients are constant. We seek a solution $x(t)$ representable as a Fourier integral. First we proceed formally and then examine the significance of the result.

If we form the Fourier transform of both sides of (5–170) and apply (5–101), we find that

$$[a_0(i\omega)^n + \cdots + a_n]\Phi[x] = \Phi[f], \tag{5–171}$$

$$\Phi[x] = Y(i\omega)\Phi[f]. \tag{5–172}$$

Therefore, if $\phi(\omega) = \Phi[f]$, then

$$x = \frac{1}{2\pi}\,(P)\int_{-\infty}^{\infty} Y(i\omega)\phi(\omega)e^{i\omega t}\,d\omega. \tag{5–173}$$

We can also write

$$\Phi^{-1}[Y(i\omega)] = W(t); \tag{5-174}$$

then (5-172) reads

$$\Phi[x] = \Phi[W]\Phi[f] = \Phi[W * f]. \tag{5-175}$$

Hence also

$$x = W * f = \int_{-\infty}^{\infty} W(u)f(t - u)\, du. \tag{5-176}$$

Thus two expressions, (5-173) and (5-176), are obtained for the solution sought.

THEOREM 19. *Let the system described by* (5-170) *have characteristic roots all with negative real parts, so that the system is stable. Let* $f(t)$ *be piecewise continuous for* $-\infty < t < \infty$ *and let* $f(t)$ *be of less than exponential growth as* $t \to -\infty$. *Then* (5-170) *has a unique solution* $x(t)$, $-\infty < t < \infty$, *which is of less than exponential growth as* $t \to -\infty$; *the inverse Fourier transform,* $W(t)$, *of* $Y(i\omega)$ *exists, and* $x(t)$ *is defined by* (5-176). *If* $f(t)$ *is absolutely integrable for* $-\infty < t < \infty$, *then* $x(t)$ *is also absolutely integrable for* $-\infty < t < \infty$ *and is represented by the Fourier integral* (5-173); $x(t)$ *is the only solution of* (5-170) *which is representable by its Fourier integral.*

Proof. First we verify existence of $W(t)$. By the assumption on the characteristic roots, $Y(s)$ has a partial fraction expansion whose general term is of form $A(s - s_j)^{-k}$ and $Y(i\omega)$ is a sum of terms of form $A(i\omega - s_j)^{-k}$, Re $(s_j) < 0$. By (5-132), each such term has an inverse; therefore, $W(t) = \Phi^{-1}[Y(i\omega)]$ is well defined. In fact W is a sum of terms of form $bt^{\alpha}e^{at}h(t)$, so that $W(t) \equiv 0$ for $t < 0$ and W is bounded for all t.

We now define a sequence of functions inductively in terms of the characteristic roots $s_1, \ldots, s_n$:

$$f_1(t) = e^{s_1 t}h(t) * \frac{f(t)}{a_0},$$

$$f_2(t) = e^{s_2 t}h(t) * f_1(t),$$

$$\vdots$$

$$f_n(t) = e^{s_n t}h(t) * f_{n-1}(t).$$

Under the hypotheses concerning $f(t)$, Theorem 16 (Section 5-12) is applicable; we conclude that $f_1(t)$ is continuous and piecewise smooth, $f_1(t)$ is of less than exponential growth as $t \to -\infty$, and, wherever f is continuous, $(D - s_1)f_1 = f(t)$. Similarly $f_2(t)$ is continuous and piecewise smooth, $f_2(t)$ is of less than exponential growth as $t \to -\infty$, and, for all t, $(D - s_2)f_2 = f_1(t)$; thus also $Df_2 = s_2 f_2 + f_1$ is continuous and piecewise smooth. Finally, $f_n(t)$ is continuous and piecewise smooth, $f_n(t)$ is of less than exponential growth as $t \to -\infty$, and $(D - s_n)f_n = f_{n-1}(t)$; $f_{n-1}(t)$ has continuous derivatives through the order $(n - 2)$, and f_n has continuous derivatives through the order $(n - 1)$. In addition,

$$(D - s_1)(D - s_2) \cdots (D - s_n)f_n = (D - s_1) \cdots (D - s_{n-1})f_{n-1}$$
$$= (D - s_1) \cdots (D - s_{n-2})f_{n-2}$$
$$\vdots$$
$$= (D - s_1)f_1 = \frac{f}{a_0}$$

wherever f is continuous. Thus $x = f_n(t)$ is the desired solution of the differential equation. By the conditions on the characteristic roots, this is the only solution of less than exponential growth as $t \to -\infty$. We can write

$$f_2(t) = e^{s_2 t}h(t) * f_1(t)$$
$$= e^{s_2 t}h(t) * \left[e^{s_1 t}h(t) * \frac{f(t)}{a_0}\right]$$
$$= [e^{s_2 t}h(t) * e^{s_1 t}h(t)] * \frac{f(t)}{a_0},$$

by Theorem 17. Therefore

$$f_2(t) = \Phi^{-1}\left[\frac{1}{i\omega - s_2} \frac{1}{i\omega - s_1}\right] * \frac{f(t)}{a_0}.$$

Similarly, by Theorem 17 (compare the remark following the proof of Theorem 17),

$$f_3(t) = e^{s_3 t}h(t) * f_2(t) = [e^{s_3 t}h(t) * e^{s_2 t}h(t)] * f_1(t)$$
$$= \{[e^{s_3 t}h(t) * e^{s_2 t}h(t)] * e^{s_1 t}h(t)\} * \frac{f(t)}{a_0}$$
$$= \Phi^{-1}\left[\frac{1}{i\omega - s_3} \cdot \frac{1}{i\omega - s_2} \cdot \frac{1}{i\omega - s_1}\right] * \frac{f(t)}{a_0},$$

and finally

$$x = f_n(t) = \Phi^{-1}\left[\frac{1}{i\omega - s_n} \cdot \frac{1}{i\omega - s_{n-1}} \cdots \frac{1}{i\omega - s_1}\right] * \frac{f(t)}{a_0}$$
$$= W(t) * f(t).$$

If $f(t)$ is absolutely integrable from $-\infty$ to ∞, then by Theorem 13 (Section 5–11) so is x and

$$\Phi[x] = \Phi[W] \cdot \Phi[f] = Y(i\omega)\Phi[f].$$

Since x is also piecewise smooth and continuous, x is representable by its Fourier integral:

$$x = \frac{1}{2\pi} (P) \int_{-\infty}^{\infty} Y(i\omega)\phi(\omega)e^{i\omega t} \, d\omega.$$

Any other solution of the differential equation differs from x by a choice of the complementary function. Since the complementary function (if not identically zero) always approaches $\pm\infty$ as $t \to -\infty$, its Fourier transform does not exist, and hence it cannot be represented by a Fourier integral. Therefore, $x = W * f$ is the only solution represented by its Fourier integral.

Remarks. Since $x, Dx, \ldots, D^{n-1}x$ are of less than exponential growth as $t \to -\infty$, we can consider the solution obtained as the only one *with "small" initial values* at $-\infty$. If $f(t) \equiv 0$ for $t < 0$, then $W * f$ is zero for $t < 0$, and $x, Dx, \ldots, D^{n-1}x$ all equal zero for $t = 0$; hence in this case the solution is $T_0[f]$, the solution with *initial values zero* at $t = 0$.

EXAMPLE 1. $(D + 2)x = t^3$. Here, by Entry 3 of Table 5–2,

$$Y(i\omega) = \frac{1}{i\omega + 2}, \qquad W = \Phi^{-1}[Y(i\omega)] = e^{-2t}h(t),$$

$$x = e^{-2t}h(t) * t^3 = \frac{-g_3(-2t)}{(-2)^4} = \frac{-1}{16}[(-2t)^3 + 3(-2t)^2 + 6(-2t) + 6],$$

$$x = \tfrac{1}{8}(4t^3 - 6t^2 + 6t - 3).$$

EXAMPLE 2. $(D + 1)(D + 2)x = h(t) - h(t - 1)$. We find, by Entry 10 of Table 5–2,

$$Y(i\omega) = \frac{1}{i\omega + 1} - \frac{1}{i\omega + 2},$$

$$W = e^{-t}h(t) - e^{-2t}h(t),$$

$$x = [e^{-t}h(t) - e^{-2t}h(t)] * [h(t) - h(t - 1)]$$

$$= \left[\frac{e^{-t} - 1}{-1} - \frac{e^{-2t} - 1}{-2}\right][h(t) - h(t - 1)]$$

$$- \left[\frac{e^{-(t-1)} - e^{-t}}{-1} - \frac{e^{-2(t-1)} - e^{-2t}}{-2}\right]h(t - 1);$$

$$x = \tfrac{1}{2} - e^{-t} + \tfrac{1}{2}e^{-2t} \quad (0 \leq t \leq 1),$$

$$x = e^{-t}(e - 1) + \tfrac{1}{2}e^{-2t}(1 - e^2) \quad (t \geq 1),$$

$$x = 0 \quad (t \leq 0).$$

In general, if $V(s) = a_0 s^n + \cdots + a_n$ has simple roots, then

$$Y(s) = \sum_{j=1}^{n} \frac{A_j}{s - s_j}, \qquad W(t) = \sum_{j=1}^{n} A_j e^{s_j t} h(t),$$

$$x = \sum_{j=1}^{n} A_j (e^{s_j t} h(t) * f). \tag{5-177}$$

Hence if a lengthy table of the form of Table 5–2 is available, computation of x is trivial once the roots of $V(s)$ are known.

If f has the special form $e^{bt} p(t)$, where Re $(b) \geqq 0$ and $p(t)$ is a polynomial, the solution $x(t)$ can be found without knowledge of the roots of $V(s)$.

THEOREM 20.   *Let all roots of* $V(s) = a_0 s^n + \cdots + a_n$ *have negative real parts. Let* $Y(s) = P(s)/V(s)$ *where* P *is a polynomial of degree less than* n. *Let* Re $(b) \geqq 0$ *and let* $p(t)$ *be a polynomial of degree* m. *Then*

$$\Phi^{-1}[Y(i\omega)] * e^{bt} p(t)$$

$$= e^{bt} \left[Y(b)p(t) + \frac{Y'(b)p'(t)}{1!} + \cdots + \frac{Y^{(m)}(b)p^{(m)}(t)}{m!} \right]. \tag{5-178}$$

Proof. By linearity it is sufficient to consider the case when $p(t) = t^m$; for the same reason, it is sufficient to consider one term in the partial fraction expansion of $Y(i\omega)$. Thus we can assume $Y(s) = (s - a)^{-k}$ $(k = 1, 2, \ldots)$, Re $(a) < 0$. Now for $k = 1$,

$$\Phi^{-1}\left[\frac{1}{i\omega - a} \right] * e^{bt} t^m = -\frac{e^{bt} g_m[(a - b)t]}{(a - b)^{m+1}}$$

$$= -e^{bt}\left[\frac{t^m}{a - b} + \frac{mt^{m-1}}{(a - b)^2} + \cdots + \frac{m!}{(a - b)^{m+1}} \right].$$

Since

$$Y(s) = (s - a)^{-1},$$
$$Y'(s) = -(a - s)^{-2},$$
$$\vdots$$
$$Y^{(l)}(s) = -l!(a - s)^{-l-1},$$
$$\vdots$$

we can write

$$\Phi^{-1}[Y(i\omega)] * e^{bt} t^m = e^{bt}\left[t^m Y(b) + mt^{m-1}\frac{Y'(b)}{1!} + \cdots + m!\frac{Y^{(m)}(b)}{m!} \right]$$

$$= e^{bt}\left[Y(b)p(t) + \frac{Y'(b)p'(t)}{1!} + \cdots \right],$$

with $p(t) = t^m$. Thus the formula is proved for $k = 1$.

To obtain the general case, we write the result for $k = 1$ as follows:

$$e^{at}h(t) * e^{bt}t^m = -e^{bt}\left[\frac{t^m}{a-b} + \frac{mt^{m-1}}{(a-b)^2} + \cdots + \frac{m!}{(a-b)^{m+1}}\right].$$

Now we differentiate both sides partially with respect to a; we justify this operation below:

$$te^{at}h(t) * e^{bt}t^m = e^{bt}\left[\frac{t^m}{(a-b)^2} + \frac{2mt^{m-1}}{(a-b)^3} + \cdots + \frac{(m+1)m!}{(a-b)^{m+2}}\right].$$

Now

$$\Phi^{-1}\left[\frac{1}{(i\omega - a)^2}\right] = te^{at}h(t),$$

and, with $Y(s) = (s-a)^{-2}$, we verify that the right-hand side is again $e^{bt}[t^m Y(b) + \cdots]$. Hence the formula follows for $k = 2$. Repeated differentiation gives it for arbitrary k.

We examine the partial differentiation in detail:

$$e^{at}h(t) * e^{bt}t^m = \int_0^\infty e^{au}e^{b(t-u)}(t-u)^m\, du.$$

Therefore,

$$\frac{\partial}{\partial a}[e^{at}h(t) * e^{bt}t^m] = \int_0^\infty ue^{au}e^{b(t-u)}(t-u)^m\, du = te^{at}h(t) * e^{bt}t^m.$$

The differentiation under the integral sign is permitted since the improper integral converges uniformly *after* differentiation (Reference 2, p. 379).

EXAMPLE 3. $(D + 2)x = t^3$.   This is the same as Example 1 above. In this instance

$$b = 0, \qquad Y(s) = \frac{1}{s+2}, \qquad Y'(s) = -\frac{1}{(s+2)^2},$$

$$Y''(s) = \frac{2}{(s+2)^3}, \qquad Y'''(s) = \frac{-6}{(s+2)^4}.$$

Hence $Y(b) = Y(0) = \frac{1}{2}$, $Y'(0) = -\frac{1}{4}$, $Y''(0) = \frac{1}{4}$, $Y'''(0) = -\frac{3}{8}$. Therefore

$$x = \Phi^{-1}[Y(i\omega)] * t^3 = \frac{1}{2}t^3 - \frac{1}{4}\cdot\frac{3t^2}{1!} + \frac{1}{4}\cdot\frac{6t}{2!} - \frac{3}{8}\cdot\frac{6}{3!}$$

$$= \frac{t^3}{2} - \frac{3t^2}{4} + \frac{3}{4}t - \frac{3}{8}.$$

EXAMPLE 4. $(D^3 + 3D^2 + D + 1)x = e^t(t^2 - 2)$.   In this case determination of the characteristic roots is awkward, but it can be verified that the system is stable. We apply Theorem 20, with $b = 1$. We have

$$Y = \frac{1}{V} = V^{-1}, \quad Y' = -V^{-2}V', \quad Y'' = 2V^{-3}V'^{2} - V^{-2}V'',$$

$$V = s^3 + 3s^2 + s + 1, \quad V' = 3s^2 + 6s + 1, \quad V'' = 6s + 6.$$

For $s = 1$, $V = 6$, $V' = 10$, $V'' = 12$, $Y = 6^{-1}$, $Y' = -6^{-2} \times 10$, $Y'' = 2 \times 6^{-3} \times 10^2 - 6^{-2} \times 12$. Hence

$$x = e^t \left[\frac{t^2 - 2}{6} - \frac{10}{36} 2t + \left(\frac{200}{216} - \frac{12}{36} \right) \left(\frac{1}{2} \right) (2) \right] = \frac{e^t}{54} (9t^2 - 30t + 14).$$

5–18 The weighting function. As in Sections 2–5 and 4–23, the function $W(t)$ can be interpreted as a *weighting function* and the solution $x = W * f$ as a weighted average over the past of f. Since $W(t) \equiv 0$ for $t < 0$,

$$x = \int_0^\infty W(u)f(t - u) \, du.$$

The total weight is

$$\int_0^\infty W(u) \, du = \int_{-\infty}^\infty W(u) \, du = Y(0);$$

for

$$Y(i\omega) = \Phi[W] = \int_{-\infty}^\infty W(t)e^{-i\omega t} \, dt$$

and $W(t) = 0$ for $t < 0$. Hence if $Y(0) = 1$, the total weight is 1 and, x is a true weighted average of $f(t)$ over the past; otherwise, x is a weighted average of $Y(0)f$.

We can compute $x(t)$ graphically as suggested in Fig. 5–7. The graph of $W(u)$ is reversed and placed above that of $f(t)$, with the origin above the point $(t, f(t))$. The value of f at $t - u$ is then multiplied by the value of W at u to obtain the graph of $W(u)f(t - u)$; the area under this curve is the value of x at t. If $f \equiv 0$ for $t < 0$, the area can be readily computed graphically. In general, we are forced to compute an approximation to the area, obtained by stopping at some negative value of t.

EXAMPLE. $(D + 1)(D + 2)x = h(t) - h(t - 1)$. This is the same as Example 2 of the preceding section. $W(t) = (e^{-t} - e^{-2t})h(t)$. The value of x at $t = \frac{1}{2}$ is computed graphically in Fig. 5–8. Since

$$Y(s) = [(s + 1)(s + 2)]^{-1},$$

then $Y(0) = \frac{1}{2}$ and x is a weighted average of $\frac{1}{2}f$. It is clear that $x = 0$ for $t < 0$; x rises almost to $\frac{1}{4}$ as t increases to 1, then continues to rise for a short time, and subsequently decreases steadily to zero as $t \to +\infty$.

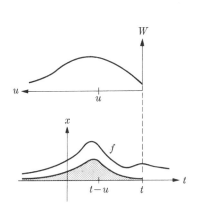

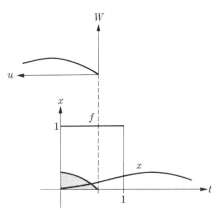

FIG. 5–7. Weighting function.   FIG. 5–8. Response to square pulse.

In general, if $Y(0) = 1$, $x(t)$ will resemble $f(t)$; that is, output $\sim$ input.
The accuracy of this approximation depends on the shape of the weighting
function. If $W(t)$ is very small except for t close to zero, then the output
will "remember" only the recent past, and the approximation is a good one.
If $W(t)$ has a sharp positive peak at $t_0 > 0$, is otherwise very small, then
$x(t) \sim f(t - t_0)$. If the graph of $W(t)$ is relatively flat, with no peaks,
then $x(t)$ will give only a smoothed-out reproduction of the input.

From the fact that

$$\Phi[W] = Y(i\omega) = \frac{1}{a_0(i\omega)^n + \cdots + a_n},$$

we can conclude that for $n \geq 2$, $W, W', \ldots, W^{(n-2)}$ are continuous, while
$W^{(n-1)}$ is piecewise continuous, and

$$\Phi[W'] = \frac{i\omega}{a_0(i\omega)^n + \cdots + a_n}, \tag{5–181}$$

$$\Phi[W^{(k)}] = \frac{(i\omega)^k}{a_0(i\omega)^n + \cdots + a_n} \quad (k = 1, \ldots, n - 1). \tag{5–182}$$

Indeed, for $n \geq 2$, the right-hand side of Eq. (5–181) is a proper rational
function of $i\omega$ and hence is the Fourier transform of a piecewise continuous
function $q(t)$. Since $\Phi[q] = i\omega\Phi[W]$, we conclude from the corollary to
Theorem 16 (Section 5–12 above) that W is continuous and $q = W'$.
Repetition of the argument establishes all the asserted continuity properties and validity of (5–182). For $n = 1$,

$$\Phi[W] = \frac{1}{a_0(i\omega) + a_1}, \qquad W = \frac{1}{a_0} e^{s_1 t} h(t),$$

where $\text{Re}\,(s_1) < 0$. Thus W is piecewise continuous, with a jump of $1/a_0$ at $t = 0$.

We can also write

$$W = \left[\frac{1}{a_0}\, e^{s_1 t} h(t) \,*\, e^{s_2 t} h(t) \,*\, \cdots\right] * e^{s_n t} h(t) = W_0(t) * e^{s_n t} h(t),$$

so that W can be considered as a solution of a differential equation

$$a_0(D - s_1)\cdots(D - s_{n-1})W = e^{s_n t} h(t). \tag{5-183}$$

This again implies continuity of $W, W', \ldots, W^{(n-2)}$, while $W^{(n-1)}$ must have a jump of $1/a_0$ at $t = 0$ (see also Problem 5 following Section 5–19). For $t > 0$, we can apply the operator $(D - s_n)$ to Eq. (5–183) and conclude that

$$a_0(D - s_1)\cdots(D - s_n)W$$
$$= (a_0 D^n + \cdots + a_n)W = 0 \quad (t > 0). \tag{5-184}$$

Thus for $t > 0$, W is a solution of the homogeneous differential equation such that $W \to 0,\ W' \to 0,\ \ldots,\ W^{(n-2)} \to 0,\ W^{(n-1)} \to 1/a_0$ as $t \to 0+$.

5–19 Response to generalized functions as inputs. In the stable differential equation

$$(a_0 D^n + \cdots + a_n)x = f(t), \tag{5-190}$$

let $f(t)$ be a generalized function. The methods of Section 1–14 tell us how to obtain solutions. It is of interest to note that a particular solution can be obtained by the methods of Section 5–17. If, for example, $f(t) = \delta(t)$, then by Section 5–16, $\Phi[f] = 1$, and

$$\Phi[x] = Y(i\omega) \cdot 1 = Y(i\omega).$$

Hence $x = \Phi^{-1}[Y(i\omega)] = W(t)$. *The weighting function is the response to the unit impulse.* The same conclusion is obtained in Section 2–8 and the two weighting functions are clearly the same (see Problem 8(a) following Section 2–8). We can also write

$$x = W * f = \int_{-\infty}^{\infty} \delta(u)W(t - u)\,du = W(t).$$

If f is chosen as $\delta^{(k)}(t)$, then we find that

$$\Phi[x] = (i\omega)^k Y(i\omega), \qquad x = W^{(k)}(t).$$

Here we have applied the rule (5–167). The same result is obtained by differentiating the equation

$$(a_0 D^n + \cdots + a_n) W(t) = \delta(t)$$

k times. It should be remarked that $W, W', \ldots, W^{(n-2)}$ are all continuous, while $W^{(n-1)}$ has a jump discontinuity at $t = 0$ (Section 5-18), and accordingly $W^{(n)}, W^{(n+1)}, \ldots$ are all generalized functions.

If f is chosen as $h(t)$, then Table 5-3 of Section 5-16 gives $\Phi[f] = (1/[i\omega]) + \pi \delta(\omega)$, so that

$$\Phi[x] = \frac{Y(i\omega)}{i\omega} + \pi \delta(\omega) Y(i\omega) = \frac{Y(i\omega)}{i\omega} + \frac{\pi}{a_n} \delta(\omega),$$

since $Y(0) = 1/a_n$. Now we can write

$$\frac{Y(s)}{s} = \frac{1}{a_n s} - \frac{a_0 s^{n-1} + \cdots + a_{n-1}}{a_n(a_0 s^n + \cdots + a_n)}$$

(the residue of $Y(s)/s$ at zero being $1/a_n$), and accordingly,

$$\frac{Y(i\omega)}{i\omega} + \frac{\pi}{a_n} \delta(\omega) = \frac{1}{a_n}\left[\frac{1}{i\omega} + \pi \delta(\omega)\right] - \frac{a_0(i\omega)^{n-1} + \cdots + a_{n-1}}{a_n[a_0(i\omega)^n + \cdots]}$$

$$= \frac{1}{a_n} \Phi[h(t)] - \frac{1}{a_n} \Phi[a_0 W^{(n-1)}(t) + a_1 W^{(n-2)}(t) + \cdots].$$

Therefore the response is

$$x = \frac{1}{a_n} [h(t) - a_0 W^{(n-1)}(t) - \cdots - a_{n-1}W(t)]. \qquad (5\text{-}191)$$

From (5-191) we see that $x = 0$ for $t < 0$. As $t \to 0+$, $W(t) \to 0, \ldots$, $W^{(n-2)}(t) \to 0$, $W^{(n-1)}(t) \to 1/a_0$, $h(t) \to 1$, so that $x \to 0$. Hence x is continuous at $t = 0$. For $t > 0$, by (5-191) and (5-184),

$$Dx = \frac{1}{a_n} [-a_0 W^{(n)}(t) - \cdots - a_{n-1}W'(t)]$$

$$= \frac{1}{a_n} a_n W(t) = W(t).$$

Therefore

$$x = D_0^{-1} W = \int_0^t W(u) \, du$$

$$= \int_0^t W(u) h(t - u) \, du = W * h,$$

in agreement with Theorem 19.

Finally, let $f = t^n$. By Entry 7 of Table 5–3 and by the rule of multiplication of Section 1–13,

$$\Phi[x] = Y(i\omega)2\pi i^n \, \delta^{(n)}(\omega)$$

$$= 2\pi i^n \left[\delta^{(n)}(\omega) \, Y(0) - \frac{n}{1!} \, \delta^{(n-1)}(\omega) i Y'(0) + \frac{n(n-1)}{2!} \right.$$

$$\left. \times \, \delta^{(n-2)}(\omega) i^2 Y''(0) + \cdots + (-1)^n \frac{n!}{n!} \, i^n \, \delta(\omega) \, Y^{(n)}(0) \right],$$

$$x = Y(0)t^n + Y'(0)nt^{n-1} + Y''(0) \frac{n(n-1)t^{n-2}}{2!} + \cdots + Y^{(n)}(0),$$

in agreement with Theorem 20.

Problems

1. For each of the following find the solution with small initial values at $t = -\infty$ (express the answer in simplest form and, if possible, as a Fourier integral):

    (a) $(D + 2)(D + 4)x = 5 \cos 3t$
    (b) $(D + 2)(D + 4)x = e^t[1 - h(t)]$
    (c) $(D + 2)(D + 4)x = t[h(t) - h(t - 1)] + (2 - t)[h(t - 1) - h(t - 2)]$

2. By Theorem 20, find the solution of each of the following with small initial values at $t = -\infty$:

    (a) $(D^3 + 5D^2 + 7D + 1)x = 2t^2 - t + 1$
    (b) $(D^2 + 3D + 1)x = e^{3t}(t^3 + 1)$
    (c) $(D^3 + 5D^2 + 7D + 1)x = t^2 \cos t$

[Hint: Since all coefficients are real, x is the real part of the response to $t^2 e^{it}$.]

3. Find the weighting function for Problem 1(c) and obtain the solution graphically.

4. Find a solution of each of the following with the aid of Fourier transforms:

    (a) $(D^2 + 3D + 2)x = 3 \, \delta(t)$
    (b) $(D^2 + 3D + 2)x = 1 - 2 \, \delta'(t) + 3 \, \delta''(t)$
    (c) $(D^2 + 3D + 2)x = h(t)$
    (d) $(D^2 + 3D + 2)x = t - th(t)$

5. Let $Y(s) = [a_0 s^n + \cdots + a_n]^{-1}$ be the transfer function of a stable system. Let $W(t) = \Phi^{-1}[Y(i\omega)]$.

    (a) Show that $W = 0$ for $t < 0$ and

$$W = \frac{1}{2\pi i} \oint_C e^{st} Y(s) \, ds \text{ for } t > 0$$

where C encloses all the poles of $Y(s)$ (see Section 5–14).

(b) From the result of part (a) show that

$$\lim_{t \to 0+} W^{(k)}(t) = -\operatorname{Res}\,[s^k Y(s),\, \infty]$$

and hence conclude that $W(t), \ldots, W^{(n-2)}(t)$ are continuous at $t = 0$, whereas $W^{(n-1)}(t)$ has a jump discontinuity.

(c) Integrate the relation

$$(a_0 D^n + \cdots + a_n) W = \delta(t)$$

to show that

$$a_n D_0^{-1}[W] = h(t) - a_0 W^{(n-1)}(t) - \cdots - a_{n-1} W(t).$$

ANSWERS

1. (a) $\dfrac{18 \sin 3t - \cos 3t}{65}$

(b) $\frac{1}{30}[2e^t\{1 - h(t)\} + \{5e^{-2t} - 3e^{-4t}\}h(t)],$

$$\frac{1}{2\pi} \int_{-\infty}^{\infty} \frac{e^{i\omega t}}{(1 - i\omega)(i\omega + 2)(i\omega + 4)}\, d\omega$$

(c) $\frac{1}{32}[(4e^{-2t} - e^{-4t} + 4t - 3)\{h(t) - h(t-1)\} + \{4(1 - 2e^2)e^{-2t}$

$\quad - (1 - 2e^4)e^{-4t} - 4t + 11\}\{h(t-1) - h(t-2)\}$

$\quad + \{4(1 - 2e^2 + e^4)e^{-2t} - (1 - 2e^4 + e^8)e^{-4t}\}h(t-2)],$

$$\frac{1}{2\pi} \int_{-\infty}^{\infty} \frac{(1 - 2e^{-i\omega} + e^{-2i\omega})e^{i\omega t}}{-\omega^2(i\omega + 2)(i\omega + 4)}\, d\omega$$

2. (a) $2t^2 - 29t + 184$ (b) $\dfrac{e^{3t}}{130{,}321}[6859t^3 - 9747t^2 + 7068t + 4537]$

(c) $\operatorname{Re}\left[e^{it}\left(\dfrac{-2 - 3i}{26}t^2 + \dfrac{280 + 4i}{26^2}t + \dfrac{-3116 + 6532i}{26^3}\right)\right]$

3. $W = \frac{1}{2}[e^{-2t} - e^{-4t}]h(t)$

4. (a) $3W = 3(e^{-t} - e^{-2t})h(t)$
(b) $\frac{1}{2} - 2W' + 3W'' = \frac{1}{2} + (5e^{-t} - 16e^{-2t})h(t) + 3\,\delta(t)$
(c) $D_0^{-1}[W] = \frac{1}{2}(1 - 2e^{-t} + e^{-2t})h(t)$
(d) $\frac{1}{4}\{2t - 3 - (2t - 3 + 4e^{-t} - e^{-2t})h(t)\}$

5-20 Application of Fourier transforms to simultaneous differential equations. We illustrate the method by an example:

$$(D + 2)x + Dy = f,$$

$$Dx + (2D + 1)y = g. \tag{5-200}$$

We replace D by s and solve formally for x and y:

$$x = \frac{2s + 1}{s^2 + 5s + 2} f + \frac{-s}{s^2 + 5s + 2} g,$$

$$y = \frac{-s}{s^2 + 5s + 2} f + \frac{s + 2}{s^2 + 5s + 2} g.$$

The coefficients of f and g form the transfer function matrix (Section 2–9). For example, the transfer function of x relative to f is

$$Y_{11}(s) = \frac{2s + 1}{s^2 + 5s + 2}.$$

We note that the roots of the denominator, which are the characteristic roots of (5–200), have negative real parts, so that the system is stable. If f is of less than exponential growth as $t \to -\infty$ and $g = 0$, then the corresponding output x is

$$x = \Phi^{-1}[Y_{11}(i\omega)] * f = \Phi^{-1}\left[\frac{2i\omega + 1}{-\omega^2 + 5i\omega + 2}\right] * f.$$

If f has an absolutely convergent integral, then x is obtained as a Fourier integral,

$$x = \frac{1}{2\pi} \int_{-\infty}^{\infty} \frac{2i\omega + 1}{2 - \omega^2 + 5i\omega} \phi(\omega) e^{i\omega t} \, d\omega,$$

where $\phi(\omega) = \Phi[f]$. If $f = 0$ but $g \neq 0$, a similar expression is obtained for the output x; if neither is zero, the output is the sum of the two outputs:

$$x = \Phi^{-1}\left[\frac{2i\omega + 1}{-\omega^2 + 5i\omega + 2}\right] * f + \Phi^{-1}\left[\frac{-i\omega}{-\omega^2 + 5i\omega + 2}\right] * g,$$

$$y = \Phi^{-1}\left[\frac{-i\omega}{-\omega^2 + 5i\omega + 2}\right] * f + \Phi^{-1}\left[\frac{i\omega + 2}{-\omega^2 + 5i\omega + 2}\right] * g.$$

When f and g are of the form $e^{bt} \cdot$ (polynomial), where Re $(b) \geqq 0$, the solution can be found by the method of Theorem 20.

The method described is applicable to any stable set of differential equations

$$(\alpha_0 D^m + \cdots)x + (\beta_0 D^n + \cdots)y = f,$$

$$(\gamma_0 D^m + \cdots)x + (\eta_0 D^n + \cdots)y = g, \tag{5–201}$$

provided

$$\begin{vmatrix} \alpha_0 & \beta_0 \\ \gamma_0 & \eta_0 \end{vmatrix} \neq 0, \tag{5–202}$$

so that the equations can be solved for the highest-order derivatives $D^m x$, $D^n y$; the method applies equally well to stable systems in more than two unknowns, provided the condition analogous to (5–202) is satisfied. In each case replacement of D by s and formal solution of the equations for x, y, ... yields the transfer function matrix. If $Y_{11}(s)$ is the transfer function of x relative to input f, then the output from this input alone is $W_{11} * f$, where $W_{11} = \Phi^{-1}[Y_{11}(i\omega)]$, provided f is of less than exponential growth as $t \to -\infty$; if f is absolutely integrable from $-\infty$ to ∞, then the output x is

$$\frac{1}{2\pi} \int_{-\infty}^{\infty} Y_{11}(i\omega)\phi(\omega)e^{i\omega t}\, d\omega,$$

where $\phi(\omega) = \Phi[f]$. Similar methods apply when f is a generalized function, as in Section 5–19.

The justification for the formal procedures described is similar to that of Sections 5–17 and 4–25 and will not be given here.

5–21 Application of Fourier transforms to integrodifferential equations. As shown in Section 1–6, integrodifferential equations can be replaced by simultaneous differential equations, so that the results of Section 5–20 are applicable. However, we can verify that exactly the same formulas are obtained if we apply the transform method to the given integrodifferential equations, with the aid of the rule:

$$\Phi[D^{-n}x] = (i\omega)^{-n}\Phi[x].$$

Here $D^{-n}x$ stands for an appropriate choice of the iterated indefinite integral. As in Section 5–20 we can describe the procedure as one of replacing D by s, solving for x, y, ... , and then interpreting the coefficients as transfer functions.

EXAMPLE 1. $(D + 3 + 2D^{-1})x = t^2$. We write

$$\left(s + 3 + \frac{2}{s}\right) x = t^2,$$

$$x = \frac{s}{s^2 + 3s + 2}\, t^2, \qquad Y(s) = \frac{s}{s^2 + 3s + 2}.$$

$$W = \Phi^{-1}[Y(i\omega)] = (2e^{-2t} - e^{-t})h(t),$$

$$x = W * t^2 = [(2e^{-2t} - e^{-t})h(t)] * t^2 = t - \tfrac{3}{2}.$$

The same result is obtained by setting $y = D^{-1}x$, so that $Dy = x$:

$$(D^2 + 3D + 2)y = t^2.$$

The solution y of less than exponential growth as $t \to -\infty$ can be obtained by Theorem 20:

$$Y(s) = V^{-1}, \qquad V = s^2 + 3s + 2,$$

$$Y' = -V^{-2}V', \qquad Y'' = 2V^{-3}V'^2 - V^{-2}V'',$$

$$V' = 2s + 3, \qquad V'' = 2,$$

$$V(0) = 2, \qquad V'(0) = 3, \qquad V''(0) = 2,$$

$$Y(0) = 2^{-1}, \qquad Y'(0) = -2^{-2} \cdot 3, \qquad Y''(0) = 2^{-2}3^2 - 2^{-2}2,$$

$$y = \frac{t^2}{2} - \frac{3}{2}t + \frac{7}{4} = D^{-1}x, \qquad x = Dy = t - \tfrac{3}{2}.$$

EXAMPLE 2.

$$(D + 2)x - (D + 1 - 2D^{-1})y = f,$$

$$(-D + 2)x + (2D + 1 + 5D^{-1})y = 0.$$

Replacement of D by s, D^{-1} by s^{-1} leads to the expression

$$x = \frac{2s^2 + s + 5}{s^3 + 6s^2 + 11s + 6} f = Y_{11}(s)f.$$

We find

$$Y_{11}(s) = \frac{3}{s + 1} - \frac{11}{s + 2} + \frac{10}{s + 3},$$

$$W_{11}(t) = (3e^{-t} - 11e^{-2t} + 10e^{-3t})h(t),$$

$$x = W_{11}(t) * f.$$

If, for example, $f = (1 + t^2)^{-1}$, then we obtain two expressions for x:

$$x = \int_{-\infty}^{t} \frac{1}{1 + u^2} [3e^{-(t-u)} - 11e^{-2(t-u)} + 10e^{-3(t-u)}] \, du;$$

$$x = \frac{1}{2} \int_{-\infty}^{\infty} \frac{-2\omega^2 + i\omega + 5}{-i\omega^3 - 6\omega^2 + 11i\omega + 6} e^{-|\omega|} e^{\omega t} \, d\omega.$$

Both formulas are difficult to simplify. However, they can be used to evaluate $x(t)$ numerically to any desired accuracy. A similar procedure yields $y(t)$.

5-22 Systems analysis by Fourier transforms. The results of the preceding sections show that the response of a stable linear system to very general inputs can be analyzed by Fourier transforms or by the related convolutions. For each input f and corresponding output x there is a transfer function $Y(s)$, which is a proper rational function of s. If f is piecewise continuous for $-\infty < t < \infty$ and of less than exponential growth as $t \to -\infty$, then $x = W * f$ is the output with "small initial values at $-\infty$." Since the system is stable, only this output will be observed after transients have died out. If $f \equiv 0$ for $t < 0$, then $x \equiv 0$ for $t < 0$ and the output obtained is $T_0[f]$, the solution with zero initial values for $t = 0$. If f has an absolutely convergent integral from $-\infty$ to ∞, then so also does the output x, and

$$x = \frac{1}{2\pi} \int_{-\infty}^{\infty} Y(i\omega)\phi(\omega)e^{i\omega t} \, d\omega,$$

where $\Phi[f] = \phi(\omega)$.

The operational method based on the Fourier transform has as its main advantage the analysis of the system over the infinite interval $-\infty < t < \infty$. If we consider only inputs and outputs which are zero for $t < 0$, then the convolution expression $W * f$ is the same as that obtained by means of the Laplace transform, to be discussed in the next chapter. If one is interested only in this case (inputs and outputs zero for $t < 0$), the Laplace transform provides considerable additional information and is to be preferred to the Fourier transform. However, as will be shown, the two transforms are very closely related.

Problems

1. For each of the following find the output with small initial values at $-\infty$:

(a) $(D + 5 + 4D^{-1})x = h(-t)$

(b) $(D^2 + 4D + 2 + D^{-1})x = t^2 - 5t + 2$

2. For each of the following obtain the solution with initial values zero at $t = 0$ in terms of convolution integrals if $f \equiv g \equiv 0$ for $t < 0$:

(a) $(D + 2)x - y = f, \qquad (D - 1)x + (3D + 5)y = g$

(b) $(D + D^{-1})x - Dy = f, \quad (5D + 4 + 5D^{-1})x + (D^2 - D + 1)y = 0$

3. Obtain a solution of the following as a Fourier integral:

$$(D + 4 + 2D^{-1} + 3D^{-2})x = \frac{\cos 2t}{t^2 + 1}$$

Answers

1. (a) $\frac{1}{3}(e^{-4t} - e^{-t})h(t)$ \qquad (b) $2t - 9$

2. (a) $x = \frac{1}{6} \int_0^t \{e^{-(t-u)}[2f(u) + g(u)] + e^{-3(t-u)}[4f(u) - g(u)]\} \, du$

$y = \frac{1}{6} \int_0^t \{e^{-(t-u)}[2f(u) + g(u)] + e^{-3(t-u)}[-4f(u) + g(u)]\} \, du$

(b) $x = \displaystyle\int_0^t e^{-u}f(t - u)(1 - 4u + 3u^2 - \tfrac{1}{2}u^3)\, du$

$\ y = \displaystyle\int_0^t e^{-u}f(t - u)(-5u + 3u^2 - u^3)\, du$

3. $x = \dfrac{1}{4}\displaystyle\int_{-\infty}^{\infty} \dfrac{-\omega^2(e^{-|\omega-2|} + e^{-|\omega+2|})}{3 - 4\omega^2 + i(2\omega - \omega^3)}\, e^{i\omega t}\, d\omega$

Suggested References

1. A. ERDÉLYI, ed., *Tables of Integral Transforms*, Vol. I (compiled by the staff of the Bateman Manuscript Project). New York: McGraw-Hill, 1954.

2. W. KAPLAN, *Advanced Calculus*. Reading, Mass.: Addison-Wesley, 1952.

3. M. J. LIGHTHILL, *An Introduction to Fourier Analysis and Generalized Functions*. Cambridge, Eng.: Cambridge University Press, 1958.

4. M. E. MUNROE, *Introduction to Measure and Integration*. Reading, Mass.: Addison-Wesley, 1953.

5. E. C. TITCHMARSH, *Introduction to the Theory of Fourier Integrals*. Oxford: Oxford University Press, 1948.

6. N. WIENER, *The Fourier Integral and Certain of Its Applications*. Cambridge, Eng.: Cambridge University Press, 1933.

CHAPTER 6

THE LAPLACE TRANSFORM

6-1 Introduction of the Laplace transform. If $f(t)$ is piecewise continuous and absolutely integrable from $-\infty$ to $+\infty$, then the theory of Chapter 5 shows that the Fourier transform $\Phi[f]$ exists. For a stable system with input $f(t)$ and transfer function $Y(s)$, the output also has a Fourier transform $Y(i\omega)\Phi[f]$ and is representable by the corresponding Fourier integral.

Now if $f(t)$ is not absolutely integrable from $-\infty$ to ∞, then in general we cannot obtain $\Phi[f]$ and the procedure described breaks down (unless generalized functions are applicable as in Section 5-16). To remedy the difficulty we can multiply f by a *damping factor* g, so that $fg = f_1$ becomes absolutely integrable. Then f_1 has a Fourier transform $\Phi[f_1]$. If f_1 is representable by its Fourier integral,

$$f_1 = \frac{1}{2\pi} (P)\int_{-\infty}^{\infty} \Phi[f_1]e^{i\omega t}\, d\omega,$$

then we can obtain something like a Fourier integral representation for f:

$$f = \frac{f_1}{g} = \frac{1}{2\pi g(t)} (P)\int_{-\infty}^{\infty} \Phi[f_1]e^{i\omega t}\, d\omega.$$

If g is of simple nature, this might prove useful in finding the output.

One natural choice of the damping factor $g(t)$ is $e^{-\sigma t}$. If $f(t)e^{-\sigma t} = f_\sigma$ is indeed absolutely integrable and representable by its Fourier integral, then we obtain the representation

$$f = \frac{1}{2\pi e^{-\sigma t}} (P)\int_{-\infty}^{\infty} \phi_\sigma(\omega)e^{i\omega t}\, d\omega$$

$$= \frac{1}{2\pi} (P)\int_{-\infty}^{\infty} \phi_\sigma(\omega)e^{(\sigma+i\omega)t}\, d\omega.$$

This is much like a Fourier integral. Furthermore,

$$\phi_\sigma(\omega) = \int_{-\infty}^{\infty} f_\sigma(t)e^{-i\omega t}\, dt = \int_{-\infty}^{\infty} f(t)e^{-(\sigma+i\omega)t}\, dt.$$

The formula for $\phi_\sigma(\omega)$ differs from that for $\Phi[f]$ only in that the exponent $i\omega t$ is replaced by $(\sigma + i\omega)t$. If we write $s = \sigma + i\omega$, then $\phi_\sigma(\omega)$ can be

301

regarded as a function of the complex variable $\sigma + i\omega$:

$$\phi_\sigma(\omega) = \psi(\sigma + i\omega) = \psi(s),$$

although we are considering σ as fixed. We can now write

$$f(t) = \frac{1}{2\pi} (P) \int_{-\infty}^{\infty} \psi(s)e^{st} \, d\omega,$$

$$\psi(s) = \int_{-\infty}^{\infty} f(t)e^{-st} \, dt,$$

(6–10)

where $s = \sigma + i\omega$. These equations will be valid if, for some choice of σ, $f(t)e^{-\sigma t}$ has an absolutely convergent integral from $-\infty$ to ∞. The function $\psi(s)$ is the *two-sided Laplace transform* of $f(t)$.

Now if $f(t)$ grows rapidly as $t \to \infty$, then multiplication by $e^{-\sigma t}$, where σ is positive, may make $f(t)e^{-\sigma t}$ absolutely integrable from 0 to $+\infty$; however, for negative t, $f(t)e^{-\sigma t}$ grows much more rapidly than f, so that we may lose absolute convergence for negative t. A similar statement holds for σ negative, with negative and positive t interchanged. In other words, the factor $e^{-\sigma t}$ can help to reduce the size of the function only on one half of the t-axis; on the other half, it makes matters worse. For functions which are very small on one half of the t-axis, it may be possible to choose σ so that $f(t)e^{-\sigma t}$ remains small on both sides. For example, if $f = e^{-t^2}h(-t) + th(t)$, then $fe^{-\sigma t}$ is absolutely integrable from $-\infty$ to ∞ if $\sigma > 0$. In particular, if $\sigma > 0$ and $f(t) \equiv 0$ for $t < 0$, then $f(t)e^{-\sigma t}$ will be small for $-\infty < t < \infty$ if only $f(t)$ does not grow too rapidly as $t \to +\infty$ (for example, if $f(t) = th(t) \sin t$).

If we concentrate on the functions which are zero for negative t, then the formulas (6–10) become

$$f(t) = \frac{1}{2\pi} (P) \int_{-\infty}^{\infty} \psi(s)e^{st} \, d\omega,$$

$$\psi(s) = \int_{0}^{\infty} f(t)e^{-st} \, dt,$$

(6–11)

where $s = \sigma + i\omega$ and σ is chosen (if possible) so that $f(t)e^{-\sigma t}$ is absolutely integrable from zero to ∞. The function $\psi(s)$ defined by (6–11) is the *one-sided Laplace transform of f*, or simply *the Laplace transform of f*.

The present chapter will be devoted to consideration of the one-sided transform. For most applications the restriction $f \equiv 0$ for $t < 0$ is a natural one. It turns out to be valuable in connection with the problem of finding solutions of differential equations with *given initial values for* $t = 0$.

6–2 Relations between the Laplace transform and the Fourier transform. Now let $f(t)$ be piecewise continuous for $t \geqq 0$ and identically zero for $t < 0$. Then the Laplace transform of f, denoted by $\mathcal{L}[f]$, is the function $\psi(s)$ of the complex variable $s = \sigma + i\omega$ defined by the equation

$$\psi(s) = \mathcal{L}[f] = \int_0^\infty f(t)e^{-st}\, dt; \qquad (6\text{–}20)$$

$\psi(s)$ is defined for all values of s for which the integral converges. Since the definition (6–20) uses only the values of f for $t \geqq 0$, we might consider it as defining a transformation on a class of functions defined only for $t \geqq 0$. However, it will be convenient *to assume always that $f(t) \equiv 0$ for $t < 0$*, even though this is not explicitly stated. Since

$$\psi(s) = \int_0^\infty f(t)e^{-(\sigma+i\omega)t}\, dt = \int_0^\infty f(t)e^{-\sigma t}e^{-i\omega t}\, dt,$$

we can consider $\psi(s)$ as the Fourier transform of $f(t)e^{-\sigma t}$:

$$\mathcal{L}[f] = \Phi[f(t)e^{-\sigma t}]. \qquad (6\text{–}21)$$

In particular, if $\Phi[f]$ exists, then

$$\Phi[f] = \mathcal{L}[f]|_{\sigma=0}. \qquad (6\text{–}22)$$

6–3 Examples of Laplace transforms. EXAMPLE 1. Let $f(t) \equiv 1$ for $t \geqq 0$; that is, $f(t) = h(t)$. Then

$$\mathcal{L}[f] = \int_0^\infty e^{-st}\, dt = \frac{e^{-st}}{-s}\bigg|_0^\infty = \lim_{t\to\infty} \frac{1 - e^{-st}}{s}.$$

Now for $s = \sigma + i\omega$,

$$e^{-st} = e^{-\sigma t}e^{-i\omega t} = e^{-\sigma t}(\cos \omega t - i \sin \omega t).$$

Hence

$$\lim_{t\to\infty} e^{-st} = 0 \quad \text{for } \sigma > 0, \qquad (6\text{–}30)$$

and there is no limit for $\sigma \leqq 0$. Therefore

$$\mathcal{L}[f] = \mathcal{L}[h(t)] = \frac{1}{s} \quad (\sigma > 0). \qquad (6\text{–}31)$$

From (6–11) we can then write

$$1 = \frac{1}{2\pi}\,(P)\int_{-\infty}^\infty \frac{1}{\sigma + i\omega}\, e^{(\sigma+i\omega)t}\, d\omega \quad (\sigma > 0),$$

for $t > 0$. For $t < 0$ the right-hand side is zero, and for $t = 0$, it yields the average value $\frac{1}{2}$.

EXAMPLE 2. Let $f(t) \equiv e^{at}$ for $t \geq 0$; that is, $f(t) = e^{at}h(t)$. Then

$$\mathcal{L}[f] = \int_0^\infty e^{(a-s)t}\, dt = \frac{1}{s-a}, \qquad (6\text{-}32)$$

provided Re $(s-a) > 0$, that is, $\sigma > $ Re (a). By inversion, for $t \neq 0$,

$$e^{at}h(t) = \frac{1}{2\pi}(P)\int_{-\infty}^\infty \frac{e^{(\sigma+i\omega)t}}{(\sigma+i\omega-a)}\, d\omega. \qquad (6\text{-}33)$$

EXAMPLE 3. Let $f(t) = t^n h(t)$, where n is a positive integer. Then

$$\mathcal{L}[f] = \int_0^\infty t^n e^{-st}\, dt$$

$$= \frac{t^n e^{-st}}{-s}\Big|_0^\infty + \frac{n}{s}\int_0^\infty t^{n-1}e^{-st}\, dt,$$

by integration by parts. As $t \to \infty$, $t^n e^{-st} \to 0$, provided $\sigma > 0$. Hence

$$\mathcal{L}[t^n h(t)] = \frac{n}{s}\int_0^\infty t^{n-1}e^{-st}\, dt = \frac{n}{s}\mathcal{L}[t^{n-1}h(t)].$$

Accordingly, by induction,

$$\mathcal{L}[t^n h(t)] = \frac{n}{s}\frac{(n-1)}{s}\cdots\frac{1}{s}\mathcal{L}[h(t)] = \frac{n!}{s^{n+1}} \quad (\sigma > 0). \quad (6\text{-}34)$$

If n is not necessarily an integer, we write $u = st$. Then

$$\mathcal{L}[f] = \frac{1}{s^{n+1}}\int_0^\infty u^n e^{-u}\, du.$$

The right-hand member is related to the *Gamma function*:

$$\Gamma(x) = \int_0^\infty u^{x-1}e^{-u}\, du \quad (x > 0). \qquad (6\text{-}35)$$

(See Chapter 2 of Reference 9, pp. 381–383 of Reference 10, and Chapter 12 of Reference 14.) We have therefore

$$\mathcal{L}[t^n h(t)] = \frac{1}{s^{n+1}}\Gamma(n+1) \quad (n > -1). \qquad (6\text{-}36)$$

When n is an integer,

$$\Gamma(n+1) = n! \qquad (6\text{-}37)$$

and (6–36) reduces to (6–34). In deriving (6–36), we have treated s as a positive real number. If s is complex, the result can be justified (by methods of Chapter 3), provided we interpret s^{n+1} as exp $[(n + 1)$ Log $s]$ (see Section 3–5) and assume $\sigma > 0$.

Additional Laplace transforms are listed in Table 6–1.

PROBLEMS

1. Verify each of the following entries in Table 6–1, either directly or as suggested. (With the aid of Table 5–1 in Section 5–10, Equations (6–21) and (6–22) can also be used as checks for parts (f), (g), (h).)

(a) No. 5 from No. 2   (b) No. 7 from No. 2
(c) No. 9 from No. 4   (d) No. 11 from Nos. 1 and 7
(e) No. 13 from No. 7   (f) No. 16
(g) No. 21   (h) No. 22

2. Express $f(t)$ in terms of $\mathcal{L}[f]$ by means of the inversion formula (6–11) for the following entries in Table 6–1:

(a) No. 5 with a real   (b) No. 11   (c) No. 16

3. Let $f(t)$ be piecewise continuous for $t \geq 0$ and periodic for $t \geq 0$: $f(t + \tau) = f(t)$ for $t \geq 0$. Let

$$\phi_0(s) = \int_0^\tau f(t)e^{-st}\, dt,$$

so that $\phi_0(in\omega) = \Phi_\tau[f]$, the finite Fourier transform of f (Section 4–14). Show that

$$\mathcal{L}[f] = \frac{\phi_0(s)}{1 - e^{-s\tau}} \quad (\sigma > 0).$$

Use this result to verify Entries 15 and 24 of Table 6–1.

ANSWERS

2. (a) $h(t) \cosh at =$

$$\frac{e^{\sigma t}}{\pi} \int_0^\infty \frac{(\sigma^2 + \omega^2 - a^2)\sigma \cos \omega t + (\sigma^2 + \omega^2 + a^2)\omega \sin \omega t}{(\sigma^2 - \omega^2 - a^2)^2 + 4\sigma^2\omega^2}\, d\omega \quad (t \neq 0)$$

(b) $h(t) \cos^2 t = \dfrac{e^{\sigma t}}{2\pi} \displaystyle\int_0^\infty \left[\dfrac{\sigma \cos \omega t + \omega \sin \omega t}{\sigma^2 + \omega^2} \right.$

$$\left. + \frac{\sigma(\sigma^2 + \omega^2 + 4) \cos \omega t + \omega(\sigma^2 + \omega^2 - 4) \sin \omega t}{(\sigma^2 - \omega^2 + 4)^2 + 4\sigma^2\omega^2} \right] d\omega \quad (t \neq 0)$$

(c) $h(t) - h(t - c) = \dfrac{1}{2\pi} (P) \displaystyle\int_{-\infty}^\infty \dfrac{1 - e^{-cs}}{s} e^{st}\, d\omega \quad (s = \sigma + i\omega,\, t \neq 0,\, t \neq c)$

TABLE 6–1 — LAPLACE TRANSFORMS

No.	$f(t)$ for $t \geqq 0$	$\mathcal{L}[f] = \phi(s) = \int_0^\infty f(t)e^{-st}\,dt$	Range of σ
1	1	$\dfrac{1}{s}$	$\sigma > 0$
2	e^{at}	$\dfrac{1}{s-a}$	$\sigma > \mathrm{Re}\,(a)$
3	$t^n,\ n > -1$	$\dfrac{n!}{s^{n+1}},\ n = 0, 1, \ldots,$ $\dfrac{\Gamma(n+1)}{s^{n+1}},$ any $n > -1$	$\sigma > 0$
4	$t^n e^{at},\ n > -1$	$\dfrac{n!}{(s-a)^{n+1}},\ n = 0, 1, \ldots,$ $\dfrac{\Gamma(n+1)}{(s-a)^{n+1}},$ any $n > -1$	$\sigma > \mathrm{Re}\,(a)$
5	$\cosh at$	$\dfrac{s}{s^2 - a^2}$	$\sigma > |\mathrm{Re}\,(a)|$
6	$\sinh at$	$\dfrac{a}{s^2 - a^2}$	$\sigma > |\mathrm{Re}\,(a)|$
7	$\cos at$	$\dfrac{s}{s^2 + a^2}$	$\sigma > |\mathrm{Im}\,(a)|$
8	$\sin at$	$\dfrac{a}{s^2 + a^2}$	$\sigma > |\mathrm{Im}\,(a)|$
9	$t^n \cos at,\ n > -1$	$\dfrac{\Gamma(n+1)}{2(s^2 + a^2)^{n+1}}[(s + ai)^{n+1} + (s - ai)^{n+1}]$	$\sigma > |\mathrm{Im}\,(a)|$

10	$t^n \sin at$, $\ n > -1$	$\dfrac{\Gamma(n+1)}{2i(s^2+a^2)^{n+1}}[(s+ai)^{n+1} - (s-ai)^{n+1}]$	$\sigma > |\text{Im}(a)|$
11	$\cos^2 t$	$\dfrac{1}{2}\left(\dfrac{1}{s} + \dfrac{s}{s^2+4}\right)$	$\sigma > 0$
12	$\sin^2 t$	$\dfrac{1}{2}\left(\dfrac{1}{s} - \dfrac{s}{s^2+4}\right)$	$\sigma > 0$
13	$\sin at \sin bt$	$\dfrac{2abs}{[s^2+(a+b)^2][s^2+(a-b)^2]}$	$\sigma > |\text{Im}(a+b)|$ and $\sigma > |\text{Im}(a-b)|$
14	$e^{at}\sin(bt+c)$, $\ a, b, c$ real	$\dfrac{(s-a)\sin c + b\cos c}{(s-a)^2 + b^2}$	$\sigma > a$
15	1 for $2nc \le t < (2n+1)c$, 0 for $(2n+1)c \le t < (2n+2)c$, $n = 0, 1, 2, \ldots, c > 0$ (square wave)	$\dfrac{1}{s(1+e^{-cs})}$	$\sigma > 0$
16	$h(t) - h(t-c)$, $\ c > 0$	$\dfrac{1-e^{-cs}}{s}$	$-\infty < \sigma < \infty$
17	$h(t-c)$, $\ c > 0$	$\dfrac{e^{-cs}}{s}$	$\sigma > 0$
18	$h(t-c_1) - h(t-c_2)$, $\ 0 < c_1 < c_2$	$\dfrac{e^{-c_1 s} - e^{-c_2 s}}{s}$	$-\infty < \sigma < \infty$
19	$e^{at}[h(t-c_1) - h(t-c_2)]$, $\ 0 < c_1 < c_2$	$\dfrac{e^{-c_1(s-a)} - e^{-c_2(s-a)}}{s-a}$	$-\infty < \sigma < \infty$

Continued

TABLE 6-1 *Continued*

No.	$f(t)$ for $t \geqq 0$	$\mathcal{L}[f] = \phi(s) = \int_0^\infty f(t)e^{-st}\, dt$	Range of σ
20	$e^{at}h(t-c), \quad c > 0$	$\dfrac{e^{-c(s-a)}}{s-a}$	$\sigma > \mathrm{Re}\,(a)$
21	$t^n h(t-c), \quad c > 0, \quad n = 1, 2, \ldots$	$\dfrac{e^{-cs}[(cs)^n + n(cs)^{n-1} + \cdots + n!]}{s^{n+1}}$	$\sigma > 0$
22	$t^n[h(t) - h(t-c)], \quad c > 0, \quad n = 1, 2, \ldots$	$\dfrac{n! - e^{-cs}[(cs)^n + n(cs)^{n-1} + \cdots + n!]}{s^{n+1}}$	$-\infty < \sigma < \infty$
23	$t[h(t) - h(t-c)] + (2c-t)[h(t-c) - h(t-2c)]$	$\dfrac{1 - 2e^{-cs} + e^{-2cs}}{s^2}$	$-\infty < \sigma < \infty$
24	$\sum\limits_{k=0}^{\infty}(t-2kc)[h(t-2kc) - h(t-2kc-c)]$ $+\sum\limits_{k=0}^{\infty}(2kc+2c-t)\cdot[h(t-2kc-c)$ $-h(t-2kc-2c)]$ (triangular wave)	$\dfrac{1 - e^{-cs}}{s^2(1 + e^{-cs})}$	$\sigma > 0$
25	$a\sum\limits_{k=0}^{\infty}(t-kc)\cdot[h(t-kc) - h(t-kc-c)]$ (sawtooth wave)	$\dfrac{a(1 + cs - e^{cs})}{s^2(1 - e^{cs})}$	$\sigma > 0$

6–4 Theory of the Laplace transform. First we formulate as a theorem the basic facts implied by the fact that the Laplace transform is a special case of a Fourier transform depending on a parameter σ.

THEOREM 1. *Let $f(t)$ be defined for $-\infty < t < \infty$, $f(t) \equiv 0$ for $t < 0$. Let $f(t)$ be piecewise continuous for $t \geqq 0$. Let the integral*

$$\int_0^\infty |f(t)|e^{-\sigma t}\,dt \tag{6–40}$$

exist for some real σ. Then for this value of σ, the Laplace transform

$$\mathcal{L}[f] = \phi(s) = \int_0^\infty f(t)e^{-st}\,dt \quad (s = \sigma + i\omega) \tag{6–41}$$

is uniformly convergent for $-\infty < \omega < \infty$ and $\phi(\sigma + i\omega)$ is continuous for all ω. If $f(t)$ is also piecewise smooth and equal to the average of left and right limits at each discontinuity, then for the same value of σ

$$f(t) = \frac{1}{2\pi}\,(P)\int_{-\infty}^\infty \phi(s)e^{st}\,d\omega. \tag{6–42}$$

We now consider the possible values of σ for which the integral (6–40) exists. It may happen that the integral exists for all σ; this will surely be so if $f(t) \equiv 0$ for $t > t_1$. Also the integral may fail to exist for all σ; this is so if $f(t) = \exp(\exp t)$, as is easily verified, since $f(t)e^{-\sigma t} \to +\infty$ as $t \to +\infty$ for every choice of σ. In Table 6–1 we have seen that the integral can exist for $\sigma > a$, for different choices of a. This is suggestive of the general rule:

THEOREM 2. *Let $f(t)$ be given as in Theorem 1 and let the integral (6–40) exist for some real σ. Then either the integral exists for all real σ or else there is a number a such that the integral exists for $\sigma > a$ and does not exist for $\sigma < a$.*

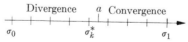

FIG. 6–1.   Abscissa of absolute convergence of Laplace transform.

The number a is called the *abscissa of absolute convergence* of the Laplace transform of f (Fig. 6–1). It is analogous to the radius of convergence of a power series. Just as for power series, there is no general rule as to convergence or divergence for the borderline case $\sigma = a$. When the integral exists for all σ, we let $a = -\infty$. In Section 6–6 we shall see that $\phi(s)$ is an *analytic function* of s for $\sigma > a$.

To prove Theorem 2, we remark that if $\sigma_1 < \sigma_2$ and the integral exists for $\sigma = \sigma_1$, then it will surely exist for $\sigma = \sigma_2$. For

$$|f(t)|e^{-\sigma_2 t} \leq |f(t)|e^{-\sigma_1 t},$$

so that we can apply the comparison test to conclude convergence of

$$\int_0^\infty |f(t)|e^{-\sigma_2 t}\, dt.$$

Now suppose the integral exists for $\sigma = \sigma_1$ but does not exist for all σ. Then there must be a value σ_0 for which the integral diverges. By the preceding reasoning, $\sigma_0 < \sigma_1$, and in general the integral diverges for $\sigma \leq \sigma_0$ and converges for $\sigma \geq \sigma_1$. If the integral diverges for all $\sigma < \sigma_1$, then the number $a = \sigma_1$ has the desired property; similarly if the integral converges for all $\sigma > \sigma_0$, then $a = \sigma_0$. If neither of these cases holds, then we can obtain the number a by a limiting process. We divide the interval from σ_0 to σ_1 into 2^k equal parts by the points

$$\sigma_0 + \frac{\sigma_1 - \sigma_0}{2^k} l \quad (l = 0, 1, 2, \ldots, 2^k).$$

Let σ_k^* be the largest value of σ among these points for which the integral diverges. As k increases, the number of subdivision points increases, and σ_k^* will move, if at all, to the right. Hence σ_k^* forms a monotone, non-decreasing, bounded sequence ($k = 1, 2, \ldots$). Therefore as $k \to +\infty, \sigma_k^*$ has a unique limit a. When $\sigma < a$ the integral must diverge. For if the integral converges for $\sigma = \sigma_2 < a$, then the integral must converge for $\sigma_2 \leq \sigma \leq a$. But $\sigma_k^* \leq a$ and $\sigma_k^* \to a$ as $k \to \infty$, so that for k sufficiently large, σ_k^* lies in the interval $\sigma_2 \leq \sigma \leq a$, and the integral *diverges* for $\sigma = \sigma_k^*$. Thus there is a contradiction. In the same way we show that the integral converges for $\sigma > a$. Hence a has the property stated and must be the only number with this property.

Remark. For certain values of $\sigma + i\omega = s$, the integral (6–41) may exist, without being absolutely convergent, i.e., without existence of (6–40). We can prove that if the integral (6–41) converges for some $s_0 = \sigma_0 + i\omega_0$, then it converges for all s such that $\sigma = \text{Re}(s) > \sigma_0$. Hence, as in Theorem 2, there is an *abscissa of convergence* c, such that (6–41) exists for $\sigma > c$ and does not exist for $\sigma < c$. We have necessarily $c \leq a$ since absolute convergence implies convergence, and it is possible that $c < a$ (Reference 4, pp. 15–19). However, for most functions considered in applications, $c = a$, and in any case it will usually be adequate for our purposes to restrict σ so that $\sigma > a$.

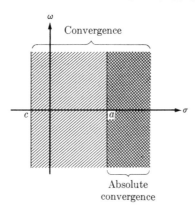

FIG. 6–2.   Convergence regions for Laplace transform.

The values of s for which the Laplace transform of f is absolutely convergent form a region in the complex s-plane. By Theorem 2, the region consists of a *half-plane* $\sigma > a$, plus perhaps the boundary line $\sigma = a$. The half-plane $\sigma > a$ (which may be the entire s-plane) is called *the half-plane of absolute convergence* of $\mathcal{L}[f]$. Similarly the region $\sigma > c$ is called the *half-plane of convergence*. These regions are pictured in Fig. 6–2.

6–5 Properties of the Laplace transform. We consider the class of functions $f(t)$ which are identically zero for $t < 0$, piecewise continuous for $t \geqq 0$, and for which the integral (6–40) exists for $\sigma > a_f$, where $-\infty \leqq a_f < \infty$. *This class is a linear space.* For if f and g are in the class, then $c_1 f + c_2 g$ is in the class and has abscissa of absolute convergence at most equal to the larger of a_f, a_g. Indeed, if $\sigma > a_f$ and $\sigma > a_g$, then both integrals

$$\int_0^\infty |f(t)| e^{-\sigma t} \, dt, \qquad \int_0^\infty |g(t)| e^{-\sigma t} \, dt$$

exist. Since

$$|c_1 f + c_2 g| \leqq |c_1| \, |f| + |c_2| \, |g|,$$

we conclude that for the same values of σ,

$$\int_0^\infty |c_1 f + c_2 g| e^{-\sigma t} \, dt$$

converges.

With σ restricted as described, one can write:

$$\mathcal{L}[c_1 f + c_2 g] = c_1 \mathcal{L}[f] + c_2 \mathcal{L}[g]. \tag{6–50}$$

This follows from the definition. Because of the restriction on σ, (6–50) is a slightly modified *linearity property* for the Laplace transform.

For applications to differential equations the following theorem is crucial.

THEOREM 3. *Let $f(t)$ be identically 0 for $t < 0$; let $f(t)$ be continuous and piecewise smooth for $t \geq 0$. Let $g(t) \equiv 0$ for $t < 0$ and let $g(t) = f'(t)$ wherever $f'(t)$ exists. Let $\mathcal{L}[f]$ and $\mathcal{L}[g]$ be absolutely convergent for $\sigma > a$. Then for $\sigma > a$,*

$$\mathcal{L}[g] = -f(0) + s\mathcal{L}[f]. \qquad (6\text{-}51)$$

Proof. We have, by integration by parts,

$$\int_0^t g(u)e^{-su}\,du = \int_0^t f'(u)e^{-su}\,du$$
$$= f(u)e^{-su}\Big|_{u=0}^{u=t} + s\int_0^t f(u)e^{-su}\,du.$$

Let Re $(s) = \sigma > a$. As $t \to +\infty$, the first integral has a limit, $\mathcal{L}[g]$, and the last integral has a limit, $\mathcal{L}[f]$. Hence $f(t)e^{-st}$ must have a limit as $t \to +\infty$. If this limit were different from zero, $\int_0^\infty f(t)e^{-st}\,dt$ could not converge; therefore the limit must be zero. Letting $t \to +\infty$, we now conclude that

$$\int_0^\infty g(u)e^{-su}\,du = -f(0) + s\int_0^\infty f(u)e^{-su}\,du,$$

so that Eq. (6–51) follows.

Remark. It is common practice to write (6–51) in the form

$$\mathcal{L}[Df] = -f(0) + s\mathcal{L}[f]. \qquad (6\text{-}51')$$

However, Df will not generally be well defined for all t. In particular, if $f(0+) \neq 0$, $|f'(0)|$ may be ∞. One way of making (6–51') consistent is to interpret Df as the *right-hand limit* of f' (or as $f'_+(t)$; see Section 4–7). For example, if $f = h(t)$, then Df would be considered to be $\equiv 0$; then

$$\mathcal{L}[Df] = \mathcal{L}[0] \equiv 0 = -f(0) + s\mathcal{L}[f]$$
$$= -1 + s \cdot \frac{1}{s} = -1 + 1.$$

Since f is assumed continuous for $t \geq 0$, $f(0)$ is also a right-hand limit: $1 = h(0) = \lim h(t)$ as $t \to 0+$.

In the subsequent work we shall use (6–51'), always recalling that $f(0)$ and Df are limiting values from the right.

COROLLARY TO THEOREM 3. *Let $f(t)$ be identically zero for $t < 0$; let $f(t)$, $f'(t), \ldots, f^{(n-1)}(t)$ be continuous for $t \geq 0$, where the values at zero are* *taken as limits to the right; let $f^{(n-1)}(t)$ be piecewise smooth for $t \geqq 0$. Let $f(t), f'(t), \ldots, f^{(n)}(t)$ all have absolutely convergent Laplace transforms for $\sigma > a$. Then for $\sigma > a$,*

$$\mathcal{L}[f'(t)] = \mathcal{L}[Df] = -f(0) + s\mathcal{L}[f],$$

$$\mathcal{L}[f''(t)] = \mathcal{L}[D^2 f] = -f'(0) - sf(0) + s^2\mathcal{L}[f],$$

$$\vdots \tag{6-52}$$

$$\mathcal{L}[f^{(n)}(t)] = \mathcal{L}[D^n f] = -[f^{(n-1)}(0) + sf^{(n-2)}(0)$$

$$+ \cdots + s^{n-1}f(0)] + s^n\mathcal{L}[f].$$

Proof. The corollary follows by repeated application of (6–51) or (6–51′):

$$\mathcal{L}[f''(t)] = \mathcal{L}[Df'] = -f'(0) + s\mathcal{L}[f']$$

$$= -f'(0) + s[-f(0) + s\mathcal{L}[f]]$$

$$= -f'(0) - sf(0) + s^2\mathcal{L}[f],$$

$$\mathcal{L}[f'''(t)] = \mathcal{L}[Df''] = -f''(0) + s\mathcal{L}[f''] = \cdots$$

If $f(0) = f'(0) = \cdots = 0$, then (6–52) reduces to the simple rule

$$\mathcal{L}[D^n f] = s^n \mathcal{L}[f]. \tag{6-52'}$$

THEOREM 4. *Let $f(t)$ be identically zero for $t < 0$ and piecewise continuous for $t \geqq 0$. Let $\mathcal{L}[f]$ be absolutely convergent for $\sigma > a \geqq 0$. Let*

$$g(t) = D_0^{-1} f = \int_0^t f(u) \, du. \tag{6-53}$$

Then for $\sigma > a$, $\mathcal{L}[g]$ is absolutely convergent, and

$$\mathcal{L}[g] = \mathcal{L}[D_0^{-1} f] = \frac{1}{s} \mathcal{L}[f]. \tag{6-54}$$

Proof. First we verify the convergence of $\mathcal{L}[g]$. Let

$$p(t) = \int_0^t |f(u)| \, du.$$

Then by integration by parts

$$\int_0^b p(t)e^{-\sigma t} \, dt = \frac{p(b)e^{-\sigma b}}{-\sigma} + \frac{1}{\sigma} \int_0^b e^{-\sigma u} |f(u)| \, du. \tag{6-55}$$

Hence for $\sigma > a \geqq 0$, and $b > 0$,

$$\int_0^b p(t)e^{-\sigma t} \, dt \leqq \frac{1}{\sigma} \int_0^b e^{-\sigma u} |f(u)| \, du.$$

Now

$$|g(t)| = \left| \int_0^t f(u)\, du \right| \le \int_0^t |f(u)|\, du = p(t).$$

Hence

$$\int_0^b |g(t)| e^{-\sigma t}\, dt \le \int_0^b p(t) e^{-\sigma t}\, dt \le \frac{1}{\sigma} \int_0^b e^{-\sigma u} |f(u)|\, du.$$

Therefore, since $\mathcal{L}[f]$ exists for this σ, $\mathcal{L}[g]$ exists; that is, $\mathcal{L}[g] = \mathcal{L}[D_0^{-1}f]$ is absolutely convergent for $\sigma > a$. Finally, by Theorem 3,

$$\mathcal{L}[g'] = -g(0) + s\mathcal{L}[g].$$

Since $\mathcal{L}[g'] = \mathcal{L}[f]$ and $g(0) = 0$, then (6–54) follows.

COROLLARY. *Under the hypotheses of Theorem 4, for $\sigma > a \ge 0$,*

$$\mathcal{L}[D_0^{-k}f] = \frac{1}{s^k}\, \mathcal{L}[f] \quad (k = 1, 2, \ldots). \tag{6–56}$$

THEOREM 5. *Let $f(t)$ be identically 0 for $t < 0$ and piecewise continuous for $t \ge 0$. Let $\mathcal{L}[f]$ be absolutely convergent for $\sigma > a$. Then for $\sigma > a$,*

$$\mathcal{L}[f(t - c)] = e^{-cs}\mathcal{L}[f] \quad (c > 0). \tag{6–57}$$

Proof. We have

$$\mathcal{L}[f(t - c)] = \int_0^\infty f(t - c)e^{-st}\, dt = \int_{-c}^\infty f(u)e^{-s(u+c)}\, du,$$

by the substitution $u = t - c$. Since $f = 0$ for $t < 0$,

$$\mathcal{L}[f(i - c)] = e^{-sc}\int_0^\infty f(u)e^{-su}\, du = e^{-sc}\mathcal{L}[f].$$

COROLLARY. *Under the hypotheses of Theorem 5, for $\sigma > a$,*

$$\mathcal{L}[f(t)h(t - c)] = e^{-sc}\mathcal{L}[f(t + c)h(t)] \quad (c > 0). \tag{6–58}$$

For if we let $g(t) = f(t + c)h(t)$, then Theorem 5 also applies to g and

$$\mathcal{L}[g(t - c)] = \mathcal{L}[f(t)h(t - c)] = e^{-sc}\mathcal{L}[g].$$

EXAMPLE 1. Let $f(t) = t^2$ for $t \ge 1$, $f = 0$ for $t < 1$. Then $f(t) = t^2 h(t - 1)$. Hence for $\sigma > 0$,

$$\mathcal{L}[f] = e^{-s}\mathcal{L}[(t + 1)^2 h(t)] = e^{-s}\mathcal{L}[(t^2 + 2t + 1)h(t)]$$
$$= e^{-s}\left[\frac{2}{s^3} + \frac{2}{s^2} + \frac{1}{s} \right].$$

EXAMPLE 2. Let $f(t) = e^{3t}$ for $1 \leqq t < 2$, $f = 0$ otherwise. Then

$$f(t) = e^{3t}[h(t-1) - h(t-2)] = e^{3t}h(t-1) - e^{3t}h(t-2).$$

Hence for $\sigma > 3$,

$$\mathcal{L}[f] = e^{-s}\mathcal{L}[e^{3(t+1)}h(t)] - e^{-2s}\mathcal{L}[e^{3(t+2)}h(t)]$$

$$= e^{-s}e^3 \frac{1}{s-3} - e^{-2s}\frac{e^6}{s-3} = \frac{e^{3-s} - e^{6-2s}}{s-3}.$$

THEOREM 6. *Let $f(t)$ be identically 0 for $t < 0$ and piecewise continuous for $t \geqq 0$. Let $\mathcal{L}[f]$ be absolutely convergent for $\sigma > a$. Then $\mathcal{L}[e^{bt}f]$ is absolutely convergent for $\sigma > a + \mathrm{Re}\ (b)$, and if $\mathcal{L}[f] = \phi(s)$,*

$$\mathcal{L}[e^{bt}f] = \phi(s - b).$$

Proof. Let $b = b_1 + ib_2$. Then $|e^{bt}| = e^{b_1 t}$, so that

$$\int_0^\infty |e^{bt}f(t)|e^{-\sigma t}\,dt = \int_0^\infty |f(t)|e^{-t(\sigma - b_1)}\,dt.$$

Since $\mathcal{L}[f]$ is absolutely convergent for $\sigma > a$, it follows that $\mathcal{L}[e^{bt}f]$ is absolutely convergent for $\sigma - b_1 > a$, that is, $\sigma > a + \mathrm{Re}\ (b)$. If σ is so chosen, then

$$\mathcal{L}[e^{bt}f] = \int_0^\infty f(t)e^{(b-s)t}\,dt = \phi(s - b).$$

THEOREM 7. *Under the hypotheses of Theorem 6, for $n = 1, 2, \ldots$, $\mathcal{L}[t^nf(t)]$ is absolutely convergent for $\sigma > a$, and*

$$\mathcal{L}[t^nf(t)] = (-1)^n \frac{d^n}{ds^n}\,\mathcal{L}[f]. \tag{6–59}$$

Proof. Let $\sigma = a + 2b$, $b > 0$. Then

$$\int_0^\infty t^n|f(t)|e^{-\sigma t}\,dt = \int_0^\infty t^n e^{-bt}|f(t)|e^{-(\sigma - b)t}\,dt.$$

Since $t^n e^{-bt} \to 0$ as $t \to \infty$, $t^n e^{-bt}$ is less than a constant; therefore the second integral is at most

$$\mathrm{const} \cdot \int_0^\infty |f(t)|e^{-(\sigma - b)t}\,dt;$$

since $\sigma - b > a$, the last integral has a finite value. Therefore $\mathcal{L}[t^nf]$ is absolutely convergent for $\sigma > a$.

The right-hand member of (6–59) involves the derivative of a complex function of a complex variable. We proceed formally and provide the details in the next section:

$$-\frac{d}{ds}\,\mathcal{L}[f] = -\frac{d}{ds}\int_0^\infty f(t)e^{-st}\,dt = \int_0^\infty tf(t)e^{-st}\,dt = \mathcal{L}[tf(t)].$$

Repeated application of the operation yields the general rule (6–59).

6–6 The Laplace transform as an analytic function. The functions of s which have appeared as Laplace transforms, $1/s$, $1/(s-a)$, $e^{-cs}/(s-a)$, $\ldots$, are all analytic functions of the complex variable s. We proceed to show that all Laplace transforms are analytic:

THEOREM 8. *Let $f(t)$ be piecewise continuous for $-\infty < t < \infty$, $f(t) = 0$ for $t < 0$. Let $\mathcal{L}[f] = \phi(s)$; let the transform be absolutely convergent for $\sigma > a$. Then $\phi(s)$ is analytic for $\sigma > a$.*

Proof. We shall show that $\phi(s)$ has derivatives of all orders in the domain $G: \sigma > a$; in particular, $\phi'(s)$ is continuous in G, so that $\phi(s)$ is analytic in G (Section 3–4).

Let s be a point of G, so that Re $(s) = \sigma = a + 2b$, $b > 0$. Then for $0 < |\Delta s| \leq b$,

$$\frac{\phi(s+\Delta s) - \phi(s)}{\Delta s} = \int_0^\infty f(t)\,\frac{e^{(-s-\Delta s)t} - e^{-st}}{\Delta s}\,dt.$$

We wish to prove that this expression approaches the derivative of $\phi(s)$ and that the derivative is

$$k = \int_0^\infty (-t)f(t)e^{-st}\,dt,$$

in accordance with Theorem 7.

Now

$$\frac{\phi(s+\Delta s) - \phi(s)}{\Delta s} - k = \int_0^\infty f(t)e^{-st}\left[\frac{e^{-t\,\Delta s} - 1}{\Delta s} + t\right]dt.$$

We need to estimate the quantity in brackets for $t \geq 0$:

$$\left|\frac{e^{-t\,\Delta s} - 1}{\Delta s} + t\right| = \left|\frac{t^2\,\Delta s}{2!} - \frac{t^3\,\overline{\Delta s}^2}{3!} + \cdots\right|$$

$$= |\Delta s|t^2\left|\frac{1}{2!} - \frac{t\Delta s}{3!} + \cdots\right|$$

$$\leq |\Delta s|t^2\left[1 + \frac{|\Delta s|t}{1!} + \frac{|\Delta s|^2 t^2}{2!} + \cdots\right]$$

$$\leq |\Delta s|t^2 e^{|\Delta s|t}.$$

Hence

$$\left| \frac{\phi(s + \Delta s) - \phi(s)}{\Delta s} - k \right| \leqq |\Delta s| \int_0^\infty t^2 |f(t)| e^{-(\sigma - |\Delta s|)t} \, dt$$

$$\leqq |\Delta s| \int_0^\infty t^2 |f(t)| e^{-(\sigma - b)t} \, dt.$$

As in the proof of Theorem 7, $t^2 f$ has an absolutely convergent Laplace transform for $\sigma > a$. Since $\sigma - b > a$, we conclude that the last integral exists. Therefore

$$\lim_{\Delta s \to 0} \left[\frac{\phi(s + \Delta s) - \phi(s)}{\Delta s} - k \right] = 0;$$

that is, $\phi'(s) = k$ as asserted. In the same way, we show that for $\sigma > a$,

$$\phi^{(n)}(s) = \int_0^\infty (-t)^n f(t) e^{-st} \, dt = \mathcal{L}[(-t)^n f],$$

so that $\phi(s)$ is analytic for $\sigma > a$. Thus the theorem is proved; at the same time, Eq. (6–59) is now fully justified.

Remark. As an analytic function, $\phi(s)$ may be prolongable by analytic continuation (Section 3–9). For example, $\mathcal{L}[h(t)] = 1/s$ for $\sigma > 0$; but $1/s$ is analytic for all s except $s = 0$. In general we agree to also call the extended function of s the Laplace transform of the given $f(t)$; that is, an analytic function $\psi(s)$ is the Laplace transform of f if $\psi(s)$ is defined where $\mathcal{L}[f] = \phi(s)$ exists and $\psi(s)$ coincides with $\phi(s)$ in the common domain.

6–7 Inverse transform. If $\mathcal{L}[f] = \phi(s)$ for $\sigma > a$, then we write

$$f = \mathcal{L}^{-1}[\phi(s)]$$

and call f an inverse transform of $\phi(s)$. If $f(t)$ is continuous and piecewise smooth and if

$$\int_0^\infty |f(t)| e^{-\sigma t} \, dt$$

converges for some σ, then by Theorem 1,

$$f = \mathcal{L}^{-1}[\phi] = \frac{1}{2\pi} (P) \int_{-\infty}^\infty \phi(s) e^{st} \, d\omega \tag{6–70}$$

for $s = \sigma + i\omega$. This shows that f is uniquely determined by ϕ, within the class of functions having the properties stated.

In point of fact, f is uniquely determined by ϕ within the class of piecewise continuous functions (except, of course, for the value of f at jump points). For let $\mathcal{L}[f_1] = \phi(s)$, $\sigma > \sigma_1$, and $\mathcal{L}[f_2] = \phi(s)$, $\sigma > \sigma_2$, where both transforms are absolutely convergent. We choose σ_0 so that $\sigma_0 > 0$, $\sigma_0 > \sigma_1$, $\sigma_0 > \sigma_2$, and let $f = f_1 - f_2$. Then $\mathcal{L}[f] = \mathcal{L}[f_1] - \mathcal{L}[f_2] = \phi(s) - \phi(s) = 0$ for $\sigma > \sigma_0 > 0$. By Theorem 4, $\mathcal{L}[D_0^{-1}f] = (1/s)\mathcal{L}[f] = 0$, $\sigma > \sigma_0$. Since $g = D_0^{-1}f$ is continuous and piecewise smooth, it is uniquely determined by its transform:

$$g = D_0^{-1}f = \mathcal{L}^{-1}[0] = 0.$$

Therefore, $g'(t) = f(t) = 0$ (except perhaps at jump points of f), and hence $f_1(t) \equiv f_2(t)$ except perhaps at jump points.

It is therefore proper to speak of *the* inverse Laplace transform of an analytic function $\phi(s)$.

An important and difficult question is that of describing the class of functions $\phi(s)$ for which there is an inverse Laplace transform. Much information concerning this topic is provided in Section 6–9. Here we remark that the class can be considered as a linear space. Let

$$\mathcal{L}^{-1}[\phi_1(s)] = f_1(t) \quad (\sigma > a_1),$$

$$\mathcal{L}^{-1}[\phi_2(s)] = f_2(t) \quad (\sigma > a_2).$$

Then

$$\mathcal{L}^{-1}[c_1\phi_1(s) + c_2\phi_2(s)] = c_1 f_1(t) + c_2 f_2(t)$$

for $\sigma > a$, where a is the larger of a_1, a_2. This follows at once from the equation

$$\mathcal{L}[c_1 f_1 + c_2 f_2] = c_1\mathcal{L}[f_1] + c_2\mathcal{L}[f_2],$$

which expresses the linearity of the operator $\mathcal{L}$ (see Section 6–5).

For applications we need the inverse transform of a proper rational function $\phi(s)$. Since $\phi(s)$ can be decomposed into partial fractions, we need only find the inverse transform of $(s - a)^{-k}$ for $k = 1, 2, \ldots$ From Entry 4 of Table 6–1, we find at once that

$$\mathcal{L}^{-1}\left[\frac{1}{(s - a)^k}\right] = \frac{t^{k-1}e^{at}}{(k - 1)!}\, h(t) \quad (\sigma > \mathrm{Re}\,(a),\ k = 1, 2, \ldots). \tag{6-71}$$

LEMMA 1. *Let* $W(t) = \mathcal{L}^{-1}[1/V(s)]$, *where* $V(s)$ *is a polynomial of degree* n: $V(s) = a_0 s^n + \cdots + a_n$. *Then* $W(0) = 0$, $W'(0) = 0, \ldots$, $W^{(n-2)}(0) = 0$, *while the right-hand derivative* $W_+^{(n-1)}(0) = 1/a_0$.

Proof. We can write

$$W(t) = \mathcal{L}^{-1}\left[\frac{1}{V(s)}\right] = \mathcal{L}^{-1}\left[\frac{1}{s^{n-1}}\,\frac{s^{n-1}}{a_0 s^n + \cdots + a_n}\right]$$

$$= D_0^{-(n-1)}\mathcal{L}^{-1}\left[\frac{s^{n-1}}{a_0 s^n + \cdots + a_n}\right],$$

by the corollary to Theorem 4. Since $D_0^{-1} = \int_0^t$, it follows that $W(0) = \cdots W^{(n-2)}(0) = 0$. To find $W_+^{(n-1)}(0)$, we write

$$W(t) = \mathcal{L}^{-1}\left[\frac{1}{V(s)}\right] = \mathcal{L}^{-1}\left[\frac{1}{a_0 s^n} - \frac{a_1 s^{n-1} + \cdots + a_n}{a_0 s^n (a_0 s^n + \cdots + a_n)}\right]$$

$$= \frac{t^{n-1}}{a_0(n-1)!}h(t) - \frac{1}{a_0}D_0^{-n}\mathcal{L}^{-1}\left[\frac{a_1 s^{n-1} + \cdots + a_n}{a_0 s^n + \cdots + s_n}\right].$$

The previous reasoning shows that the second term has all derivatives zero through order $n - 1$, while the first term has right-hand $(n - 1)$-st derivative $1/a_0$ at $t = 0$. Hence the conclusion follows (see Section 5–18).

LEMMA 2. *Under the hypotheses of Lemma 1,*

$$\mathcal{L}^{-1}\left[\frac{b_0 s^{n-1} + b_1 s^{n-2} + \cdots + b_{n-1}}{a_0 s^n + \cdots + a_n}\right]$$

$$= b_0 W^{(n-1)} + b_1 W^{(n-2)} + \cdots + b_{n-1}W. \qquad (6\text{–}72)$$

For since $W(0) = 0, \ldots, W^{(n-2)}(0) = 0$, then

$$\mathcal{L}[DW] = s\mathcal{L}[W] = \frac{s}{V}, \quad \mathcal{L}[D^2W] = \frac{s^2}{V}, \ldots, \quad \mathcal{L}[D^{n-1}W] = \frac{s^{n-1}}{V},$$

by the corollary to Theorem 3.

EXAMPLE.

$$\mathcal{L}^{-1}\left[\frac{1}{s^2 - 1}\right] = \mathcal{L}^{-1}\left[\frac{1}{2}\left(\frac{1}{s-1} - \frac{1}{s+1}\right)\right]$$

$$= \tfrac{1}{2}[e^t - e^{-t}]h(t) = h(t)\sinh t.$$

Hence for $t \geqq 0$,

$$\mathcal{L}^{-1}\left[\frac{s}{s^2 - 1}\right] = D\sinh t = \cosh t.$$

Since

$$\frac{s}{s^2 - 1} = \frac{1}{2}\left(\frac{1}{s-1} + \frac{1}{s+1}\right),$$

the result is seen to be correct. Furthermore, since $\sinh 0 = 0$,

$$\mathcal{L}[D\sinh t] = s\mathcal{L}[\sinh t] = \frac{s}{s^2 - 1}$$

by (6–51′).

A table of inverse transforms can be built up with the aid of Table 6–1, the preceding remarks, Theorems 3–7, and the convolution formulas presented later in this chapter. Some useful inverses are presented in Table 6–2.

TABLE 6–2

INVERSE LAPLACE TRANSFORMS

No.	$\phi(s)$	$f(t) = \mathcal{L}^{-1}[\phi]$ for $t \geqq 0$
1	$\dfrac{c}{as+b}$, $\quad a \neq 0$	$\dfrac{c}{a} e^{-(b/a)t}$
2	$\dfrac{1}{(s+\alpha)(s+\beta)}$, $\quad \alpha \neq \beta$	$\dfrac{e^{-\alpha t} - e^{-\beta t}}{\beta - \alpha}$
3	$\dfrac{ps+q}{(s+\alpha)(s+\beta)}$, $\quad \alpha \neq \beta$	$\dfrac{(q - p\alpha)e^{-\alpha t} - (q - p\beta)e^{-\beta t}}{\beta - \alpha}$
4	$\dfrac{1}{(s+\alpha)^2}$	$te^{-\alpha t}$
5	$\dfrac{ps+q}{(s+\alpha)^2}$	$e^{-\alpha t}[p + t(q - \alpha p)]$
6	$\dfrac{1}{as^2 + bs + c}$, $b^2 - 4ac > 0$, $\quad a \neq 0$	$\dfrac{e^{-\beta t} - e^{-\alpha t}}{\mu}$, $\quad \alpha = \dfrac{b + \mu}{2a}$, $\beta = \dfrac{b - \mu}{2a}$, $\quad \mu = \sqrt{b^2 - 4ac}$
7	$\dfrac{ps+q}{as^2 + bs + c}$, $b^2 - 4ac > 0$, $\quad a \neq 0$	$\dfrac{(q - p\beta)e^{-\beta t} - (q - p\alpha)e^{-\alpha t}}{\mu}$, α, β, μ as in No. 6
8	$\dfrac{1}{as^2 + bs + c}$, $b^2 - 4ac = 0$, $\quad a \neq 0$	$\dfrac{te^{-\alpha t}}{a}$, $\quad \alpha = \dfrac{b}{2a}$
9	$\dfrac{ps+q}{as^2 + bs + c}$, $b^2 - 4ac = 0$, $\quad a \neq 0$	$\dfrac{e^{-\alpha t}[p + t(q - \alpha p)]}{a}$, $\quad \alpha = \dfrac{b}{2a}$
10	$\dfrac{1}{as^2 + bs + c}$, $b^2 - 4ac < 0$, $\quad a \neq 0$	$\dfrac{2e^{-\alpha t}}{\mu} \sin \dfrac{\mu}{2a} t$, $\mu = \sqrt{4ac - b^2}$, $\quad \alpha = \dfrac{b}{2a}$
11	$\dfrac{ps+q}{as^2 + bs + c}$, $b^2 - 4ac < 0$, $\quad a \neq 0$	$e^{-\alpha t}\left[\dfrac{p}{a} \cos \dfrac{\mu}{2a} t + \dfrac{2q - 2\alpha p}{\mu} \sin \dfrac{\mu}{2a} t\right]$ $\alpha = \dfrac{b}{2a}$, $\quad \mu = \sqrt{4ac - b^2}$

TABLE 6–2 *Continued*

No.	$\phi(s)$	$f(t) = \mathcal{L}^{-1}[\phi]$ for $t \geqq 0$
12	$\dfrac{1}{(s+\alpha)(s+\beta)(s+\gamma)}$, α, β, γ distinct	$-\left(\dfrac{e^{-\alpha t}}{BC} + \dfrac{e^{-\beta t}}{AC} + \dfrac{e^{-\gamma t}}{AB}\right)$, $A = \beta - \gamma, \ B = \gamma - \alpha, \ C = \alpha - \beta$
13	$\dfrac{1}{(s+\alpha)^2(s+\beta)}$, $\quad \alpha \neq \beta$	$\dfrac{e^{-\beta t}}{(\beta-\alpha)^2} + \left[\dfrac{t}{\beta-\alpha} - \dfrac{1}{(\beta-\alpha)^2}\right] e^{-\alpha t}$
14	$\dfrac{1}{(s+\alpha)^3}$	$\dfrac{t^2}{2} e^{-\alpha t}$
15	$\dfrac{1}{(s+\alpha)(as^2+bs+c)}$, $N = a\alpha^2 - b\alpha + c \neq 0, \ a \neq 0$	$\dfrac{e^{-\alpha t}}{N} + \dfrac{1}{N}\mathcal{L}^{-1}\left[\dfrac{-as + a\alpha - b}{as^2+bs+c}\right]$
16	$\dfrac{1}{(as^2+bs+c)(As^2+Bs+C)}$, $as^2 + bs + c$ and $As^2 + Bs + C$ having no common roots, $a \neq 0, \ A \neq 0$.	$\mathcal{L}^{-1}\left[\dfrac{p_0 s + q_0}{as^2+bs+c}\right]$ $\quad + \mathcal{L}^{-1}\left[\dfrac{p_1 s + q_1}{As^2+Bs+C}\right]$; let $\beta = aB - bA, \ \gamma = aC - cA$, $\delta_0 = a\gamma^2 - b\beta\gamma + c\beta^2$, $\delta_1 = A\gamma^2 - B\beta\gamma + C\beta^2$. Then $p_0 = \dfrac{-a^2\beta}{\delta_0}, \ q_0 = \dfrac{a^2\gamma - ab\beta}{\delta_0}$, $p_1 = \dfrac{A^2\beta}{\delta_1}, \quad q_1 = \dfrac{-A^2\gamma + AB\beta}{\delta_1}$.
17	$\dfrac{1}{(as^2+bs+c)^2}$, $a \neq 0, \ b^2 - 4ac < 0$	$\dfrac{e^{-\alpha t}}{2a^2\beta^3}(\sin\beta t - \beta t \cos\beta t)$, $\alpha = \dfrac{b}{2a}, \quad \beta = \dfrac{\sqrt{4ac - b^2}}{2a}$
18	$\dfrac{e^{-cs}}{(s-a)^k}$, $c \geqq 0, \ k = 1, 2, \ldots$	$\dfrac{(t-c)^{k-1} e^{a(t-c)} h(t-c)}{(k-1)!}$

PROBLEMS

1. Obtain Entry 3 of Table 6–1 from Entry 1 by applying (6–59); assume $n = 1, 2, \ldots$

2. Obtain Entry 4 of Table 6–1 from Entry 3 by applying Theorem 6.

3. Obtain Entry 7 of Table 6–1 from Entry 8 by applying (6–51′).

4. Obtain Entry 17 of Table 6–1 by applying (6–57) to Entry 1.

5. Obtain Entry 19 of Table 6–1 from Entry 18 by applying Theorem 6.

6. Find the Laplace transforms of the following functions with the aid of Table 6–1 and the rules of Section 6–5:

(a) $h(t - \pi) \sin t$ (b) $te^{2t}h(t - 1)$

(c) $h(t) - 2h(t - 1) + 2h(t - 2) - 2h(t - 3) + \cdots$

(d) $\int_0^t f(u)\, du$, where f is the function of part (c)

(e) $(t^3 + 1)[h(t - 2) - h(t - 3)]$ (f) $\cos\left(t - \dfrac{\pi}{2}\right) h\left(t - \dfrac{\pi}{2}\right)$

(g) $h(t)t^2 \cos^2 t$ (h) $h(t)te^t \sin t$

7. Verify the following entries of Table 6–2 as suggested:

(a) No. 1 from Entry 2 of Table 6–1

(b) No. 2 by partial fractions

(c) No. 3 from No. 2 by Lemma 2

(d) No. 5 from No. 4 by Lemma 2   (e) No. 6 from No. 2

(f) No. 7 from No. 6 by Lemma 2   (g) No. 8 from No. 4

(h) No. 12 by partial fractions   (i) No. 18 from Entry 4 of Table 6–1

8. By Table 6–2 and Lemma 2, find the inverse Laplace transforms of the following:

(a) $\dfrac{1}{(s + 2)^2(s + 1)}$ (b) $\dfrac{s^2 - s + 5}{(s + 2)^2(s + 1)}$

(c) $\dfrac{s}{(s^2 + 1)(s^2 + 4)}$ (d) $\dfrac{e^{-s}}{s^2} + \dfrac{e^{-2s}}{(s - 1)^3}$

9. Prove the following rule: If $f(t)$ is piecewise continuous for $t \geqq 0$ and $\mathcal{L}[f] = F(s)$ is absolutely convergent for $\sigma > a$, then $f(bt)$, where b is a positive constant, has an absolutely convergent Laplace transform for $\sigma > ab$ and $\mathcal{L}[f(bt)] = (1/b)F(s/b)$.

ANSWERS

6. (a) $-e^{-\pi s}/(s^2 + 1)$ (b) $e^{-(s-2)}(s - 1)(s - 2)^{-2}$

(c) $(1 - e^{-s})s^{-1}(1 + e^{-s})^{-1}$ (d) $(1 - e^{-s})s^{-2}(1 + e^{-s})^{-1}$

(e) $e^{-2s}s^{-4}(6 + 12s + 12s^2 + 9s^3) - e^{-3s}s^{-4}(6 + 18s + 27s^2 + 28s^3)$

(f) $se^{-\pi s/2}(s^2 + 1)^{-1}$ (g) $s^{-3} + s(s^2 - 12)(s^2 + 4)^{-3}$

(h) $(2s - 2)(s^2 - 2s + 2)^{-2}$

8. (a) $e^{-t} - (1 + t)e^{-2t}$ (b) $7e^{-t} - (6 + 11t)e^{-2t}$

(c) $(\cos t - \cos 2t)/3$

(d) $(t - 1)h(t - 1) + \tfrac{1}{2}(t - 2)^2 e^{t-2}h(t - 2)$

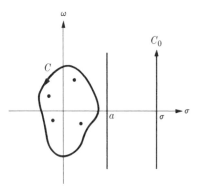

FIG. 6–3.   Inverse Laplace transform as a line integral.

6–8 Evaluation of inverse transforms by residues.   The reasoning of Section 5–14 carries over at once to Laplace transforms. If $\mathcal{L}[f] = \phi(s)$, then $f(t) = 0$ for $t < 0$ and (for proper choice of f at jumps)

$$f(t) = \frac{1}{2\pi} (P) \int_{-\infty}^{\infty} \phi(s)e^{st}\, d\omega \quad (t > 0), \tag{6–80}$$

where $s = \sigma + i\omega$ and σ has some fixed value greater than the abscissa of absolute convergence. We can interpret (6–80) as a complex line integral (Fig. 6–3) of the *analytic function* $\phi(s)e^{st}$ along the corresponding line $\sigma = \text{const}$:

$$f(t) = \frac{1}{2\pi i} \int_{C_0} \phi(s)e^{st}\, ds, \tag{6–81}$$

where C_0 is the line in question, directed upwards. Exactly the same reasoning as that of Section 5–14 then permits us to replace C_0 by a path C enclosing the singularities of $\phi(s)$, which must lie to the left of the path C_0:

THEOREM 9.   *Let $f(t)$ be piecewise continuous for $0 \leqq t < \infty$ and let $\mathcal{L}[f] = \phi(s)$ be absolutely convergent for $\sigma > a$. Let $\phi(s)$ be analytic for all s except for singularities $s_1, \ldots, s_k$ and let $\phi(s)$ have a zero at ∞. Then for $t > 0$,*

$$f(t) = \frac{1}{2\pi i} \oint_{C} \phi(s)e^{st}\, ds = \sum_{j=1}^{k} \text{Res}\,[\phi(s)e^{st}, s_j], \tag{6–82}$$

$$f(0) = \tfrac{1}{2} \lim_{t\to 0+} f(t) = \sum_{j=1}^{k} \text{Res}\,[\phi(s), s_j] + \tfrac{1}{2}\,\text{Res}\,[\phi(s), \infty], \tag{6–83}$$

where C is a path enclosing all the s_j.

EXAMPLE 1. $\mathscr{L}[f] = 1/s$,

$$\text{for } t > 0, \quad f(t) = \frac{1}{2\pi i} \oint_C \frac{e^{st}}{s} \, ds = \operatorname{Res}\left[\frac{e^{st}}{s}, 0\right] = 1;$$

$$f(0) = \operatorname{Res}\left[\frac{1}{s}, 0\right] + \tfrac{1}{2}\operatorname{Res}\left[\frac{1}{s}, \infty\right] = 1 - \tfrac{1}{2} = \tfrac{1}{2}.$$

EXAMPLE 2. $\mathscr{L}[f] = 1/(s-1)^3$. We find, for $t > 0$,

$$f(t) = \frac{1}{2\pi i} \oint_C \frac{e^{st}}{(s-1)^3} \, ds = \operatorname{Res}\left[\frac{e^{st}}{(s-1)^3}, 1\right] = \frac{t^2 e^t}{2!}.$$

For $t = 0$ the value must be zero.

6–9 Laplace transforms analytic at infinity. Let $\mathscr{L}[f] = \phi(s)$ for $\sigma > a$. If $\phi(s)$ is analytic at ∞ (Fig. 6–4) and has a zero there, then we can reason as in Section 6–8 to conclude that

$$f(t) = \frac{1}{2\pi i} \oint_C \phi(s)e^{st} \, ds \quad (t > 0), \tag{6–90}$$

without any assumption about the singularities of $\phi(s)$. The path C must simply be chosen as a circle $|s| = R$, on and outside of which ϕ is analytic. The integral (6–90) can be evaluated by residues if $\phi(s)$ has only a finite number of singularities inside C. But there may be infinitely many. In that case, we can still use the integral expression (6–90).

In (6–90) we can use the equation to define $f(t)$ for negative t and in fact for *complex t*. If we do so, it will be seen that $f(t)$ is an analytic function of the complex variable t for all t; such a function, analytic for all (finite)

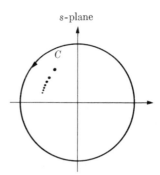

s-plane

FIG. 6–4.   Representation of Laplace transform analytic at infinity as a line integral.

values, is called an *entire* function. Furthermore, $f(t)$ is of *exponential type*. That is, $f(t)$ satisfies an equality

$$|f(t)| < ce^{b|t|} \quad (b, c \text{ const})$$

for all complex t. Thus $f(t)$ grows no more rapidly than an exponential function.

THEOREM 10. *Let $\phi(s)$ be analytic at ∞ and have a zero at ∞:*

$$\phi(s) = \sum_{n=1}^{\infty} \frac{a_n}{s^n} \quad (|s| > R_0). \tag{6–91}$$

For arbitrary complex t, let

$$f(t) = \sum_{n=0}^{\infty} a_{n+1} \frac{t^n}{n!}. \tag{6–92}$$

Then $\mathcal{L}[f(t)h(t)] = \phi$, the function $f(t)$ is an entire function of exponential type, and

$$f(t) = \frac{1}{2\pi i} \oint_C e^{st} \phi(s) \, ds, \tag{6–93}$$

where C is a circle: $|s| = R > R_0$.

Proof. The series for ϕ is a Laurent series. Hence by Section 3–10

$$a_n = \frac{1}{2\pi i} \oint_C \phi(s) s^{n-1} \, ds,$$

where C is a circle: $|s| = R > R_0$. Therefore, if M is the maximum of $|\phi|$ on the circle C, then

$$|a_n| \leqq \frac{1}{2\pi} M R^{n-1} \cdot 2\pi R = M R^n.$$

Accordingly,

$$\left| \frac{a_{n+1} t^n}{n!} \right| \leqq M R \frac{R^n |t|^n}{n!}.$$

Since the series $\sum R^n |t|^n / n!$ converges to $\exp(R|t|)$, we conclude by the comparison test that the series

$$\sum_{n=0}^{\infty} \frac{a_{n+1} t^n}{n!}$$

converges for all complex t, so that its sum $f(t)$ is an entire function. The comparison test shows that

$$|f(t)| \leqq M R \, e^{R|t|}. \tag{6–94}$$

Hence $f(t)$ is of exponential type. As the sum of a power series, $f(t)$ is continuous for $t \geq 0$. It remains to show that $\mathcal{L}[f(t)h(t)] = \phi$ and that (6–93) holds.

From (6–94) it follows that $\mathcal{L}[f(t)h(t)]$ is absolutely convergent for $\sigma > R$. Furthermore, for $\sigma > R$,

$$\mathcal{L}[f(t)h(t)] = \lim_{b \to \infty} \int_0^b e^{-st}f(t)\, dt = \lim_{b \to \infty} \int_0^b e^{-st}\sum a_{n+1}\frac{t^n}{n!}\, dt$$

$$= \lim_{b \to \infty} \sum_{n=0}^{\infty} \frac{a_{n+1}}{n!} \int_0^b e^{-st}t^n\, dt$$

$$= \lim_{b \to \infty} \sum_{n=0}^{\infty} \frac{a_{n+1}}{n!} \left\{ \int_0^{\infty} e^{-st}t^n\, dt - \int_b^{\infty} e^{-st}t^n\, dt \right\}$$

$$= \sum_{n=0}^{\infty} \frac{a_{n+1}}{s^{n+1}} - \lim_{b \to \infty} \sum_{n=0}^{\infty} \frac{a_{n+1}}{n!} \int_b^{\infty} e^{-st}t^n\, dt.$$

But the last limit is zero. For when $t > 0$,

$$\frac{t^n}{n!} < \frac{e^{Rt}}{R^n} = \frac{1}{R^n}\left(1 + Rt + \cdots + \frac{R^n t^n}{n!} + \cdots\right).$$

Hence

$$\left| \sum_{n=0}^{\infty} a_{n+1} \int_b^{\infty} e^{-st}\frac{t^n}{n!}\, dt \right| \leq \sum_{n=0}^{\infty} \frac{|a_{n+1}|}{R^n} \int_b^{\infty} e^{-\sigma t}e^{Rt}\, dt.$$

The last integral is independent of n, and approaches zero as $b \to \infty$, for $\sigma > R$. The series $\sum |a_{n+1}|R^{-n}$ converges by the hypothesis on ϕ. Hence for $\sigma > R$ the limit in question is zero, and

$$\mathcal{L}[f(t)h(t)] = \sum_{n=0}^{\infty} \frac{a_{n+1}}{s^{n+1}} = \phi(s) \quad (\sigma > R).$$

Since R is an arbitrary number greater than R_0, it follows that equality holds for $\sigma > R_0$.

Since $\mathcal{L}[f(t)h(t)] = \phi$, Theorem 9 shows that

$$f(t) = \frac{1}{2\pi i} \oint_C \phi(s)e^{st}\, ds \quad (t > 0).$$

If we insert the series $\sum(a_n s^{-n})$ for $\phi(s)$ and integrate term by term, we see that the equality holds for all *complex* t. Hence Theorem 10 is proved.

EXAMPLE 1. Let

$$\phi(s) = e^{1/s} - 1 = \frac{1}{s} + \frac{1}{2s^2} + \cdots$$

Then

$$\mathcal{L}^{-1}[\phi] = \sum_{n=0}^{\infty} \frac{t^n}{n!(n+1)!} = \frac{1}{2\pi i} \oint_C e^{st}(e^{1/s} - 1)\, ds = \frac{1}{2\pi i} \oint_C e^{st}e^{1/s}\, ds,$$

where C is a circle: $|s| = R > 0$. The -1 can be omitted because, by Cauchy's theorem, $\int e^{st}\, ds = 0$ on C. From either form of the answer, $f(t)$ can be computed numerically.

It is natural to ask whether all entire functions $f(t)$ of exponential type appear as inverse transforms of functions $\phi(s)$ of form (6–91). The answer is affirmative.

THEOREM 11. *Let* $f(t) = \sum_{n=0}^{\infty} a_{n+1} t^n / n!$ *for all complex t and let, for all t,*

$$|f(t)| < c \cdot e^{b|t|}, \tag{6–95}$$

where b and c are positive constants. Then

$$\mathcal{L}[f(t)h(t)] = \phi(s) = \sum_{n=1}^{\infty} \frac{a_n}{s^n} \quad (\sigma > b). \tag{6–96}$$

The series for $\phi(s)$ converges at least for $|s| > b$.

Proof. We reason as in the proof of Theorem 10 to conclude that $\mathcal{L}[f(t)h(t)]$ exists for $\sigma > b$. Also

$$\frac{a_{n+1}}{n!} = \frac{1}{2\pi i} \oint_{|t|=k} \frac{f(t)}{t^{n+1}}\, dt \quad (k > 0),$$

by Section 3–9. Hence

$$\left| \frac{a_{n+1}}{n!} \right| < \frac{c}{2\pi} e^{bk} \frac{2\pi k}{k^{n+1}} = \frac{ce^{bk}}{k^n}.$$

Therefore, for $k > 0$,

$$\left| \frac{a_n}{s^n} \right| \leqq \frac{ce^{bk}(n-1)!}{|s|^n k^{n-1}}.$$

In particular, for $k = n/b$ and $n = 1, 2, \ldots$,

$$\left| \frac{a_n}{s^n} \right| \leqq \frac{ce^n(n-1)!}{bn^{n-1}} \left(\frac{b}{|s|} \right)^n = \frac{c}{b} c_n \left(\frac{b}{|s|} \right)^n,$$

$$c_n = \frac{e^n(n-1)!}{n^{n-1}}.$$

Now we can verify by the ratio test (Problem 4 below) that $\sum c_n r^n$ converges for $|r| < 1$. Hence for $|s| > b$ the series $\sum_0^\infty a_n s^{-n}$ converges to $\phi(s)$. By Theorem 10, $\mathcal{L}[f(t)h(t)] = \phi$.

EXAMPLE 2. Let

$$f(t) = \cosh \sqrt{t} = 1 + \frac{t}{2} + \cdots + \frac{t^n}{(2n)!} + \cdots$$

We can verify (Problem 3 below) that $|f(t)| \leq e^{|t|}$. Hence the series for $\phi(s)$ converges at least for $|s| > 1$:

$$\mathcal{L}[f] = \phi(s) = \sum_{n=0}^\infty \frac{n!}{(2n)!s^{n+1}}.$$

We can show (Problem 5 below) that the series for $\phi(s)$ converges except for $s = 0$. Hence as in the proof of Theorem 10, $|f(t)| < Ce^{b|t|}$ for every $b > 0$ (for appropriate choice of C).

Remark. Theorem 10 provides a class of analytic functions which are Laplace transforms. Larger classes are described on pp. 126–128 of Reference 4.

Inverse Laplace transforms of rational functions. If $a_0 \neq 0$, then

$$Y(s) = \frac{1}{a_0 s^n + \cdots + a_n} = \frac{1}{s^n \left(a_0 + \frac{a_1}{s} + \cdots + \frac{a_n}{s^n} \right)}$$

has a zero of order n at ∞:

$$Y(s) = \frac{1}{a_0 s^n} - \frac{a_1}{a_0^2} \frac{1}{s^{n+1}} + \frac{a_1^2 - a_0 a_2}{a_0^3} \frac{1}{s^{n+2}}$$

$$+ \frac{2a_0 a_1 a_2 - a_0^2 a_3 - a_1^3}{a_0^4} \frac{1}{s^{n+3}} + \cdots$$

Hence by Theorem 10, if $W(t) = \mathcal{L}^{-1}[Y(s)]$, then

$$W(t) = \frac{1}{a_0} \frac{t^{n-1}}{(n-1)!} - \frac{a_1}{a_0^2} \frac{t^n}{n!}$$

$$+ \frac{a_1^2 - a_0 a_2}{a_0^3} \frac{t^{n+1}}{(n+1)!} + \frac{2a_0 a_1 a_2 - a_0^2 a_3 - a_1^3}{a_0^4} \frac{t^{n+2}}{(n+2)!} + \cdots \qquad (6\text{-}97)$$

for $t > 0$, while we consider $W(t)$ to be zero for $t < 0$. The poles of $Y(s)$ are at zeros of the polynomial $a_0 s^n + \cdots + a_n$. If all these zeros lie in the circle $|s| \leq R_0$, then the series for $Y(s)$ converges for all s *outside* the circle, while the series for $W(t)$ converges for all t.

From the expansion (6–97) we conclude at once that

$$W(0) = 0, \quad W'(0) = 0, \ldots, W^{(n-2)}(0) = 0, \quad W^{(n-1)}(0) = \frac{1}{a_0},$$

$$W^{(n)}(0) = -\frac{a_1}{a_0^2}, \quad W^{(n+1)}(0) = \frac{a_1^2 - a_0 a_2}{a_0^3},$$

$$W^{(n+2)}(0) = \frac{2a_0 a_1 a_2 - a_0^2 a_3 - a_1^3}{a_0^4}, \ldots \quad (6\text{–}98)$$

The derivatives $W^{(n-1)}(0)$, $W^{(n)}(0)$, ... are considered as derivatives to the *right* at zero, since W is considered as zero for $t < 0$. The first n relations (6–98) were proved by other means in Section 6–7.

Since

$$s^k Y(s) = \frac{1}{a_0 s^{n-k}} - \frac{a_1}{a_0^2} \frac{1}{s^{n+1-k}} + \cdots,$$

we conclude that for $k = 0, 1, 2, \ldots, n - 1$,

$$\mathcal{L}^{-1}[s^k Y(s)] = \frac{1}{a_0} \frac{t^{n-k-1}}{(n-k-1)!} - \frac{a_1}{a_0^2} \frac{t^{n-k}}{(n-k)!} + \cdots$$

This is identical with $W^{(k)}(t)$, as follows from (6–97). The same fact was also deduced in Section 6–7. We can now add the information, that if $g(t) = \mathcal{L}^{-1}[s^k Y(s)]$, then

$$g(0) = 0, \quad g'(0) = 0, \ldots, g^{n-k-2}(0) = 0, \quad g^{(n-k-1)}(0) = \frac{1}{a_0},$$

$$g^{(n-k)}(0) = -\frac{a_1}{a_0^2}, \quad g^{(n-k+1)}(0) = \frac{a_1^2 - a_0 a_2}{a_0^3},$$

$$g^{(n-k+2)}(0) = \frac{2a_0 a_1 a_2 - a_0^2 a_3 - a_1^3}{a_0^4}, \ldots, \quad (6\text{–}99)$$

where again the derivatives are right-hand values.

PROBLEMS

1. Evaluate by residues the inverse Laplace transforms of the following functions:

(a) $\dfrac{1}{s^2 - 1}$

(b) $\dfrac{1}{s^3 + 1}$

(c) $\dfrac{1}{(s - 1)^2}$

(d) $\dfrac{s^2}{(s^2 - 1)^2}$

(e) $\dfrac{s}{(s + 1)(s + 2)}$

(f) $\dfrac{s^5}{s^6 - 1}$

2. For each of the following find the inverse Laplace transforms as power series and as contour integrals:

(a) $se^{1/s} - 1 - s$ (b) $\sin \dfrac{1}{s}$ (c) $\displaystyle\sum_{n=1}^{\infty} \dfrac{n^2 + 1}{s^n}$

3. For each of the following prove that $f(t)$ is of exponential type, and represent $f(t)$ as an integral as in (6–90):

(a) t^2 (b) te^t (c) $e^t - e^{2t}$

(d) $\displaystyle\sum_{n=0}^{\infty} \dfrac{t^n}{n! + n^2}$ (e) $\cosh \sqrt{t}$ (Example 2 above)

4. Prove that the series

$$\sum_{n=1}^{\infty} c_n r^n = \sum_{n=1}^{\infty} \frac{e^n (n-1)!}{n^{n-1}} r^n$$

converges for $|r| < 1$.

5. Prove that the series

$$\sum_{n=0}^{\infty} \frac{n!}{(2n)! s^{n+1}}$$

converges for $s \neq 0$.

6. Find $f(0), f'(0), f''(0), f'''(0)$ for the inverse Laplace transform $f(t)$ of each of the following functions of s:

(a) $\displaystyle\sum_{n=1}^{\infty} \dfrac{2^n}{(n! + 1)s^n}$ (b) $\dfrac{e^{1/s} - 1}{s}$

(c) $\dfrac{1}{s^3 + 2s + 1}$ (d) $\dfrac{s^2 - 1}{s^4 + s + 1}$

ANSWERS

1. (a) $\frac{1}{2}(e^t - e^{-t})$ (b) $\frac{1}{3}[e^{-t} - 2\,\mathrm{Re}\,(ce^{ct})]$, $c = \exp(i\pi/3)$ (c) te^t
(d) $\frac{1}{4}[e^t(t+1) + e^{-t}(t-1)]$ (e) $2e^{-2t} - e^{-t}$
(f) $\frac{1}{3}(\cosh t + \cosh ct + \cosh \bar{c}t)$, $c = \exp(i\pi/3)$

2. (a) $\displaystyle\sum_{n=0}^{\infty} \dfrac{t^n}{n!(n+2)!}$, $\dfrac{1}{2\pi i}\displaystyle\oint_{|s|=R} e^{st} se^{1/s}\,ds$ $(R>0)$

(b) $\displaystyle\sum_{n=0}^{\infty} \dfrac{(-1)^n t^{2n}}{(2n)!\cdot(2n+1)!}$, $\dfrac{1}{2\pi i}\displaystyle\oint_{|s|=R} e^{st} \sin(1/s)\,ds$ $(R>0)$

(c) $\displaystyle\sum_{n=0}^{\infty} \dfrac{n^2 + 2n + 2}{n!} t^n$, $\dfrac{1}{2\pi i}\displaystyle\oint_{|s|=R} e^{st} \sum_{1}^{\infty} \dfrac{n^2+1}{s^n}\,ds$ $(R>1)$

3. (a) $\dfrac{1}{2\pi i} \displaystyle\oint_{|s|=R} \dfrac{2e^{st}}{s^3}$ $(R > 0)$ (b) $\dfrac{1}{2\pi i} \displaystyle\oint_{|s|=R} \dfrac{e^{st}}{(s-1)^2}\, ds$ $(R > 1)$

(c) $-\dfrac{1}{2\pi i} \displaystyle\oint_{|s|=R} \dfrac{e^{st}\, ds}{s^2 - 3s + 2}$ $(R > 2)$

(d) $\dfrac{1}{2\pi i} \displaystyle\oint_{|s|=R} e^{st} \sum_{n=0}^{\infty} \dfrac{n!}{(n!+n^2)s^{n+1}}\, ds$ $(R > 1)$

(e) $\dfrac{1}{2\pi i} \displaystyle\oint_{|s|=R} e^{st} \sum_{n=0}^{\infty} \dfrac{n!}{(2n)!s^{n+1}}\, ds$ $(R > 0)$

6. (a) $1, \frac{4}{3}, \frac{8}{7}, \frac{16}{25}$ (b) $0, 1, \frac{1}{2}, \frac{1}{6}$
   (c) $0, 0, 1, 0$ (d) $0, 1, 0, -1$

6–10 Expansion in terms of Laguerre polynomials.   The Laguerre polynomials $L_n(t)$ $(n = 0, 1, 2, \ldots)$ are defined as follows:

$$L_n(t) = \frac{1}{n!}\, e^t D^n(e^{-t}t^n) \quad (n = 0, 1, \ldots) \tag{6–100}$$

As pointed out in Section 4–12, the functions $e^{-t/2}L_n(t)$ form a complete orthogonal sequence for the interval $0 \leq t < \infty$; hence, under appropriate assumptions, a function $f(t)$ defined on that interval can be represented as a series $\sum c_n e^{-t/2}L_n(t)$.

Here we obtain such an expansion, slightly modified, for the functions $f(t)$ of Section 6–9, that is, the functions f whose Laplace transforms $F(s)$ are analytic at infinity. If $F(s)$ is given, the procedure described may be easier to carry out and may give more useful results than the Taylor series of Section 6–9.

First we list the first few Laguerre polynomials in Table 6–3.

Next we note that these polynomials have Laplace transforms as follows:

$$\mathcal{L}[L_n(t)] = \frac{(s-1)^n}{s^{n+1}} \quad (n = 0, 1, \ldots). \tag{6–101}$$

The proof is left as an exercise (Problem 1 below). From (6–101) we deduce (by Theorem 6, Section 6–5) that

$$\mathcal{L}[e^{-kt}L_n(t)] = \frac{(s+k-1)^n}{(s+k)^{n+1}}. \tag{6–102}$$

Let $F(s)$ be given, analytic at ∞ and having a zero at ∞. Then we can choose a circle of center s_0 (s_0 real) and radius ρ outside of which $F(s)$ is

TABLE 6–3

LAGUERRE POLYNOMIALS

n	$L_n(t)$
0	1
1	$1 - t$
2	$1 - 2t + \frac{1}{2}t^2$
3	$1 - 3t + \frac{3}{2}t^2 - \frac{1}{6}t^3$
4	$1 - 4t + 3t^2 - \frac{2}{3}t^3 + \frac{1}{24}t^4$
5	$1 - 5t + 5t^2 - \frac{5}{3}t^3 + \frac{5}{24}t^4 - \frac{1}{120}t^5$
6	$1 - 6t + \frac{15}{2}t^2 - \frac{10}{3}t^3 + \frac{5}{8}t^4 - \frac{1}{20}t^5 + \frac{1}{720}t^6$
7	$1 - 7t + \frac{21}{2}t^2 - \frac{35}{6}t^3 + \frac{35}{24}t^4 - \frac{7}{40}t^5 + \frac{7}{720}t^6 - \frac{1}{5040}t^7$
8	$1 - 8t + 14t^2 - \frac{28}{3}t^3 + \frac{35}{12}t^4 - \frac{7}{15}t^5 + \frac{7}{180}t^6 - \frac{1}{40320}t^8$

analytic. We now let

$$z = \frac{s + k - 1}{s + k}, \tag{6–103}$$

where k is a real number to be adjusted so that, as s varies over the exterior of the circle $|s - s_0| = \rho$, z varies over the interior of the circle $|z| = r$. We verify that this condition is satisfied when

$$\rho = \frac{r}{r^2 - 1}, \qquad s_0 = -\frac{1}{r^2 - 1} - k \tag{6–104}$$

with $r > 1$, so that

$$r = \frac{1 + \sqrt{1 + 4\rho^2}}{2\rho}, \qquad k = \frac{-2\rho^2}{1 + \sqrt{1 + 4\rho^2}} - s_0 \tag{6–105}$$

(Problem 2 below). From (6–103) we obtain

$$s = \frac{k - 1 - kz}{z - 1}, \tag{6–103'}$$

so that we can form $F[s(z)]$. As z varies over the circle $|z| < r$, s varies over the region $|s - s_0| > \rho$, with $z = 1$ corresponding to $s = \infty$. Since $F(s)$ is analytic and, by (6–103'), $s(z)$ is analytic except perhaps at $z = 1$, we conclude that $F[s(z)]$ is analytic for $|z| < r$, except perhaps at $z = 1$. As $z \to 1$, $s \to \infty$ and $F(s) \to 0$; since F has a limit as $z \to 1$, we conclude (see Section 3–11) that $F[s(z)]$ remains analytic at $z = 1$. The function $(s + k)F(s)$ is also analytic at $s = \infty$, and hence the same argument applies to this function; that is, when s is expressed in terms of z by (6–103′), $(s + k)F(s)$ is analytic for $|z| < r$. Accordingly,

$$(s + k)F(s) = \sum_{n=0}^{\infty} c_n z^n \quad (|z| < r),$$

and hence by (6–103)

$$(s + k)F(s) = \sum_{n=0}^{\infty} c_n \left(\frac{s + k - 1}{s + k} \right)^n,$$

$$F(s) = \sum_{n=0}^{\infty} c_n \frac{(s + k - 1)^n}{(s + k)^{n+1}} \quad (|s - s_0| > \rho). \tag{6–106}$$

By (6–102) the general term of the last series is $\mathcal{L}[c_n e^{-kt} L_n(t)]$. Hence we expect that $f(t)$, the inverse Laplace transform of $F(s)$, is given by the series

$$f(t) = \sum_{n=0}^{\infty} c_n e^{-kt} L_n(t) \quad (0 \leq t < \infty). \tag{6–107}$$

For a proof refer to pp. 136–138 of Reference 4.

The series (6–107) differs from the one referred to in the first paragraph of this section in that e^{-kt} appears instead of $e^{-t/2}$. However, we can write (6–107) as

$$e^{[k-(1/2)]t} f(t) = \sum_{n=0}^{\infty} c_n e^{-(1/2)t} L_n(t),$$

which shows that the function $e^{[k-(1/2)]t} f(t)$ is being expanded in the series of orthogonal functions.

EXAMPLE. $F(s) = (s + 1)/(s^2 + 2s + 2)$. Here $f(t) = e^{-t} \cos t$, so that we can compare our result with a known result. First we choose a circle: $|s - s_0| = \rho$ outside of which $F(s)$ is analytic. Here there is considerable freedom, but experience shows that it is advantageous to make the circle as small as possible. We here choose $s_0 = -2$, $\rho = \sqrt{2}$. (This is not quite the smallest circle, but it happens to yield simple formulas.) By (6–105) we find $r = \sqrt{2}$, $k = 1$, so that by (6–103′) $s = -z/(z - 1)$. Now

$$(s + k)F(s) = (s + 1)F(s) = \frac{(s + 1)^2}{(s + 1)^2 + 1} = \frac{1}{z^2 - 2z + 2},$$

and we are to expand the last function in powers of z for $|z| < \sqrt{2}$.

We can write

$$\frac{1}{z^2 - 2z + 2} = \frac{1}{2[1 - (z - \frac{1}{2}z^2)]} = \frac{1}{2} \sum_{n=0}^{\infty} (z - \frac{1}{2}z^2)^n$$

$$= \frac{1}{2} \sum_{n=0}^{\infty} z^n (1 - \frac{1}{2}z)^n = \frac{1}{2} \sum_{n=0}^{\infty} z^n \sum_{r=0}^{n} \binom{n}{r} (-\frac{1}{2}z)^r$$

$$= \frac{1}{2} \sum_{n=0}^{\infty} z^n \left[\binom{n}{0} + \binom{n-1}{1}\left(\frac{-1}{2}\right) + \binom{n-2}{2}\frac{1}{2^2} + \cdots \right.$$

$$\left. + \binom{n - p_n}{p_n} \frac{(-1)^{p_n}}{2^{p_n}} \right],$$

where p_n is the largest integer such that $2p_n \leqq n$. Accordingly,

$$\frac{1}{z^2 - 2z + 2} = \frac{1}{2}\left\{ 1 + \binom{1}{0}z + \left[\binom{2}{0} + \binom{1}{1}(-\frac{1}{2})\right]z^2 \right.$$

$$\left. + \left[\binom{3}{0} + \binom{2}{1}(-\frac{1}{2})\right]z^3 + \cdots \right\}$$

$$= \frac{1}{2}(1 + z + \frac{1}{2}z^2 + 0z^3 - \frac{1}{4}z^4 - \frac{1}{4}z^5 + \cdots).$$

Therefore, by (6–107),

$$f(t) = \frac{1}{2} \{ e^{-t}L_0(t) + e^{-t}L_1(t) + \frac{1}{2}e^{-t}L_2(t) - \frac{1}{4}e^{-t}L_4(t) - \frac{1}{4}e^{-t}L_5(t) + \cdots$$

$$+ \left[\binom{n}{0} + \binom{n-1}{1}\left(\frac{-1}{2}\right) + \cdots \right.$$

$$\left. + \binom{n - p_n}{p_n} \frac{(-1)^{p_n}}{2^{p_n}} \right] e^{-t}L_n(t) + \cdots \},$$

and we know that $f(t) = e^{-t} \cos t$. If we use the terms through $L_n(t)$, we obtain an approximate expression $S_n(t)$. Thus

$$S_0(t) = \frac{1}{2}e^{-t}, \qquad S_1(t) = e^{-t}(1 - \frac{1}{2}t), \qquad S_2(t) = e^{-t}\left(\frac{t^2}{8} - t + \frac{5}{4}\right),$$

$$S_4(t) = e^{-t}\left(\frac{9}{8} - \frac{t}{2} - \frac{t^2}{4} + \frac{t^3}{12} - \frac{t^4}{192}\right).$$

In Fig. 6–5 the graphs of $e^{-t} \cos t$, $S_2(t)$, $S_4(t)$ are shown, as well as the partial sum $1 - t + (t^3/3)$ of the power series for $e^{-t} \cos t$. It is clear that the partial sum of the power series is much better for t close to zero, but it is extremely poor for $t \geqq 2$; whereas $S_2(t)$ is a fairly good approximation out to $t = 8$, and $S_4(t)$ is fairly good out to $t = 11$.

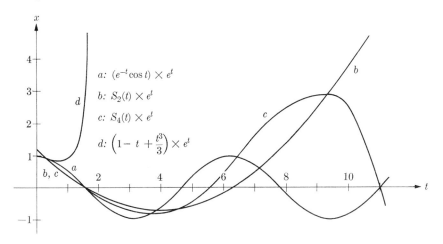

Fig. 6–5.   Expansion of $e^{-t} \cos t$ in Laguerre polynomials. (All four functions are multiplied by e^t.)

The example illustrates the advantage of an expansion of $f(t)$ in Laguerre polynomials; the function can be well approximated over a large interval by relatively few terms of the series.

Problems

1. Prove from the definition (6–100) that (6–101) is valid, and hence obtain (6–102).

2. Prove that if $r > 1$ and (6–104) holds, then, by (6–103), as s varies over the region $|s - s_0| > \rho$, z varies over the region $|z| < r$. [*Hint:* Let $w = s - s_0$ and show that $z = (w - r\rho)/[w - (\rho/r)]$. Then write

$$|z|^2 = z\bar{z} = \frac{w - r\rho}{w - (\rho/r)} \frac{\bar{w} - r\rho}{\bar{w} - (\rho/r)}$$

and verify that the inequality $|z|^2 < r^2$ implies the inequality $|w\bar{w}| = |s - s_0|^2 > \rho^2$, and conversely.]

3. For each of the following expand $f(t)$ for $t \geqq 0$ in a series of functions $e^{-kt}L_n(t)$, with k as indicated:

(a) $f(t) = \mathcal{L}^{-1}\left[\dfrac{s}{s^2 + 1}\right] = \cos t$, $k = 0$ (use $s_0 = -1$)

(b) $f(t) = \mathcal{L}^{-1}\left[\dfrac{1}{s^2 + 3s + 2}\right] = e^{-t} - e^{-2t}$, $k = 1$ (use $s_0 = -2$)

(c) $f(t) = \mathcal{L}^{-1}\left[\dfrac{e^{1/(s+1)}}{s + 1}\right]$, $k = 1$ (use $s_0 = -\tfrac{3}{2}$)

4. Prove from the definition (6–100) that

$$L_n(t) = \sum_{k=0}^{n} \binom{n}{k} (-1)^k \frac{t^k}{k!}.$$

ANSWERS

3. (a) $\frac{1}{2}L_0(t) + \frac{1}{2}L_1(t) + \cdots$

$$+ \frac{1}{2}\left[\binom{n}{0} + \binom{n-1}{1}(-\tfrac{1}{2}) + \cdots + \binom{n-p_n}{p_n}(-2)^{-p_n}\right]L_n(t) + \cdots,$$

where p_n is the largest integer such that $2p_n \leqq n$

(b) $e^{-t}\left(\frac{1}{2} - \sum_{n=1}^{\infty} L_n(t)2^{-n-1}\right)$ (c) $\sum_{n=0}^{\infty} [(-1)^n e/n!]e^{-t}L_n(t)$

6–11 Initial- and final-value theorems. For functions $f(t)$ whose Laplace transforms are analytic at $s = \infty$, the behavior of $f(t)$ for t near zero is related to the behavior of $F(s) = \mathcal{L}[f]$ near $s = \infty$. Indeed,

$$f(t) = \sum_{n=0}^{\infty} a_{n+1} \frac{t^n}{n!}, \qquad F(s) = \sum_{n=1}^{\infty} \frac{a_n}{s^n}, \qquad (6\text{--}110)$$

so that (for the right-hand limits)

$$f(0) = a_1, f'(0) = a_2, f''(0)/2! = a_3, \ldots,$$

while

$$sF(s) = s\left(\frac{a_1}{s} + \frac{a_2}{s^2} + \cdots\right) = a_1 + \frac{a_2}{s} + \cdots,$$

so that

$$f(0) = a_1 = \lim_{s\to\infty} sF(s). \qquad (6\text{--}111)$$

Similarly,

$$f'(0) = a_2 = \lim_{s\to\infty} s[sF(s) - a_1].$$

If $a_1, \ldots, a_k$ are 0, but $a_{k+1} \neq 0$, so that for small t the dominant term in the series for $f(t)$ is $a_{k+1}t^k/k!$, then for large $|s|$ the dominant term in the series for $F(s)$ is a_{k+1}/s^{k+1}; more precisely, we verify that

$$\lim_{t\to 0+} \frac{f(t)}{a_{k+1}t^k/k!} = 1, \qquad \lim_{s\to\infty} \frac{F(s)}{a_{k+1}/s^{k+1}} = 1. \qquad (6\text{--}112)$$

We now proceed to show that the *initial-value theorem* (6–111) remains valid without assumption of analyticity of $F(s)$, provided $s \to \infty$ through positive real values.

THEOREM 12. *Let $f(t)$ be piecewise continuous for $t \geqq 0$ and let $f(t) \to B$ as $t \to 0+$. Let $\mathcal{L}[f]$ be absolutely convergent to $F(s)$ for $\sigma > a$. Then as s approaches $+\infty$ through real values,*

$$\lim_{s \to +\infty} sF(s) = B.$$

Proof. Let $F(s) = F_1(s) + F_2(s)$, where

$$F_1(s) = \int_0^1 f(t)e^{-st}\,dt, \qquad F_2(s) = \int_1^\infty f(t)e^{-st}\,dt.$$

Let s be real and $> a + 1$, so that $s = a + 1 + \alpha$, where $\alpha > 0$. Then

$$|F_2(s)| \leqq \int_1^\infty |f(t)|e^{-st}\,dt$$

$$= \int_1^\infty |f(t)|e^{-(a+1)t}e^{-\alpha t}\,dt \leqq e^{-\alpha} \int_1^\infty |f(t)|e^{-(a+1)t}\,dt = Ke^{-\alpha},$$

where K is a constant. Hence

$$|sF_2(s)| \leqq |a + 1 + \alpha| Ke^{-\alpha}.$$

As s tends to $+\infty$, α also tends to $+\infty$, and hence $sF_2(s) \to 0$.

Next, since $\int_0^\infty e^{-st}\,dt = 1/s$,

$$sF_1(s) - B = s\int_0^1 f(t)e^{-st}\,dt - Bs\int_0^\infty e^{-st}\,dt$$

$$= s\int_0^1 [f(t) - B]e^{-st}\,dt - Bs\int_1^\infty e^{-st}\,dt.$$

The second term approaches zero as $s \to +\infty$, as above for $sF_2(s)$. We show that the first term also approaches zero. To this end, we choose T so small that $|f(t) - B| < \epsilon$ for $0 < t \leqq T$, so that, for $s > 0$,

$$\left| s\int_0^1 [f(t) - B]e^{-st}\,dt \right| = \left| s\int_0^T [f(t) - B]e^{-st}\,dt + s\int_T^1 [f(t) - B]e^{-st}\,dt \right|$$

$$\leqq s\epsilon \int_0^T e^{-st}\,dt + s\int_T^\infty |f(t) - B|e^{-st}\,dt.$$

The second term approaches zero as $s \to +\infty$, as above, and the first term is

$$s\epsilon \frac{1 - e^{-sT}}{s} = \epsilon(1 - e^{-sT}) \leqq \epsilon.$$

We conclude that

$$sF(s) - B = [sF_1(s) - B] + sF_2(s)$$

can be made less than 2ϵ by choosing s real and sufficiently large. Thus $sF(s) - B$ approaches zero as $s \to +\infty$, as asserted.

Remark. We can verify from the proof that the result remains valid if $s \to \infty$ in a sector $|\arg s| \leq \beta$, $0 < \beta < \pi/2$. We can also extend the result as follows [as suggested by (6–112)]: if $f(t)/(Bt^\gamma)$ approaches 1 as $t \to 0+$, $\gamma > -1$, then $sF(s)/[B\Gamma(\gamma + 1)s^{-\gamma}] \to 1$ as $s \to \infty$ in such a sector. For a proof see pp. 200–201 of Reference 4.

It is now natural to ask whether the behavior of $f(t)$ for large positive t is related to that of $F(s)$ for s near 0. In general, $F(s)$ is analytic only in a half-plane $\sigma > a$, of which $s = 0$ need be neither an interior point nor a boundary point. However, if $f(t)$ behaves like a power of t for large positive t, then we can verify that $s = 0$ is either in or on the boundary of the half-plane. In particular, if $f(t)$ has limit A as $t \to +\infty$ (so that $f(t)$ behaves like At^0 for large t), then $sF(s)$ has limit A as s approaches $0+$ through real values:

$$\lim_{t \to +\infty} f(t) = \lim_{s \to 0+} sF(s). \tag{6–113}$$

This is the *final-value theorem.*

THEOREM 13. *Let $f(t)$ be piecewise continuous for $t \geq 0$ and let $f(t)$ approach A as $t \to +\infty$. Then $F(s) = \mathcal{L}[f]$ is absolutely convergent for $\sigma > 0$ and, as $s \to 0+$ on the real axis,*

$$\lim_{s \to 0+} sF(s) = A.$$

Proof. In each finite interval of the t-axis, $t \geq 0$, $f(t)$ has only a finite number of jump discontinuities and hence $|f(t)|$ is bounded. For t sufficiently large, $f(t)$ is as close to A as desired, hence $|f(t)|$ is bounded. Therefore, for some constant M, $|f(t)| < M$ for all positive t. Therefore, for $\sigma > 0$,

$$\int_0^\infty |f(t)e^{-st}|\, dt \leq M \int_0^\infty e^{-\sigma t}\, dt < \infty,$$

and hence $\mathcal{L}[f]$ is absolutely convergent.

We can write $f(t) = A + g(t)$, where $g(t)$ approaches zero as $t \to \infty$. Then for given $\epsilon > 0$, we choose T so large that $|g(t)| < \epsilon$ for $t \geq T$. Now

$$F(s) = \int_0^\infty (A + g(t))e^{-st}\, dt = \frac{A}{s} + \int_0^T g(t)e^{-st}\, dt + \int_T^\infty g(t)e^{-st}\, dt,$$

so that

$$sF(s) - A = s\int_0^T g(t)e^{-st}\, dt + s\int_T^\infty g(t)e^{-st}\, dt.$$

We can write

$$\left| \int_T^\infty g(t)e^{-st}\, dt \right| \leq \int_T^\infty |g(t)|e^{-\sigma t}\, dt < \int_0^\infty \epsilon e^{-\sigma t}\, dt = \frac{\epsilon}{\sigma}.$$

Also $|g(t)| = |f(t) - A| \leq |f(t)| + |A| < M + |A|$, so that

$$\left| \int_0^T g(t)e^{-st}\, dt \right| < (M + |A|) \int_0^T e^{-\sigma t}\, dt < (M + |A|)T.$$

Hence for s real, so that $s = \sigma > 0$,

$$|sF(s) - A| = \sigma \left| \int_0^T g(t)e^{-st}\, dt + \int_T^\infty g(t)e^{-st}\, dt \right|$$
$$< \sigma \left(\frac{\epsilon}{\sigma} + (M + |A|)T \right) = \epsilon + (M + |A|)\sigma T.$$

Hence, for $\sigma < \epsilon[(M + |A|)T]^{-1}$, $|sF(s) - A| < 2\epsilon$, so that

$$\lim_{s \to 0+} |sF(s) - A| = 0.$$

Remark. By similar methods we can show that the conclusion holds with s allowed to approach zero in a sector $|\arg s| \leq \alpha$, where $0 < \alpha < \pi/2$. Furthermore, as in (6–112), the theorem can be generalized as follows: if $\gamma > -1$ and $f(t)/[At^\gamma]$ approaches 1 as $t \to +\infty$, then $sF(s)/[A\Gamma(\gamma + 1)s^{-\gamma}]$ approaches 1 as $s \to 0$ in such a sector. For a proof, see pp. 188–189 of Reference 4.

Problems

1. Verify the initial-value theorem for the following choices of $f(t)$:

(a) e^t (b) $\cos t$ (c) $h(t) - h(t - 1)$ (d) $e^{2t}h(t - 2)$

2. Verify that the relations (6–112) are satisfied, with s approaching $+\infty$ through real values, for appropriate choice of k and a_{k+1}:

(a) $\sin t$ (b) $1 - \cos t$ (c) $\sin^3 t$ (d) $e^t - 1 - t$

3. Verify the final-value theorem for the following choices of $f(t)$:

(a) $1 - e^{-t}$ (b) $t[h(t) - h(t - 1)] + 2h(t - 1)$

4. Prove that if b is real and $f(t)/(Ae^{bt})$ approaches unity as $t \to +\infty$, then $\mathcal{L}[f]$ is absolutely convergent to $F(s)$ for $\sigma > b$, and $(s - b)F(s)/A$ approaches unity as $s \to b+$ through real values. [*Hint:* Consider $g(t) = f(t)e^{-bt}$].

5. For each of the following verify that, for appropriate choices of A and k, $f(t)/(At^k)$ approaches one as $t \to +\infty$ and $sF(s)/(Ak!s^{-k})$ approaches one as $s \to 0+$ through real values:

(a) $5t^2 - 3t + 1$ (b) $t^3(1 - e^{-t}) - t \sin t$

6-12 Convolution. The convolution is defined as in Section 5–11. However, since here all functions are zero for $t < 0$, the definition can be simplified.

DEFINITION. Let $f(t)$ and $g(t)$ be identically zero for $t < 0$ and piecewise continuous for $t \geq 0$. Then the convolution of f and g, denoted by $f * g$, is the function of t:

$$f * g = \int_0^t f(u)g(t - u)\, du. \tag{6–120}$$

THEOREM 14. *With the definition above, $f * g$ is identically zero for $t < 0$, and is continuous for $-\infty < t < \infty$. Furthermore,*

$$f * g = g * f, \tag{6–121}$$

$$f * (cg) = (cf) * g = c(f * g), \tag{6–122}$$

$$f * (g_1 + g_2) = f * g_1 + f * g_2, \tag{6–123}$$

$$e^{at}(f * g) = e^{at}f * e^{at}g. \tag{6–124}$$

The continuity of $f * g$ was shown in Section 2–8. The remaining proofs are left as an exercise (Problem 7 following Section 6–13).

THEOREM 15. *Let $f(t)$ and $g(t)$ be identically 0 for $t < 0$ and piecewise continuous for $t \geq 0$. Let $\mathcal{L}[f]$ and $\mathcal{L}[g]$ be absolutely convergent for $\sigma > a$. Then, for $\sigma > a$, $\mathcal{L}[f * g]$ is absolutely convergent and*

$$\mathcal{L}[f * g] = \mathcal{L}[f] \cdot \mathcal{L}[g]. \tag{6–125}$$

This is a special case of the rule for Fourier transforms, and the proof is included in that of Section 5–11. However, we recall the essential steps:

$$\mathcal{L}[f * g] = \int_0^\infty (f * g)_t \, e^{-st} \, dt = \int_0^\infty \int_0^t f(u)g(t - u)e^{-st} \, du \, dt$$

$$= \int_0^\infty \int_u^\infty f(u)g(t - u)e^{-st} \, dt \, du,$$

the integration being over the sector of Fig. 6–6 in the ut-plane. Accordingly,

$$\mathcal{L}[f * g] = \int_0^\infty f(u)e^{-su} \left(\int_u^\infty g(t - u)e^{-s(t-u)} \, dt \right) du.$$

If we replace $t - u$ by v in the inner integral, it no longer depends on t:

$$\mathcal{L}[f * g] = \int_0^\infty f(u)e^{-su} \left(\int_0^\infty g(v)e^{-sv} \, dv \right) du = \mathcal{L}[f]\mathcal{L}[g].$$

The justification of the steps and the proof that $\mathcal{L}[f * g]$ is absolutely convergent for the value of σ considered are the same as for Fourier trans-

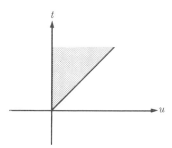

Fig. 6–6.   Proof of convolution theorem.

forms. In fact we can deduce the formula from the corresponding one for Fourier transforms by means of (6–124):

$$\mathcal{L}[f * g] = \Phi[(f * g)e^{-\sigma t}] = \Phi[(e^{-\sigma t}f) * (e^{-\sigma t}g)]$$
$$= \Phi[e^{-\sigma t}f]\Phi[e^{-\sigma t}g] = \mathcal{L}[f]\mathcal{L}[g].$$

COROLLARY (Associative law). *Let* $f_1(t)$, $f_2(t)$, $f_3(t)$ *be identically* 0 *for* $t < 0$ *and piecewise continuous for* $t \geqq 0$. *Then*

$$f_1 * (f_2 * f_3) = (f_1 * f_2) * f_3. \tag{6–126}$$

Proof. If f_1, f_2, f_3 have absolutely convergent Laplace transforms for $\sigma > a$, then (6–126) is a consequence of (6–125). For

$$\mathcal{L}[f_1 * (f_2 * f_3)] = \mathcal{L}[f_1]\mathcal{L}[f_2 * f_3] = \mathcal{L}[f_1]\mathcal{L}[f_2]\mathcal{L}[f_3]$$
$$= \mathcal{L}[f_1 * f_2]\mathcal{L}[f_3] = \mathcal{L}[(f_1 * f_2) * f_3].$$

Since a function is uniquely determined by its Laplace transform, we conclude that

$$f_1 * (f_2 * f_3) = (f_1 * f_2) * f_3.$$

We reduce the general case to the special case just considered. Let c be a fixed positive number and set, for $j = 1, 2, 3$,

$$g_j(t) = f_j(t) \quad (t \leqq c), \qquad g_j(t) = 0 \quad (t > c).$$

Then

$$g_1 * g_2 = \int_0^t g_1(u)g_2(t - u)\, du$$

$$= \int_0^t f_1(u)f_2(t - u)\, du = f_1 * f_2 \quad (t \leqq c),$$

and similarly,

$$(g_1 * g_2) * g_3 = (f_1 * f_2) * f_3 \quad (t \leqq c),$$
$$g_1 * (g_2 * g_3) = f_1 * (f_2 * f_3) \quad (t \leqq c). \tag{6–127}$$

Now the functions g_1, g_2, g_3 have absolutely convergent Laplace transforms for all σ. Hence by the special case considered above, for all t,

$$g_1 * (g_2 * g_3) = (g_1 * g_2) * g_3.$$

Hence for $t \leqq c$, by (6–127),

$$(f_1 * f_2) * f_3 = f_1 * (f_2 * f_3). \qquad (6\text{–}128)$$

But c is arbitrary; therefore (6–128) holds for all t.

A second proof of the associative law is given in Section 2–8.

THEOREM 16. *Let $f(t)$, $g(t)$ have absolutely convergent Laplace transforms $\phi(s)$, $\psi(s)$ for $\sigma > a$. For each $\sigma > a$ let $\phi(\sigma + i\omega)$, $\psi(\sigma + i\omega)$ be absolutely integrable with respect to ω from $-\infty$ to ∞. Then, for $\sigma > 2a$, $f \cdot g$ has an absolutely convergent Laplace transform, and*

$$\mathcal{L}[f \cdot g] = \frac{1}{2\pi} \int_{-\infty}^{\infty} \phi(\sigma_1 + i\lambda)\psi(s - (\sigma_1 + i\lambda))\, d\lambda, \qquad (6\text{–}129)$$

where $a < \sigma_1 < \sigma - a$.

Proof. By definition,

$$\mathcal{L}[f] = \phi(\sigma + i\omega) = \Phi[fe^{-\sigma t}] \quad (\sigma > a),$$
$$\mathcal{L}[g] = \psi(\sigma + i\omega) = \Phi[ge^{-\sigma t}] \quad (\sigma > a),$$
$$\mathcal{L}[f \cdot g] = \Phi[fge^{-\sigma t}]$$
$$= \Phi[fe^{-\sigma_1 t}ge^{-\sigma_2 t}],$$

where $\sigma_1 + \sigma_2 = \sigma$ and σ_1, σ_2 are to be determined. If $\sigma_1 > a, \sigma_2 > a$, then we can apply the corollary to Theorem 13 of Section 5–11,

$$\Phi[fe^{-\sigma_1 t} \cdot ge^{-\sigma_2 t}] = \frac{1}{2\pi} \Phi[fe^{-\sigma_1 t}] * \Phi[ge^{-\sigma_2 t}],$$

where the convolution is with respect to ω. Hence

$$\Phi[fe^{-\sigma_1 t} \cdot ge^{-\sigma_2 t}] = \frac{1}{2\pi} \phi(\sigma_1 + i\omega) * \psi(\sigma_2 + i\omega)$$
$$= \frac{1}{2\pi} \int_{-\infty}^{\infty} \phi(\sigma_1 + i\lambda)\psi(\sigma_2 + i(\omega - \lambda))\, d\lambda$$
$$= \frac{1}{2\pi} \int_{-\infty}^{\infty} \phi(\sigma_1 + i\lambda)\psi(\sigma - \sigma_1 + i(\omega - \lambda))\, d\lambda$$
$$= \frac{1}{2\pi} \int_{-\infty}^{\infty} \phi(\sigma_1 + i\lambda)\psi(s - (\sigma_1 + i\lambda))\, d\lambda.$$

If now $\sigma > 2a$ and $a < \sigma_1 < \sigma - a$, then

$$\sigma_2 = \sigma - \sigma_1 > a,$$

and the above steps are valid.

EXAMPLE. Let $f = th(t)$, $g = te^th(t)$. Then $\mathfrak{L}[f] = s^{-2}$, and $\mathfrak{L}[g] = (s - 1)^{-2}$, both integrals converging absolutely for $\sigma > 1$. Furthermore for fixed $\sigma > 1$, we find that $\phi(s) = (\sigma + i\omega)^{-2}$ and $\psi(s) = (\sigma - 1 + i\omega)^{-2}$ are absolutely integrable with respect to ω for $-\infty < \omega < \infty$, as is easily verified. Hence for $\sigma > 2$,

$$\mathfrak{L}[fg] = \mathfrak{L}[t^2e^th(t)] = \frac{1}{2\pi} \int_{-\infty}^{\infty} \frac{1}{(\sigma_1 + i\lambda)^2} \frac{1}{(s - 1 - \sigma_1 - i\lambda)^2} \, d\lambda,$$

where $1 < \sigma_1 < \sigma - 1$. Thus if $\sigma > 4$, we can choose $\sigma_1 = 2$:

$$\mathfrak{L}[fg] = \frac{1}{2\pi} \int_{-\infty}^{\infty} \frac{1}{(2 + i\lambda)^2} \frac{1}{(s - 3 - i\lambda)^2} \, d\lambda.$$

The integral on the right can be evaluated most easily by residues (see Reference 10, p. 577). As expected, the value is found to be $2/(s - 1)^3$.

6–13 Special convolutions. As in Section 5–12, the convolution $e^{st}h(t) * f$ is of importance. Since $f = 0$ for $t < 0$ here, we conclude at once from Theorem 15 of Section 5–12 that

$$e^{st}h(t) * f(t) = e^{st}F_t(s),$$

$$F_t(s) = \int_0^t f(u)e^{-su} \, du. \tag{6-130}$$

Hence $e^{st}h(t) * f(t)$ is simply e^{st} times the truncated Laplace transform of f. From Table 5–2 of Section 5–12 we at once obtain a table of these convolutions (Table 6–4). In Table 6–4, s must be treated as a complex *constant*.

The other rules of Section 5–12 can also be repeated:

$$(D - s)[e^{st}h * f] = f, \tag{6-131}$$

$$e^{st}h(t) * f(t - c) = g_s(t - c) \qquad (c > 0), \tag{6-132}$$

where $g_s(t) = e^{st}h(t) * f$. From (6–124) we obtain the rule

$$e^{st}h(t) * e^{at}f(t) = e^{at}g_{s-a}(t). \tag{6-133}$$

In (6–131), $e^{st}h * f$ is continuous and piecewise smooth, and equality holds except at the discontinuities of f.

TABLE 6-4

CONVOLUTIONS

No.	$f(t), \quad t \geq 0 \quad (f = 0 \text{ for } t < 0)$	$e^{st}h(t) * f(t), \quad t \geq 0, \quad s = \text{const}$
1	$1, \quad s \neq 0$	$\dfrac{e^{st} - 1}{s}$
2	$t, \quad s \neq 0$	$\dfrac{e^{st} - 1 - st}{s^2}$
3	$t^n, \quad n = 1, 2, \ldots, \quad s \neq 0$	$\dfrac{n!e^{st} - g_n(st)}{s^{n+1}}, \quad g_n(x) = x^n + nx^{n-1} + \cdots + n!$
4	$e^{at}, \quad a \neq s$	$\dfrac{e^{at} - e^{st}}{a - s}$
5	e^{st}	te^{st}
6	$t^n e^{at}, \quad a \neq s, \quad n = 1, 2, \ldots$	$\dfrac{n!e^{st} - e^{at}g_n[(s - a)t]}{(s - a)^{n+1}}$
7	$t^n e^{st}, \quad n = 1, 2, \ldots$	$\dfrac{e^{st} t^{n+1}}{n+1}$

8	$h(t-c)$, $c > 0$, $s \neq 0$	$\dfrac{e^{s(t-c)} - 1}{s} h(t - c)$
9	$(t-c)h(t-c)$, $c > 0$, $s \neq 0$	$\dfrac{e^{s(t-c)} - 1 - s(t-c)}{s^2} h(t - c)$
10	$t[h(t) - h(t-c)]$, $c > 0$, $s \neq 0$	$\dfrac{e^{st} - 1 - st}{s^2}[h(t) - h(t-c)] + \dfrac{e^{st}[1 - (1+cs)e^{-sc}]}{s^2} h(t - c)$
11	$t[h(t) - h(t-c)] + (2c-t)[h(t-c) - h(t-2c)]$, $c > 0$, $s \neq 0$	$\dfrac{e^{st} - 1 - st}{s^2}[h(t) - h(t-c)]$ $+ \dfrac{e^{st} - 2e^{s(t-c)} + 1 + st - 2cs}{s^2} \cdot [h(t-c) - h(t-2c)]$ $+ \dfrac{e^{st}(1 - 2e^{-cs} + e^{-2cs})}{s^2} h(t - 2c)$
12	$e^{at}\cos bt$, $s \neq a \pm bi$	$\dfrac{e^{at}[(a-s)\cos bt + b\sin bt] - (a-s)e^{st}}{(a-s)^2 + b^2}$
13	$e^{at}\sin bt$, $s \neq a \pm bi$	$\dfrac{e^{at}[(a-s)\sin bt - b\cos bt] + be^{st}}{(a-s)^2 + b^2}$

If $Y(s)$ is a proper rational function of s, then we can evaluate $\mathcal{L}^{-1}[Y] * f$ by decomposing $Y(s)$ into partial fractions; then we have to compute terms of the form

$$\mathcal{L}^{-1}\left[\frac{A}{(s - s_j)^k}\right] * f.$$

If $k = 1$, this becomes

$$\mathcal{L}^{-1}\left[\frac{A}{s - s_j}\right] * f = Ae^{s_jt}h(t) * f,$$

and therefore is of the special type considered above. If $k = 2$, we can write

$$\mathcal{L}^{-1}\left[\frac{A}{(s - s_j)^2}\right] * f = A\mathcal{L}^{-1}\left[\frac{1}{s - s_j} \cdot \frac{1}{s - s_j}\right] * f$$

$$= A[e^{s_jt}h(t) * e^{s_jt}h(t)] * f$$

$$= Ae^{s_jt}h(t) * [e^{s_jt}h(t) * f].$$

Hence again the operation is built out of convolutions of the special type. A similar reasoning applies to higher values of k.

When $f(t)$ is of the form $e^{bt}p(t)h(t)$, where $p(t)$ is a polynomial, then we can obtain $\mathcal{L}^{-1}[Y] * f$ in other ways, for example, by *partial fractions*. Thus

$$\mathcal{L}^{-1}[Y] * e^{bt}p(t)h(t) = \mathcal{L}^{-1}[Y] * \mathcal{L}^{-1}[\phi(s)],$$

where $\phi(s) = \mathcal{L}[e^{bt}p(t)h(t)]$. Now $\phi(s)$ is also a proper rational function. Hence

$$\mathcal{L}^{-1}[Y] * e^{bt}p(t)h(t) = \mathcal{L}^{-1}[Y(s)\phi(s)],$$

where $Y(s)\phi(s)$ is a proper rational function. We can expand $Y(s)\phi(s)$ in partial fractions (or use Table 6–2) and thereby obtain the inverse transform.

The convolution $\mathcal{L}^{-1}[Y] * e^{bt}p$ can also be expressed in other forms. The results to be described are related to Theorem 20 of Section 5–17.

THEOREM 17. *Let*

$$Y(s) = \sum_{j=1}^{n} \frac{A_j}{s - s_j}.$$

Let $b \neq s_j$ for $j = 1, \ldots, n$. Let $p(t)$ be a polynomial and let $\phi(s) = \mathcal{L}[e^{bt}p(t)]$. Then for $t \geq 0$,

$$\mathcal{L}^{-1}[Y] * e^{bt}p(t)h(t)$$

$$= \sum_{j=1}^{n} A_j\phi(s_j)e^{s_jt} + e^{bt}\left[Y(b)p(t) + \frac{Y'(b)p'(t)}{1!} + \cdots\right]. \qquad (6\text{–}134)$$

Proof. By linearity it is sufficient to prove the rule when $p(t)$ reduces to one term t^k, so that $\phi(s) = k!/(s - b)^{k+1}$. Then $Y(s)\phi(s)$ is a proper rational function with poles of first order at $s_1, \ldots, s_n$ and of order at most $k + 1$ at b. Since $\phi(s)$ is analytic at s_j, and $Y(s)$ has residue A_j, it follows that $Y(s)\phi(s)$ has residue $A_j\phi(s_j)$ (Section 3–13). At the pole at $s = b$, $Y(s)\phi(s)$ has principal part

$$\frac{k!}{(s - b)^{k+1}}\left[Y(b) + \frac{Y'(b)}{1!}(s - b) + \cdots + \frac{Y^{(k)}(b)}{k!}(s - b)^k \right].$$

Therefore, as in Section 3–16, $Y(s)\phi(s)$ has the partial fraction expansion

$$\sum_{j=1}^{n} \frac{A_j\phi(s_j)}{s - s_j} + k!\left[\frac{Y(b)}{(s - b)^{k+1}} + \frac{Y'(b)}{1!(s - b)^k} + \cdots + \frac{Y^{(k)}(b)}{k!(s - b)} \right].$$

Hence $\mathcal{L}^{-1}[Y(s)\phi(s)]$ is

$$\sum_{j=1}^{n} A_j\phi(s_j)e^{s_jt} + e^{bt}\left[Y(b)t^k + \frac{Y'(b)}{1!}kt^{k-1} + \cdots + \frac{Y^{(k)}(b)}{k!}k! \right].$$

Therefore the theorem is proved.

EXAMPLE 1. Let $Y(s) = (s + 7)/(s^2 + 4s + 3)$. Then to evaluate

$$g(t) = \mathcal{L}^{-1}[Y] * \{e^{2t}(t^2 + 5t + 1)h(t)\},$$

we compute

$$Y'(s) = \frac{-s^2 - 14s - 25}{(s^2 + 4s + 3)^2}, \qquad Y''(s) = \frac{2s^3 + 42s^2 + 150s + 158}{(s^2 + 4s + 3)^3},$$

so that $Y(2) = 3/5$, $Y'(2) = -19/75$, $Y''(2) = 214/1125$. Furthermore,

$$Y(s) = \frac{3}{s + 1} - \frac{2}{s + 3},$$

$$\phi(s) = \frac{2}{(s - 2)^3} + \frac{5}{(s - 2)^2} + \frac{1}{s - 2} = \frac{s^2 + s - 4}{(s - 2)^3},$$

so that $\phi(-1) = 4/27$, $\phi(-3) = -2/125$. Accordingly, for $t \geq 0$,

$$g(t) = 3(\tfrac{4}{27})e^{-t} - 2(-\tfrac{2}{125})e^{-3t}$$
$$+ e^{2t}[\tfrac{3}{5}(t^2 + 5t + 1) - \tfrac{19}{75}(2t + 5) + \tfrac{107}{1125}(2)].$$

When multiple roots appear, the rule has to be modified.

THEOREM 18. *Let*

$$Y(s) = \sum_{j=1}^{n} \frac{A_j}{s - s_j} + \frac{B_1}{s - c} + \cdots + \frac{B_m}{(s - c)^m} \quad (B_m \neq 0).$$

Let $b \neq s_j$ for $j = 1, \ldots, n$, $b \neq c$. Let $p(t)$ be a polynomial and let $\phi(s) = \mathcal{L}[e^{bt}p(t)]$. Then for $t \geqq 0$,

$$\mathcal{L}^{-1}[Y] * e^{bt}p(t)h(t) = \sum_{j=1}^{n} A_j\phi(s_j)e^{s_jt}$$

$$+ e^{bt}\left[Y(b)p(t) + \frac{Y'(b)}{1!} p'(t) + \cdots \right]$$

$$+ \frac{e^{ct}}{(m - 1)!}\left[Y_c(c)t^{m-1} + \frac{Y_c'(c)}{1!} (m - 1)t^{m-2} + \cdots \right], \quad (6\text{-}135)$$

where

$$Y_c(s) = (s - c)^m Y(s)\phi(s). \quad (6\text{-}136)$$

Remark. Under the assumptions made, $Y(s)$ can be written in the form

$$Y(s) = \frac{Z(s)}{(z - c)^m},$$

where $Z(s)$ is a rational function, with $z - c$ not a factor of its denominator. Thus $(z - c)^m Y(s) = Z(s)$, and (6-136) is understood to mean

$$Y_c(s) = Z(s)\phi(s).$$

Proof. The proof of Theorem 17 can be repeated except for the fact that we must add the principal part of $Y(s)\phi(s)$ at $s = c$ to the partial fraction expansion. We can write

$$Y(s)\phi(s) = \frac{Z(s)\phi(s)}{(s - c)^m} = \frac{Y_c(s)}{(s - c)^m}.$$

Therefore the principal part in question is

$$\frac{1}{(s - c)^m}\left[Y_c(c) + \frac{Y_c'(c)}{1!} (s - c) + \cdots + \frac{Y_c^{(m-1)}(c)}{(m - 1)!} (s - c)^{m-1} \right].$$

The inverse transform of this function provides the terms in e^{ct} in Eq. (6-135).

When there are several multiple roots of the denominator of Y, we simply add terms for each as in (6-135).

EXAMPLE 2. Let

$$Y(s) = \frac{1}{(s+1)(s+2)^2(s+3)^2}, \qquad b = 1, p = t^2 + t.$$

Then

$$\log Y = -\log(s+1) - 2\log(s+2) - 2\log(s+3),$$

$$\frac{Y'}{Y} = -\frac{1}{s+1} - \frac{2}{s+2} - \frac{2}{s+3} = \psi(s),$$

$$Y' = \psi(s)Y,$$

$$Y'' = \psi Y' + \psi'Y,$$

$$\psi' = \frac{1}{(s+1)^2} + \frac{2}{(s+2)^2} + \frac{2}{(s+3)^2}.$$

Hence $\psi(1) = -\frac{1}{2} - \frac{2}{3} - \frac{2}{4} = -\frac{5}{3}$, $\psi'(1) = \frac{1}{4} + \frac{2}{9} + \frac{2}{16} = \frac{43}{72}$, $Y(1) = \frac{1}{288}$, $Y'(1) = \psi(1)Y(1) = -\frac{5}{864}$, $Y''(1) = \psi(1)Y'(1) + \psi'(1)Y(1) = \frac{3}{256}$. Accordingly,

$$e^{bt}\left[Y(b)p(t) + \frac{Y'(b)p'(t)}{1!} + \cdots\right]$$

$$= e^t\left[\frac{t^2+t}{288} + \left(-\frac{5}{864}\right)(2t+1) + \frac{3}{256}\frac{(2)}{2}\right] = e^t\left[\frac{t^2}{288} - \frac{7t}{864} + \frac{41}{6912}\right].$$

Next $\phi(s) = \mathcal{L}[e^t(t^2 + t)] = [2/(s-1)^3] + [1/(s-1)^2]$. For the factor $(s+2)^2$, we form

$$Y_{-2}(s) = (s+2)^2 Y(s)\phi(s) = \frac{1}{(s+1)(s+3)^2}\left(\frac{2}{(s-1)^3} + \frac{1}{(s-1)^2}\right).$$

Since m is 2, we need only $Y_{-2}(-2)$, $Y'_{-2}(-2)$. We find easily $Y_{-2}(-2) = -\frac{1}{27}$, $Y'_{-2}(-2) = \frac{1}{27}$. Hence for $c = -2$,

$$\frac{e^{ct}}{(m-1)!}[Y_c(c)t^{m-1} + \cdots] = \frac{e^{-2t}}{1}[-\tfrac{1}{27}t + \tfrac{1}{27}].$$

Similarly,

$$Y_{-3}(s) = (s+3)^2 Y(s)\phi(s) = \frac{1}{(s+1)(s+2)^2}\left(\frac{2}{(s-1)^3} + \frac{1}{(s-1)^2}\right),$$

$$Y_{-3}(-3) = \frac{-1}{64}, \qquad Y'_{-3}(-3) = \frac{-11}{256},$$

$$\frac{e^{-3t}}{(1)!}[Y_{-3}(-3)t + \cdots] = e^{-3t}\left[\frac{-t}{64} - \frac{11}{256}\right].$$

The term corresponding to the simple factor $(s+1)$ has yet to be evaluated. It can be done in the same way, with $m = 1$. Thus

$$Y_{-1}(s) = (s+1)Y(s)\phi(s) = \frac{1}{(s+2)^2(s+3)^2}\left(\frac{2}{(s-1)^3} + \frac{1}{(s-1)^2}\right).$$

Since $Y_{-1}(-1) = 0$, this term is missing. Thus finally, for $t \geq 0$,

$$\mathcal{L}^{-1}\left[\frac{1}{(s+1)(s+2)^2(s+3)^2}\right] * e^t(t^2 + t)h(t)$$

$$= e^t\left(\frac{t^2}{288} - \frac{7t}{864} + \frac{41}{6912}\right) + e^{-3t}\left(-\frac{t}{64} - \frac{11}{256}\right) + e^{-2t}\left(-\frac{t}{27} + \frac{1}{27}\right).$$

Neither Theorem 17 nor Theorem 18 takes care of the case in which $s - b$ appears in the denominator of $Y(s)$. This case is covered by the following rule:

THEOREM 19. Let $Y(s)$ be a proper rational function whose denominator is $(\text{const})(s - b)^m(s - c_1)^{m_1}(s - c_2)^{m_2}\cdots$ Then for $t \geq 0$,

$$\mathcal{L}^{-1}[Y] * e^{bt}p(t) = e^{bt}\left[B(b)q(t) + \frac{B'(b)q'(t)}{1!} + \cdots\right]$$

$$+ \frac{e^{c_1 t}}{(m_1 - 1)!}[Y_{c_1}(c_1)t^{m_1-1} + \cdots]$$

$$+ \frac{e^{c_2 t}}{(m_2 - 1)!}[Y_{c_2}(c_2)t^{m_2-1} + \cdots] + \cdots, \qquad (6\text{–}137)$$

where $B(s) = (s - b)^m Y(s)$, $q(t) = D_0^{-m}p(t)$,

$$Y_{c_j}(s) = (s - c_j)^{m_j}Y(s)\phi(s), \qquad \phi(s) = \mathcal{L}[e^{bt}p].$$

Thus the terms in e^{bt} are modified, but the other terms are computed as before.

Proof. The proof of Theorem 17 can again be repeated with appropriate modification of the principal parts in the partial fraction expansion of $Y(s)\phi(s)$. The principal parts at $c_1, c_2, \ldots$ are computed as for Theorem 18, while the principal part at $s = b$ has to be modified. We write

$$Y(s)\phi(s) = \frac{B(s)\phi(s)}{(s-b)^m} = \frac{B(s)k!}{(s-b)^{k+m+1}}.$$

Hence the principal part at $s = b$ is

$$\frac{k!}{(s-b)^{k+m+1}}\left[B(b) + \frac{B'(b)}{1!}(s-b) + \cdots + \frac{B^{(k+m)}(b)}{(k+m)!}(s-b)^{k+m}\right];$$

its inverse Laplace transform is

$$k!e^{bt}\left[B(b)\,\frac{t^{k+m}}{(k+m)!} + \frac{B'(b)}{1!}\,\frac{t^{k+m-1}}{(k+m-1)!} + \cdots\right],$$

in agreement with the terms in e^{bt} in Eq. (6–137), with

$$q(t) = D_0^{-m}t^k = \frac{t^{k+m}}{(k+m)(k+m-1)\cdots(k+1)} = \frac{k!t^{k+m}}{(k+m)!}.$$

Remark. If we wish the complete expression for $\mathcal{L}^{-1}[Y] * e^{bt}p$, then Theorems 17 through 19 do not provide a significant saving in labor over the method of partial fractions. However, they are advantageous if we seek only some of the terms, not all. In particular, we can find the terms in e^{bt} *without knowing any of the roots of the denominator of* $Y(s)$. This is of great value, as will be seen.

PROBLEMS

1. Evaluate the following Laplace transforms (all functions zero for $t < 0$):

(a) $\mathcal{L}[\sin t * \cos t]$

(b) $\mathcal{L}[te^t * h(t-1)]$

(c) $\mathcal{L}[\int_0^t u^2(t-u-1)^2\,du]$

(d) $\mathcal{L}[e^{2t}\int_0^t \sin u \; e^{-2u}\,du]$

2. Express the following inverse Laplace transforms as convolutions:

(a) $\mathcal{L}^{-1}\left[\dfrac{s}{(s^2-1)}\cdot\dfrac{1}{s^2-1}\right]$

(b) $\mathcal{L}^{-1}\left[\dfrac{e^{-s}}{s}\,\dfrac{1-e^{-2s}}{s}\right]$

3. (a) Prove that $e^{at}h(t) * e^{at}h(t) * \cdots * e^{at}h(t)$ (n factors) equals $t^{n-1}e^{at}/(n-1)!$.

(b) Prove that $t^m e^{at}h(t) * t^n e^{at}h(t) = \dfrac{m!n!}{(m+n+1)!}\,t^{m+n+1}e^{at}$.

(c) Prove that $\mathcal{L}[e^{a_1 t}h(t) * e^{a_2 t}h(t) * \cdots * e^{a_n t}h(t)]$ equals $[(s-a_1)(s-a_2)\cdots(s-a_n)]^{-1}$.

4. For each of the following express $\mathcal{L}[f \cdot g]$ as an integral as in Theorem 16:

(a) $f = h(t)\sin t, \quad g = h(t)\sinh t$

(b) $f = h(t)\sin t, \quad g = t^2 h(t)$

5. (a) Prove that $e^{at}h(t) * e^{at}f(t) = e^{at}\int_0^t f(t)\,dt$.

(b) Prove that if p is a polynomial and $a \neq b$, then

$$e^{at}h(t) * e^{bt}p(t)h(t) = \phi(a)e^{at} + e^{bt}\left[Y(b)p + \frac{Y'(b)}{1!}\,p' + \cdots\right],$$

where $\phi(s) = \mathcal{L}[e^{bt}p(t)h(t)]$, and $Y(s) = 1/(s-a)$.

(c) Prove that $t^k h(t) * f(t) = k!D_0^{-(k+1)}f, \quad k = 0, 1, 2, \ldots$

6. Evaluate the following convolutions:

(a) $\mathcal{L}^{-1}\left[\dfrac{1}{(s-1)(s-2)}\right] * h(t-3)$ (b) $\mathcal{L}^{-1}\left[\dfrac{1}{(s+1)(s+2)^2}\right] * t^2 h(t)$

7. (a) Prove (6–121). (b) Prove (6–122). (c) Prove (6–123). (d) Prove (6–124). (e) Prove that $(f * g) * (f_1 * g_1) = (f * f_1) * (g * g_1)$.

8. Evaluate the following convolutions:

(a) $\mathcal{L}^{-1}\left[\dfrac{1}{(s+1)(s-2)}\right] * e^{3t}(2t-1)h(t)$

(b) $\mathcal{L}^{-1}\left[\dfrac{1}{s(s+1)^2}\right] * e^t(t^2-t)h(t)$

(c) $\mathcal{L}^{-1}\left[\dfrac{s}{(s-1)(s-2)}\right] * e^t(t+1)h(t)$

9. Find the term in e^{2t} for the following convolutions:

(a) $\mathcal{L}^{-1}\left[\dfrac{1}{s^3+s+1}\right] * e^{2t}h(t)$

(b) $\mathcal{L}^{-1}\left[\dfrac{1}{s^5-s^3+1}\right] * e^{2t}(t^2+t-1)h(t)$

(c) $\mathcal{L}^{-1}\left[\dfrac{1}{(s-2)(s^3+s^2+1)}\right] * e^{2t}(t-2)h(t)$

10. Prove that the Laguerre polynomials $L_n(t)$ (Section 6–10) satisfy the relations

$$L_n(t)h(t) * L_m(t)h(t) = D_0^{-1}L_{n+m}(t)h(t).$$

Answers

1. (a) $\dfrac{s}{(s^2+1)^2}$ (b) $\dfrac{e^{-s}}{s(s-1)^2}$

   (c) $\dfrac{2}{s^3}\left(\dfrac{2}{s^3}-\dfrac{2}{s^2}+\dfrac{1}{s}\right)$ (d) $\dfrac{1}{(s-2)(s^2+1)}$

2. (a) $\cosh t * \sinh t$ (b) $h(t-1) * [h(t)-h(t-2)]$

4. (a) $\dfrac{1}{2\pi}\displaystyle\int_{-\infty}^{\infty}\dfrac{1}{(\sigma_1+i\lambda)^2+1}\dfrac{1}{[s-(\sigma_1+i\lambda)]^2-1}\,d\lambda$

   $(\sigma > 2, 1 < \sigma_1 < \sigma - 1)$

   (b) $\dfrac{1}{2\pi}\displaystyle\int_{-\infty}^{\infty}\dfrac{1}{(\sigma_1+i\lambda)^2+1}\dfrac{2}{[s-(\sigma_1+i\lambda)]^3}\,d\lambda \quad (0 < \sigma_1 < \sigma)$

6. (a) $(\frac{1}{2}e^{2t-6} - e^{t-3} + \frac{1}{2})h(t - 3)$

(b) $-2e^{-t} + e^{-2t}\left(\dfrac{t}{4} + \dfrac{5}{8}\right) + \dfrac{t^2}{4} - t + \dfrac{11}{8}$

8. (a) $e^{2t} - \dfrac{e^{-t}}{8} + e^{3t}\left(\dfrac{t}{2} - \dfrac{7}{8}\right)$

(b) $-3 + e^{t}\left(\dfrac{t^2}{4} - \dfrac{5}{4}t + \dfrac{15}{8}\right) + e^{-t}\left(\dfrac{t}{2} + \dfrac{9}{8}\right)$

(c) $4e^{2t} - \frac{1}{2}e^{t}(t^2 + 6t + 8)$

9. (a) $\dfrac{e^{2t}}{11}$ (b) $e^{2t}\left(\dfrac{t^2}{25} - \dfrac{111}{625}t + \dfrac{3223}{15{,}625}\right)$

(c) $e^{2t}\left(\dfrac{t^2}{26} - \dfrac{42}{169}t + \dfrac{581}{2197}\right)$

6-14 Laplace transforms of generalized functions. As in Section 5–16 we can define transforms of generalized functions by using the theory of Section 1–13. In particular,

$$\mathcal{L}\left[\delta(t)\right] = \int_{-\infty}^{\infty} \delta(t)e^{-st}\,dt = 1; \tag{6-140}$$

the limits of integration are from $-\infty$ to ∞, since we consider $\delta(t)$ to be zero for $t < 0$. Similarly,

$$\mathcal{L}\left[\delta^{(k)}(t)\right] = \int_{-\infty}^{\infty} \delta^{(k)}(t)e^{-st}\,dt$$

$$= (-1)^{k}\frac{d^{k}}{dt^{k}}e^{-st}\bigg|_{t=0} = s^{k}; \tag{6-141}$$

$$\mathcal{L}\left[\delta^{(k)}(t - c)\right] = (-1)^{k}\frac{d^{k}}{dt^{k}}e^{-st}\bigg|_{t=c} = s^{k}e^{-cs}, \tag{6-142}$$

for $c \geqq 0$. The definition can be extended to each generalized function

$$g(t) = p(t) + a_1\,\delta(t - c_1) + \cdots + b_1\,\delta'(t - d_1) + \cdots$$

by linearity, provided $p(t)$ is piecewise continuous and has a Laplace transform in the ordinary sense. We see at once that the terms added because of δ, δ', ... are *analytic* for all s. Hence $\mathcal{L}[g]$ is, as before, analytic in a half-plane $\sigma > a$. The new analytic functions thus obtained include all polynomials and all functions $P(s)e^{-cs}$, where $P(s)$ is a polynomial and $c \geqq 0$. Thus inverse transforms of these functions are now defined.

If $f(t)$ is piecewise smooth for $t \geq 0$, $f(t) = 0$ for $t < 0$, and f has only a finite number of jump discontinuities, then $f'(t)$ is a generalized function such as $g(t)$. With the above definitions we have the simplified rule

$$\mathcal{L}[f'] = s\mathcal{L}[f], \qquad (6\text{--}143)$$

provided only that $\mathcal{L}[f]$ and $\mathcal{L}[f']$ exist for $\sigma > a$. For example, if $f(t) = F(t)h(t)$, where $F(t)$ has a continuous derivative for all t, then [by Eq. (6–51')]

$$f'(t) = F'(t)h(t) + F(0)\,\delta(t),$$

$$\mathcal{L}[f'] = \mathcal{L}[F'] + F(0) \cdot 1$$

$$= s\mathcal{L}[F] - F(0) + F(0)$$

$$= s\mathcal{L}[f].$$

The general case is established in similar fashion. As a particular case we take $f = h(t)$, $f' = \delta(t)$; then

$$\mathcal{L}[f'] = 1, \qquad \mathcal{L}[f] = \frac{1}{s},$$

so that (6–143) is satisfied.

The rule

$$\mathcal{L}[D_0^{-1}f] = \frac{1}{s}\,\mathcal{L}[f] \qquad (6\text{--}144)$$

continues to hold for generalized functions, provided we interpret $D_0^{-1}f$ as the solution $x(t)$ of the equation $Dx = f$ such that $x(t) \to 0$ as $t \to 0-$. Then in particular $D_0^{-1}\,\delta(t) = h(t)$, $D_0^{-1}\,\delta^{(k)}(t - c) = \delta^{(k-1)}(t - c)$; Eq. (6–144) now follows from (6–140), (6–141), and (6–142). The rules

$$\mathcal{L}[f(t - c)] = e^{-cs}\mathcal{L}[f] \quad (c > 0), \qquad (6\text{--}145)$$

$$\mathcal{L}[e^{bt}f] = \phi(s - b), \quad \text{where } \phi(s) = \mathcal{L}[f], \qquad (6\text{--}146)$$

also continue to hold, where, in accordance with Eq. (1–132),

$$e^{bt}\,\delta^{(k)}(t - c) = (-1)^k e^{bc}\left[b^k\,\delta(t - c) - \frac{k}{1!}\,b^{k-1}\,\delta'(t - c) \right.$$

$$\left. + \frac{k(k-1)}{2!}\,b^{k-2}\,\delta''(t - c) + \cdots + (-1)^k\,\delta^{(k)}(t - c) \right]. \qquad (6\text{--}147)$$

The proofs are left as an exercise (Problem 5 below).

We can define the convolution $f * g$, where f and g are both generalized functions. First we define, for $k = 0, 1, 2, \ldots$ and $l = 0, 1, 2, \ldots$,

$$\delta^{(k)}(t - a) * \delta^{(l)}(t - b) = \delta^{(k+l)}(t - a - b) \quad (a \geqq 0, b \geqq 0),$$

$$g(t) * \delta(t - c) = \delta(t - c) * g(t) = g(t - c), \qquad (6\text{–}148)$$

$$g(t) * \delta^{(k)}(t - c) = \delta^{(k)}(t - c) * g(t) = g^{(k)}(t - c),$$

where $g(t)$ is an ordinary function and $g^{(k)}(t)$ is defined as a generalized function. The last definition agrees with that obtained from the integral

$$\int_{-\infty}^{\infty} \delta^{(k)}(u - c)g(t - u)\, du$$

if g has a continuous ordinary derivative $g^{(k)}(t)$ for all t. The general convolution $f * g$, when f and g are generalized functions, is then obtained by writing

$$f = f_1(t) + a_1\, \delta(t - c_1) + b_1\, \delta'(t - c_2) + \cdots,$$

$$g = g_1(t) + a_2\, \delta(t - k_1) + b_2\, \delta'(t - k_2) + \cdots,$$

and by applying linearity. For example,

$$[th(t) + \delta(t)] * [h(t) + \delta'(t - 1) + \delta''(t)]$$

$$= th(t) * h(t) + \delta(t) * h(t)$$

$$+ th(t) * \delta'(t - 1) + \delta(t) * \delta'(t - 1)$$

$$+ th(t) * \delta''(t) + \delta(t) * \delta''(t)$$

$$= \frac{t^2}{2} h(t) + h(t) + h(t - 1)$$

$$+ \delta'(t - 1) + \delta(t) + \delta''(t).$$

From (6–148) we can now verify that the rule

$$\mathcal{L}[f * g] = \mathcal{L}[f]\mathcal{L}[g] \qquad (6\text{–}149)$$

continues to hold, provided f and g have transforms (Problem 6 below). We note the particular cases

$$\mathcal{L}[f * \delta(t)] = \mathcal{L}[f(t)] = \mathcal{L}[f] \cdot \mathcal{L}[\delta],$$

$$\mathcal{L}[f * \delta^{(k)}(t)] = \mathcal{L}[f^{(k)}(t)] = \mathcal{L}[f] \cdot \mathcal{L}\,[\delta^{(k)}(t)] = s^k \mathcal{L}[f].$$

Problems

1. Evaluate the following Laplace transforms and inverse transforms:

(a) $\mathcal{L}[2h(t) - \delta(t-2) + \delta''(t-3)]$

(b) $\mathcal{L}[\delta(t) + \delta(t-1) + \cdots + \delta(t-n)]$

(c) $\mathcal{L}^{-1}[(1+2s)e^{-3s} + s^3 + 1]$

(d) $\mathcal{L}^{-1}\left[1 + s + s^2 - s^3 + \dfrac{s^4}{2!} + \cdots + (-1)^n \dfrac{s^{n+2}}{n!} + \cdots\right]$

2. Verify that $\mathcal{L}[f'] = s\mathcal{L}[f]$, $\mathcal{L}[f''] = s^2\mathcal{L}[f]$ for the following choices of f:

(a) $f = e^t h(t) \cdot$

(b) $f = t[h(t) - h(t-1)]$

3. Verify that $\mathcal{L}[D_0^{-1}f] = s^{-1}\mathcal{L}[f]$ for the following choices of f:

(a) $\delta(t) - \delta(t-1)$

(b) $e^t h(t) + \delta(t) + 2\,\delta'(t)$

4. Evaluate the following convolutions:

(a) $t^2 h(t) * \delta(t)$

(b) $t^2 h(t) * \delta'''(t)$

(c) $\delta''(t-1) * \delta'''(t-2)$

(d) $th(t-1) * [e^t h(t) + 2\,\delta''(t)]$

5. Prove that the rules (6–145) and (6–146) hold when $f = \delta^{(k)}(t-a)$, $a \geqq 0$, and hence conclude that they hold for an arbitrary generalized function having a Laplace transform.

6. Prove (6–149) on the basis of (6–148).

Answers

1. (a) $2s^{-1} - e^{-2s} + s^2 e^{-3s}$ (b) $(e^s - e^{-ns})/(e^s - 1)$

(c) $\delta(t-3) + 2\,\delta'(t-3) + \delta'''(t) + \delta(t)$

(d) $\delta(t) + \delta'(t) + \delta''(t-1)$

4. (a) $t^2 h(t)$ (b) $2\,\delta(t)$ (c) $\delta^{(v)}(t-3)$

(d) $(2e^{t-1} - t - 1)h(t-1) + 2\,\delta(t-1) + 2\,\delta'(t-1)$

6–15 Application of Laplace transforms to differential equations. The applications are based on the rule (6–52), which we restate for a function $x(t)$ with initial values $x_0, x_0', \ldots, x_0^{(n-1)}$ at $t = 0$:

$$\mathcal{L}[Dx] = -x_0 + s\mathcal{L}[x],$$

$$\mathcal{L}[D^2 x] = -x_0' - sx_0 + s^2\mathcal{L}[x],$$

$$\vdots$$

$$\mathcal{L}[D^n x] = -[x_0^{(n-1)} + sx_0^{(n-2)} + \cdots + s^{n-1}x_0] + s^n\mathcal{L}[x].$$

$$(6\text{--}150)$$

Here $x(t)$ is assumed to be an ordinary function, equal to 0 for $t < 0$, and such that $x, Dx, \ldots D^{n-1}x$ are continuous for $t > 0$, with limiting values $x_0, x_0', \ldots, x_0^{(n-1)}$ as $t \to 0+$, while $D^n x$ is piecewise continuous for $t \geqq 0$; it is assumed that $\mathcal{L}[x], \mathcal{L}[Dx], \ldots, \mathcal{L}[D^n x]$ are absolutely convergent in some half-plane $\sigma > a$.

Given a differential equation with constant coefficients

$$(a_0 D^n + \cdots + a_n)x = f(t) \quad (a_0 \neq 0), \tag{6–151}$$

we then proceed formally by taking Laplace transforms on both sides:

$$a_0\{s^n \mathcal{L}[x] - (x_0^{(n-1)} + sx_0^{(n-2)} + \cdots + s^{n-1}x_0)\} + \cdots$$

$$+ a_{n-1}\{s\mathcal{L}[x] - x_0\} + a_n\mathcal{L}[x] = \mathcal{L}[f].$$

Hence

$$(a_0 s^n + a_1 s^{n-1} + \cdots + a_{n-1}s + a_n)\mathcal{L}[x] = \mathcal{L}[f] + Q(s), \tag{6–152}$$

$$Q(s) = x_0 a_0 s^{n-1} + (x_0 a_1 + x_0' a_0)s^{n-2} + \cdots + (x_0 a_{n-1} + \cdots + x_0^{(n-1)}a_0).$$

Thus $Q(s)$ is a polynomial in s of degree at most $n - 1$, and depends on the initial values $x_0, x_0', \ldots, x_0^{(n-1)}$. In terms of the characteristic function $V(s)$ and transfer function $Y(s) = 1/V(s)$, we can then write

$$V(s)\mathcal{L}[x] = \mathcal{L}[f] + Q(s),$$

$$\mathcal{L}[x] = \frac{\mathcal{L}[f]}{V(s)} + \frac{Q(s)}{V(s)},$$

$$\mathcal{L}[x] = Y(s)\mathcal{L}[f] + Y(s)Q(s).$$

If all initial values $x_0, \ldots, x_0^{(n-1)}$ are zero, then $Q(s)$ is identically zero and

$$\mathcal{L}[x] = Y(s)\mathcal{L}[f],$$

$$\mathcal{L}[x] = \mathcal{L}[W]\mathcal{L}[f] = \mathcal{L}[W * f],$$

$$x = W * f,$$

where $W = \mathcal{L}^{-1}[Y(s)]$. On the other hand, if $f \equiv 0$, then $\mathcal{L}[f] \equiv 0$ and

$$\mathcal{L}[x] = Y(s)Q(s) = \frac{Q(s)}{V(s)}. \tag{6–152'}$$

Here $Q(s)$ is a polynomial of degree at most $(n - 1)$ while $V(s)$ has degree n. Hence Q/V is a rational function whose inverse transform is easily computed:

$$x = \mathcal{L}^{-1}\left[\frac{Q}{V}\right].$$

This is the solution of the homogeneous equation, for $t \geq 0$, with given initial conditions at $t = 0$.

Thus far we have proceeded formally. Now we state and prove a general theorem:

THEOREM 20. *Let $f(t)$ be piecewise continuous for $t \geqq 0$, $f \equiv 0$ for $t < 0$. Let $V(s) = a_0 s^n + \cdots + a_n$, $Y(s) = 1/V(s)$, $W(t) = \mathcal{L}^{-1}[Y(s)]$. Then*

$$x = W * f \tag{6-153}$$

is the unique solution of the differential equation (6–151), for $-\infty < t < \infty$, such that $x = 0, \ldots, x^{(n-1)} = 0$ for $t = 0$. If f has an absolutely convergent Laplace transform for $\sigma > a$ and for every root s_j of $V(s)$ we have $\mathrm{Re}\ (s_j) < a$, then $\mathcal{L}[x]$ is absolutely convergent for $\sigma > a$ and

$$\mathcal{L}[x] = Y(s)\mathcal{L}[f]. \tag{6-154}$$

If $Q(s)$ is defined by (6–152), then $\mathcal{L}^{-1}[Q/V]$ is the unique solution of the homogeneous equation

$$(a_0 D^n + \cdots + a_n)x = 0 \quad (t \geqq 0), \tag{6-155}$$

with initial values $x_0, x_0', \ldots, x_0^{(n-1)}$ for $t = 0$. Finally $x = W * f + \mathcal{L}^{-1}[Q/V]$ is the unique solution of the nonhomogeneous equation for $t \geqq 0$, with the initial values $x_0, x_0', \ldots, x_0^{(n-1)}$ for $t = 0$.

Proof. Let

$$Y(s) = \frac{1}{a_0(s - s_1) \cdots (s - s_n)} = \frac{1}{a_0} \frac{1}{s - s_1} \cdots \frac{1}{s - s_n}.$$

Then

$$W = \mathcal{L}^{-1}[Y] = \frac{1}{a_0} \mathcal{L}^{-1}\left[\frac{1}{s - s_1}\right] \cdots \mathcal{L}^{-1}\left[\frac{1}{s - s_n}\right]$$

$$= \frac{1}{a_0} e^{s_1 t} h * e^{s_2 t} h * \cdots * e^{s_n t} h,$$

by Theorem 15; no parentheses are needed by the associative law (6–126). Hence

$$W * f = \frac{1}{a_0} e^{s_1 t} h * e^{s_2 t} h * \cdots * e^{s_n t} h * f,$$

$$(D - s_1)(W * f) = \frac{1}{a_0} (D - s_1)\{e^{s_1 t} h * (e^{s_2 t} * \cdots)\}$$

$$= \frac{1}{a_0} e^{s_2 t} * e^{s_3 t} h * \cdots * f,$$

by (6–131). Repeating the process we conclude that

$$(D - s_n)(D - s_{n-1}) \cdots (D - s_1)(W * f) = \frac{f}{a_0},$$

wherever f is continuous. The same reasoning shows that $W * f$ has continuous derivatives through the order $(n - 1)$ and that $D^{n-1}(W * f)$ is piecewise smooth. Hence $W * f$ is a solution of the differential equation

$$(a_0 D^n + \cdots + a_n)x \equiv a_0(D - s_n) \cdots (D - s_1)x = f$$

for all t. As a convolution, $W * f = 0$ for $t = 0$; since $(D - s_1)(W * f)$ is a convolution, it is also zero for $t = 0$; hence $D(W * f)$ is zero for $t = 0$. Similarly, $D^2(W * f)$, $\ldots$, $D^{(n-1)}(W * f)$ are zero for $t = 0$. Thus $x = W * f$ is the solution of (6–151) with initial values zero for $t = 0$.

Now let $\mathcal{L}[f]$ be absolutely convergent for $\sigma > a$, and let Re $(s_j) < a$ for every j. Then $W(t)$ is a sum of terms of form $e^{s_j t} \cdot$ (polynomial) and thus has an absolutely convergent Laplace transform for $\sigma > a$. Hence, by Theorem 15, $\mathcal{L}[W * f]$ converges absolutely for $\sigma > a$, and $\mathcal{L}[W * f] = Y(s)\mathcal{L}[f]$.

Next we consider the homogeneous equation. We know that the general solution is a sum of terms of the form (polynomial) $\cdot e^{bt}$; hence each solution has a Laplace transform. Furthermore there is a unique solution with prescribed initial values of x, Dx, $\ldots$, $D^{n-1}x$ at $t = 0$. Hence the steps leading to Eq. (6–152′) are valid, and we conclude that $\mathcal{L}^{-1}[Q/V]$ is the solution of the homogeneous equation with the prescribed initial values. Since $W * f$ is the solution of the nonhomogeneous equation with the initial values zero, $x = W * f + \mathcal{L}^{-1}[Q/V]$ is the solution of the nonhomogeneous equation for $t \geqq 0$ with prescribed initial values at $t = 0$.

Remarks. We can write

$$Q(s) = x_0 Q_0(s) + x_0' Q_1(s) + \cdots + x_0^{(n-1)} Q_{n-1}(s),$$
$$Q_0(s) = a_0 s^{n-1} + a_1 s^{n-2} + \cdots + a_{n-1},$$
$$Q_1(s) = a_0 s^{n-2} + a_1 s^{n-3} + \cdots + a_{n-2}, \qquad (6\text{--}156)$$
$$\vdots$$
$$Q_{n-1}(s) = a_0,$$

so that

$$x = \mathcal{L}^{-1}\left[\frac{Q}{V}\right] = x_0 \mathcal{L}^{-1}\left[\frac{Q_0}{V}\right] + x_0' \mathcal{L}^{-1}\left[\frac{Q_1}{V}\right] + \cdots + x_0^{(n-1)} \mathcal{L}^{-1}\left[\frac{Q_{n-1}}{V}\right]$$
$$= x_0 P_0(t) + x_0' P_1(t) + \cdots + x_0^{(n-1)} P_{n-1}(t),$$

where $P_k(t) = \mathcal{L}^{-1}[Q_k/V]$.

The solution of the nonhomogeneous equation with general initial values can then be written as

$$x = W * f + x_0 P_0(t) + x_0' P_1(t) + \cdots + x_0^{(n-1)} P_{n-1}(t)$$

$$= T_0[f] + x_0 P_0(t) + x_0' P_1(t) + \cdots + x_0^{(n-1)} P_{n-1}(t). \quad (6\text{--}157)$$

Thus the linear operator T_0, which yields the solution with zero initial conditions (Section 2–3), is now identified as the convolution with $W = \mathcal{L}^{-1}[Y(s)]$:

$$T_0[f] = \int_0^t W(u)f(t - u)\,du. \quad (6\text{--}158)$$

By Lemma 2 of Section 6–7, we can also write

$$x = W * f + \mathcal{L}^{-1}\left[\frac{Q}{V}\right]$$

$$= W * f + \mathcal{L}^{-1}\left[\frac{x_0 a_0 s^{n-1} + (x_0 a_1 + x_0' a_0)s^{n-2} + \cdots}{V}\right]$$

$$= (W * f) + x_0 a_0 W^{(n-1)} + (x_0 a_1 + x_0' a_0)W^{(n-2)} + \cdots$$

$$+ (x_0 a_{n-1} + \cdots + x_0^{(n-1)} a_0)W. \quad (6\text{--}159)$$

This shows that, once $W(t)$ is known, the determination of the output x is reduced to one integration (6–158) and $n - 1$ differentiations.

6–16 Examples. The following examples illustrate the general method of the preceding section, and also the variations available for its application.

EXAMPLE 1. $(D^2 + 4)x = 0$; $x = 0$, $x' = 1$ for $t = 0$. We take Laplace transforms, and applying (6–150), we find

$$\mathcal{L}[(D^2 + 4)x] = \mathcal{L}[D^2 x] + 4\mathcal{L}[x] = -1 + s^2\mathcal{L}[x] + 4\mathcal{L}[x] = 0,$$

$$\mathcal{L}[x] = \frac{1}{s^2 + 4}, \qquad x = \tfrac{1}{2}\sin 2t,$$

by Entry 8 of Table 6–1.

EXAMPLE 2. $(D^2 + 4D + 3)x = 0$; $x = 1$, $x' = 2$ for $t = 0$. As in Example 1, $-2 - s + s^2\mathcal{L}[x] + 4(-1 + s\mathcal{L}[x]) + 3\mathcal{L}[x] = 0$,

$$\mathcal{L}[x] = \frac{6 + s}{s^2 + 4s + 3} = \frac{5}{2}\frac{1}{s + 1} - \frac{3}{2}\frac{1}{s + 3},$$

$$x = \tfrac{5}{2}e^{-t} - \tfrac{3}{2}e^{-3t}.$$

We could also find the inverse transform by Entries 2 or 3 of Table 6–2.

Thus

$$\mathcal{L}^{-1}\left[\frac{1}{s^2 + 4s + 3}\right] = \mathcal{L}^{-1}\left[\frac{1}{(s+1)(s+3)}\right] = \frac{e^{-t} - e^{-3t}}{2} = W(t),$$

$$x = 6W + W' = 6\left(\frac{e^{-t} - e^{-3t}}{2}\right) + \left(\frac{-e^{-t} + 3e^{-3t}}{2}\right)$$

$$= \tfrac{5}{2}e^{-t} - \tfrac{3}{2}e^{-3t}.$$

EXAMPLE 3. $(D^2 + 4D + 3)x = e^t(t^3 + t^2 - 1)$. We seek only a particular solution. The solution with initial values zero is $W * f$, where $W = \mathcal{L}^{-1}[Y(s)]$, $Y(s) = 1/(s^2 + 4s + 3)$. Now by Theorem 17,

$$\mathcal{L}^{-1}[Y] * e^t(t^3 + t^2 - 1)$$

$$= e^t\left[Y(1)(t^3 + t^2 - 1) + \frac{Y'(1)}{1!}(3t^2 + 2t) + \cdots\right]$$

plus terms in e^{-t} and e^{-3t}. The extra terms are clearly part of the complementary function. Hence $e^t[Y(1)(\cdots) + \cdots]$ is a particular solution. Since $Y(1) = \tfrac{1}{8}$, $Y'(1) = -\tfrac{3}{32}$, $Y''(1) = \tfrac{7}{64}$, and $Y'''(1) = -\tfrac{45}{256}$, we find

$$x = e^t[\tfrac{1}{8}(t^3 + t^2 - 1) - \tfrac{3}{32}(3t^2 + 2t) + \tfrac{7}{128}(6t + 2) - \tfrac{45}{256}]$$

$$= e^t\left[\frac{t^3}{8} - 5\frac{t^2}{32} + 9\frac{t}{64} - \frac{49}{256}\right].$$

EXAMPLE 4. $(D^2 + 4D + 3)x = t[h(t) - h(t - 2)]$. We seek the solution with initial values zero: $x = W * f$. Since

$$W = \mathcal{L}^{-1}\left[\frac{1}{s^2 + 4s + 3}\right] = \tfrac{1}{2}(e^{-t} - e^{-3t})h(t),$$

$$x = \tfrac{1}{2}[e^{-t}h * f - e^{-3t}h * f].$$

By Entry 10 of Table 6–4,

$$x = \frac{1}{2}\left\{\frac{e^{-t} - 1 + t}{1} - \frac{e^{-3t} - 1 + 3t}{9}\right\}[h(t) - h(t - 2)]$$

$$+ \frac{1}{2}\left\{\frac{e^{-t}[1 - (-1)e^2]}{1} - \frac{e^{-3t}[1 - (-5)e^6]}{9}\right\}h(t - 2).$$

EXAMPLE 5. $(D^3 + 4D^2 + D + 1)x = e^t \sin t$. We seek the solution with initial values zero. Hence

$$x = \mathcal{L}^{-1}\left[\frac{\mathcal{L}[e^t \sin t]}{s^3 + 4s^2 + s + 1}\right] = \mathcal{L}^{-1}\left[\frac{1}{(s^2 - 2s + 2)(s^3 + 4s^2 + s + 1)}\right]$$

$$= \mathcal{L}^{-1}[\phi(s)].$$

Evaluation of the roots of the cubic polynomial is awkward. If we do not wish to carry this out, we write

$$x = \frac{1}{2\pi} (P) \int_{-\infty}^{\infty} \frac{e^{st} \, d\omega}{(s^2 - 2s + 2)(s^3 + 4s^2 + s + 1)},$$

where $s = \sigma + i\omega$ and σ is chosen larger than the real part of the roots of the denominator (it can be verified that this means $\sigma > 1$). We can also apply Theorem 10 (Section 6–9):

$$x = \frac{1}{2\pi i} \oint_C e^{st} \phi(s) \, ds,$$

where C can be chosen, for example, as the circle $|s| = 5$. If $\phi(s)$ is expanded in a Laurent series at ∞, we obtain a power series for x:

$$\phi(s) = \frac{1}{s^5} - \frac{2}{s^6} + \frac{9}{s^7} + \cdots,$$

$$x = \frac{t^4}{24} - \frac{t^5}{60} + \frac{t^6}{80} + \cdots$$

EXAMPLE 6. $(D^4 + D^3 + 1)x = t \sin t$. We seek a particular solution. Since the right-hand side can be written as

$$\tfrac{1}{2}ite^{-it} - \tfrac{1}{2}ite^{it},$$

we can apply Theorem 17, retaining only the terms in e^{it} and e^{-it}.

$$x = \tfrac{1}{2}ie^{-it}[tY(-i) + Y'(-i)] - \frac{i}{2} \cdot e^{it}[tY(i) + Y'(i)]$$

$$= \operatorname{Re} \{ie^{-it}[tY(-i) + Y'(-i)]\}$$

$$= \operatorname{Re} \left(ie^{-it} \left[t\frac{2-i}{5} - \frac{7+24i}{25} \right] \right)$$

$$= \sin t \left(\frac{2t}{5} - \frac{7}{25} \right) + \cos t \left(\frac{t}{5} + \frac{24}{25} \right).$$

EXAMPLE 7. $(D^4 + 4D^3 + 5D^2 + 3D + 1)x = e^{-t}(2t - 3)$. Again we seek a particular solution. If we proceed as in Example 3 or 6 we find that $Y(-1)$ is undefined. This corresponds to the fact that -1 is a root of $V(s)$:

$$Y(s) = \frac{1}{(s + 1)(s^3 + 3s^2 + 2s + 1)}.$$

We apply Theorem 19, retaining only the terms in e^{-t}. With $B(s) =$

$1/(s^3 + 3s^2 + 2s + 1)$, $q = t^2 - 3t$, we obtain

$$x = e^{-t}[B(-1)(t^2 - 3t) + B'(-1)(2t - 3) + B''(-1)] = e^{-t}(t^2 - t - 1).$$

We summarize the procedures suggested by these examples in the following theorem:

THEOREM 21. *To compute the solution $x = W * f + \mathcal{L}^{-1}[Q/V]$ described in Theorem 20 we have the following alternatives:*

(a) *Expand $Y(s)$ in partial fractions and compute $W(t)$ as the sum of the inverse transforms of the terms of the expansion. If $V(s)$ has only simple roots, then*

$$Y(s) = \sum_{j=1}^{n} \frac{A_j}{s - s_j}, \qquad W(t) = \sum_{j=1}^{n} A_j e^{s_j t} h(t).$$

(b) *Compute $W * f$ from the result of (a). In the case of simple roots,*

$$W * f = \sum_{j=1}^{n} A_j [e^{s_j t} h(t) * f].$$

The individual terms here can be computed by integration or with the aid of Table 6–4.

(c) *If the roots of $V(s)$ are not known, we can in any case write*

$$W(t) = \mathcal{L}^{-1}[Y] = \frac{1}{2\pi i} \oint_C Y(s) e^{st} \, ds,$$

where C is any simple closed path enclosing all roots of $V(s)$. Then $W * f$ is represented as an iterated integral:

$$W * f = \frac{1}{2\pi i} \int_0^t \oint_C f(t - u) e^{su} Y(s) \, ds \, du.$$

(d) *If $\mathcal{L}[f] = \phi(s)$ is absolutely convergent for $\sigma > a$, then*

$$x = \mathcal{L}^{-1}[Y(s)\phi(s)] = \frac{1}{2\pi} (P) \int_{-\infty}^{\infty} Y(s)\phi(s) e^{st} \, d\omega.$$

This can be computed with the aid of Table 6–2. If $\phi(s)$ is rational, $Y(s)\phi(s)$ can be expanded in partial fractions and the inverse computed term by term. If $\phi(s)$ is analytic at ∞, then

$$x = \frac{1}{2\pi i} \oint_C Y(s)\phi(s) e^{st} \, ds,$$

where C is any simple closed path outside of which $Y(s)\phi(s)$ is analytic.

(e) *If $f = e^{bt}p(t)$, where p is a polynomial, $W * f$ can be computed by Theorems 17 through 19. In particular, if b is not a root of $V(s)$, then*

$$e^{bt}\left[Y(b)p(t) + \frac{Y'(b)}{1!}p'(t) + \frac{Y''(b)}{1!}p''(t) + \cdots\right]$$

is a particular solution of Eq. (6–151) for $t \geqq 0$. If b is a root of $V(s)$ of multiplicity m, then

$$e^{bt}\left[B(b)q(t) + \frac{B'(b)}{1!}q'(t) + \cdots\right]$$

is a particular solution, where $B(s) = (s - b)^m Y(s)$, $q = D_0^{-m}p(t)$.

(f) *To compute $\mathcal{L}^{-1}[Q/V]$ one can expand in partial fractions and evaluate the inverse term by term. If $W = \mathcal{L}^{-1}[Y(s)]$ has already been found, one can use Lemma 2 of Section 6–7.*

6–17 The equation for forced vibrations. As a further application of the Laplace transform method we give an analysis of the equation for forced vibrations treated in Section 1–10:

$$(mD^2 + 2qD + k^2)x = A\cos\lambda t. \tag{6–170}$$

We shall assume $q > 0$, the case $q = 0$ (including resonance) being left to Problem 6 below. We assume the initial conditions:

$$x = x_0, \qquad x' = v_0 \quad \text{for } t = 0. \tag{6–171}$$

Here

$$V(s) = ms^2 + 2qs + k^2,$$

$$Y(s) = \frac{1}{ms^2 + 2qs + k^2}, \qquad \mathcal{L}[f] = \frac{As}{s^2 + \lambda^2}.$$

Accordingly,

$$\mathcal{L}[x] = Y(s)\mathcal{L}[f] + \frac{Q}{V}, \qquad Q = mx_0 s + 2qx_0 + mv_0.$$

Now

$$\mathcal{L}^{-1}[Y(s)\mathcal{L}[f]] = \mathcal{L}^{-1}\left[\frac{1}{ms^2 + 2qs + k^2} \cdot \frac{As}{s^2 + \lambda^2}\right]$$

$$= A\frac{d}{dt}\mathcal{L}^{-1}\left[\frac{1}{ms^2 + 2qs + k^2}\frac{1}{s^2 + \lambda^2}\right]$$

$$= \frac{A}{\Delta}\frac{d}{dt}\left\{\mathcal{L}^{-1}\left[\frac{2qms + m(\lambda^2 m - k^2) + 4q^2}{ms^2 + 2qs + k^2}\right]\right.$$

$$\left. + \mathcal{L}^{-1}\left[\frac{-2qs - (\lambda^2 m - k^2)}{s^2 + \lambda^2}\right]\right\} \quad [\Delta = (\lambda^2 m - k^2)^2 + 4q^2\lambda^2],$$

by Entry 16 of Table 6–2 and Lemma 2 of Section 6–7. If we write

$$W(t) = \mathcal{L}^{-1}[Y(s)] = \mathcal{L}^{-1}\left[\frac{1}{ms^2 + 2qs + k^2}\right],$$

$$G(t) = \mathcal{L}^{-1}\left[\frac{1}{s^2 + \lambda^2}\right] = \frac{\sin \lambda t}{\lambda} \quad (G' = \cos \lambda t, \; G'' = -\lambda \sin \lambda t),$$

then we can write

$$x = \frac{A}{\Delta}\left[2qmW'' + \{m(\lambda^2 m - k^2) + 4q^2\}W' - 2qG'' - (\lambda^2 m - k^2)G'\right]$$

$$+ mx_0 W' + (2qx_0 + mv_0)W$$

$$= \frac{2qmA}{\Delta} W'' + \left[(m(\lambda^2 m - k^2) + 4q^2)\frac{A}{\Delta} + mx_0\right]W'$$

$$+ (2qx_0 + mv_0)W + \frac{2qA}{\Delta}\lambda \sin \lambda t - \frac{(\lambda^2 m - k^2)}{\Delta} A \cos \lambda t.$$

The form of $W(t)$ depends on the discriminant of the quadratic:

Case I. $q^2 > mk^2$. Entry 6 of Table 6–2 applies;

$$W = \frac{e^{-\beta t} - e^{-\alpha t}}{\mu}, \quad \left(\alpha = \frac{2q + \mu}{2m}, \; \beta = \frac{2q - \mu}{2m}, \; \mu = 2\sqrt{q^2 - mk^2}\right),$$

$$W' = \frac{-\beta e^{-\beta t} + \alpha e^{-\alpha t}}{\mu}, \qquad W'' = \frac{\beta^2 e^{-\beta t} - \alpha^2 e^{-\alpha t}}{\mu}.$$

Case II. $q^2 = mk^2$. Entry 8 of Table 6–2 applies;

$$W = \frac{t}{m} e^{-(q/m)t}, \qquad W' = \frac{m - tq}{m^2} e^{-(q/m)t},$$

$$W'' = \frac{tq^2 - 2qm}{m^3} e^{-(q/m)t}.$$

Case III. $q^2 < mk^2$. Entry 10 applies, and we find

$$W = \frac{1}{m\beta} e^{-\alpha t} \sin \beta t, \quad \left(\alpha = \frac{q}{m}, \; \beta = \frac{\sqrt{mk^2 - q^2}}{m}\right),$$

$$W' = \frac{1}{m\beta} e^{-\alpha t}(\beta \cos \beta t - \alpha \sin \beta t),$$

$$W'' = \frac{1}{m\beta} e^{-\alpha t}[(\alpha^2 - \beta^2) \sin \beta t - 2\alpha\beta \cos \beta t].$$

Problems

1. For each of the following find a solution for $t \geq 0$ satisfying the given initial conditions:

(a) $(D + 3)x = 0$; $x = 1$ for $t = 0$
(b) $(D^2 + 5D + 6)x = 0$; $x = 1$, $x' = 0$ for $t = 0$
(c) $(D^2 + 9)x = 0$; $x = 0$, $x' = 1$ for $t = 0$
(d) $(D^2 + 9)x = e^{2t}$; $x = 1$, $x' = 1$ for $t = 0$

2. For each of the following find a solution with initial values zero for $t = 0$:

(a) $(D^2 + 4D + 4)x = h(t) \sin t$
(b) $(D^3 + 4D^2 + 5D + 2)x = e^{-3t} h(t)$

3. Find a particular solution for $t \geq 0$ of each of the following:

(a) $(D^2 + 5D + 5) = e^t(t^2 + 2t)$
(b) $(D^2 + 3)x = e^{-2t}(t^3 - 1)$
(c) $(D^4 + 2)x = t^3$
(d) $(D^3 + 4D^2 + 5D + 2)x = e^{-2t}(t^2 + 2t)$

4. For each of the following find a solution with initial values zero in the form of a convolution integral:

(a) $(D^2 + 9)x = e^{-t^2}$
(b) $(D^2 + 6D + 8)x = \log(1 + t^2)$

5. From the definitions (6–156) prove that $x = P_k(t) = \mathcal{L}^{-1}[Q_k/V]$ is the solution of the homogeneous equation such that at $t = 0$ the initial values of x, Dx, ..., $D^{n-1}x$ are 0 except for $D^k x$ which has initial value 1.

6. Let $q = 0$ in (6–170), and obtain the solution with initial conditions (6–171).

Answers

1. (a) e^{-3t} (b) $3e^{-2t} - 2e^{-3t}$ (c) $(\sin 3t)/3$
   (d) $(3e^{2t} + 36 \cos 3t + 11 \sin 3t)/39$

2. (a) $\frac{1}{25}[3 \sin t - 4 \cos t + e^{-2t}(5t + 4)]h(t)$
   (b) $\{e^{-2t} - \frac{1}{4}[e^{-3t} + e^{-t}(3 - 2t)]\} h(t)$

3. (a) $e^t \left(\dfrac{t^2}{11} + \dfrac{8}{121} t - \dfrac{78}{1331} \right)$ (b) $e^{-2t} \left(\dfrac{t^3}{7} + \dfrac{12t^2}{49} + \dfrac{54t}{343} - \dfrac{295}{2401} \right)$

   (c) $\frac{1}{2}t^3$ (d) $\dfrac{e^{-2t}}{3} (t^3 + 9t^2 + 30t + 12)$

4. (a) $\frac{1}{3}\int_0^t e^{-u^2} \sin 3(t - u) \, du$
   (b) $\frac{1}{2}\int_0^t \log(1 + u^2)[e^{-2(t-u)} - e^{-4(t-u)}] \, du$

6. If $k^2 \neq m\lambda^2$, then
   $$x = A(\lambda^2 m - k^2)^{-1}(\cos \gamma t - \cos \lambda t) + x_0 \cos \gamma t + v_0 \gamma^{-1} \sin \gamma t,$$
   $$\gamma = k/\sqrt{m};$$
   if $k^2 = m\lambda^2$, then $x = A(2m\lambda)^{-1}t \sin \lambda t + x_0 \cos \lambda t + v_0\lambda^{-1} \sin \lambda t.$

6–18 Weighting function.   Response to generalized functions. As in Sections 2–8, 4–23, and 5–18, the convolution integral can be interpreted as a weighted average. In fact the interpretation is identical with that of Section 5–18, with the same weighting function $W(t)$; here $f(t)$ is restricted to be zero for negative t. If we assume, as in Section 5–18, that all roots of $V(s)$ have negative real parts, then the total weight $\int_0^\infty W(t)\,dt$ is well defined and

$$\int_0^\infty W(t)\,dt = Y(0) = \frac{1}{a_n}. \tag{6–180}$$

The weighting function itself can be regarded as the output $x(t)$ with initial conditions zero as $t \to 0-$, corresponding to a delta function input. For we have

$$(a_0 D^n + \cdots + a_n)x = \delta(t),$$

and therefore as in Section 1–14, $x = Dy$ where

$$(a_0 D^n + \cdots + a_n)y = h(t) \quad (y(0) = 0,\ y'(0) = 0, \ldots).$$

The transform method yields

$$\mathcal{L}[y] = Y(s) \cdot \mathcal{L}[h] = \frac{Y(s)}{s}.$$

Hence as in Section 6–14,

$$\mathcal{L}[x] = s\mathcal{L}[y] = Y(s),$$

so that

$$x = \mathcal{L}^{-1}[Y(s)] = W(t).$$

If we treat input and output as generalized functions, so that all the rules of Section 6–14 are applicable, then we obtain a simpler approach to the transform method. For all functions are considered to have initial values zero (regarded as limits as $t \to 0-$) and

$$\mathcal{L}[D^k f] = s^k \mathcal{L}[f].$$

Thus the solution of

$$(a_0 D^n + \cdots + a_n)x = f$$

is given by

$$\mathcal{L}[x] = Y(s) \cdot \mathcal{L}[f], \qquad x = W * f.$$

If $f = \delta(t)$, then $\mathcal{L}[f] = 1$ and $\mathcal{L}[x] = Y(s)$, $x = W(t)$ as above. If $f = \delta^{(k)}(t)$, then $\mathcal{L}[f] = s^k$ and

$$\mathcal{L}[x] = s^k Y(s),\ x = W^{(k)}(t).$$

The generalized functions can also be applied to obtain solutions with prescribed initial values as $t \to 0+$. For a first order equation

$$(a_0 D + a_1)x = f$$

with initial value x_0 given, and $f(t)$ piecewise continuous, we consider the modified equation with generalized function input:

$$(a_0 D + a_1)x = f(t) + a_0 x_0 \, \delta(t).$$

Solving as above we find

$$\mathcal{L}[x] = \frac{1}{a_0 s + a_1} \, \mathcal{L}[f] + \frac{a_0 x_0}{a_0 s + a_1},$$

$$x = W * f + x_0 e^{-(a_1/a_0)t} h(t).$$

The first term has initial value zero at $t = 0$; the second has initial value zero as $t \to 0-$, but initial value x_0 as $t \to 0+$. Thus the delta-function input provides a jump in x from 0 to x_0 at $t = 0$.

For a second order equation, terms in $\delta(t)$ and $\delta'(t)$ are needed; as above,

$$(a_0 D^2 + a_1 D + a_2)x = f + (a_0 x_0' + a_1 x_0) \, \delta(t) + a_0 x_0 \, \delta'(t)$$

has a solution $x(t)$ for which x jumps from 0 to x_0, and x' jumps from 0 to x_0' at $t = 0$. For the equation of order n the solution of the homogeneous equation with initial values $x_0, x_0', \ldots, x^{(n-1)}$ as $t \to 0+$ is given by (6–159) as

$$x = x_0 a_0 W^{(n-1)} + (x_0 a_1 + x_0' a_0) W^{(n-2)} + \cdots$$
$$+ (x_0 a_{n-1} + \cdots + x_0^{(n-1)} a_0) W,$$

so that

$$\mathcal{L}[x] = x_0 a_0 s^{n-1} Y(s) + (x_0 a_1 + x_0' a_0) s^{n-2} Y(s) + \cdots$$
$$+ (x_0 a_{n-1} + \cdots + x_0^{(n-1)} a_0) Y(s);$$

hence x is the output corresponding to the generalized function input

$$x_0 a_0 \, \delta^{(n-1)}(t) + (x_0 a_1 + x_0' a_0) \, \delta^{(n-2)}(t) + \cdots$$
$$+ (x_0 a_{n-1} + \cdots + x_0^{(n-1)} a_0) \, \delta(t).$$

The advantage gained by using generalized functions lies not in a shortening of the process of computing particular solutions, but rather in the conceptual simplification: solutions with all possible initial values at $t = 0$ are obtained if we allow generalized functions as inputs and then proceed as if all initial values at $t = 0$ are zero. Thus *the initial conditions are absorbed into the input*. If the complete input is f, the output is simply $W * f$ or (if the transform of f exists) $\mathcal{L}^{-1}[Y(s)\mathcal{L}[f]]$.

Remark. If the system considered is stable, all characteristic roots lie to the left of some line Re $s = a$, $a < 0$, so that $Y(s) = \mathcal{L}[W]$ is analytic for Re $s > a$, and

$$\int_0^\infty |W(t)|\, dt = \int_0^\infty |W(t)|e^{0t}\, dt$$

is finite. If the system is unstable, then $Y(s)$ has a pole for some s_0, Re $s_0 \geq 0$, so that $\int_0^\infty |W(t)|\, dt$ must be infinite. (See Section 2–10.)

6–19 Application of Laplace transforms to simultaneous differential equations. First we imitate the procedures of Section 6–15. The application of generalized functions is considered at the end of the section.

EXAMPLE. We consider the system

$$(2D - 1)x + (D + 13)y = 16e^{2t}, \qquad (2D - 3)x + (-D + 7)y = 0,$$

with initial values $x_0 = 1$, $y_0 = 0$ for $t = 0$. Then

$$2s\mathcal{L}[x] - 2 - \mathcal{L}[x] + s\mathcal{L}[y] + 13\mathcal{L}[y] = \frac{16}{s - 2},$$

$$2s\mathcal{L}[x] - 2 - 3\mathcal{L}[x] - s\mathcal{L}[y] + 7\mathcal{L}[y] = 0;$$

$$(2s - 1)\mathcal{L}[x] + (s + 13)\mathcal{L}[y] = \frac{16}{s - 2} + 2 = \frac{12 + 2s}{s - 2},$$

$$(2s - 3)\mathcal{L}[x] + (-s + 7)\mathcal{L}[y] = 2.$$

If we solve for $\mathcal{L}[x]$, $\mathcal{L}[y]$, we find

$$\mathcal{L}[x] = \begin{vmatrix} \dfrac{12 + 2s}{s - 2} & s + 13 \\[2mm] 2 & -s + 7 \end{vmatrix} \div V, \qquad \mathcal{L}[y] = \begin{vmatrix} 2s - 1 & \dfrac{12 + 2s}{s - 2} \\[2mm] 2s - 3 & 2 \end{vmatrix} \div V,$$

$$V(s) = \begin{vmatrix} 2s - 1 & s + 13 \\ 2s - 3 & -s + 7 \end{vmatrix} = -4s^2 - 8s + 32;$$

$$\mathcal{L}[x] = \frac{s^2 + 5s - 34}{(s + 4)(s - 2)^2}, \qquad \mathcal{L}[y] = \frac{7s - 10}{(s + 4)(s - 2)^2};$$

$$x = (D^2 + 5D - 34)\tfrac{1}{36}[e^{-4t} + (6t - 1)e^{2t}]$$

$$= \tfrac{1}{18}[-19e^{-4t} + (37 - 60t)e^{2t}],$$

$$y = (7D - 10)\tfrac{1}{36}[e^{-4t} + (6t - 1)e^{2t}]$$

$$= \tfrac{1}{18}[-19e^{-4t} + (19 + 12t)e^{2t}].$$

For the same system with general inputs f and g,

$$(2D - 1)x + (D + 13)y = f,$$

$$(2D - 3)x + (-D + 7)y = g,$$

and initial values zero, the same procedure leads to the equations

$$\mathcal{L}[x] = \frac{(-s + 7)\mathcal{L}[f] - (s + 13)\mathcal{L}[g]}{-4s^2 - 8s + 32},$$

$$\mathcal{L}[y] = \frac{(-2s + 3)\mathcal{L}[f] + (2s - 1)\mathcal{L}[g]}{-4s^2 - 8s + 32}.$$

These are of the form

$$\mathcal{L}[x] = Y_{11}(s)\mathcal{L}[f] + Y_{12}(s)\mathcal{L}[g],$$

$$\mathcal{L}[y] = Y_{21}(s)\mathcal{L}[f] + Y_{22}(s)\mathcal{L}[g], \qquad (6\text{--}190)$$

where

$$Y_{11}(s) = \frac{s - 7}{4s^2 + 8s - 32}, \cdots$$

are the elements of the transfer function matrix of the system. Accordingly, the solution is obtained in the form

$$x = W_{11} * f + W_{12} * g,$$

$$y = W_{21} * f + W_{22} * g, \qquad (6\text{--}191)$$

where $W_{11} = \mathcal{L}^{-1}[Y_{11}], \ldots$ Even if $\mathcal{L}[f], \mathcal{L}[g]$ do not exist, we expect x, y to be obtainable by (6–191).

We now justify the procedure described for a general case of two equations:

$$(\alpha_0 D^m + \cdots)x + (\beta_0 D^n + \cdots)y = f,$$

$$(\gamma_0 D^m + \cdots)x + (\eta_0 D^n + \cdots)y = g, \qquad (6\text{--}192)$$

where $m > 0$, $n > 0$, and

$$\kappa = \begin{vmatrix} \alpha_0 & \beta_0 \\ \gamma_0 & \eta_0 \end{vmatrix} \neq 0. \qquad (6\text{--}193)$$

For simplicity we assume all initial values to be zero for $t = 0$ [the general case is actually included in this if we allow f and g to include terms in $\delta(t), \delta'(t), \ldots$].

We write

$$V_{11}(s) = \alpha_0 s^m + \cdots, \qquad V_{12}(s) = \beta_0 s^n + \cdots,$$
$$V_{21}(s) = \gamma_0 s^m + \cdots, \qquad V_{22}(s) = \eta_0 s^n + \cdots, \tag{6–194}$$

$$V(s) = \begin{vmatrix} V_{11}(s) & V_{12}(s) \\ V_{21}(s) & V_{22}(s) \end{vmatrix}, \tag{6–195}$$

$$Y_{11}(s) = \frac{V_{22}(s)}{V(s)}, \qquad Y_{12}(s) = \frac{-V_{12}(s)}{V(s)},$$

$$Y_{21}(s) = \frac{-V_{21}(s)}{V(s)}, \qquad Y_{22}(s) = \frac{V_{11}(s)}{V(s)}. \tag{6–196}$$

Then application of Laplace transforms leads to the equations (6–190) for $\mathcal{L}[x]$, $\mathcal{L}[y]$, and hence leads to (6–191). First we justify (6–191); if $\mathcal{L}[f]$, $\mathcal{L}[g]$ are absolutely convergent for $\sigma > a$, then (6–190) immediately follows.

We can further assume $g = 0$, $f \neq 0$. For the solution (6–191) is the sum of the solutions for given f, with $g = 0$, and for given g, with $f = 0$. Accordingly,

$$x = \mathcal{L}^{-1}\left[\frac{V_{22}}{V}\right] * f = \mathcal{L}^{-1}\left[\frac{\eta_0 s^n + \cdots}{\kappa s^{n+m} + \cdots}\right] * f.$$

As shown in Problem 5 below, we can also write

$$x = (\eta_0 D^n + \cdots)\left(\mathcal{L}^{-1}\left[\frac{1}{\kappa s^{n+m} + \cdots}\right] * f\right).$$

Similarly,

$$y = \mathcal{L}^{-1}\left[\frac{-V_{21}}{V}\right] * f = -(\gamma_0 D^m + \cdots)\left(\mathcal{L}^{-1}\left[\frac{1}{V}\right] * f\right).$$

Now

$$u = \mathcal{L}^{-1}\left[\frac{1}{V}\right] * f$$

is the solution of the differential equation

$$\begin{vmatrix} \alpha_0 D^m + \cdots & \beta_0 D^n + \cdots \\ \gamma_0 D^m + \cdots & \eta_0 D^n + \cdots \end{vmatrix} u = f,$$

with initial values zero. Hence

$$(\alpha_0 D^m + \cdots)(\eta_0 D^n + \cdots)u - (\beta_0 D^n + \cdots)(\gamma_0 D^m + \cdots)u = f,$$
$$(\alpha_0 D^m + \cdots)x + (\beta_0 D^n + \cdots)y = f.$$

Thus the first differential equation is satisfied. Similarly the second differential equation is satisfied. Furthermore x and y are zero for $t = 0$, since they are convolutions. Similarly

$$Dx = \mathcal{L}^{-1}\left[\frac{\eta_0 s^{n+1} + \cdots}{\kappa s^{n+m} + \cdots}\right] * f, \ldots, \quad D^{m-1}x = \mathcal{L}^{-1}\left[\frac{\eta_0 s^{m+n-1} + \cdots}{\kappa s^{n+m} + \cdots}\right] * f$$

are zero for $t = 0$, and $Dy, \ldots D^{n-1}y$ are zero for $t = 0$. Thus the formulas (6–191) are fully justified, as are the expressions (6–190), under the appropriate hypotheses.

If for equations (6–192) the initial values are not assumed to be zero, then application of the Laplace transform leads to equations

$$V_{11}(s)\mathcal{L}[x] + V_{12}(s)\mathcal{L}[y] = \mathcal{L}[f] + Q_1(s),$$
$$V_{21}(s)\mathcal{L}[x] + V_{22}(s)\mathcal{L}[y] = \mathcal{L}[g] + Q_2(s),$$

where $Q_1(s)$, $Q_2(s)$ are polynomials of degree less than max (m, n). Hence $Q_1(s)$, $Q_2(s)$ are Laplace transforms of generalized functions, which are linear combinations of $\delta(t)$, $\delta'(t)$, $\ldots$ Hence imposition of initial conditions (as $t \to 0+$) is equivalent to adding appropriate generalized functions to the inputs and solving for initial values zero as $t \to 0-$.

6–20 Application of Laplace transforms to integrodifferential equations. The method of the preceding section extends to integrodifferential equations without change. The formulas (6–56) are now needed, in addition to the formulas (6–52). The proofs are similar to those of Section 6–19.

EXAMPLE. We consider the system

$$(D + D^{-1})x + (D + 26D^{-1})y = f,$$
$$(D - D^{-1})x + (2D + 10D^{-1})y = g,$$

and assume that all initial values are zero (or equivalently that all initial values are absorbed in f and g). Then application of Laplace transforms leads to the equations

$$(s + s^{-1})\mathcal{L}[x] + (s + 26s^{-1})\mathcal{L}[y] = \mathcal{L}[f],$$

$$(s - s^{-1})\mathcal{L}[x] + (2s + 10s^{-1})\mathcal{L}[y] = \mathcal{L}[g],$$

$$\mathcal{L}[x] = \frac{(2s + 10s^{-1})\mathcal{L}[f] - (s + 26s^{-1})\mathcal{L}[g]}{(s + s^{-1})(2s + 10s^{-1}) - (s - s^{-1})(s + 26s^{-1})}$$

$$= \frac{(2s^3 + 10s)\mathcal{L}[f] - (s^3 + 26s)\mathcal{L}[g]}{s^4 - 13s^2 + 36} = Y_{11}(s)\mathcal{L}[f] + Y_{12}(s)\mathcal{L}[g].$$

Hence

$$x = W_{11}(t) * f + W_{12}(t) * g,$$

where

$$W_{11}(t) = \mathcal{L}^{-1}[Y_{11}] = (2D^3 + 10D)\mathcal{L}^{-1}\left[\frac{1}{s^4 - 13s^2 + 36}\right]$$

$$= (2D^3 + 10D)\tfrac{1}{60}(2e^{3t} - 2e^{-3t} - 3e^{2t} + 3e^{-2t})$$

$$= \tfrac{1}{5}(14e^{3t} + 14e^{-3t} - 9e^{2t} - 9e^{-2t}),$$

$$W_{12}(t) = \mathcal{L}^{-1}[Y_{12}] = -(D^3 + 26D)\tfrac{1}{60}(2e^{3t} - 2e^{-3t} - 3e^{2t} + 3e^{-2t})$$

$$= \tfrac{1}{2}(-7e^{3t} - 7e^{-2t} + 6e^{2t} + 6e^{-2t}).$$

Similar expressions are obtained for y.

6–21 System analysis by means of Laplace transforms. We restrict attention to systems described by a set of linear integrodifferential equations with constant coefficients, with inputs zero for $t < 0$ and with all initial values zero for $t = 0$. We further assume for simplicity that each input has an absolutely convergent Laplace transform for σ sufficiently large.

Then the description of the system can be carried out completely in terms of Laplace transforms. For each input f_j there is a transform $\mathcal{L}[f_j] = \phi_j(s)$; for each output x_k there is a transform $\mathcal{L}[x_k] = \psi_k(s)$. There is a transfer function $Y_{kj}(s)$ relating these, and the total output $\psi_k(s)$ is

$$\psi_k(s) = \sum_j Y_{kj}(s)\phi_j(s).$$

The functions $Y_{kj}(s)$ are rational functions of the complex variable s. The $\phi_j(s)$ are functions of s which lie in the class of Laplace transforms of piecewise continuous functions plus possible linear combinations of delta functions. The $\psi_k(s)$ are similarly restricted.

It is very natural to try to think *in terms of the s language alone.* This can be done if we correlate each $\phi_j(s)$ and $\psi_k(s)$ with a corresponding physical phenomenon. Thus $(s + a)^{-1}$, with a real and > 0, represents *exponential decay;* $A\lambda(s^2 + \lambda^2)^{-1}$ represents a *sinusoidal oscillation* of frequency λ and amplitude A; a rational function is a linear combination (via the partial fraction decomposition) of such functions and their generalizations; $e^{-cs}\phi(s)$, if ϕ is rational, represents a similar function *delayed by time c;* 1 represents a *unit impulse.*

For each such function $\phi(s)$, multiplication by a rational transfer function $Y(s)$ converts ϕ into some other function ψ whose physical meaning depends on that of ϕ and the nature of $Y(s)$. Thus, for any $\phi(s)$, $s^k\phi(s)$

represents the kth derivative ($k = 1, 2, \ldots$), while $s^{-1}\phi(s)$ represents the integral (from 0 to t), and $s^{-k}\phi(s)$, the k-fold integral. If Y is split into partial fractions, multiplication by Y can be interpreted in terms of the basic operation $e^{at}h * f$ of Table 6–4.

Problems

1. Express in terms of the weighting function W the output, for zero initial conditions, corresponding to the following inputs:

(a) $2\,\delta(t) - 3\,\delta'(t)$

(b) $h(t)$

(c) $h(t - 1)$

(d) $th(t)$

(e) $t^n h(t)$ ($n = 1, 2, \ldots$)

(f) $e^{at} = \sum a^n t^n/n!$

2. Show that the solution of $(a_0 D^n + \cdots + a_n)x = 0$ with initial values $x_0^{(j)} = 0$ for $j \ne k$, $x_0^{(k)} = 1$ for $0 \le k < n$, is the same as the solution of $(a_0 D^n + \cdots + a_n)x = a_0\,\delta^{(n-k-1)}(t) + a_1\,\delta^{(n-k-2)}(t) + \cdots + a_{n-k-1}\,\delta(t)$, with initial values zero.

3. For each of the following find the solution for $t \ge 0$ satisfying the given initial conditions:

(a) $(D - 1)x - 2y = t$, $-12x + (D + 1)y = 0$, $x = 0$ and $y = 0$ for $t = 0$

(b) $(D - 1)x + y = 0$, $(D - 1)y + z = 0$, $x + (D - 1)z = 0$, $x = 1$, $y = 1$, $z = 1$ for $t = 0$

(c) $(D^2 - 1)x + y = e^t$, $x + (D^2 - 1)y = 0$, $x = 0$, $y = 0$, $x' = 1$, $y' = 1$ for $t = 0$

(d) $(D - 1)x - 2y = a\,\delta(t)$, $-12x + (D + 1)y = b\,\delta(t)$, a, b constants, $x \to 0$, $y \to 0$ for $t \to 0-$ (What initial values are approached as $t \to 0+$?)

(e) $(3D + 5)x + (2D - 2)y = 1$, $(D + 1)x + (D - D_0^{-1})y = 0$, $x = 0$, $y = 1$ for $t = 0$

(f) $(2D + 1 - 3D^{-1})x + (D + 5 - 6D^{-1})y = 0$, $(D - D^{-1})x + (D + 1 - 2D^{-1})y = 0$, $x = 1$, $y = 2$, $x^{(-1)} = 2$, $y^{(-1)} = 3$ for $t = 0$

(g) $(2D_0^{-1} + 3D_0^{-2})x + (D_0^{-1} + 6D_0^{-2})y = 1 + 2t^3$, $(D_0^{-1} + D_0^{-2})x + (D_0^{-1} + 2D_0^{-2})y = 0$

4. For each of the following find the solution for $t \ge 0$ satisfying the given initial conditions:

(a) $(3D + 5 + 2D^{-1})x = 0$, $x_0 = 1$, $x^{(-1)} = 0$ for $t = 0$

(b) $(D^2 + D + 1 + D_0^{-1})x = e^{-t}h(t)$, $x = 1$, $x' = 0$ for $t = 0$

(c) $2x + D_0^{-1}x = t[h(t) - h(t - 1)]$

(d) $(D + D_0^{-1})x = h(t) + \delta(t)$, $x \to 0$ as $t \to 0-$

5. Prove that if $a_0 \neq 0$ and $n < m$, then

$$\mathcal{L}^{-1}\left[\frac{s^n}{a_0 s^m + a_1 s^{m-1} + \cdots + a_m}\right] * f = D^n\left\{\mathcal{L}^{-1}\left[\frac{1}{a_0 s^m + \cdots + a_m}\right] * f\right\}.$$

[*Hint:* Write the right-hand member as

$$D^n\left\{\mathcal{L}^{-1}\left[\frac{1}{s^n} \cdot Y(s)\right] * f\right\} = D^n\left\{\mathcal{L}^{-1}\left[\frac{1}{s}\right] * \mathcal{L}^{-1}\left[\frac{1}{s}\right] * \cdots * Y(s) * f\right\}$$

$$= D^n\{h(t) * h(t) * \cdots * Y(s) * f\}$$

and show that $h(t) * g = D_0^{-1}g$].

ANSWERS

1. (a) $2W - 3W'$ (b) $D_0^{-1}W$
   (c) $g(t - 1)$, where $g = D_0^{-1}W$ (d) $D_0^{-2}W$
   (e) $n!D_0^{-n-1}W$ (f) $\sum a^n D_0^{-(n+1)}W$
3. (a) $x = \frac{1}{125}(3e^{5t} + 2e^{-5t} - 5t - 5)$, $y = \frac{1}{125}(6e^{5t} - 6e^{-5t} - 60t)$
   (b) $x = y = z = 1$
   (c) $x = -\frac{1}{2}(1 - t) + \frac{1}{8}[(2 + a)e^{at} + (2 - a)e^{-at}]$,
        $y = -\frac{1}{2}(1 - t) - \frac{1}{8}[(2 + a)e^{at} + (2 - a)e^{-at}] + e^t$, $a = \sqrt{2}$
   (d) $x = \frac{1}{5}[(3a + b)e^{5t} + (2a - b)e^{-5t}]$,
        $y = \frac{1}{5}[(6a + 2b)e^{5t} + (3b - 6a)e^{-5t}]$; $t = 0+$, $x \to a$, $y \to b$
   (e) $x = \frac{1}{5} + \frac{1}{2}e^{-t} - \frac{7}{10}e^{-5t}$, $y = \frac{1}{12}(3e^{-t} + 7e^{-5t} + 2e^t)$
   (f) $x = -6 + 7e^t$, $y = 3 - e^t$
   (g) $x = \delta(t) + 2 + 6t^2 + 4t^3$, $y = -(\delta(t) + 1 + 6t^2 + 2t^3)$
4. (a) $(3e^{-t} - 2e^{-2t/3})h(t)$ (b) $(-\frac{1}{2}te^{-t} + \cos t + \frac{1}{2}\sin t)h(t)$
   (c) $(1 - e^{-(1/2)t})[h(t) - h(t - 1)] + (\frac{1}{2}e^{(1-t)/2} - e^{-(1/2)t})h(t - 1)$
   (d) $(\sin t + \cos t)h(t)$

6-22 The z-transform. In many systems of importance, input and output functions are considered at discrete values of t, usually at values nT, $n = 0, 1, 2, \ldots$, where T is a fixed positive number. Such systems are called *sampled-data* systems. The term *sampling* refers to replacement of a function $f(t)$ by the function $f(nT)$, as suggested in Fig. 6-7.

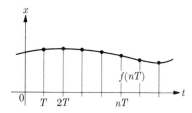

FIG. 6-7. Sampling.

The study of such systems is simplified by means of the *z-transform*, to be described here. The *z*-transform also finds application in the study of *difference equations*, as illustrated below.

DEFINITION. Let T be a fixed positive number. Let $f(t)$ be defined for $t \geq 0$. Then the *z*-transform of $f(t)$ is the function

$$Z[f] = F(z) = \sum_{n=0}^{\infty} f(nT)z^{-n} \tag{6-220}$$

of the complex variable z.

Since only the values c_n of f at nT are used, the *z*-transform is actually defined for such sequences $\{c_n\}$:

$$Z[\{c_n\}] = \sum_{n=0}^{\infty} c_n z^{-n} = F(z). \tag{6-221}$$

The function F is the sum of a power series in $1/z$ and hence the study of the *z*-transform is essentially the study of the relation between the sum of a power series and its coefficients. From the theory of power series (Section 3–8), we know that the series has a *radius of convergence* ρ such that the series converges for $|1/z| < \rho$ (that is, $|z| > 1/\rho$) and diverges for $|z| < \rho$. It may happen that $\rho = 0$, in which case $F(z)$ is defined only for $1/z = 0$, that is, for $z = \infty$. Actually, much of the theory has no relation to convergence questions, and we can always consider the series $\sum c_n z^{-n}$ as a "formal series," to be manipulated in certain ways and not necessarily to be summed.

The function $f(t)$ of the definition above need satisfy no continuity requirement. However, ambiguity can arise if $f(t)$ has a jump discontinuity at a value nT. We shall always interpret $f(nT)$ as the limit of $f(t)$ as $t \to nT+$, and shall assume existence of this limit, for $n = 0, 1, 2, \ldots$, for all $f(t)$ considered.

Relation to Laplace transform. The function $f(nT)$ of Fig. 6–7 suggests a train of pulses $f(0)\,\delta(t) + f(T)\,\delta(t-T) + \cdots + f(nT)\,\delta(t-nT) + \cdots$ The Laplace transform of such a generalized function is

$$\sum_{n=0}^{\infty} f(nT)e^{-nTs}.$$

If we set $z = e^{sT}$, so that $s = T^{-1} \log z$, this becomes $\sum f(nT)z^{-n}$, the *z*-transform of f. We can write

$$f^*(t) = \sum_{n=0}^{\infty} f(nT)\,\delta(t-nT), \qquad F^*(s) = \mathcal{L}[f^*] = \sum_{n=0}^{\infty} f(nT)e^{-nTs},$$

$$F(z) = Z[f] = F^*(s)|_{s=T^{-1}\log z}. \tag{6-222}$$

It should be remarked that the sampled function $f(nT)$ of Fig. 6–7 can be considered as a sum of rectangular pulses of area $\epsilon f(nT)$, where ϵ is the width of the rectangle. Such a sum approximates $\sum \epsilon f(nT)\, \delta(t - nT)$, since $\delta(t - nT)$ should be considered as the limit of a pulse of *unit area*. Hence a scale factor $1/\epsilon$ is needed to convert the sampled function $f(nT)$ into the train of pulses $f^*(t)$.

Examples of z-transforms. 1. If $f(t) = h(t) = 1$, then

$$F(z) = 1 + \frac{1}{z} + \cdots + \frac{1}{z^n} + \cdots = \frac{1}{1 - (1/z)} = \frac{z}{z - 1} \quad (|z| > 1).$$

2. If $f(t) = a^t$, $a > 0$, then

$$F(z) = \sum_{n=0}^{\infty} a^{nT} z^{-n}$$

$$= \sum_{n=0}^{\infty} (a^T/z)^n$$

$$= \frac{1}{1 - (a^T/z)} = \frac{z}{z - a^T} \quad (|z| > a^T).$$

3. If $f(t) = \left[\Gamma\left(\frac{t}{T} + 1\right)\right]^{-1}$, so that $f(nT) = [\Gamma(n + 1)]^{-1} = 1/n!$, then

$$F(z) = \sum_{n=0}^{\infty} \frac{1}{n!} \frac{1}{z^n} = e^{1/z} \quad (|z| > 0).$$

Other transforms are listed in Table 6–5.

Linearity of the z-transform. If we consider $F(z)$ always as a formal power series, then $Z[f]$ is defined in the linear space of all functions $f(t)$ $(t \geqq 0)$ and is a *linear operator* in that space:

$$Z[c_1 f_1 + c_2 f_2] = \sum_{n=0}^{\infty} [c_1 f_1(nT) + c_2 f_2(nT)]z^{-n}$$

$$= c_1 \sum_{n=0}^{\infty} f_1(nT)z^{-n} + c_2 \sum_{n=0}^{\infty} f_2(nT)z^{-n}$$

$$= c_1 Z[f_1] + c_2 Z[f_2].$$

If we restrict to functions $f(t)$ such that $F(z)$ converges for $|z| > R$ for some finite R (depending on f), then again we obtain a linear space and Z is again a linear operator.

TABLE 6–5

z-TRANSFORMS*

$$Z[f] = \sum_{n=0}^{\infty} f(nT)z^{-n} = F(z), \quad |z| > R = \frac{1}{\rho}$$

No.	$f(t), \quad t = 0, T, 2T, \ldots$	$F(z)$	R
1	1	$\dfrac{z}{z-1}$	1
2	t	$\dfrac{zT}{(z-1)^2}$	1
3	t^2	$\dfrac{T^2 z(z+1)}{(z-1)^3}$	1
4	e^{ct}	$\dfrac{z}{z - e^{cT}}$	$|e^{cT}|$
5	$t^k e^{ct}$	$\dfrac{\partial^k}{\partial c^k} \dfrac{z}{z - e^{cT}}$	$|e^{cT}|$
6	$\cos bt$	$\dfrac{z(z - \cos bT)}{z^2 - 2z \cos bT + 1}$	1
7	$\sin bt$	$\dfrac{z \sin bT}{z^2 - 2z \cos bT + 1}$	1
8	$e^{ct} \cos bt$	$\dfrac{z(z - e^{cT} \cos bT)}{z^2 - 2z e^{cT} \cos bT + e^{2cT}}$	e^{cT}
9	$e^{ct} \sin bt$	$\dfrac{z e^{cT} \sin bT}{z^2 - 2z e^{cT} \cos bT + e^{2cT}}$	e^{cT}
10	$\cosh bt$	$\dfrac{z(z - \cosh bT)}{z^2 - 2z \cosh bT + 1}$	$e^{|b|T}$
11	$\sinh bt$	$\dfrac{z \sinh bT}{z^2 - 2z \cosh bT + 1}$	$e^{|b|T}$
12	$h(t - T)$	$\dfrac{1}{z-1}$	1
13	$h(t) - h(t - T)$	1	0

* *Note.* In Entries 6 through 11 the formulas are valid for complex b and c, but the range of z is given only for real b and c. In Entries 15 through 17, a is assumed real and $\neq 0$; however, the transforms remain valid for complex a (not 0), if we choose a definite value for the exponential a^z, for example, $\exp(z \log a)$ as in Section 3–5.

TABLE 6–5 *Continued*

$$Z[f] = \sum_{n=0}^{\infty} f(nT)z^{-n} = F(z), \quad |z| > R = \frac{1}{\rho}$$

No.	$f(t), \quad t = 0, T, 2T, \ldots$	$F(z)$	R
14	$h(t - kT), \quad k = 1, 2, \ldots$	$\dfrac{1}{z^{k-1}(z - 1)}$	1
15	$a^{(t-T)/T}h(t - T)$	$\dfrac{1}{z - a}$	$|a|$
16	$\dfrac{1}{T}(t - T)a^{(t-2T)/T}h(t - T)$	$\dfrac{1}{(z - a)^2}$	$|a|$
17	$\dfrac{(t - T)(t - 2T)\cdots(t - kT + T)}{(k - 1)!T^{k-1}}$ $\times\, a^{(t-kT)/T}h(t - T)$	$\dfrac{1}{(z - a)^k}$	$|a|$
18	$h(t - T) - h(t - 2T)$	$\dfrac{1}{z}$	0
19	$h(t - kT) - h[t - (k + 1)T]$	$\dfrac{1}{z^k}$	0
20	$\dfrac{1}{\Gamma\left(\dfrac{t}{T} + 1\right)}$	$e^{1/z}$	0
21	$\dfrac{T}{t}h(t - T)$	$\log\dfrac{z}{z - 1} = \displaystyle\sum_{n=1}^{\infty}\dfrac{z^{-n}}{n}$	1
22	$f(nT) = \dfrac{k(k - 1)\cdots(k - n + 1)}{n!}$	$\left(1 + \dfrac{1}{z}\right)^k$	1
23	$\dfrac{(t + T)(t + 2T)\cdots(t + kT - T)}{(k - 1)!T^{k-1}}$	$\left(1 - \dfrac{1}{z}\right)^{-k}, \quad k = 2, 3, \ldots$	1
24	f has period NT, $f = t$ for $0 \leqq t < NT$	$\dfrac{Tz}{z - 1}\left(\dfrac{1}{z - 1} - \dfrac{N}{z^N - 1}\right)$	1
25	f has period NT, $f = 1$ for $0 \leqq t \leqq MT, \quad M < N$	$\dfrac{z^{M+1} - 1}{z^{M-N}(z - 1)(z^N - 1)}$	1
26	f has period NT, $f_1(t) = f(t)[1 - h(t - NT)]$	$\dfrac{z^N}{z^N - 1}F_1(z), \quad F_1 = Z[f_1]$	1

It is clear from the definition that $Z[f_1] = Z[f_2]$ whenever f_1 and f_2 agree at the values nT ($n = 0, 1, 2, \ldots$), so that Z is not a one-to-one transformation. For example, if $T = \pi$, then $f_1(t) = t$ and $f_2(t) = t + \sin t$ have the same z-transform. However, if $F(z)$ is given, either as an analytic function for $|z| > R$ or as a formal power series, then the coefficients c_n are uniquely determined. Thus Z is a one-to-one transformation between the sequences $\{c_n\}$ and the functions $F(z)$. Hence given $F(z)$, we can always determine $f(nT)$ (but we have no information on $f(t)$ for other values of t).

Inversion. The problem of finding $f(nT)$ from $F(z)$ is that of evaluating the inverse z-transform $Z^{-1}[F]$. When F is given as a power series $\sum c_n z^{-n}$, we can read off the result: $f(nT) = c_n$. When F is given as a function analytic for $|z| > R$ (and at $z = \infty$), we need only expand $F(z)$ in such a series or, equivalently, expand $F(1/z)$ in its Taylor series in z.

EXAMPLE 4. Let

$$F(z) = \frac{z + 3}{z - 2} \quad \text{for } |z| > 2.$$

Then

$$F\left(\frac{1}{z}\right) = \frac{1 + 3z}{1 - 2z} = (1 + 3z)(1 + 2z + \cdots + 2^n z^n + \cdots)$$

$$= 1 + 5z + 10z^2 + \cdots + 5 \cdot 2^{n-1} z^n + \cdots,$$

$$F(z) = 1 + \frac{5}{z} + \frac{10}{z^2} + \cdots + \frac{5 \cdot 2^{n-1}}{z^n} + \cdots,$$

so that

$$f(0) = 1, f(T) = 5, \ldots, f(nT) = 5 \cdot 2^{n-1} \quad (n = 1, 2, \ldots).$$

We can also represent the coefficients as complex integrals. Since $F(z)$ can be regarded as a Laurent series (Section 3–10),

$$c_n = \frac{1}{2\pi i} \oint_C F(z) z^{n-1} \, dz \quad (n = 0, 1, 2, \ldots), \tag{6–223}$$

where C is a circle $|z| = R_0$, $R_0 > R$, or any simple closed path on and outside of which F is analytic.

The familiar Maclaurin series for a function of z analytic at $z = 0$ becomes, on replacement of z by $1/z$, a series $\sum c_n z^{-n}$ for a function $F(z)$ having inverse z-transform $\{c_n\}$. For example, the binomial series

$$(1 + z)^k$$
$$= 1 + kz + \frac{k(k-1)}{2!} z^2 + \cdots + \frac{k(k-1) \cdots (k - n + 1)}{n!} z^n + \cdots,$$

valid for $|z| < 1$, yields

$$F(z) = \left(1 + \frac{1}{z}\right)^k = 1 + \frac{k}{z} + \frac{k(k-1)}{2!}\frac{1}{z^2} + \cdots$$

$$+ \frac{k(k-1)\cdots(k-n+1)}{n!}\frac{1}{z^n} + \cdots,$$

so that

$$Z^{-1}\left[\left(1 + \frac{1}{z}\right)^k\right] = \left\{\frac{k(k-1)\cdots(k-n+1)}{n!}\right\}.$$

If k is a positive integer, $c_n = 0$ for $n > k$. Figure 6–8 shows a corresponding $f(t)$ for $k = 7$. It is well known that, as $k \to \infty$, the shape of the graph approaches that of the error function (see Reference 7, pp. 170–171).

If $F(z)$ is a rational function of z, analytic at ∞, then $F(1/z)$ is also a rational function, analytic at $z = 0$. We can expand $F(1/z)$ in partial fractions and then use the expansion

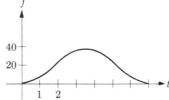

$$\frac{1}{(z-a)^m} = (-a)^{-m}\left(1 - \frac{z}{a}\right)^{-m}$$

FIG. 6–8. Function whose z-transform is $[1 + (1/z)]^7$.

$$= (-a)^{-m}\left(1 + \frac{mz}{a} + \frac{m(m+1)}{2!}\frac{z^2}{a^2} + \cdots\right),$$

obtainable from the binomial expansion given above, or we can use Entries 15 through 17 of Table 6–5.

EXAMPLE 5. Let

$$F(z) = \frac{z^3 - 9z^2 + 5z - 1}{4z^3 - 8z^2 + 5z - 1}.$$

Then

$$F\left(\frac{1}{z}\right) = \frac{z^3 - 5z^2 + 9z - 1}{z^3 - 5z^2 + 8z - 4} = 1 + \frac{z+3}{(z-1)(z-2)^2}$$

$$= 1 + \frac{4}{z-1} - \frac{4}{z-2} + \frac{5}{(z-2)^2}$$

$$= 1 - 4(1 + z + z^2 + \cdots + z^n + \cdots)$$

$$+ 2\left(1 + \frac{z}{2} + \cdots + \frac{z^n}{2^n} + \cdots\right)$$

$$+ \frac{5}{4}\left(1 + z + \cdots + \frac{(n+1)z^n}{2^n} + \cdots\right).$$

Thus, if $Z[f] = F$, then

$$f(0) = \tfrac{1}{4}, \qquad f(nT) = -4 + 2^{-n+1} + 5(n+1)2^{-n-2} \quad (n > 0).$$

Alternatively, by Entries 12, 13, 15, and 16,

$$F(z) = \frac{1}{4} - \frac{4}{z-1} + \frac{9/4}{z-(1/2)} + \frac{5/16}{[z-(1/2)]^2}$$

$$= \tfrac{1}{4}Z[h(t) - h(t - T)] - 4Z[h(t - T)]$$

$$+ \tfrac{9}{4}Z[(\tfrac{1}{2})^{(t-T)/T}h(t - T)] + \tfrac{5}{16}Z\left[\frac{1}{T}(t - T)(\tfrac{1}{2})^{(t-2T)/T}h(t - T)\right].$$

Hence

$$f(t) = \tfrac{1}{4}[h(t) - h(t - T)] + \left(\frac{13T + 5t}{4T}2^{-t/T} - 4\right)h(t - T),$$

in agreement with the previous answer at $t = nT$. Finally, we can use long division:

$$
\begin{array}{r}
\frac{1}{4} - \frac{7}{4z} - \frac{41}{16z^2}, \quad \text{etc.} \\
4z^3 - 8z^2 + 5z - 1 \ \overline{)z^3 - 9z^2 + 5z \quad - 1} \\
z^3 - 2z^2 + \frac{5}{4}z \quad - \frac{1}{4} \\
\hline
-7z^2 + \frac{15}{4}z - \frac{3}{4} \\
-7z^2 + 14z - \frac{35}{4} + \frac{7}{4z} \\
\hline
-\frac{41}{4}z + 8 \quad - \frac{7}{4z}
\end{array}
$$

We read off: $f(0) = \tfrac{1}{4}$, $f(T) = -\tfrac{7}{4}$, $f(2T) = -\tfrac{41}{16}$, ... This last method provides no general formula, but it does eliminate the necessity of finding the roots of a polynomial and expansion in partial fractions.

Translation. Since only the values of f at multiples of T are concerned, we consider only translations through mT $(m = 1, 2, \ldots)$ and assume $f(t) = 0$ for $t < 0$. We find

$$Z[f(t - mT)] = z^{-m}Z[f],$$

$$Z[f(t + mT)] = z^m Z[f] - \{z^m f(0) + z^{m-1}f(T) \qquad (6\text{-}224)$$

$$+ \cdots + zf[(m - 1)T]\}$$

(Problem 5 following Section 6–24).

Multiplication by e^{at} and by t. Let $Z[f] = F = \sum c_n z^{-n}$. Then

$$Z[e^{at}f(t)] = F(e^{-aT}z),$$

$$Z[tf(t)] = -zTF'(z),$$

(6–225)

where $F'(z) = \sum (-n)c_n z^{-n-1}$ (Problem 6 below).

EXAMPLE 6. Let $f(t) = 1$, so that $F(z) = z/(z - 1)$. Then

$$Z[t] = Z[tf(t)] = -zTF'(z) = zT(z - 1)^{-2}.$$

Convolution. We define

$$f_1 *_T f_2 = \sum_{k=0}^{n} f_1(kT)f_2[(n - k)T]$$

(6–226)

to be the z-convolution of $f_1(t), f_2(t)$. The convolution depends only on the values $c_n = f_1(nT)$, $d_n = f_2(nT)$ and is itself a sequence $\{e_n\}$:

$$e_n = \sum_{k=0}^{n} c_k d_{n-k}.$$

We verify (Problem 7 following Section 6–24) that

$$Z[f_1 *_T f_2] = Z[f_1] \cdot Z[f_2].$$

(6–227)

Transform of product. If $Z[f] = F(z) = \sum c_n z^{-n}$ and $Z[g] = G(z) = \sum d_n z^{-n}$, then $f(t)g(t)$ has value $c_n d_n$ at nT, so that $Z[fg] = \sum c_n d_n z^{-n} = H(z)$. It is of interest to note that when F and G are analytic for $|z| > R_0$, the function $H(z)$ is obtainable from F and G as follows:

$$H(z) = \frac{1}{2\pi i} \oint_{|\zeta|=R_1} F(\zeta)G\left(\frac{z}{\zeta}\right) \frac{d\zeta}{\zeta}$$

for $|z| > R_0 R_1$, where $R_1 > R_0$. (The path of integration can also be chosen as a simple closed curve C). To verify this relation, we write

$$F(\zeta) = \sum_{n=0}^{\infty} c_n \zeta^{-n}, \qquad G\left(\frac{z}{\zeta}\right) = \sum_{m=0}^{\infty} d_m z^{-m} \zeta^{m},$$

$$F(\zeta)G\left(\frac{z}{\zeta}\right) = \sum_{n=0}^{\infty} \sum_{m=0}^{\infty} c_n d_m z^{-m} \zeta^{m-n}.$$

The two power series are absolutely and uniformly convergent on the circle $|\zeta| = R_1$. It follows that the double series for the product can be arranged in any order to form a simple series and can be integrated term by term on the circle $|\zeta| = R_1$. But

$$\frac{1}{2\pi i} \oint_{|\zeta|=R_1} \zeta^{m-n} \frac{d\zeta}{\zeta} = \begin{cases} 0 & (m \neq n) \\ 1 & (m = n). \end{cases}$$

Hence only the terms for which $m = n$ give a contribution, and we find as asserted that

$$\frac{1}{2\pi i} \oint_{|\zeta|=R_1} F(\zeta)G\left(\frac{z}{\zeta}\right) \frac{d\zeta}{\zeta} = \sum_{n=0}^{\infty} c_n d_n z^{-n} = H(z).$$

EXAMPLE 7. By Example 6, $Z[t] = zT(z-1)^{-2}$, $|z| > 1$. Hence

$$Z[t^2] = Z[t \cdot t] = \frac{1}{2\pi i} \oint \frac{\zeta T}{(\zeta-1)^2} \frac{zT/\zeta}{\left(\frac{z}{\zeta}-1\right)^2} \frac{d\zeta}{\zeta}$$

$$= \frac{T^2}{2\pi i} \oint \frac{z\zeta}{(\zeta-1)^2(\zeta-z)^2} d\zeta,$$

where z is outside the path of integration and $\zeta = 1$ is inside. Hence the integral reduces to a residue at $\zeta = 1$:

$$Z[t^2] = T^2 z \operatorname{Res}[\zeta(\zeta-1)^{-2}(\zeta-z)^{-2}, \zeta = 1] = T^2 \frac{z(z+1)}{(z-1)^3}.$$

Initial and final values. Since $f(0)$ is the constant term in the series for $Z[f] = F(z)$, we conclude at once that

$$f(0) = \lim_{z\to\infty} F(z). \tag{6-228}$$

Thus the behavior of $f(t)$ at $t = 0$ is related to that of $F(z)$ at $z = \infty$. However, the behavior of $f(t)$ at ∞ is related to that of $F(z)$ at $z = 1$. In particular, if $f(t) \to A$ as $t \to +\infty$, then $F(z)$ is analytic for $|z| > 1$ and

$$\lim_{z\to 1+} (z-1)F(z) = A, \tag{6-229}$$

where $z \to 1+$ through real values. To prove this, we write $f(nT) = A + b_n$, where $b_n \to 0$ as $n \to \infty$. Then

$$F(z) = \sum_{n=0}^{\infty} \frac{A}{z^n} + \sum_{n=0}^{\infty} \frac{b_n}{z^n} = \frac{Az}{z-1} + \sum_{n=0}^{\infty} \frac{b_n}{z^n}.$$

The series $\sum A z^{-n}$ is a geometric series, converging for $|z| > 1$. Since $b_n \to 0$, the second series converges for $|z| > 1$ by comparison with a geometric series Br^n $(r = 1/|z|)$. Hence $F(z)$ is analytic for $|z| > 1$. Since

$$\lim_{z \to 1+} (z - 1) \frac{Az}{z - 1} = A,$$

we need only show that

$$\lim_{z \to 1+} (z - 1) \sum_{n=0}^{\infty} \frac{b_n}{z^n} = 0.$$

Given $\epsilon > 0$, we choose $N > 1$ and so large that $|b_n| < \frac{1}{2}\epsilon$ for $n \geq N$. Then for z real and > 1,

$$\left| (z - 1) \sum_{n=N}^{\infty} \frac{b_n}{z^n} \right| \leq (z - 1) \frac{\epsilon}{2} \sum_{n=N}^{\infty} \frac{1}{z^n}$$

$$= (z - 1) \frac{\epsilon}{2} \cdot \frac{1}{z^N} \cdot \frac{z}{z - 1} \leq \frac{\epsilon}{2z^{N-1}} < \frac{\epsilon}{2}.$$

Furthermore, since $(z - 1)(b_0 + b_1 z^{-1} + \cdots + b_{N-1} z^{-(N-1)})$ is continuous and equal to zero at $z = 1$,

$$\left| (z - 1) \sum_{n=0}^{N-1} \frac{b_n}{z^n} \right| < \frac{\epsilon}{2} \quad \text{for } |z - 1| < \delta,$$

for δ sufficiently small. Hence for z real, $z > 1$, $|z - 1| < \delta$,

$$\left| (z - 1) \sum_{n=0}^{\infty} \frac{b_n}{z^n} \right| \leq \left| (z - 1) \sum_{n=0}^{N-1} \frac{b_n}{z^n} \right|$$

$$+ \left| (z - 1) \sum_{n=N}^{\infty} \frac{b_n}{z^n} \right| < \frac{\epsilon}{2} + \frac{\epsilon}{2} = \epsilon.$$

Therefore (6–229) is proved.

6–23 Application of the z-transform to difference equations. We consider here difference equations relating the values of functions defined only at the discrete values $t = nT$ $(n = 0, 1, 2, \ldots)$. By a *linear difference equation* we mean an equation

$$a_0 x(t + NT) + a_1 x[t + (N - 1)T] + \cdots + a_N x(t) = f(t), \quad (6\text{–}230)$$

where t is restricted to the values nT. The coefficients a_k may depend on t, but here we restrict attention to the case of *constant* coefficients a_k. Furthermore we assume $a_0 \neq 0$, $a_N \neq 0$, $N \geq 1$. Equation (6–230) can be considered as a *recursion formula*, defining the value of x at each t in terms of the value of f at t and of x at the N preceding values of t. Thus, given

$x(0)$, $x(T)$, ..., $x[(N-1)T]$, Eq. (6–230) permits us to calculate successively $x(NT)$, $x[(N+1)T]$, ...

The z-transform is generally applicable to Eq. (6–230), and for simple choices of $f(t)$ yields a useful form of the solution. We illustrate the procedure.

EXAMPLE 1. $x(t+2) + 3x(t+1) + 2x(t) = 0$, $x(0) = 1$, $x(1) = 2$. Application of the z-transform and the second translation formula (6–224), with $T = 1$ and $X(z) = Z[x]$, gives

$$z^2 X(z) - (z^2 + 2z) + 3[zX(z) - z] + 2X(z) = 0,$$

$$X(z) = \frac{z^2 + 5z}{z^2 + 3z + 2} = 1 + \frac{6}{z+2} - \frac{4}{z+1},$$

$$x(t) = h(t) - h(t-1) + [6(-2)^{(t-1)} - 4(-1)^{(t-1)}]h(t-1).$$

We verify that for all $t = 0, 1, 2, \ldots$ we can write

$$x = 4(-1)^t - 3(-2)^t,$$

so that $x(0) = 1$, $x(1) = 2$, $x(2) = -8$, ...

EXAMPLE 2. $x(t+2) + 3x(t+1) + 2x(t) = t$; $x(0) = c_1$, $x(1) = c_2$ unspecified. Applying the z-transform, with $T = 1$, we obtain

$$(z^2 + 3z + 2)X(z) - c_1(z^2 + 3z) - c_2 z = z(z-1)^{-2},$$

$$X(z) = \frac{1}{z^2 + 3z + 2} \frac{z}{(z-1)^2} + \frac{c_1(z^2 + 3z)}{z^2 + 3z + 2} + \frac{c_2 z}{z^2 + 3z + 2}.$$

If we take inverse transforms, we obtain the solution in the form

$$x = x_p(t) + c_1 x_1(t) + c_2 x_2(t),$$

where $x_p(t)$ is a particular solution and $c_1 x_1(t) + c_2 x_2(t)$ is the general solution of the related homogeneous equation (obtained by replacing the forcing function t by zero).

To evaluate the particular solution

$$x_p(t) = Z^{-1}\left[\frac{1}{z^2 + 3z + 2} \frac{z}{(z-1)^2}\right],$$

we can introduce a weighting function

$$w(t) = Z^{-1}\left[\frac{1}{z^2 + 3z + 2}\right].$$

Then

$$Z[x_p] = Z[w] \cdot Z[t]$$

so that, for $t = n$,

$$x_p(t) = w(t) *_T t = \sum_{k=0}^{n} w(k)(n - k).$$

We find, as in Example 1, $w(t) = (-1)^t(2^{t-1} - 1)h(t - 1)$, so that

$$x_p(n) = \sum_{k=1}^{n} (-1)^k(2^{k-1} - 1)(n - k).$$

We could also expand $Z[x_p]$ in partial fractions and get the result in a simpler form (without a summation). We can also write

$$Z[x_p] = Z[w] \cdot Z[t]$$

$$= \sum_{n=1}^{\infty} (-1)^n(2^{n-1} - 1)z^{-n} \sum_{n=1}^{\infty} z^{-n}$$

$$= \left(\frac{1}{z^2} - \frac{3}{z^3} + \frac{7}{z^4} + \cdots\right)\left(\frac{1}{z} + \frac{1}{z^2} + \frac{1}{z^3} + \cdots\right)$$

$$= \frac{1}{z^3} - \frac{1}{z^4} + \frac{4}{z^5} + \cdots,$$

so that $x_p(0) = 0$, $x_p(1) = 0$, $x_p(2) = 0$, $x_p(3) = 1$, $x_p(4) = -1$, $x_p(5) = 4, \ldots$

To find the "complementary function," we can write

$$Z[x_1(t)] = \frac{z^2 + 3z}{z^2 + 3z + 2} = \frac{z^2}{z^2 + 3z + 2} + 3\frac{z}{z^2 + 3z + 2}$$

$$= Z[w(t + 2) + 3w(t + 1)],$$

$$x_1(t) = w(t + 2) + 3w(t + 1),$$

$$Z[x_2(t)] = \frac{z}{z^2 + 3z + 2} = Z[w(t + 1)],$$

$$x_2(t) = w(t + 1),$$

by the translation rule (6–224), since $w(0) = 0$, $w(1) = 0$. We note that $w(t)$ is an "impulse" response; that is, the solution of the equation

$$x(n + 2) + 3x(n + 1) + 2x(n) = \delta_{0n},$$

where δ_{0n} (Kronecker delta) is 1 for $n = 0$, 0 otherwise, and $w(0) = w(1) = 0$.

The procedures illustrated can easily be generalized to Eq. (6–230). The general solution has the form

$$x = x_p(t) + c_1 x_1(t) + \cdots + c_N x_N(t) \quad (t = 0, T, 2T, \ldots),$$

where $x_p(t)$ is a particular solution and $c_1 x_1(t) + \cdots + c_N x_N(t)$ is the general solution of the related homogeneous equation. We can choose

$$x_p(t) = w(t) *_T f(t),$$

where $w(t)$ is the weighting function or impulse response,

$$w(t) = Z^{-1}\left[\frac{1}{a_0 z^N + \cdots + a_N}\right],$$

and $w(0) = w(T) = \cdots = w[(N-1)T] = 0$; equivalently,

$$x_p(t) = Z^{-1}\left[\frac{F(z)}{a_0 z^N + \cdots + a_N}\right],$$

where $F(z) = Z[f]$. The arbitrary constants $c_1, \ldots, c_N$, can be chosen as the initial values $x(0), x(T), \ldots, x[(N-1)T]$. If the "characteristic equation"

$$a_0 z^N + \cdots + a_N = 0$$

has distinct roots $z_1, \ldots, z_N$, we can alternatively choose

$$x_1(t) = z_1^t, \ldots, \qquad x_N(t) = z_N^t.$$

6–24 Sampled-data systems. The main unit in a sampled-data system is a component which transforms an input $f(t)$ into a train of pulses which somehow represent the values of f at the sampling instants $0, T, 2T, \ldots$ The description is made definite by postulating that the sampler converts $f(t)$ into

$$f^*(t) = \sum_{n=0}^{\infty} f(nT)\, \delta(t - nT).$$

In practice, we can only approximate $f^*(t)$ by, for example, a sequence of rectangular pulses of area $f(nT)$. As in Section 6–22, we interpret $f(nT)$ as $\lim f(t)$, as $t \to nT+$, if nT is a discontinuity point of f. For example, if $f(t) = h(t) - 2h(t-1) + 3h(t-2)$, then for $T = 1$

$$f^*(t) = \delta(t) - \delta(t-1) + 2\,\delta(t-2) + \cdots$$

We call a function $f^*(t)$, or any function $\sum c_n \delta(t - nT)$, a *sampled function*.

Let $f(t)$ have Laplace transform $\mathcal{L}[f] = F(s)$. Then $f^*(t)$ has Laplace transform

$$F^*(s) = \sum_{k=0}^{\infty} f(nT)e^{-nTs} = Z[f]\big|_{z=e^{sT}}.$$

Given $F(s)$, we can consider $F^*(s)$ as a new function of s, obtained by applying the Laplace transform to the sampled inverse Laplace transform of $F(s)$:

$$F^*(s) = \mathcal{L}[\{\mathcal{L}^{-1}[F]\}^*] = Z[\mathcal{L}^{-1}[F]]\big|_{z=e^{sT}}.$$

It should be noted that $F(s)$ and $F^*(s)$ are defined (if at all) in half-planes $\sigma > a,\ \sigma > a^*$, which are not necessarily the same. In particular, $F(s)$ may be a Laplace transform, without $F^*(s)$ being defined for any s; conversely, $f(t)$ may have no Laplace transform $F(s)$, even though $f^*(t)$ has a transform $F^*(s)$. In the following, we shall tacitly assume that all the transforms do exist.

The star operation is itself a linear operator [either on functions $f(t)$ or on functions $F(s)$]. It has the following basic property:

$$[F(s)G^*(s)]^* = F^*(s)G^*(s). \tag{6–240}$$

To prove this, we evaluate the left and right sides separately. If $F(s) = \mathcal{L}[f]$, $G^*(s) = \mathcal{L}[g^*] = \sum_{k=0}^{\infty} c_k e^{-ksT}$, then

$$F(s)G^*(s) = \mathcal{L}[f] \sum_{k=0}^{\infty} c_k e^{-ksT}$$

$$= \sum_{k=0}^{\infty} c_k \mathcal{L}[f] e^{-ksT}$$

$$= \sum_{k=0}^{\infty} c_k \mathcal{L}[f(t - kT)]$$

$$= \mathcal{L}\left[\sum_{k=0}^{\infty} c_k f(t - kT)\right];$$

we assume existence of the transform of the series and that $\mathcal{L}$ can be applied term by term. Hence

$$[F(s)G^*(s)]^* = \sum_{n=0}^{\infty}\sum_{k=0}^{\infty} c_k f(nT - kT)e^{-nsT}$$

$$= \sum_{n=0}^{\infty}\sum_{k=0}^{n} c_k f[(n - k)T]e^{-nsT},$$

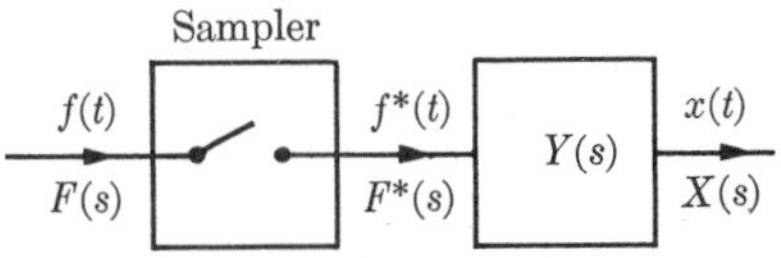

FIG. 6-9. System with one sampler.

since $f(t) = 0$ for $t < 0$. For the right side we write $F_1(z) = \sum_{n=0}^{\infty} f(nT)z^{-n} = Z[f]$, $G_1(z) = \sum_{k=0}^{\infty} g(nT)z^{-n} = Z[g]$, $g(nT) = c_n$. Then

$$F^*(s) = F_1(e^{sT}), \qquad G^*(s) = G_1(e^{sT}),$$

$$F^*(s)G^*(s) = F_1(z)G_1(z) \quad (z = e^{sT})$$

$$= Z[f]Z[g] \quad (z = e^{sT})$$

$$= Z[f *_T g] \quad (z = e^{sT})$$

$$= Z\left[\sum_{k=0}^{n} c_k f[(n-k)T]\right] \quad (z = e^{sT})$$

$$= \sum_{n=0}^{\infty} \sum_{k=0}^{n} c_k f[(n-k)T]e^{-nsT}$$

This is identical with the expression for the left side. Hence (6–240) is proved.

EXAMPLE 1. Let $T = 1$, let $F(s) = \mathcal{L}[h(t) - h(t-1)] = (1 - e^{-s})/s$, $G^*(s) = \sum_{k=0}^{\infty} e^{-ks} = 1/(1 - e^{-s})$. Then

$$F(s)G^*(s) = \frac{1}{s} = \mathcal{L}[h(t)],$$

$$[F(s)G^*(s)]^* = \sum_{k=0}^{\infty} e^{-ks}, \qquad F^*(s) = \sum_{k=0}^{\infty} [h(k) - h(k-1)]e^{-ks} = 1,$$

so that

$$F^*(s)G^*(s) = G^*(s) = [F(s)G^*(s)]^*.$$

We now consider the effect of combining one or more samplers with normal components of a linear system. In Fig. 6–9, there is one sampler and one normal component, described by a transfer function $Y(s) = \mathcal{L}[W(t)]$. To study the relation between input $f(t)$ and output $x(t)$, we use Laplace transforms $F(s) = \mathcal{L}[f]$, $X(s) = \mathcal{L}[x]$. The sampler transforms $f(t)$ to $f^*(t)$ and $F(s)$ to $F^*(s)$, so that

$$X(s) = Y(s)F^*(s), \qquad x(t) = \mathcal{L}^{-1}[Y(s)F^*(s)] = W(t) * f^*(t).$$

(We have assumed zero initial values of x at $t = 0$; general initial values

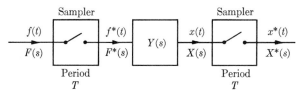

FIG. 6–10.   System with two samplers.

can be taken care of by an appropriate generalized function input to the normal component.)

We now ask for the values of the output x at the sampling instants. This can be measured by inserting a second sampler, as in Fig. 6–10. The sampled output has Laplace transform

$$X^*(s) = [Y(s)F^*(s)]^* = Y^*(s)F^*(s), \qquad (6\text{–}241)$$

by the rule (6–240). The result becomes even simpler if we use z-transforms. Following custom, we write $F(z)$, $X(z)$, $Y(z)$ for the z-transforms of f, x, W, even though these are not the same as $F(s)$, $X(s)$, $Y(s)$, with s replaced by z. In fact

$$F(z) = F^*(s)\big|_{z=e^{sT}}, \quad X(z) = X^*(s)\big|_{z=e^{sT}}, \quad Y(z) = Y^*(s)\big|_{z=e^{sT}}.$$

Now equation (6–241) becomes

$$X(z) = Y(z)F(z). \qquad (6\text{–}242)$$

Thus, in the z-language, $Y(z)$ acts as a transfer function analogous to the usual one. For this reason, $Y(z)$ is termed the *sampled transfer function*. We can also write (6–242) thus:

$$Z[x] = Z[W] \cdot Z[f],$$

so that

$$x = W *_T f. \qquad (6\text{–}243)$$

EXAMPLE 2.   In Fig. 6–10, let $T = 1$, $f(t) = t$, $Y(s) = 1/(s+1) = \mathcal{L}[e^{-t}]$. Then

$$F(z) = Z[f] = \frac{z}{(z-1)^2}, \qquad Y(z) = Z[e^{-t}] = \frac{z}{z - e^{-1}},$$

so that

$$X(z) = \frac{z}{z - e^{-1}} \cdot \frac{z}{(z-1)^2}$$

$$= (e-1)^{-2}\left[\frac{1}{z - e^{-1}} + \frac{e^2 - 2e}{z - 1} + \frac{e^2 - e}{(z-1)^2} \right],$$

$$x(t) = (e-1)^{-2}[e^{-(t-1)} + e^2 - 2e + (e^2 - e)(t-1)]h(t-1)$$

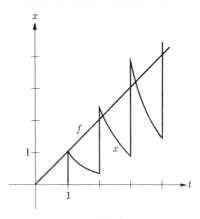

FIG. 6–11.   Response of sampled-data system to ramp input.

for $t = 0, 1, 2, \ldots$ This expression is *not* correct for other values of t. Indeed, as above,

$$X(s) = Y(s)F^*(s) = \frac{1}{s+1} \sum_{n=0}^{\infty} ne^{-ns} = \frac{e^{-s}}{s+1} + \frac{2e^{-2s}}{s+1} + \cdots,$$

so that for $t \geq 0$,

$$x(t) = e^{-(t-1)}h(t-1) + 2e^{-(t-2)}h(t-2) + \cdots$$
$$= e^{-t}[eh(t-1) + 2e^2h(t-2) + 3e^3h(t-3) + \cdots].$$

We can verify that the two expressions agree for $t = 0, 1, 2, \ldots$ Input f and output x are shown in Fig. 6–11.

For further information on sampled-data systems refer to Reference 13 and Chapter 10 of Reference 2.

PROBLEMS

1. Verify the following entries of Table 6–5: (a) No. 4   (b) No. 12   (c) No. 13 (d) No. 15   (e) No. 18   (f) No. 21   (g) No. 24

2. Show that if

$$Z[f(t, a)] = F(z, a), \qquad \text{then} \qquad Z\left[\frac{\partial f}{\partial a}(t, a)\right] = \frac{\partial F}{\partial a};$$

assume all continuity conditions needed.

3. Verify the following entries of Table 6–5 as suggested:

(a) No. 5, with the aid of Problem 2
(b) No. 6, with the aid of No. 4
(c) No. 8, with the aid of No. 6
(d) No. 16, with the aid of No. 15 and Problem 2
(e) No. 23, with the aid of No. 22
(f) No. 26
(g) No. 24, with the aid of No. 26

4. Find $Z^{-1}[F]$ for the following choices of F:

(a) $\dfrac{z^2 + 1}{z^2 - 1}$

(b) $\dfrac{2z}{(z + 1)(z + 3)}$

(c) $\dfrac{z^2 + 1}{(z - 1)^3}$

(d) $\dfrac{z}{z^2 + 1}$

5. Prove the translation formulas (6–224).
6. Prove the formulas (6–225).
7. (a) Prove the convolution rule (6–227).
(b) Prove that $f_1 *_T f_2 = f_2 *_T f_1$.
(c) Prove that $f_1 *_T (f_2 *_T f_3) = (f_1 *_T f_2) *_T f_3$.
(d) Prove that $f_1 *_T (f_2 + f_3) = f_1 *_T f_2 + f_1 *_T f_3$.

8. Evaluate $f(0)$ and $\lim_{n \to \infty} f(nT)$ for the following choices of $F(z) = Z[f]$:

(a) $\dfrac{2z^2 - 1}{2z^2 - 3z + 1}$

(b) $e^{1/z}$

(c) $\log \dfrac{z}{z - 1} = \sum_{n=1}^{\infty} \dfrac{z^{-n}}{n}$

9. Solve the following difference equations with the aid of the z-transform:
(a) $x(t + 2) - 4x(t) = 0$, $x(0) = 1$, $x(1) = -1$
(b) $x(t + 2) + 2x(t + 1) + x(t) = 0$, $x(0) = 1$, $x(1) = 0$
(c) $x(t + 1) - x(t) = 2t + 1$, $x(0) = c_1$

10. Evaluate $F^*(s)$ for the following choices of $F(s)$, with $T = 1$:

(a) $\dfrac{1}{s^2 - 1}$

(b) $\dfrac{e^{-s}}{s^2}$

(c) $\dfrac{s}{s^2 + 1}$

11. Verify the rule (6–240) for $T = 1$ and the following choices of $F(s)$, $G(s)$:

(a) $F(s) = \dfrac{1}{s}$, $G(s) = \dfrac{1}{s^2}$

(b) $F(s) = \dfrac{1}{s^2}$, $G(s) = \dfrac{e^{-s}}{s}$

12. In Fig. 6–10 let $f(t) = \sin t$, $T = \pi/2$, $Y(s) = s[(s + 1)(s + 2)]^{-1}$. Find the output $x(t)$ for initial values zero at $t = 0$. Use the z-transform to find $x(nT)$. Graph $f(t)$ and $x(t)$.

13. Figure 6–12 shows a feedback system with two samplers. Analyze the system with the aid of z-transforms and show that

$$X(z) = \frac{Y_1(z)F(z)}{Y_1(z)Y_2(z) + 1}$$

14. *Difference-differential equations.* An equation such as

$$x'(t) + 2x(t - 1) + x(t) = f(t)$$

is a difference-differential equation. Here t ranges over all real values. We can obtain a solution by Laplace transforms if f has a transform $F(s)$. Show that if $x(t)$ is a solution having a Laplace transform $X(s)$ and if $x \equiv 0$ for $t \leq 0$, then

$$X(s) = \frac{F(s)}{s + 1 + 2e^{-s}}.$$

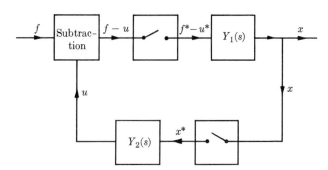

FIG. 6–12. Feedback system with two samplers.

Write

$$(s + 1 + 2e^{-s})^{-1} = (s + 1)^{-1}[1 + \{2e^{-s}/(s + 1)\}]^{-1}$$

$$= (s + 1)^{-1} \sum_{n=0}^{\infty} (-1)^n \left(\frac{2e^{-s}}{s + 1} \right)^n$$

and obtain a series solution

$$x = \sum_{n=0}^{\infty} (-1)^n 2^n p_n(t - n), \qquad p_n(t) = \left(\frac{t^n}{n!} e^{-t} h(t) \right) * f(t),$$

so that $x = p_0(t)$ for $0 \le t \le 1$, $x = p_0(t) - 2p_1(t - 1)$ for $1 \le t \le 2, \ldots$
We can verify directly that the series solution is valid, provided only that $f(t)$
is piecewise continuous.

For the general theory of difference-differential equations refer to Reference 11.

ANSWERS

4. (a) $f(0) = 1, f(nT) = 1 + (-1)^n$ $(n = 1, 2, \ldots)$
(b) $f(t) = (-1)^{t/T} - (-3)^{t/T}$
(c) $f(0) = 0, f(t) = (t/T)^2 - (t/T) + 1$ $(t = T, 2T, \ldots)$
(d) $f(t) = \sin (\tfrac{1}{2}\pi t/T)$

8. (a) $f(0) = 1, f \to 1$ as $t \to \infty$ (b) $f(0) = 1, f \to 0$ as $t \to \infty$
(c) $f(0) = 0, f \to 0$ as $t \to \infty$

9. (a) $[2^t + 3(-2)^t]/4$ (b) $(-1)^t(1 - t)$ (c) $c_1 + t^2$

10. (a) $(e^s \sinh 1)/[e^{2s} - 2e^s \cosh 1 + 1]$
(b) $(e^s - 1)^{-2}$ (c) $e^s(e^s - \cos 1)/[e^{2s} - 2e^s \cos 1 + 1]$

12. $x = \sum_{n=0}^{\infty} \sin \frac{n\pi}{2} (2e^{-2(t - n\pi/2)} - e^{-(t - n\pi/2)})h(t - n\pi/2);$

$$x(0) = 0, \quad x(t) = \alpha\beta^2 e^{-2t} - \beta\gamma e^{-t} + (\alpha - \gamma) \sin t + (\beta\gamma - \beta^2\alpha) \cos t,$$

$$\beta = e^{-\pi/2}, \quad \alpha = 2(\beta^4 + 1)^{-1}, \quad \gamma = (\beta^2 + 1)^{-1}, \quad \text{for } t = T, 2T, \ldots$$

6–25 Hilbert transforms. In applications of systems analysis to electrical networks, the problem arises of finding the frequency response function when only its real part or imaginary part is known. Mathematically the problem is that of studying the relations between real functions $\alpha(\omega)$ and $\beta(\omega)$, when $\phi(\omega) = \alpha(\omega) + i\beta(\omega)$ is the Fourier transform of a function $f(t)$ which is zero for $t < 0$. We assume $f(t)$ to be absolutely integrable on the interval $0 \leq t < \infty$. Accordingly, the Laplace transform $F(s) = \mathcal{L}[f]$ is absolutely convergent for $\sigma \geq 0$, and the problem becomes that of relating the real and imaginary parts of $f(s)$ along a line $\sigma = 0$; more generally, we could seek relations on any line $\sigma = $ const.

DEFINITION. *Real functions* $\alpha(\omega)$, $\beta(\omega)$ *defined for* $-\infty < \omega < \infty$ *are a pair of* conjugate functions *if there exists a piecewise continuous function* $f(t)$, *equal to 0 for* $t \leq 0$ *and absolutely integrable from* $t = 0$ *to* $t = +\infty$, *such that* $\Phi[f] = \alpha(\omega) + i\beta(\omega)$.

The function $f(t)$ is permitted to be complex-valued. Since $-if$ has Fourier transform $\beta(\omega) - i\alpha(\omega)$, we conclude that $\beta(\omega)$ and $-\alpha(\omega)$ are also a pair of conjugate functions. Thus the relation between $\alpha(\omega)$ and $\beta(\omega)$ is not symmetric.

By Theorem 2 of Section 5–4, we know that $\alpha(\omega)$ and $\beta(\omega)$ must be continuous and that

$$f(t) = \frac{1}{2\pi} (P) \int_{-\infty}^{\infty} [\alpha(\omega) + i\beta(\omega)]e^{i\omega t} \, d\omega, \qquad (6\text{--}250)$$

for proper assignment of value to $f(t)$ at discontinuities. Setting $t = 0$, we conclude, not that $\alpha(\omega)$ and $\beta(\omega)$ are absolutely integrable, but that the principal values

$$(P)\int_{-\infty}^{\infty} \alpha(\omega) \, d\omega = \lim_{b \to \infty} \int_{-b}^{b} \alpha(\omega) \, d\omega,$$

$$(P)\int_{-\infty}^{\infty} \beta(\omega) \, d\omega$$

exist.

THEOREM 22. *Let* $\alpha(\omega)$, $\beta(\omega)$ *be a pair of conjugate functions. Let* $\alpha(\omega)$, $\beta(\omega)$ *be absolutely integrable from* $-\infty$ *to* ∞, *and let* $\alpha'(\omega)$, $\beta'(\omega)$ *exist for all* ω. *Then*

$$\beta(\omega) = \frac{1}{\pi} \int_{0}^{\infty} \frac{\alpha(\omega + v) - \alpha(\omega - v)}{v} \, dv,$$

$$\alpha(\omega) = -\frac{1}{\pi} \int_{0}^{\infty} \frac{\beta(\omega + v) - \beta(\omega - v)}{v} \, dv. \qquad (6\text{--}251)$$

Proof. By assumption, $\alpha(\omega) + i\beta(\omega) = \Phi[f]$. Under the hypotheses made, we can drop the (P) in Eq. (6–250) and write

$$f(t) = \frac{1}{2\pi} \int_{-\infty}^{\infty} [\alpha(\omega) + i\beta(\omega)] e^{i\omega t} \, d\omega$$

$$= \frac{1}{2\pi} \int_{-\infty}^{\infty} [\alpha(\omega) \cos \omega t - \beta(\omega) \sin \omega t] \, d\omega$$

$$+ \frac{i}{2\pi} \int_{-\infty}^{\infty} [\alpha(\omega) \sin \omega t + \beta(\omega) \cos \omega t] \, d\omega.$$

Since $f(t) = 0$ for $t \leq 0$, $f(-t) = 0$ for $t \geq 0$; hence for $t \geq 0$,

$$\int_{-\infty}^{\infty} [\alpha(\omega) \cos \omega t + \beta(\omega) \sin \omega t] \, d\omega = 0,$$

$$\int_{-\infty}^{\infty} [-\alpha(\omega) \sin \omega t + \beta(\omega) \cos \omega t] \, d\omega = 0.$$

Therefore we can write, for $t \geq 0$,

$$f(t) = \frac{1}{\pi} \int_{-\infty}^{\infty} \alpha(\omega) \cos \omega t \, d\omega + \frac{i}{\pi} \int_{-\infty}^{\infty} \alpha(\omega) \sin \omega t \, d\omega,$$

$$f(t) = \frac{1}{\pi} \int_{-\infty}^{\infty} \alpha(\omega) e^{i\omega t} \, d\omega = \frac{1}{\pi} \int_{-\infty}^{\infty} \alpha(u) e^{iut} \, du. \tag{6–252}$$

Now, by definition,

$$\beta(\omega) = \mathrm{Im}[\Phi[f]] = \mathrm{Im} \int_{0}^{\infty} f(t) e^{-i\omega t} \, dt.$$

Therefore, by (6–252),

$$\beta(\omega) = \mathrm{Im} \, \frac{1}{\pi} \int_{0}^{\infty} \int_{-\infty}^{\infty} \alpha(u) e^{iut} e^{-i\omega t} \, du \, dt$$

$$= \frac{1}{\pi} \int_{0}^{\infty} \int_{-\infty}^{\infty} \alpha(u) \sin t(u - \omega) \, du \, dt$$

$$= \frac{1}{\pi} \lim_{b \to \infty} \int_{0}^{b} \int_{-\infty}^{\infty} \alpha(u) \sin t(u - \omega) \, du \, dt$$

$$= \frac{1}{\pi} \lim_{b \to \infty} \int_{-\infty}^{\infty} \int_{0}^{b} \alpha(u) \sin t(u - \omega) \, dt \, du,$$

the interchange of order of integration being permitted by Theorem 5 of

Section 5–6. Accordingly,

$$
\beta(\omega) = \frac{1}{\pi} \lim_{b \to \infty} \int_{-\infty}^{\infty} \alpha(u) \, \frac{1 - \cos b(u - \omega)}{u - \omega} \, du
$$

$$
= \frac{1}{\pi} \lim_{b \to \infty} \int_{-\infty}^{\infty} \alpha(v + \omega) \, \frac{1 - \cos bv}{v} \, dv
$$

$$
= \frac{1}{\pi} \lim_{b \to \infty} \left[\int_{-\infty}^{0} \alpha(v + \omega) \, \frac{1 - \cos bv}{v} \, dv + \int_{0}^{\infty} \alpha(v + \omega) \, \frac{1 - \cos bv}{v} \, dv \right]
$$

$$
= \frac{1}{\pi} \lim_{b \to \infty} \int_{0}^{\infty} \frac{\alpha(\omega + v) - \alpha(\omega - v)}{v} \, (1 - \cos bv) \, dv,
$$

by replacing v by $-v$ in the integral from $-\infty$ to 0. As in Section 5–8, we show that

$$
\lim_{b \to \infty} \int_{0}^{\infty} \frac{\alpha(\omega + v) - \alpha(\omega - v)}{v} \cos bv \, dv = 0
$$

wherever $\alpha'(\omega)$ exists. Hence

$$
\beta(\omega) = \frac{1}{\pi} \int_{0}^{\infty} \frac{\alpha(\omega + v) - \alpha(\omega - v)}{v} \, dv,
$$

as asserted. The second equation (6–251) is proved in the same way, since $\beta(\omega)$, $-\alpha(\omega)$ are a pair of conjugate functions.

Remark. Equations (6–251) are valid under considerably more general conditions than those stated. For a full discussion refer to Chapter V of Reference 12.

Hilbert transforms. We can write the formulas (6–251) in other ways. In particular, we can write

$$
\beta(\omega) = \frac{1}{\pi} \, (P_\omega) \int_{-\infty}^{\infty} \frac{\alpha(u)}{u - \omega} \, du,
$$

$$
\alpha(\omega) = -\frac{1}{\pi} \, (P_\omega) \int_{-\infty}^{\infty} \frac{\beta(u)}{u - \omega} \, du,
$$

(6–253)

where the (P_ω) refers to the discontinuity at $u = \omega$:

$$
(P_\omega) \int_{-\infty}^{\infty} \frac{\alpha(u)}{u - \omega} \, du = \lim_{\epsilon \to 0+} \left[\int_{-\infty}^{\omega - \epsilon} \frac{\alpha(u)}{u - \omega} \, du + \int_{\omega + \epsilon}^{\infty} \frac{\alpha(u)}{u - \omega} \, du \right].
$$

The equations (6–253) define $\beta(\omega)$ as the *Hilbert transform* of $\alpha(\omega)$, or $\alpha(\omega)$

as the Hilbert transform of $-\beta(\omega)$. In symbols,

$$\beta(\omega) = \Im[\alpha(\omega)],$$
$$\alpha(\omega) = \Im[-\beta(\omega)].$$

The operator $\Im$ is defined in the linear space of functions $\alpha(\omega)$ which are differentiable and absolutely integrable from $-\infty$ to ∞, and is a linear operator in that class. Of course, equations (6–251) could equally well be used to define the Hilbert transform.

To show the equivalence of (6–253) and (6–251), we write the first of (6–251) as follows:

$$\beta(\omega) = \lim_{\epsilon \to 0+} \frac{1}{\pi} \int_{\epsilon}^{\infty} \frac{\alpha(\omega + v) - \alpha(\omega - v)}{v}\, dv$$

$$= \lim_{\epsilon \to 0+} \left[\frac{1}{\pi} \int_{\epsilon}^{\infty} \frac{\alpha(\omega + v)}{v}\, dv - \frac{1}{\pi} \int_{\epsilon}^{\infty} \frac{\alpha(\omega - v)}{v}\, dv \right]$$

$$= \lim_{\epsilon \to 0+} \left[\frac{1}{\pi} \int_{\epsilon}^{\infty} \frac{\alpha(\omega + v)}{v}\, dv + \frac{1}{\pi} \int_{-\infty}^{-\epsilon} \frac{\alpha(\omega + v)}{v}\, dv \right]$$

$$= \lim_{\epsilon \to 0+} \left[\frac{1}{\pi} \int_{\omega+\epsilon}^{\infty} \frac{\alpha(u)}{u - \omega}\, du + \frac{1}{\pi} \int_{-\infty}^{\omega-\epsilon} \frac{\alpha(u)}{u - \omega}\, du \right]$$

$$= \frac{1}{\pi} (P_\omega) \int_{-\infty}^{\infty} \frac{\alpha(u)}{u - \omega}\, du.$$

The steps can be reversed, so that the relations are equivalent.

We can also write

$$\beta(\omega) = \frac{1}{\pi} \int_{-\infty}^{\infty} \frac{\alpha(u) - \alpha(\omega)}{u - \omega}\, du, \tag{6–254}$$

where a principal value is understood at both ω and ∞:

$$(P_{\omega,\infty}) \int_{-\infty}^{\infty} \cdots d\omega = \lim_{\substack{b \to \infty \\ \epsilon \to 0+}} \left[\int_{-b}^{\omega-\epsilon} \cdots + \int_{\omega+\epsilon}^{b} \right].$$

Since

$$(P_{\omega,\infty}) \int_{-\infty}^{\infty} \frac{du}{u - \omega} = 0,$$

as is easily verified (Problem 4 below), Eq. (6–254) follows from the first of (6–253). If $\alpha(\omega)$ has a derivative, we can verify that the principal value at ω is not needed.

From (6–251), we conclude that if $\alpha(\omega)$ is an odd function, then $\beta(\omega)$ is even. For

$$\beta(-\omega) = \frac{1}{\pi} \int_0^\infty \frac{\alpha(-\omega + v) - \alpha(-\omega - v)}{v} \, dv$$

$$= \frac{1}{\pi} \int_0^\infty \frac{-\alpha(\omega - v) + \alpha(\omega + v)}{v} \, dv = \beta(\omega).$$

Under these conditions we can write

$$\beta(\omega) = -\frac{2}{\pi} \int_0^\infty \frac{u\alpha(u)}{u^2 - \omega^2} \, du, \tag{6–255}$$

where a principal value is needed at $u = |\omega|$. This is known as the *Kronig-Kramers relation*. To verify it, we write, temporarily disregarding principal values,

$$\beta(\omega) = \frac{1}{\pi} \int_{-\infty}^\infty \frac{\alpha(u)}{u - \omega} \, du = \frac{1}{\pi} \int_{-\infty}^\infty \frac{\alpha(-u)}{\omega - u} \, du = \frac{1}{\pi} \int_{-\infty}^\infty \frac{\alpha(u)}{u + \omega} \, du$$

$$= \frac{1}{2\pi} \int_{-\infty}^\infty \alpha(u) \left(\frac{1}{u - \omega} + \frac{1}{u + \omega} \right) du = \frac{1}{\pi} \int_{-\infty}^\infty \frac{u\alpha(u)}{u^2 - \omega^2} \, du$$

$$= \frac{2}{\pi} \int_0^\infty \frac{u\alpha(u)}{u^2 - \omega^2} \, du.$$

If principal values are carried along, we verify that, when $\omega > 0$, for example, the integral can be computed as the limit, as $\epsilon \to 0+$, of the sum of the integrals from 0 to $\omega - \epsilon$ and from $\omega + \epsilon$ to ∞; for $\omega = 0$, the integral is simply the usual improper integral at 0 (and at ∞). If $\alpha(\omega)$ is even, then $\beta(\omega)$ is odd, and a similar relation is found:

$$\beta(\omega) = \frac{2\omega}{\pi} \int_0^\infty \frac{\alpha(u)}{u^2 - \omega^2} \, du. \tag{6–256}$$

Examples of conjugate functions. We easily obtain examples of pairs from Table 5–1 (Section 5–10) by taking real and imaginary parts of $\phi(\omega)$ for entries $f(t)$ which are zero for $t < 0$. From Entry 4, with a real, we find

$$\alpha(\omega) = -\frac{a}{a^2 + \omega^2}, \qquad \beta(\omega) = \frac{-\omega}{\omega^2 + a^2}.$$

From Entry 16, with $c = 1$, we find

$$\alpha(\omega) = \frac{1 - \cos\omega - \omega\sin\omega}{-\omega^2}, \qquad \beta(\omega) = \frac{\sin\omega - \omega\cos\omega}{-\omega^2}.$$

Further examples can be obtained from Table 6–1 (Section 6–3), since $\mathcal{L}[f] = \Phi[f(t)e^{-\sigma t}]$ with $f(t) = 0$ for $t < 0$. Thus, from Entry 3,

$$\alpha(\omega) = \text{Re } \frac{1}{(\sigma + i\omega)^n}, \qquad \beta(\omega) = \text{Im } \frac{1}{(\sigma + i\omega)^n}$$

are conjugate functions for each $\sigma > 0$. For further examples, see the table of Hilbert transforms in Chapter XV of Reference 6.

PROBLEMS

1. With the aid of Table 6–1, verify that the following are pairs of conjugate functions:

(a) $\alpha(\omega) = \dfrac{\sigma(\sigma^2 + \omega^2 + 1)}{(\sigma^2 + \omega^2)^2 + 2(\sigma^2 - \omega^2) + 1}$,

$\beta(\omega) = \dfrac{\omega(1 - \sigma^2 - \omega^2)}{(\sigma^2 + \omega^2)^2 + 2(\sigma^2 - \omega^2) + 1} \quad (\sigma > 0)$

(b) $\dfrac{\sigma \cos \omega - \omega \sin \omega}{\sigma^2 + \omega^2}$, $\dfrac{-\omega \cos \omega - \sigma \sin \omega}{\sigma^2 + \omega^2} \quad (\sigma > 0)$

2. Let $\alpha(\omega) = (\omega^2 + 1)^{-1}$, $\beta(\omega) = -\omega(\omega^2 + 1)^{-1}$. Verify that each of the following sets of relations is correct:

(a) (6–251)   (b) (6–253)   (c) (6–254)   (d) (6–256)

3. Prove that if $\Phi[f] = \alpha(\omega) + i\beta(\omega)$, $f = 0$ for $t \leq 0$, then $\alpha(\omega)$ is even if and only if f is real-valued, and $\alpha(\omega)$ is odd if and only if f has pure imaginary values. [*Hint:* Use (6–252).]

4. Prove that

$$(P_{\omega,\infty}) \int_{-\infty}^{\infty} \frac{du}{u - \omega} = 0.$$

5. Prove the following properties of the Hilbert transform:

(a) $\mathcal{H}[\alpha(\omega + c)] = \beta(\omega + c)$
(b) $\mathcal{H}[\alpha(c\omega)] = \beta(c\omega) \quad (c > 0)$
(c) $\mathcal{H}[\alpha(-c\omega)] = -\beta(-c\omega) \quad (c > 0)$

6. Prove that

$$\mathcal{H}[\omega\alpha(\omega)] = \omega\beta(\omega) + \frac{1}{\pi} \int_{-\infty}^{\infty} \alpha(\omega) \, d\omega.$$

[*Hint:* Use the first of (6–251), or use (6–254).]

7. Prove that $\mathcal{H}[\alpha'(\omega)] = \beta'(\omega)$. [*Hint:* Use (6–251) and assume conditions on $\alpha(\omega)$ such that differentiation and integration may be interchanged. See p. 379 of Reference 10.]

8. Prove that if $\alpha(\omega)$ is even, then

(a) $\beta(\omega) = \dfrac{2\omega}{\pi} \displaystyle\int_0^\infty \dfrac{\alpha(u) - \alpha(\omega)}{u^2 - \omega^2} \, du$ [*Hint:* Use (6–254).]

(b) $\beta(\omega) = \dfrac{1}{\pi} \displaystyle\int_0^\infty \alpha'(u) \log \left| \dfrac{\omega + u}{\omega - u} \right| du$ [*Hint:* Integrate by parts in the result of part (a).]

(c) $\beta(\omega) = \dfrac{1}{\pi} \displaystyle\int_{-\infty}^\infty \log \left| \coth \dfrac{u}{2} \right| \dfrac{d}{dv} \alpha(\omega e^v) \, dv$ [*Hint:* Set $u = |\omega| e^v$ in the result of part (a) and then integrate by parts.]

A discussion of the physical meaning of these formulas is given on pp. 313–320 of Reference 1.

9. The *Stieltjes transform* of $f(t)$ is $\mathcal{S}[f] = g(s)$, and is defined as follows:

$$\mathcal{S}[f] = g(s) = \int_0^\infty \frac{f(t)}{t + s} \, dt.$$

Thus $f(t)$ is to be considered 0 for negative t. The variable s is a complex variable. Prove, under appropriate hypotheses, that

(a) $\mathcal{S}[f] = \mathcal{L}[\mathcal{L}[f]]$ (b) $\mathcal{S}[f'(t)] = -s^{-1} f(0) - g'(s)$

For relations between Stieltjes and Hilbert transforms, see Chapter XIV of Reference 6.

Suggested References

1. H. W. Bode, *Network Analysis and Feedback Amplifier Design.* New York: Van Nostrand, 1945.

2. D. K. Cheng, *Analysis of Linear Systems.* Reading, Mass.: Addison-Wesley, 1959.

3. R. V. Churchill, *Operational Mathematics*, 2nd ed. New York: McGraw-Hill, 1958.

4. G. Doetsch, *Theorie und Anwendung der Laplace Transformation.* Berlin: Springer, 1937.

5. G. Doetsch, *Handbuch der Laplace Transformation*, Vols. 1, 2, 3. Basel: Birkhäuser, 1950, 1955, 1956.

6. A. Erdélyi, editor, *Tables of Integral Transforms*, Vols. I, II (compiled by the staff of the Bateman Manuscript Project). New York: McGraw-Hill, 1954.

7. W. Feller, *Introduction to Probability Theory*, Vol. 1, 2nd ed. New York: John Wiley, 1957.

8. D. Holl and B. Vinograde, *Introduction to Laplace Transforms.* New York: Appleton-Century-Crofts, 1959.

9. E. Jahnke and F. Emde, *Tables of Functions*, 4th ed. New York: Dover, 1945.

10. W. Kaplan, *Advanced Calculus*. Reading, Mass.: Addison-Wesley, 1952.

11. E. Pinney, *Ordinary Difference-Differential Equations*. Berkeley: University of California Press, 1959.

12. E. C. Titchmarsh, *Introduction to the Theory of Fourier Integrals*. Oxford: Oxford University Press, 1948.

13. J. Tou, *Sampled-data Control Systems*. New York: McGraw-Hill, 1959.

14. E. T. Whittaker and G. N. Watson, *A Course of Modern Analysis*, 4th ed. Cambridge, Eng.: Cambridge University Press, 1940.

15. D. V. Widder, *The Laplace Transform*. Princeton: Princeton University Press, 1941.

CHAPTER 7

STABILITY

7-1 Introduction. In this chapter we consider the stability of systems governed by simultaneous linear differential, or integrodifferential, equations with constant coefficients. In Section 2–10 it is shown that *the system is stable precisely when all characteristic roots have negative real parts.* Evaluation of the characteristic roots requires solution of an algebraic equation, which may be of high degree. It is in general very time-consuming to solve this equation with precision. If only stability is being investigated, we need not find the roots exactly, for we need to know only the signs of the real parts of the roots. We shall see that there are several methods which provide the desired information without solving for the roots.

For brevity we shall call a linear differential equation, a system of linear differential equations, an algebraic equation, or a polynomial *stable* if all roots or characteristic roots have negative real parts; otherwise we call the differential equation, etc., *unstable.*

7-2 Signs of the coefficients. The following theorem provides a simple necessary condition for stability.

THEOREM 1. *If the equation*

$$a_0 s^n + \cdots + a_n = 0 \tag{7-20}$$

has real coefficients and the equation is stable, then all coefficients have the same sign; if $a_0 > 0$, then

$$a_1 > 0, \qquad a_2 > 0, \qquad \ldots, \qquad a_n > 0. \tag{7-21}$$

Thus the following differential equations are *unstable:*

$$(D^2 - 2D + 3)x = f(t),$$
$$(D^3 + 2D^2 + 1)x = f(t).$$

In the first, one coefficient is negative; in the second, one coefficient is zero.

Theorem 1 can be proved by induction. It is true for $n = 1$, for $a_0 s + a_1 = 0$ has a negative root (with $a_0 > 0$) only if $a_1 > 0$. If it is true for $n = 1, \ldots, N$, then let the equation have degree $N + 1$:

$$a_0 s^{N+1} + \cdots + a_{N+1} = 0 \quad (a_0 > 0). \tag{7-22}$$

If Eq. (7–22) has a real root s_1, then the equation can be factored as

$$a_0(s - s_1)(s^N + \cdots) = 0.$$

Now by assumption s_1 is negative, so that $-s_1$ is positive. The third factor is a polynomial $s^N + \cdots$ whose roots have negative real parts. Hence by the induction assumption all coefficients in this factor are positive. Accordingly, the original polynomial $a_0 s^{N+1} + \cdots + a_{N+1}$ is a product of factors with positive coefficients, and must itself have positive coefficients.

If Eq. (7–22) has no real root, then it must have a pair of conjugate complex roots $\alpha \pm \beta i$, with $\alpha < 0$. Hence

$$a_0 s^{N+1} + \cdots + a_{N+1} = a_0(s - \alpha - \beta i)(s - \alpha + \beta i)(s^{N-1} + \cdots)$$
$$= a_0(s^2 - 2\alpha s + \alpha^2 + \beta^2)(s^{N-1} + \cdots).$$

Since $\alpha < 0$, again the factors have positive coefficients and all coefficients of $a_0 s^{N+1} + \cdots$ are positive. Thus truth of the theorem for $n = 1, \ldots, N$ implies its truth for $n = N + 1$. Hence the induction is complete and the theorem is true for all n.

Remark. It should be stressed that failure of (7–21) implies instability, but validity of (7–21) does not imply stability (the condition is necessary, but not sufficient). For example, $s^3 + s^2 + s + 6$ has all coefficients positive, but has roots $-2, \frac{1}{2}(1 \pm \sqrt{11}\, i)$, of which two have positive real parts. If all coefficients are positive, we can conclude that there are no *real* roots with positive real parts, but there may be imaginary ones.

7–3 Direct method. We can of course attack an equation directly by solving for the roots (References 2 and 5). If all coefficients are positive, then the only roots with positive real parts must be complex. For $n \leqq 4$, we can obtain an equation for the real parts of the complex roots as illustrated in the following example:

$$s^3 + 2s^2 + s + 4 = 0, \qquad s = x + iy; \qquad (7\text{–}30)$$
$$(x + iy)^3 + 2(x + iy)^2 + (x + iy) + 4 = 0;$$
$$x^3 - 3xy^2 + 2x^2 - 2y^2 + x + 4 + i(3x^2 y - y^3 + 4xy + y) = 0.$$

Hence real and imaginary parts must be zero:

$$x^3 - 3xy^2 + 2x^2 - 2y^2 + x + 4 = 0,$$
$$y(3x^2 - y^2 + 4x + 1) = 0.$$

The second equation gives $y = 0$ (that is, real roots) or

$$3x^2 - y^2 + 4x + 1 = 0.$$

Solving for y^2 and substituting in the first equation, we obtain an equation for x:

$$8x^3 + 16x^2 + 11x - 2 = 0. \qquad (7\text{--}31)$$

Now x must be real, and we need only determine whether (7–31) has any positive real roots. By the Descartes rule of signs (Reference 2), (7–31) has precisely one positive real root. Hence (7–30) has two complex roots $x \pm iy$ with positive real parts and Eq. (7–30) is unstable.

PROBLEMS

1. Test the following for stability:

(a) $\dfrac{d^2x}{dt^2} + \dfrac{dx}{dt} + 4x = F(t)$ $\qquad$ (b) $\dfrac{d^3x}{dt^3} - \dfrac{d^2x}{dt^2} + \dfrac{dx}{dt} - x = F(t)$

2. Test the following for stability $(D = d/dt)$:

(a) $(D^3 + D^2 + 4)x = F(t)$
(b) $(D^3 + 2D^2 + D + 1)x = F(t)$
(c) $(D^4 + D^3 + 2D^2 + D)x = F(t)$
(d) $(D^3 + 2D^2 + 3D + 4)x = F(t)$
(e) $(D^4 + 2D^3 + D^2 + D + 1)x = F(t)$

3. Prove that an equation of second order with real coefficients

$$(a_0 D^2 + a_1 D + a_2)x = F(t) \quad (a_0 > 0)$$

is stable if and only if $a_1 > 0$, $a_2 > 0$.

4. Prove that an equation of third order with real coefficients

$$(a_0 D^3 + a_1 D^2 + a_2 D + a_3)x = F(t) \quad (a_0 > 0)$$

is stable if and only if $a_1 > 0$, $a_2 > 0$, $a_3 > 0$ and $a_0 a_3 < a_1 a_2$. [*Hint:* We can write

$$a_0 s^3 + a_1 s^2 + a_2 s + a_3 = a_0(s + a)(s^2 + bs + c)$$

and hence relate a_0, a_1, a_2, a_3 to a, b, c. Show that the stated conditions on the a_k are equivalent to the conditions $a > 0$, $b > 0$, $c > 0$ and hence to all roots having negative real parts.]

ANSWERS

1. (a) stable   (b) unstable
2. (a) unstable   (b) stable   (c) unstable   (d) stable   (e) unstable

7–4 Hurwitz-Routh criterion. As pointed out above, the fact that $a_0 > 0$, $a_1 > 0$, ..., $a_n > 0$ does not by itself imply that all roots of Eq. (7–20) have negative real parts. We can ask what additional conditions the coefficients must satisfy to ensure that all the roots s_j have negative real parts. The answer is provided by the following theorem:

THEOREM 2. (Hurwitz-Routh criterion). *Let*

$$\Delta_1 = a_1, \quad \Delta_2 = \begin{vmatrix} a_1 & a_0 \\ a_3 & a_2 \end{vmatrix}, \quad \Delta_3 = \begin{vmatrix} a_1 & a_0 & 0 \\ a_3 & a_2 & a_1 \\ a_5 & a_4 & a_3 \end{vmatrix}, \ldots,$$

$$\Delta_k = \begin{vmatrix} a_1 & a_0 & 0 & \cdots & \cdots & 0 \\ a_3 & a_2 & a_1 & a_0 & \cdots & 0 \\ \vdots & & & & & \vdots \\ a_{2k-1} & a_{2k-2} & \cdots & \cdots & \cdots & a_k \end{vmatrix}, \ldots \qquad (7\text{--}40)$$

For the equation with real coefficients of degree n,

$$a_0 s^n + a_1 s^{n-1} + \cdots + a_{n-1} s + a_n = 0 \quad (a_0 > 0), \qquad (7\text{--}41)$$

form the determinants $\Delta_1, \ldots, \Delta_n$, *with* a_m *replaced by* 0 *for* $m > n$. *The roots all have negative real parts if and only if*

$$\Delta_1 > 0, \qquad \Delta_2 > 0, \qquad \ldots, \qquad \Delta_n > 0. \qquad (7\text{--}42)$$

If the coefficients $a_1, \ldots, a_n$ are positive, then the Hurwitz-Routh criterion provides at most $n - 2$ additional conditions. For $\Delta_1 > 0$ is the same as $a_1 > 0$, and for the equation of degree n,

$$\Delta_n = \begin{vmatrix} a_1 & a_0 & 0 & \cdots & \cdots & 0 \\ a_3 & a_2 & a_1 & a_0 & \cdots & 0 \\ \vdots & & & & & \vdots \\ 0 & 0 & 0 & 0 & \cdots & a_n \end{vmatrix} = a_n \Delta_{n-1},$$

so that $\Delta_n > 0$ tells no more than $\Delta_{n-1} > 0$. Accordingly it is sufficient to consider only whether $\Delta_2 > 0, \ldots, \Delta_{n-1} > 0$.

As illustrations, the test is written out in full for equations of degrees 2 to 5 in Table 7–1.

The proof of Theorem 2 is given in the next section. Here we consider a few examples.

EXAMPLE 1. $s^4 + 2s^3 + s^2 + 3s + 2 = 0$.   Here $a_0 = 1$, $a_1 = 2$, $a_2 = 1, a_3 = 3, a_4 = 2, \Delta_2 = -1, \Delta_3 = -11$. The equation is unstable.

EXAMPLE 2. $s^5 + 5s^4 + 11s^3 + 14s^2 + 10s + 4 = 0$. Here $a_0 = 1$, $a_1 = 5$, $a_2 = 11$, $a_3 = 14$, $a_4 = 10$, $a_5 = 4$, $\Delta_2 = 41$, $\Delta_3 = 344$, $\Delta_4 = 1820$. The equation is stable.

EXAMPLE 3. $s^4 + as^3 + 6s^2 + 4s + 1 = 0$, where a is unspecified. Hence $a_0 = 1$, $a_1 = a$, $a_2 = 6$, $a_3 = 4$, $a_4 = 1$, $\Delta_2 = 6a - 4$, $\Delta_3 =$

Table 7–1

Equation	Condition for stability
$a_0 s^2 + a_1 s + a_2 = 0$	all $a_j > 0$
$a_0 s^3 + a_1 s^2 + a_2 s + a_3 = 0$	all $a_j > 0$ and $\Delta_2 = a_1 a_2 - a_0 a_3 > 0$
$a_0 s^4 + a_1 s^3 + a_2 s^2 + a_3 s + a_4 = 0$	all $a_j > 0$, $\Delta_2 = a_1 a_2 - a_0 a_3 > 0$, and $\Delta_3 = a_3 \Delta_2 - a_4 a_1^2 > 0$
$a_0 s^5 + a_1 s^4 + a_2 s^3 + a_3 s^2 + a_4 s$ $+ a_5 = 0$	all $a_j > 0$, $\Delta_2 = a_1 a_2 - a_0 a_3 > 0$, $\Delta_3 = a_3 \Delta_2 - a_1(a_1 a_4 - a_0 a_5) > 0$, and $\Delta_4 = a_4 \Delta_3 - a_2 a_5 \Delta_2$ $+ a_5 a_0 (a_1 a_4 - a_0 a_5) > 0$

Table 7–2

Stability Test

							Explanation
a_0	a_2	a_4	a_6	$\ldots$	0	0	$b_1 = a_1 a_2 - a_0 a_3,$
a_1	a_3	a_5	$\ldots$	$\ldots$	0	0	$b_2 = a_1 a_4 - a_0 a_5, \ldots,$
b_1	b_2	b_3	$\ldots$	$\ldots$	0	0	$c_1 = b_1 a_3 - a_1 b_2,$
c_1	c_2	c_3	$\ldots$	$\ldots$	0	0	$c_2 = b_1 a_5 - a_1 b_3, \ldots,$
d_1	d_2	d_3	$\ldots$	$\ldots$	0	0	$d_1 = c_1 b_2 - b_1 c_2, \ldots$

$-a^2 + 24a - 16$. Therefore the equation is stable if $6a - 4 > 0$ and $-a^2 + 24a - 16 > 0$. The first condition holds when $a > \frac{2}{3}$, the second when $12 - 8\sqrt{2} < a < 12 + 8\sqrt{2}$. Hence the equation is stable if $12 - 8\sqrt{2} < a < 12 + 8\sqrt{2}$.

Tabular form of test. The computation of the determinants can be considerably simplified by the following procedure. We write the coefficients a_0, a_2, a_4, $\ldots$ on one row and below them the coefficients a_1, a_3, a_5, $\ldots$ The rows are considered to consist of zeros after the last entry, as in Table 7–2. We then compute the next row b_1, b_2, b_3, $\ldots$ as directed, so that the b's are actually second order determinants with a_1, a_0 as elements of the first column. In general,

$$b_j = a_1 a_{2j} - a_0 a_{2j+1},$$

where missing a's are replaced by 0's. The fourth row, c_1, c_2, . . . , is computed from the preceding two in the same way:

$$c_j = b_1 a_{2j+1} - a_1 b_{j+1};$$

the fifth row, d_1, d_2, . . . , is computed from the preceding two in analogous fashion. It will be seen that the numbers in the first column of the table are equal to the determinants $\Delta_0 = a_0$, $\Delta_1 = a_1$, . . . , Δ_k, . . . multiplied by powers of preceding numbers of the column:

$$a_1 = \Delta_1,\ b_1 = \Delta_2,\ c_1 = \Delta_3,\ d_1 = a_1\,\Delta_4,\ e_1 = a_1 b_1\,\Delta_5,\ f_1 = a_1^2 b_1 c_1\,\Delta_6,\ \ldots$$

The general rule is given in the next section. Hence if the first n numbers in the first column are all positive, then Δ_1, . . . , Δ_n are positive and conversely.

Examples 1 and 2, considered above, lead to the following results:

+1	1	2		1	11	10	0
+2	3	0		5	14	4	0
−1	4	0		41	46	0	
−11	0			344	164	0	
−44	0			9100	0		

The minus signs in the first column show instability for Example 1; the plus signs show stability for Example 2.

7–5 Proof of Hurwitz-Routh criterion. The plan of the proof is as follows. It is based on induction. The theorem is true if $n = 1$. It will be assumed true for $n - 1$, and then proved true for n ($n \geqq 2$). To carry out the last step, it will first be assumed that $f(s) = a_0 s^n + \cdots$ has all its zeros in the left half-plane. A new polynomial $f_1(s)$ of degree $n - 1$ will then be constructed, related to $f(s)$, and it will be shown that $f_1(s)$ also has all its zeros in the left half-plane. By the induction assumption, $f_1(s)$ therefore satisfies the criterion; that is, certain determinants δ_1, δ_2, . . . are positive. Because of the relationship between $f(s)$ and $f_1(s)$, the determinants Δ_1, Δ_2, . . . for f are simply related to δ_1, δ_2, . . . , and we can conclude that Δ_1, Δ_2, . . . are also positive. This proves the necessity of the criterion. To prove the sufficiency, we use the same induction. If Δ_1, Δ_2, . . . are positive, then so are δ_1, δ_2, . . . Hence, by induction assumption, $f_1(s)$ has its zeros in the left half-plane, and it will be shown that this implies that $f(s)$ has its zeros in the left half-plane.

It will be shown that $f(s)$ is a stable polynomial if and only if a certain *rational function* $\phi(s) = g(s)/q(s)$ has the property that its real part is positive in the right half-plane and negative in the left half-plane. This restatement of the property will prove to simplify the study of the relationship between $f(s)$ and $f_1(s)$.

Now let $f(s) = a_0 s^n + a_1 s^{n-1} + \cdots$, $(a_0 > 0)$, be an arbitrary polynomial of degree n with real coefficients. Then we can write

$$f(s) = g(s) + q(s),$$
$$g(s) = a_0 s^n + a_2 s^{n-2} + \cdots, \tag{7-50}$$
$$q(s) = a_1 s^{n-1} + a_3 s^{n-3} + \cdots$$

Since

$$(-1)^n f(-s) = a_0 s^n - a_1 s^{n-1} + a_2 s^{n-2} - a_3 s^{n-3} + \cdots = g - q,$$

it follows that

$$g(s) = \tfrac{1}{2}[f(s) + (-1)^n f(-s)],$$
$$\tag{7-51}$$
$$q(s) = \tfrac{1}{2}[f(s) - (-1)^n f(-s)].$$

We now associate with $f(s)$ the rational function

$$\phi(s) = \frac{g(s)}{q(s)} = \frac{f(s) + (-1)^n f(-s)}{f(s) - (-1)^n f(-s)}.$$

If

$$f(s) = a_0(s - s_1)(s - s_2) \cdots (s - s_n),$$

then

$$(-1)^n f(-s) = a_0(s + s_1)(s + s_2) \cdots (s + s_n),$$

so that

$$\phi(s) = \frac{a_0(s - s_1) \cdots (s - s_n) + a_0(s + s_1) \cdots (s + s_n)}{a_0(s - s_1) \cdots (s - s_n) - a_0(s + s_1) \cdots (s + s_n)} = \frac{A(s) + B(s)}{A(s) - B(s)},$$

$$A(s) = (s - s_1) \cdots (s - s_n), \qquad B(s) = (s + s_1) \cdots (s + s_n). \tag{7-52}$$

Now suppose $f(s)$ is a stable polynomial; that is, all roots of $f(s)$ satisfy $\operatorname{Re} s < 0$. Then we assert that $\phi(s)$ *is irreducible and* that

$$\operatorname{Re} \phi(s) < 0 \quad \text{for } \operatorname{Re} s < 0,$$
$$\tag{7-53}$$
$$\operatorname{Re} \phi(s) > 0 \quad \text{for } \operatorname{Re} s > 0.$$

To prove this, we first show that $\phi(s) = g(s)/q(s)$ is irreducible, that is, that $g(s)$ and $q(s)$ have no common zeros. If there were such a common zero s, then by (7-51),

$$f(s) + (-1)^n f(-s) = 0, \qquad f(s) - (-1)^n f(-s) = 0,$$

so that $f(s) = 0$ and $f(-s) = 0$; this is impossible, since all zeros of $f(s)$

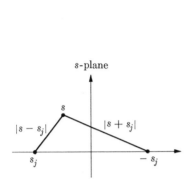

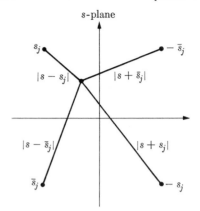

FIG. 7-1. Proof that $|A(s)| < |B(s)|$ for Re $s < 0$.

FIG. 7-2. Proof that $|A(s)| < |B(s)|$ for Re $s < 0$.

are in the left half-plane, while those of $f(-s)$ are in the right half-plane. Hence $\phi(s)$ is irreducible.

Next let s be in the left half-plane. We then assert that $|A(s)| < |B(s)|$. Indeed, if s_j is a real root of f, then $|s - s_j| < |s + s_j|$, as in Fig. 7-1. If s_j is a complex root, then $\bar{s}_j$ is also a root, and $(s - s_j)(s - \bar{s}_j)$ is a factor of $A(s)$, $(s + s_j)(s + \bar{s}_j)$ is a factor of $B(s)$. Now $|s - s_j| < |s + \bar{s}_j|$, $|s - \bar{s}_j| < |s + s_j|$, as in Fig. 7-2, so that

$$|(s - s_j)(s - \bar{s}_j)| < |(s + s_j)(s + \bar{s}_j)|.$$

Thus $A(s)$ is made up of factors which are in absolute value less than the corresponding factors of $B(s)$ and, accordingly, $|A(s)| < |B(s)|$. A parallel argument shows that $|A(s)| > |B(s)|$ when Re $s > 0$. We note that the poles of $\phi(s)$ occur where $A(s) = B(s)$; hence these must lie on the imaginary axis. Furthermore,

$$\text{Re } \phi(s) = \text{Re } \frac{A + B}{A - B} \frac{\bar{A} - \bar{B}}{\bar{A} - \bar{B}}$$

$$= \text{Re } \frac{A\bar{A} - B\bar{B} + \bar{A}B - A\bar{B}}{|A - B|^2} = \frac{|A|^2 - |B|^2}{|A - B|^2}, \quad (7\text{-}54)$$

since $\bar{A}B - A\bar{B}$ is of form $z - \bar{z}$ and hence is pure imaginary. We now conclude that Re $\phi(s) < 0$ for Re $s < 0$, where $|A| < |B|$, and Re $\phi(s) > 0$ for Re $s > 0$, where $|A| > |B|$. Thus (7-53) is proved.

Conversely, if $\phi(s)$ is irreducible and satisfies (7-53), then $f(s)$ is a stable polynomial. For, by virtue of (7-54), (7-53) implies that $|A(s)| < B(s)$ for Re $s < 0$, and $|A(s)| > |B(s)|$ for Re $s > 0$. Furthermore $A(s)$ and $B(s)$ have no common zeros; for otherwise, $\phi(s)$ would be reducible. Now if $f(s)$ had a root s_j in the right half-plane, then $A(s_j) = 0$, $B(s_j) \neq 0$, so that $|A(s_j)| < |B(s_j)|$, which is impossible. If $f(s)$ had a root s_j on the imaginary axis, then again $|A(s_j)| = 0 < |B(s_j)|$; hence by continuity $|A(s)| < |B(s)|$ in a neighborhood of s_j. Such a neighborhood includes points in the right half-plane, so that again there is a contradiction.

Now let $f(s)$ be a stable polynomial, so that (7–53) holds and $\phi(s)$ is irreducible. Then

$$\phi(s) = \frac{g}{q} = \frac{a_0 s^n + a_2 s^{n-2} + \cdots}{a_1 s^{n-1} + a_3 s^{n-3} + \cdots}$$

has $n - 1$ poles $\beta_1, \ldots, \beta_{n-1}$. All the poles must lie on the imaginary axis, as noted above. Also each pole must be simple, for near a pole of order k, Re $[\phi(s)]$ behaves like Re $([s - \beta]^{-k})$ and hence changes sign in alternate sectors of angle π/k; for the same reason, the principal part of $\phi(s)$ at each pole β must be of the form $\lambda/(s - \beta)$, where λ is real and positive (see Problem 4 below). Hence $\phi(s)$ has a partial fraction expansion:

$$\phi(s) = \frac{a_0}{a_1} s + \alpha + \sum_{j=1}^{n-1} \frac{\lambda_j}{s - \beta_j} \quad (\lambda_j > 0).$$

Here $\alpha = 0$, for

$$\phi(s) - \frac{a_0}{a_1} s = \frac{(a_1 a_2 - a_0 a_3) s^{n-2} + \cdots}{a_1 (a_1 s^{n-1} + \cdots)} \rightarrow 0 \quad \text{as} \quad s \rightarrow \infty.$$

Hence

$$\phi_0(s) = \phi(s) - \frac{a_0}{a_1} s = \sum_{j=1}^{n-1} \frac{\lambda_j}{s - \beta_j} = \frac{g}{q} - \frac{a_0}{a_1} s = \frac{a_1 g - a_0 s q}{a_1 q}$$

satisfies (7–53); for each term of the summation satisfies (7–53); also $\phi_0(s)$ is irreducible. Next

$$\phi_1(s) = \frac{1}{\phi_0(s)} = \frac{a_1 q}{a_1 g - a_0 s q} \tag{7–55}$$

is irreducible and satisfies (7–53). For

$$\text{Re} \, [\phi_1(s)] = \frac{\text{Re} \, [\phi_0]}{|\phi_0|^2}.$$

Now ϕ_1 is related to a polynomial $f_1(s)$ of degree $n - 1$ in the same way in which $\phi(s)$ is related to $f(s)$. We set

$$f_1(s) = (a_1 - a_0 s) q(s) + a_1 g(s)$$

$$= a_1^2 s^{n-1} + \cdots$$

$$= a_1 q(s) + [a_1 g(s) - a_0 s q(s)],$$

where the first term contains the terms of degree $n - 1, n - 3, \ldots$, and the second, those of degree $n - 2, n - 4, \ldots$ Hence

$$g_1(s) = a_1 q(s), \qquad q_1(s) = a_1 g(s) - a_0 s q(s),$$

and the associated rational function is

$$\frac{g_1(s)}{q_1(s)} = \frac{a_1 q}{a_1 g - a_0 s q} = \phi_1(s),$$

in agreement with (7–55).

Since $\phi_1(s)$ satisfies (7–53) and is irreducible, we conclude that $f_1(s)$ is a stable polynomial. Conversely, if $f_1(s)$ *is a stable polynomial*, then $\phi_1(s)$ satisfies (7–53) and is irreducible; hence so does $\phi_0(s)$, and thus if $a_1 > 0$, so does

$$\phi(s) = \phi_0(s) + \frac{a_0}{a_1} s.$$

Accordingly, $f(s)$ is a stable polynomial. Thus we have proved that *if* $a_0 > 0$ *and* $a_1 > 0$, *then* $f(s)$ *is a stable polynomial of degree n if and only if* $f_1(s)$ *is a stable polynomial of degree* $n - 1$.

We now study the relationship between the Hurwitz determinants of $f(s)$ and $f_1(s)$. We have

$$f_1(s) = a_1^2 s^{n-1} + (a_1 a_2 - a_0 a_3) s^{n-2} + a_1 a_3 s^{n-3}$$
$$+ (a_1 a_4 - a_0 a_5) s^{n-4} + a_1 a_5 s^{n-5}$$
$$+ (a_1 a_6 - a_0 a_7) s^{n-6} + a_1 a_7 s^{n-7} + \cdots$$

The Hurwitz determinants of $f_1(s)$ are

$$\delta_1 = a_1 a_2 - a_0 a_3, \qquad \delta_2 = \begin{vmatrix} a_1 a_2 - a_0 a_3 & a_1^2 \\ a_1 a_4 - a_0 a_5 & a_1 a_3 \end{vmatrix},$$

$$\delta_3 = \begin{vmatrix} a_1 a_2 - a_0 a_3 & a_1^2 & 0 \\ a_1 a_4 - a_0 a_5 & a_1 a_3 & a_1 a_2 - a_0 a_3 \\ a_1 a_6 - a_0 a_7 & a_1 a_5 & a_1 a_4 - a_0 a_5 \end{vmatrix}, \cdots$$

Now $a_0 a_1 \delta_j$ can be written as follows:

$$a_0 a_1 \delta_j = \begin{vmatrix} a_0 a_1 & 0 & 0 & 0 & \cdots \\ a_0 a_3 & a_1 a_2 - a_0 a_3 & a_1^2 & 0 & \cdots \\ a_0 a_5 & a_1 a_4 - a_0 a_5 & a_1 a_3 & a_1 a_2 - a_0 a_3 & \cdots \\ \cdots & \cdots & \cdots & \cdots & \cdots \end{vmatrix}.$$

Here a new first column and first row have been added. If we now add the first column to the second, the third times a_0/a_1 to the fourth, and so on, then we obtain

$$a_0 a_1 \, \delta_j = \begin{vmatrix} a_0 a_1 & a_1 a_0 & 0 & 0 & \cdots \\ a_0 a_3 & a_1 a_2 & a_1^2 & a_1 a_0 & \cdots \\ a_0 a_5 & a_1 a_4 & a_1 a_3 & a_1 a_2 & \cdots \\ \cdots & \cdots & \cdots & \cdots & \cdots \end{vmatrix},$$

so that

$$a_0 a_1 \, \delta_1 = \begin{vmatrix} a_0 a_1 & a_1 a_0 \\ a_0 a_3 & a_1 a_2 \end{vmatrix} = a_0 a_1 \begin{vmatrix} a_1 & a_0 \\ a_3 & a_2 \end{vmatrix} = a_0 a_1 \, \Delta_2,$$

$$a_0 a_1 \, \delta_2 = \begin{vmatrix} a_0 a_1 & a_1 a_0 & 0 \\ a_0 a_3 & a_1 a_2 & a_1^2 \\ a_0 a_5 & a_1 a_4 & a_1 a_3 \end{vmatrix} = a_0 a_1^2 \begin{vmatrix} a_1 & a_0 & 0 \\ a_3 & a_2 & a_1 \\ a_5 & a_4 & a_3 \end{vmatrix} = a_0 a_1^2 \, \Delta_3,$$

and in general

$$a_0 a_1 \, \delta_k = a_0 a_1^k \, \Delta_{k+1} \quad (k = 1, 2, \ldots, n - 1). \tag{7-56}$$

Now the induction can be carried out. The theorem is true for $n = 1$. If it is true for $n - 1$, then $f_1(s)$ is a stable polynomial if and only if $\delta_1, \ldots, \delta_{n-1}$ are positive. But if $a_1 > 0$, then $f(s)$ is a stable polynomial if and only if $f_1(s)$ is a stable polynomial, and hence by (7-56) if and only if $\Delta_2, \Delta_3, \ldots, \Delta_n$ are positive. Since $\Delta_1 = a_1$, we conclude that $f(s)$ is a stable polynomial if and only if $\Delta_1 > 0, \ldots, \Delta_n > 0$. Thus truth of the theorem for $n - 1$ implies its truth for n, and it is valid for all positive integers.

Justification of tabular form of test. We refer to Table 7-2 of Section 7-4. The first two rows are formed of the coefficients of $f(s)$ (those of even index on the first row, those of odd index in the second row). We can consider the second and third rows as coefficients of a polynomial of degree $n - 1$:

$$\begin{aligned} G_2(s) &= a_1 s^{n-1} + b_1 s^{n-2} + a_3 s^{n-3} + \cdots \\ &= a_1 s^{n-1} + (a_1 a_2 - a_0 a_3) s^{n-2} + a_3 s^{n-3} \\ &\quad + (a_1 a_4 - a_0 a_5) s^{n-4} + \cdots \end{aligned}$$

Hence $G_2(s)$ is related to $f_1(s)$; $f_1(s)$ is obtained from $G_2(s)$ by multiplying every other coefficient by a_1. If $\delta_k^{(2)}$ denotes the kth Hurwitz determinant for $G_2(s)$, we readily verify that

$$\delta_k^{(2)} = \delta_k \div a_1^{[k/2]},$$

where $[k/2]$ denotes the largest integer contained in $k/2$: $[2/2] = 1$, $[3/2] = 1$, $[4/2] = 2, \ldots$ Hence by (7-56),

$$\delta_k^{(2)} = a_1^{k-1-[k/2]} \, \Delta_{k+1}$$

or

$$\delta_k^{(2)} = a_1^{k^*} \, \Delta_{k+1}, \qquad k^* = k - 1 - [k/2].$$

A similar relationship holds between each successive pair of rows. Thus if we denote the elements of the first column as $A_0 = a_0$, $A_1 = a_1$, $A_2 = b_1$, $A_3 = c_1$, ..., then A_m is the second coefficient (in the "a_1 position") for a corresponding polynomial $G_m(s)$. Thus $G_1(s) = a_0 s^n + a_1 s^{n-1} + \cdots = f(s)$; $G_2(s)$ is as above; $G_3(s) = b_1 s^{n-2} + c_1 s^{n-3} + \cdots = A_2 s^{n-2} + A_3 s^{n-3} + \cdots$ If $\delta_k^{(m)}$ denotes the kth Hurwitz determinant for $G_m(s)$, then, as above,

$$\delta_k^{(m)} = A_{m-1}^{k^*} \, \delta_{k+1}^{(m-1)}.$$

Furthermore $\delta_1^{(m)} = A_m$ (just as $\Delta_1 = a_1$), so that

$$A_m = \delta_1^{(m)} = A_{m-1}^{1^*} \delta_2^{(m-1)} = A_{m-1}^{1^*} A_{m-2}^{2^*} \delta_3^{(m-2)} = \cdots$$

$$= A_{m-1}^{1^*} A_{m-2}^{2^*} \cdots A_1^{(m-1)^*} \delta_m^{(1)}$$

or, since $G_1(s) = f(s)$,

$$A_m = A_{m-1}^{1^*} A_{m-2}^{2^*} \cdots A_1^{(m-1)^*} \Delta_m.$$

Since $1^* = 0$, $2^* = 0$, $3^* = 1$, $4^* = 1$, $5^* = 2$, $6^* = 2$, $A_1 = a_1$, $A_2 = b_1$, $A_3 = c_1$, $A_4 = d_1$, $A_5 = e_1$, ..., we obtain the rules

$$b_1 = \Delta_2, \quad c_1 = \Delta_3, \quad d_1 = a_1 \Delta_4,$$

$$e_1 = a_1 b_1 \Delta_5, \quad f_1 = a_1^2 b_1 c_1 \Delta_6, \ldots$$

PROBLEMS

1. Test the following for stability:

(a) $(D^4 + 3D^3 + D^2 + 2D + 1)x = F(t)$

(b) $(D^5 + 7D^4 + 11D^3 + 17D^2 + 16D + 3)x = F(t)$

2. Determine the range of b for which each of the following is stable:

(a) $(D^4 + D^3 + bD^2 + 2D + 4)x = F(t)$

(b) $(D^5 + 5D^4 + 10D^3 + 10D^2 + bD + 1)x = F(t)$

3. Find $\phi(s)$ and $f_1(s)$ for each of the following choices of $f(s)$:

(a) $s^3 + 2s^2 + 3s + 1$ (b) $s^4 + 3s^3 + 2s^2 + s + 1$

4. Show in detail that if $\phi(s)$ is a rational function satisfying (7–53), then at each pole β of $\phi(s)$ (necessarily on the imaginary axis), $\phi(s)$ has principal part $\lambda/(s - \beta)$, where λ is real and positive. [*Hint:* Let $s - \beta = \rho e^{i\psi}$ and consider the behavior of $\arg \phi(s)$ on a ray $\psi = \text{const.}$ Write

$$\phi(s) = (s - \beta)^{-k} \cdot [\lambda + \lambda_1(s - \beta) + \cdots],$$

where $\lambda \neq 0$; let $\gamma = \arg \lambda$ and show that, on the ray, $\arg \phi(s)$ approaches $\gamma - k\psi$ as $\rho \to 0$. Show by choosing $\psi = 0$, $\psi = (\gamma \pm \pi)/k$ that (7–53) is contradicted unless $k = 1$ and $\gamma = 0$.]

1. (a) unstable   (b) stable

2. (a) $b > 4$ (b) $1.04 < b < 15.3$

3. (a) $\phi = (s^3 + 3s)/(2s^2 + 1)$, $f_1 = 4s^2 + 5s + 2$

(b) $\phi = \dfrac{s^4 + 2s^2 + 1}{3s^3 + s}$, $f_1 = 9s^3 + 5s^2 + 3s + 3$

7–6 Nyquist criterion. Polynomial case. Since every polynomial
$V(s) = a_0 s^n + \cdots$ is an analytic function of the complex variable s, the
number N_0 of zeros of $V(s)$ in a region bounded by a curve C can be found
by the argument principle (Section 3–15):

$$N_0 = \frac{1}{2\pi i} \oint_C \frac{V'(s)}{V(s)}\, ds = \frac{1}{2\pi} \cdot [\text{total increment in arg } V \text{ about } C]. \quad (7\text{–}60)$$

To study stability we wish to determine whether $V(s)$ has any zeros in the
right half-plane, Re $(s) \geqq 0$. To this end we choose C as a semicircle of
radius R:

$$\text{Re } s = 0, \quad -R \leqq \text{Im } s \leqq R; \qquad |s| = R, \quad -\frac{\pi}{2} \leqq \text{arg } s \leqq \frac{\pi}{2}$$

(see Fig. 7–3). If R is sufficiently large, C will enclose all roots of $V(s)$ in
the right half-plane. If, in accordance with (7–60), the total increment of
arg V about C is zero, then there are no zeros of V in the right half-plane,
and the corresponding differential equation $V(D)x = f(t)$ is stable. (It
should be noted that if V has a zero on the imaginary axis, then arg V
becomes undefined at this point; because of the root with Re $(s) = 0$, the
system must be *unstable*.)

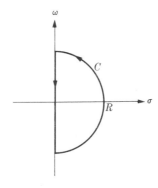

FIG. 7–3. Application of argument principle to stability.

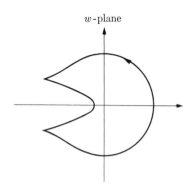

w-plane

arg $V(i\omega)$

FIG. 7–4. Nyquist diagram for $V(s) = s^2 + 2s - 3$.

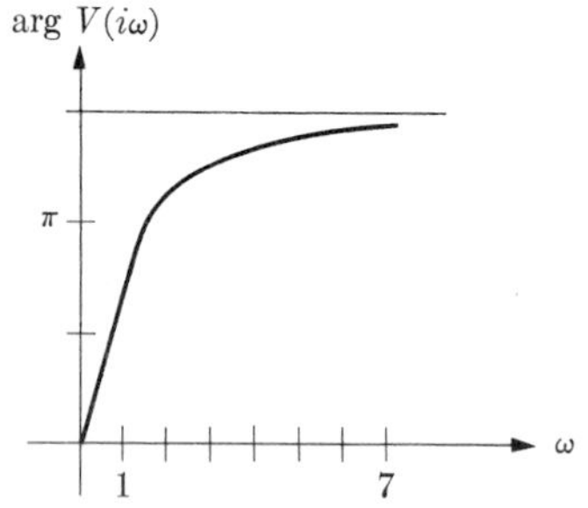

FIG. 7–5. Graph of arg $V(i\omega)$, $V(s) = s^3 + 2s^2 + s + 1$.

It is important to know how large R must be chosen to ensure that all roots of V in the right half-plane lie inside C. The following theorem provides an answer.

THEOREM 3. *The polynomial* $V(s) = a_0 s^n + \cdots + a_n$ *has no roots for which* $|s| \geq R$, *if* $R = 1 + M/|a_0|$ *and* M *is the largest of* $|a_1|, \ldots, |a_n|$.

The proof is left as an exercise (Problem 1 below).

In computing the total change of arg V on C we can reason as in Section 3–15, that as s traces C, $w = V(s)$ traces a curve C_w in the w-plane. The number of zeros inside C is then equal to the number of times C_w encircles the origin of the w-plane in the positive direction. Hence it is sufficient to graph C_w and determine whether it encircles $w = 0$. In graphing C_w we can simplify the work by noting that along the circular part of C, $V(s)$ is approximately equal to $a_0 s^n = a_0 R^n e^{in\theta}$, for R large. Hence this part of C corresponds to an approximately circular arc in the w-plane, along which arg w varies from

$$-n\frac{\pi}{2} \quad \text{to} \quad +n\frac{\pi}{2} \cdot$$

This need not be graphed in detail, but can be joined to the graph of the part of C_w corresponding to one diameter of C to give a closed path in the w-plane.

EXAMPLE. $V = s^2 + 2s - 3$. Here of course we immediately verify that the system is unstable. Along the diameter of C, $V = -3 - \omega^2 + 2i\omega$ and the path $w = V(s)$ is easily traced (Fig. 7–4). Along the circular part, $w = s^2$ approximately, and accordingly we trace most of the circle $|w| = R^2$, $-\pi < \arg w < \pi$. In accordance with Theorem 3, R is chosen as 4. The resulting curve C_w encircles the origin $w = 0$ once; hence there is one zero inside C, and the system is unstable.

A diagram such as that of Fig. 7–4 is known as a *Nyquist diagram*, and the determination of stability by determining how many times C_w encircles $w = 0$ is known as the *Nyquist criterion*. The essential part of the Nyquist diagram is a graph of $V(i\omega)$. Since $V(i\omega) = 1/Y(i\omega)$, this is essentially a graph of the *frequency response function*. Indeed we can graph $w = Y(s) = 1/V(s)$ for s on C, and can determine stability in the same way. For $\arg Y = -\arg V$, and if the total change of $\arg Y$ is zero on C, then the total change of $\arg V$ is zero. If $Y(s)$, rather than $V(s)$, is graphed, then the circular part of C corresponds to a *small* circular arc $w = 1/(a_0 R^n e^{in\theta})$, rather than a large one.

7–7 Stability determined from graph of arg $V(i\omega)$. The analysis can be considerably simplified by noting that all the information is contained in the graph of $V(i\omega)$ or, even better, in the graph of $\arg V(i\omega)$. We formulate a precise form of the rule as follows:

THEOREM 4. *Let $V(s) = a_0 s^n + \cdots + a_0$ be a polynomial with real coefficients, with $a_0 > 0$. Let M be the largest of $|a_1|, |a_2|, \ldots, |a_n|$; let $R = 1 + 3M/|a_0|$. Let $\arg V(i\omega)$ be defined as a continuous function of ω for $\omega \geqq 0$. Let α be the total change in $\arg V(i\omega)$ as ω goes from 0 to R:*

$$\alpha = \arg V(iR) - \arg V(0).$$

Let k be the integer closest to $\alpha \div (\pi/2)$. Then $-n \leqq k \leqq n$. If $k = n$, $V(s)$ has all its roots in the left half-plane. If $k < n$, then $V(s)$ has $\frac{1}{2}(n - k)$ roots in the right half-plane.

Remark. As pointed out above, when V has a root on the imaginary axis, the procedure breaks down and the system is unstable.

Before proving Theorem 4, we consider an example: $V(s) = s^3 + 2s^2 + 2s + 1$. Here $V(i\omega) = 1 - 2\omega^2 + i(2\omega - \omega^3)$, and it is clear that $V(i\omega)$ is never zero, so that

$$\arg V(i\omega) = \arctan \frac{2\omega - \omega^3}{1 - 2\omega^2}$$

can be defined as a continuous function of ω. Here $M = 2$, $a_0 = 1$, $R = 7$, and we need to graph $\arg V(i\omega)$ in the interval $0 \leqq \omega \leqq 7$. For $\omega = 0$, $V = 1$, and $\arg V(i\omega)$ can be chosen as zero. Continuity then dictates the choice of $\arg V(i\omega)$ for $0 \leqq \omega \leqq 7$. The graph is shown in Fig. 7–5. Hence α is very close to $3\pi/2$, so that $2\alpha/\pi$ is very close to 3; $k = 3 = n$. Hence $V(s)$ has no roots in the right half-plane.

Proof of Theorem 4. We return to Fig. 7–3 and seek the number N_0 of roots of $V(s)$ inside C by determining the total change in $\arg V(s)$ about C. It follows from Theorem 3 that if R is chosen as $1 + 3M/|a_0|$, then V has no roots in the right half-plane outside C. Hence with this choice of R, N_0 equals the number of roots of V in the right half-plane.

We first show that arg $V(s)$ changes by approximately $n \cdot \pi$ on the semicircular part of C, with an error less than $\pi/2$. Indeed,

$$\text{arg } V(s) = \text{arg } [a_0 s^n + a_1 s^{n-1} + \cdots + a_n]$$

$$= \text{arg} \left[(a_0 s^n) \left(1 + \frac{a_1}{a_0 s} + \cdots + \frac{a_n}{a_0 s^n} \right) \right]$$

$$= \text{arg } (a_0 s^n) + \text{arg} \left(1 + \frac{a_1}{a_0 s} + \cdots + \frac{a_n}{a_0 s^n} \right)$$

$$= \text{arg } (a_0 s^n) + \beta(s),$$

$$\beta(s) = \text{arg} \left(1 + \frac{a_1}{a_0 s} + \cdots + \frac{a_n}{a_0 s^n} \right).$$

The term arg $(a_0 s^n)$ = arg $(a_0 R^n e^{in\theta})$ changes by $n\pi$ on the semicircle, as θ goes from $-\pi/2$ to $+\pi/2$. It will be shown that, because of the choice of R, the second term changes by less than $\pi/2$. To this end, we remark that on the semicircle,

$$\left| \frac{a_1}{a_0 s} + \cdots + \frac{a_n}{a_0 s^n} \right| \leqq \frac{1}{|a_0|} \left(\frac{|a_1|}{R} + \cdots + \frac{|a_n|}{R^n} \right)$$

$$\leqq \frac{M}{|a_0|} \left(\frac{1}{R} + \cdots + \frac{1}{R^n} \right)$$

$$< \frac{M}{|a_0| R} \left(1 + \frac{1}{R} + \cdots + \frac{1}{R^n} + \cdots \right) = \frac{M}{|a_0|(R-1)}$$

$$< \frac{M}{|a_0|} \cdot \frac{1}{3M/|a_0|} = \frac{1}{3},$$

since $R > 1$ and $R - 1 = 3M/|a_0|$. Hence the complex number

$$\zeta = 1 + \left(\frac{a_1}{a_0 s} + \cdots + \frac{a_n}{a_0 s^n} \right)$$

lies inside the circle of radius $\frac{1}{3}$ about 1 (Fig. 7–6). From this, simple geometry shows that $\beta(s) = $ arg ζ can be chosen to lie between $-\arcsin \frac{1}{3}$ and $+\arcsin \frac{1}{3}$. That is,

$$|\beta(s)| < \arcsin \frac{1}{3} < \frac{\pi}{4} = \arcsin \frac{1}{\sqrt{2}}.$$

Hence the total change of $\beta(s)$ on the semicircle must be less than $\pi/2$. We can now conclude that arg $V(s)$ changes by $n\pi + \epsilon$ on the semicircle, with $|\epsilon| < \pi/2$.

ζ -plane

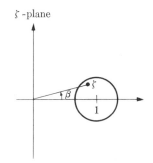

Fig. 7–6.   Proof of Theorem 4.

Next we consider the change in arg $V(i\omega)$ as ω goes from R to $-R$. Since $V(s)$ is a polynomial with real coefficients, $V(-i\omega)$ is the conjugate of $V(i\omega)$; that is, the graph of $V(i\omega)$ exhibits symmetry in the real axis (compare Fig. 7–4). Hence the change of arg $V(i\omega)$ from $\omega = R$ to $\omega = -R$ is twice the change from $\omega = R$ to 0. By definition of α, the change of arg $V(i\omega)$ from $\omega = R$ to $\omega = -R$ is thus -2α. Hence

$$N_0 = \frac{1}{2\pi}\,(-2\alpha + n\pi + \epsilon) = \frac{1}{2}\left(-\frac{\alpha}{\pi/2} + n + \frac{\epsilon}{\pi}\right).$$

It follows that

$$\frac{\alpha}{\pi/2} - \frac{\epsilon}{\pi} = n - 2N_0$$

must be an integer. Since $|\epsilon/\pi| < 1/2$, this must be the integer k nearest to $\alpha \div (\pi/2)$. Hence $k = n - 2N_0$ and

$$N_0 = \tfrac{1}{2}(n - k).$$

Since $0 \leqq N_0 \leqq n$, $-n \leqq k \leqq n$. If $k = n$, $N_0 = 0$ and there are no roots in the right half-plane; if $k < n$, there are $\frac{1}{2}(n - k)$ roots in the right half-plane.

7–8 Nyquist criterion.   Rational function case.   In various problems the transfer function $Y(s)$ appears as a general rational function of s, rather than as $1/V(s)$, where V is a polynomial. In these cases $Y = P(s)/Q(s)$, where P and Q are polynomials. The characteristic equation is $Q(s) = 0$, and stability is determined by whether its roots all lie in the left half-plane.

We can apply the methods of the preceding sections to determine whether the roots of Q lie in the left half-plane. We can also apply the procedure of Section 7–6 to $Y(s)$ rather than to $Q(s)$. The roots of Q now correspond to *poles* of $Y(s)$ and we wish to know how many poles $Y(s)$ has in the right half-plane. If N_0, N_p are the number of zeros and poles of Y inside

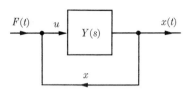

Fig. 7-7.   System with feedback.

C, then by the argument principle,

$$N_0 - N_p = \frac{1}{2\pi i} \oint_C \frac{Y'(s)}{Y(s)} \, ds = \frac{1}{2\pi} \cdot [\text{total increment of arg } Y(s) \text{ about } C].$$

(7–80)

If R is chosen large enough (see Theorem 3 above), all zeros and poles of $Y(s)$ in the right half-plane lie inside C. However, formula (7–80) will permit us to determine only the difference $N_0 - N_p$. If N_0 is known, then N_p can be found; if N_0 is known to be zero, then

$$N_p = -\frac{1}{2\pi} \cdot [\text{total increment of arg } Y(s)].$$

In any case, if $N_0 - N_p$ is negative, $Y(s)$ must have poles in the right half-plane, so that the system is unstable.

In graphing $w = Y(s)$ for s on C, we can again obtain the image of the semicircular portion of C as an approximately circular arc. Thus if

$$Y(s) = \frac{a_0 s^n + \cdots + a_n}{b_0 s^m + \cdots + b_m},$$

then Y is approximately $(a_0/b_0 s^{n-m})$ for large R. In general, n will be less than m, so that the arc will be on a circle of small radius $|a_0/b_0| R^{n-m}$. Along the diameter of C, $Y(s) = Y(i\omega)$, the frequency response function. Accordingly, it is essentially the graph of the frequency response function which determines whether or not the system is stable.

In studying physical systems with "feedback," the point of view under consideration leads us to important applications. In the system illustrated in Fig. 7–7, there is an input $F(t)$ and an output $x(t)$. The output is compared to the input to yield the "error" $u(t) = F(t) - x(t)$. The error is then fed as input to a mechanism with transfer function $Y(s)$, and x is the corresponding output. Therefore, all initial values being zero,

$$\mathcal{L}[u] = \mathcal{L}[F] - \mathcal{L}[x], \qquad \mathcal{L}[x] = Y(s)\mathcal{L}[u],$$

so that

$$\mathcal{L}[x] = Y(s)(\mathcal{L}[F] - \mathcal{L}[x]), \qquad \mathcal{L}[x] = \frac{Y(s)}{1 + Y(s)} \mathcal{L}[F].$$

Hence the final transfer function is

$$Y_1(s) = \frac{Y(s)}{1 + Y(s)}.$$

This equation gives a relation between the "open-loop" transfer function $Y(s)$ and the "closed-loop" transfer function $Y_1(s)$. In many cases the open-loop transfer function is well known, and in particular the Nyquist diagram of $Y(s)$ for the curve C may be known. Finding the number of poles of $Y_1(s)$ inside C is now reduced to finding the number of zeros of $1 + Y(s)$ inside C. The argument principle can be applied to the graph of $Y(s)$ translated 1 unit to the right, and we can reason as before; or we can simply count the total number of times the graph of $Y(s)$ encloses the point -1. This gives the number of zeros of $1 + Y(s)$ inside C minus the number of poles of $Y(s)$ inside C. In many cases it is known that $Y(s)$ has no poles in the right half-plane, that is, that the open loop by itself is stable. In such a case, the closed loop is stable if and only if the point -1 is not enclosed by the graph of $Y(s)$, s on C.

PROBLEMS

1. Prove Theorem 3. [*Hint:* Suppose s is a root and $|s| > 1$. Show that

$$1 = -\left(\frac{a_1}{a_0 s} + \frac{a_2}{a_0 s} + \cdots + \frac{a_n}{a_0 s^n}\right)$$

and hence, as in the proof of Theorem 4,

$$\left. 1 < \frac{M}{|a_0|(|s| - 1)}.\right]$$

2. Find the image of the curve C of Fig. 7–3 under the transformation $w = s^3 + 2$ for the cases $R = 1, 2, 3$. Use these graphs to determine the number of roots of $s^3 + 2 = 0$ in the right half-plane.

3. Graph $\arg V(i\omega)$ against ω for $\omega \geqq 0$, and use the results to determine stability for the following choices of V:

(a) $s^3 + 2s^2 + 3s + 1$ (b) $s^4 + 3s^3 + s^2 + s + 8$

4. Let the system of Fig. 7–7 have open-loop transfer function

$$Y(s) = \frac{s^2 + s + 7}{s^3 + 2s^2 + s + 1}.$$

Show that the open loop is stable, but the closed loop is unstable.

ANSWERS

2. Two roots in the right half-plane.   3. (a) stable   (b) unstable

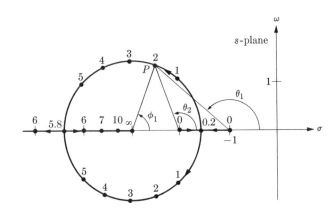

Fig. 7–8. Root locus for $(s + 1)(s + 2) + K(s + 3) = 0$. Values of K are shown above the locus points.

7–9 Root-locus method. In a variety of applications, the characteristic equation has at least one adjustable parameter; we are concerned with the influence of such a parameter on the stability of the system. An important case is that in which the equation has form

$$p(s) + Kq(s) = 0, \tag{7–90}$$

where $p(s)$ and $q(s)$ are polynomials with known zeros and K is a real constant, the adjustable parameter. For each value of K, we can in principle find all roots $s_1, \ldots, s_n$ of Eq. (7–90). As K varies, the n roots vary and hence trace n curves in the complex plane. These curves together form the *root locus*.

For example, in the system of Fig. 7–7 (Section 7–8), we might have

$$Y(s) = \frac{Kq(s)}{p(s)}$$

as the open-loop transfer function. The constant K affects the amplification or *gain* $|Y(i\omega)|$, and is hence called the *gain constant* or simply the *gain*. The closed-loop transfer function is

$$Y_1(s) = \frac{Y(s)}{1 + Y(s)} = \frac{Kq(s)}{p(s) + Kq(s)}.$$

Hence the characteristic equation is (7–90), and a study of the root locus yields information on the effect upon stability of varying the open-loop gain, K.

In this section and the following section we consider the principal features of the root locus associated with Eq. (7–90). The general approach is applicable to equations in which the parameter enters in other ways, and to equations with several parameters. We shall normally assume that $p(s)$ and $q(s)$ have all coefficients positive or zero and all zeros in the left half-plane or on the imaginary axis, and that $K > 0$. The root locus will be plotted for $0 < K < \infty$. The limiting values $K = 0$ and $K = +\infty$ will also be of interest.

EXAMPLE 1. $(s + 1)(s + 2) + K(s + 3) = 0$. Since the equation is quadratic, we can solve explicitly:

$$s^2 + (3 + K)s + 2 + 3K = 0,$$

$$s = \frac{-3 - K \pm \sqrt{K^2 - 6K + 1}}{2}.$$

The locus is shown in Fig. 7–8; arrows point in the direction of increasing K. We note that when $K \to 0$, the roots become -1, -2; these are the zeros of $p(s)$, in accordance with Eq. (7–90). When $K \to \infty$, the roots approach $-\infty$ and -3; this could also be anticipated from the relation

$$K = -\frac{p(s)}{q(s)} = -\frac{(s + 1)(s + 2)}{(s + 3)}. \tag{7–91}$$

For each K, there are two distinct values of s, except when the discriminant $K^2 - 6K + 1$ is 0; the corresponding values are $K = 5.8$ and $K = 0.2$, for which $s = -4.4$ and -1.6, respectively; these are self-intersections, or double-points, of the locus.

The root locus has one fundamental geometrical property, which is the basis of a purely graphical method of constructing the locus. We write

$$p(s) = A(s - a_1)(s - a_2) \cdots (s - a_n),$$

$$q(s) = B(s - b_1)(s - b_2) \cdots (s - b_m),$$

where $A > 0$ and $B > 0$. Our equation (7–90) can be written

$$\frac{A(s - a_1) \cdots (s - a_n)}{B(s - b_1) \cdots (s - b_m)} = -K.$$

We now take the argument of both sides and find (see Section 1–7), since $\arg A = \arg B = 0$,

$$\arg (s - a_1) + \cdots + \arg (s - a_n)$$
$$-[\arg (s - b_1) + \cdots + \arg (s - b_m)] = \pi + 2k\pi,$$

424 STABILITY [CHAP. 7

where $k = 0, \pm 1, \ldots$ If we write $\theta_j = \arg(s - a_j)$, $\phi_j = \arg(s - b_j)$, then our condition reads

$$\theta_1 + \theta_2 + \cdots + \theta_n - (\phi_1 + \phi_2 + \cdots + \phi_m) = (2k + 1)\pi. \quad (7\text{--}92)$$

This geometric condition is satisfied at every point of the root locus. Conversely, if (7–92) holds, we conclude that $p(s)/q(s)$ has argument $(2k + 1)\pi$, so that $p(s)/q(s)$ equals some negative real number $-K$. Thus Eq. (7–92) completely characterizes the root locus.

In Example 1, $\theta_1 = \arg(s + 1)$, $\theta_2 = \arg(s + 2)$, $\phi_1 = \arg(s + 3)$, and our condition is $\theta_1 + \theta_2 - \phi_1 = (2k + 1)\pi$. The three angles are shown at the point P in Fig. 7–8; with the aid of a protractor, we find $\theta_1 = 139°, \theta_2 = 111°, \phi_1 = 70°$, so that $\theta_1 + \theta_2 - \phi_1 = 250° - 70° = 180° = \pi$ radians. If we now try to move P so as to maintain the condition $\theta_1 + \theta_2 - \phi_1 = \pi$, we easily discover that P must trace, at least approximately, the locus shown. For example, if P moves upward from the position marked, $\theta_1 + \theta_2$ decreases more rapidly than ϕ_1 increases, so that no locus points occur.

The portions of the locus on the real axis are found at once from the condition $\theta_1 + \theta_2 - \phi_1 = (2k + 1)\pi$. For $s > -1$ all angles can be taken to be 0, so that no locus points occur; between -2 and -1 we have $\theta_1 = \pi, \theta_2 = 0, \phi_1 = 0$, so that the locus condition is satisfied; between -2 and -3, $\theta_1 = \pi, \theta_2 = \pi, \phi_1 = 0$, so that

$$\theta_1 + \theta_2 - \phi_1 = 2\pi \neq (2k + 1)\pi,$$

and again there is no locus; between -3 and $-\infty$, $\theta_1 = \theta_2 = \phi_1 = \pi$, so that

$$\theta_1 + \theta_2 - \phi_1 = \pi,$$

and again the locus appears. In general, as we move from right to left, the expression $\theta_1 + \theta_2 - \phi_1$ jumps by π each time we pass a zero of p or q, so that we obtain alternating locus and nonlocus intervals; a multiple zero of p or q must here be counted according to multiplicity, so that the alternation occurs only when we pass a zero of *odd* multiplicity.

EXAMPLE 2. $(s^2 + 2s + 2)(s + 2) + K = 0$. Here $p = (s + 1 - i) \cdot (s + 1 + i)(s + 2)$, $q(s) = 1$, so that $a_1 = -1 + i$, $a_2 = -1 - i$, $a_3 = -2$, and our locus condition is $\theta_1 + \theta_2 + \theta_3 = (2k + 1)\pi$, where $\theta_1, \theta_2, \theta_3$ are the angles of Fig. 7–9. For s real, $\theta_1 + \theta_2 = 0$, so that we require $\theta_3 = (2k + 1)\pi$; the condition is satisfied only for $s < -2$. As we move upward from the real axis near a point on the locus interval,

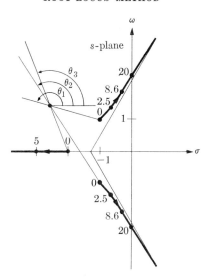

FIG. 7–9.   Root locus for $(s^2 + 2s + 2)(s + 2) + K = 0$.

$\theta_1 + \theta_2$ becomes negative, while θ_3 also decreases; hence there are no complex locus points near this part of the real axis; a similar exploration shows that there are no complex locus points near the rest of the real axis. At the point $-1 + i$ we have $\theta_2 = \pi/2$, $\theta_3 = \pi/4$, while θ_1 is undefined. However, if we move away from this point in the direction $\theta_1 = \pi/4$, the locus condition will be satisfied near the point, and we discover by measurement that we can continue to move upward and to the right, approaching a straight line along which θ_1, θ_2, θ_3 are all approximately $\pi/3$.

Since all coefficients in our equation are real, the values of s satisfying it come in conjugate pairs. Thus the root locus is always *symmetric in the real axis.* Corresponding to the curve moving upward from $-1 + i$, we hence obtain a curve moving downward from $-1 - i$. Values of K can easily be assigned at the locus points found thus far by writing

$$K = -(s^2 + 2s + 2)(s + 2) = -(s + 1 - i)(s + 1 + i)(s + 2)$$

or, by taking absolute values ($K > 0$),

$$K = |s + 1 - i|\,|s + 1 + i|\,|s + 2| = |s - a_1|\,|s - a_2|\,|s - a_3|.$$

Thus the value of K at s is simply the product of the distances of s from the zeros a_1, a_2, a_3. Typical values are shown in Fig. 7–9; they indicate that we have the complete locus, since we have three values of s for each K.

In this example, the locus can easily be obtained exactly. For our locus equation can be written $s^3 + 4s^2 + 6s + 4 + K = 0$ or, with $s = \sigma + i\omega$,

$$\sigma^3 + 3i\sigma^2\omega - 3\sigma\omega^2 - i\omega^3 + 4(\sigma^2 + 2i\sigma\omega - \omega^2)$$
$$+ 6(\sigma + i\omega) + 4 + K = 0.$$

We take real and imaginary parts:

$$\sigma^3 - 3\sigma\omega^2 + 4\sigma^2 - 4\omega^2 + 6\sigma + 4 + K = 0, \qquad (7\text{--}93)$$

$$3\sigma^2\omega - \omega^3 + 8\sigma\omega + 6\omega = 0. \qquad (7\text{--}94)$$

The second equation itself describes the root locus; more precisely, it describes those points s for which the locus equation is satisfied for some real K, positive or negative. Hence if we plot the locus (7–94) and reject the points for which K is negative, we have our root locus. The equation (7–93) gives the value of K at each point:

$$K = -\sigma^3 + 3\sigma\omega^2 - 4\sigma^2 + 4\omega^2 - 6\sigma - 4. \qquad (7\text{--}93')$$

Now Eq. (7–94) is satisfied when $\omega = 0$ (that is, *along the real axis* of the s-plane) or when

$$3\sigma^2 - \omega^2 + 8\sigma + 6 = 0. \qquad (7\text{--}95)$$

This second degree equation represents a hyperbola:

$$\frac{\omega^2}{2/3} - \frac{[\sigma + (4/3)]^2}{2/9} = 1;$$

the asymptotes, obtained by replacing the 1 by 0, are the straight lines

$$\omega = \pm\sqrt{3}\,(\sigma + \tfrac{4}{3}).$$

From Eq. (7–93') we conclude that, when $\omega = 0$,

$$K = -\sigma^3 - 4\sigma^2 - 6\sigma - 4 = -(\sigma^2 + 2\sigma + 2)(\sigma + 2),$$

so that K is positive only for $\sigma < -2$. From (7–95) we find

$$\omega^2 = 3\sigma^2 + 8\sigma + 6,$$

so that along the hyperbola, by (7–93'),

$$K = -\sigma^3 + 3\sigma(3\sigma^2 + 8\sigma + 6) - 4\sigma^2 + 4(3\sigma^2 + 8\sigma + 6) - 6\sigma - 4$$
$$= 8\sigma^3 + 32\sigma^2 + 44\sigma + 20$$
$$= 4(\sigma + 1)(2\sigma^2 + 6\sigma + 5).$$

The last factor is always positive, so that K is positive only when $\sigma > -1$.

Thus we obtain the portions of the hyperbola shown in Fig. 7–9. The values of K can also be readily computed from the last equation.

The procedure illustrated here can be applied to an equation $p(s) + Kq(s) = 0$ for which $q(s) \equiv 1$ and $p(s)$ is at most of fourth degree; if $q(s)$ is of first degree or higher or if $p(s)$ is of degree greater than four, the method is applicable but may be cumbersome.

7–10 Properties of the root locus. The examples of Section 7–9 suggest a number of properties valid for all root loci. We formulate these here as a theorem. Throughout we assume that

$$p(s) = A_0 s^n + A_1 s^{n-1} + \cdots + A_n = A_0(s - a_1) \cdots (s - a_n),$$

$$q(s) = B_0 s^m + B_1 s^{m-1} + \cdots + B_m = B_0(s - b_1) \cdots (s - b_m),$$

$$A_0 > 0, \qquad B_0 > 0, \qquad K > 0, \tag{7–100}$$

that all coefficients of $p(s)$ and $q(s)$ are real, and that $p(s)$ and $q(s)$ have no common zeros. The root locus is always formed of all complex s such that $p(s) + Kq(s) = 0$ for some real K, $K > 0$. We can further assume that $n \geqq m$, for otherwise we can achieve the same effect by writing the equation as $q(s) + (1/K)p(s) = 0$.

THEOREM 5. *The root locus has the following properties:*

I. *The locus is formed of exactly n curves (branches), each of which leads from a zero a_k of $p(s)$ to a zero b_l of $q(s)$, or to ∞, as K goes from 0 to $+\infty$.*

II. *The branches do not meet except at points s of the locus such that*

$$pq' - qp' = 0. \tag{7–101}$$

At such points the branches meet at right angles, unless $pq'' - qp'' = 0$.

III. *The locus is symmetric with respect to the real axis.*

IV. *The portion of the locus on the real axis consists of those real s such that an odd number of a_k and b_l lie to the right of s on the real axis.*

V. *If $n = m$, the whole root locus lies inside the circle $|s| = 1 + R$, where*

$$R = \max \left(\frac{|A_1|}{A_0}, \frac{|A_2|}{A_0}, \ldots, \frac{|A_n|}{A_0}, \frac{|B_1|}{B_0}, \ldots, \frac{|B_m|}{B_0} \right).$$

If $n > m$, then precisely $n - m$ branches tend to ∞ as $K \to +\infty$; these branches have as asymptotes the rays meeting at

$$s = s_0 = \frac{1}{n - m} \left[\frac{B_1}{B_0} - \frac{A_1}{A_0} \right],$$

with angles of inclination α, 3α, 5α, $\ldots$, $[2(n - m) - 1]\alpha$, where $\alpha = \pi/(n - m)$.

VI. *Let* $\gamma_i = \sum_{l=1}^{m} \arg (a_i - b_l) - \sum_{j,\,a_i \neq a_j} \arg (a_i - a_j) + \pi$, *and let* $\eta_l = \sum_{i=1}^{n} \arg (b_l - a_i) - \sum_{j,\,b_j \neq b_l} \arg (b_l - b_j) + \pi$, *for some choice of the angle in each case. If a_i is a simple zero of $p(s)$, then exactly one branch of the locus starts at a_i, with angle of inclination γ_i; if b_l is a simple zero of $q(s)$, then precisely one branch of the locus starts at b_l, with angle of inclination η_l. If a_i is a zero of $p(s)$ of multiplicity $\mu > 1$, then μ branches start at a_i with angles of inclination $(\gamma_i + 2k\pi)/\mu$, where $k = 0, 1, \ldots, \mu - 1$; if b_l is a zero of $q(s)$ of multiplicity $\mu > 1$, then μ branches start at b_l, with angles of inclination $(\eta_l + 2k\pi)/\mu$, $k = 0, 1, \ldots, \mu - 1$.*

Proof for I *and* II. For each K the equation $p(s) + Kq(s) = 0$ is an algebraic equation with n roots; the roots are distinct unless $p'(s) + Kq'(s) = 0$ at a root s. Since in general the roots vary continuously with the coefficients, it is clear that n curves are obtained, meeting only where $p(s) + Kq(s) = 0$ and $p'(s) + Kq'(s) = 0$. At these points we can eliminate K to obtain Eq. (7–101):

$$p(s)q'(s) - p'(s)q(s) = 0.$$

Thus they satisfy an algebraic equation of degree at most $n + m - 1$ and hence are finite in number. Conversely, if s is a point of the locus and (7–101) holds, then we deduce that

$$Kq(s)q'(s) + p'(s)q(s) = 0,$$

so that $q(s) = 0$ or $p'(s) + Kq'(s) = 0$; but $q(s) = 0$ and $p(s) + Kq(s) = 0$ imply that $p(s) = 0$, contrary to the assumption that p and q have no common zeros; thus $q(s) \neq 0$ and hence $p'(s) + Kq'(s) = 0$, so that the point s is a multiple point of the locus.

When $K = 0$, our n values of s are given by $a_1, \ldots, a_n$; by continuity these must be the limiting values as $K \to 0$. To analyze the behavior of the locus for large K, we first write the equation thus: $(1/K)p(s) + q(s) = 0$. Then for $1/K = 0$ we obtain the m values $b_1, \ldots, b_m$; hence for K large positive there must be m values close to these. If $m < n$, then $n - m$ values are unaccounted for. To find these values, we set $z = 1/s$ and write the equation as follows:

$$\frac{1}{K}(A_0 + \cdots + A_n z^n) + B_0 z^{n-m} + \cdots + B_m z^n = 0. \quad (7\text{–}102)$$

Now for $1/K = 0$ we obtain exactly $n - m$ roots $z = 0, \ldots, 0$; hence for K large plus there are $n - m$ locus points s close to ∞. Therefore $n - m$ branches lead to $s = \infty$.

To study the locus near a point c of the intersection of two branches, we write the equation as follows:

$$K = -\frac{p}{q}\bigg|_c + \left(\frac{-p}{q}\right)'\bigg|_c (s - c) + \frac{1}{2!}\left(\frac{-p}{q}\right)''\bigg|_c (s - c)^2 + \cdots$$

$$= -\frac{p(c)}{q(c)} - \frac{p'(c)q(c) - p(c)q'(c)}{[q(c)]^2}(s - c) + \cdots,$$

by a Taylor series expansion. At a multiple point c, (7-101) holds, so that $p'(c)q(c) - p(c)q'(c) = 0$, and we can write the equation as follows:

$$K = K_c + \frac{1}{2}\frac{p(c)q''(c) - p''(c)q(c)}{[q(c)]^2}(s - c)^2 + \cdots,$$

where $K_c = -p(c)/q(c)$. If the coefficient of the second degree term is not zero, the behavior of the locus near $s = c$ can be found by neglecting higher-degree terms and we obtain

$$s = c \pm [b(K - K_c)]^{1/2}, \qquad b = \frac{2q^2}{pq'' - p''q}\bigg|_c.$$

As K varies near K_c, s varies on two perpendicular lines through c; for example, if b is real, s varies on lines parallel to the real and imaginary axes. Hence, in general, the tangents to the branches meet at right angles at a multiple point, provided $pq'' - qp'' \neq 0$ at the point. When $pq'' - qp'' = 0$, we are led to a similar expression for s, with the square root replaced by an lth root, for some $l > 2$; we conclude that l branches meet at the point, the angle between adjacent branches being π/l. (The case in which $p = 0$ or $q = 0$ at c is discussed in the proof of property VI.)

Remark. The preceding discussion is not fully rigorous. A complete analysis requires greater use of complex function theory. In brief, the function $w = p(s)/q(s)$ is an algebraic function mapping the extended s-plane onto an n-sheeted Riemann surface over the w-plane, and we are interested in the branches of the inverse function along the negative real axis, $w = -K$. That these branches do correspond to curves as described follows at once from known properties of algebraic Riemann surfaces (Vol. II of Reference 4).

Proof for III *and* IV. The symmetry follows at once from the fact that $p(s)$, $q(s)$ have real coefficients and K is real. To study the locus for real s, we write our equation as

$$K = -\frac{p(s)}{q(s)}.$$

For large positive s, $p(s)$ and $q(s)$ are positive, so that a positive K cannot be obtained and there is no locus. The ratio $p(s)/q(s)$ changes sign each time a real root of $p(s)$ or $q(s)$ of odd multiplicity is passed. We thus obtain alternating locus and nonlocus intervals, whose end points are roots of $p(s)$ or $q(s)$; an odd number of such roots must lie to the right of a point in a locus interval, by the sign-change property. At each root of even multiplicity, we may have locus intervals to the left and right, so that the root appears as an interior point of a locus interval; however, it is more proper to consider the intervals to the left and right as different branches of the locus, as in Property I. We may also have nonlocus intervals to the left and right. In the latter case, branches of the locus must lead from the root to the upper and lower half-planes, as in Property VI.

If the two end points of a locus interval are roots of $p(s)$, then $p(s)/q(s)$ is continuous in the interval, and by Rolle's theorem the derivative $[p(s)/q(s)]' = (qp' - pq')/q^2$ must equal zero at least once in the interval. Thus one of the crossing points (7–101) lies in the interval. As in Properties I and II we conclude that from each such point branches lead into the upper and lower half-planes; when $pq'' - p''q \neq 0$ at the point, there are two such branches meeting the real axis at right angles. A similar analysis applies if both end points are roots of $q(s)$.

Proof for V. If $n = m$, we write our equation as

$$(A_0 + KB_0)s^n + (A_1 + KB_1)s^{n-1} + \cdots + (A_n + KB_n) = 0.$$

By Theorem 3 (Section 7–6), for each fixed K, each root s satisfies the inequality

$$|s| < 1 + \frac{M}{A_0 + KB_0}, \qquad M = \max(|A_1 + KB_1|, \ldots, |A_n + KB_n|).$$

Hence also

$$|s| < 1 + \max\left(\frac{|A_1 + KB_1|}{A_0 + KB_0}, \ldots, \frac{|A_n + KB_n|}{A_0 + KB_0}\right). \qquad (7\text{–}103)$$

From this relation we deduce the conclusion: $|s| < 1 + R$, where R is given as in the theorem (Problem 4 below).

When $n > m$, we have seen that there are $n - m$ branches leading to $s = \infty$ as K approaches ∞. It is convenient to describe these by Eq. (7–102), in which $z = 1/s$, so that $z = 0$ corresponds to $s = 0$. We write also $t = [KB_0/A_0]^{1/(n-m)}$, so that $0 < t < \infty$, and write Eq. (7–102) thus:

$$\frac{1}{t^{n-m}} = -z^{n-m}\frac{B_0 + B_1 z + \cdots + B_m z^m}{A_0 + A_1 z + \cdots + A_n z^n} \cdot \frac{A_0}{B_0}$$

$$= -z^{n-m}(1 + (n - m)s_0 z + \cdots).$$

The expression in parentheses is analytic for $z = 0$ and reduces to 1 for $z = 0$. Hence the $(n - m)$-th root of this function can be chosen to be analytic (Problem 5 below), and we find

$$\frac{1}{t} = \omega_k z(1 + s_0 z + \cdots) \quad (k = 0, 1, \ldots, n - m - 1),$$

where $\omega_0, \ldots, \omega_{n-m-1}$ are the $(n - m)$-th roots of -1, $\omega_k = \exp\left[(2k + 1)\pi i/(n - m)\right]$. These are the $n - m$ branches leading to ∞ in the s-plane. If we set $\zeta = 1/(\omega_k t)$, our equation reads

$$\zeta = z + s_0 z^2 + \cdots;$$

this defines ζ as an analytic function of z near $z = 0$, where $\zeta = 0$. We can obtain a series for the inverse function by writing $z = z(\zeta) = \zeta + c_2 \zeta^2 + c_3 \zeta^3 + \cdots$ (see page 184 of Reference 3). We find, for $|\zeta|$ sufficiently small,

$$z = \zeta - s_0 \zeta^2 + \cdots = \zeta(1 - s_0 \zeta + \cdots).$$

Hence

$$s = \frac{1}{z} = \frac{1}{\zeta(1 - s_0 \zeta + \cdots)} = \frac{1}{\zeta}(1 + s_0 \zeta + \cdots)$$

$$= \omega_k t\left(1 + \frac{s_0}{\omega_k t} + \frac{\text{const}}{\omega_k^2 t^2} + \cdots\right)$$

$$= \omega_k t + s_0 + \frac{\text{const}}{\omega_k t} + \cdots$$

The first two terms on the right describe a straight line, approaching ∞ as $t \to \infty$: $s = \omega_k t + s_0$. Since the remaining terms approach zero as $t \to \infty$, the line describes the desired asymptote to the kth branch. The asymptote is clearly a ray through s_0, with angle of inclination $\arg \omega_k = (2k + 1)\alpha$, with $\alpha = \pi/(n - m)$. Hence V is proved.

Proof for VI. For simplicity we take $i = 1$ and assume first that a_1 is a simple zero of $p(s)$, so that $p(a_1) = 0$, $p'(a_1) \neq 0$. Then we proceed as above to obtain a series expansion:

$$K = -\frac{p(s)}{q(s)} = -\frac{p(a_1)}{q(a_1)} - \left(\frac{p}{q}\right)'\bigg|_{a_1}(s - a_1) + \cdots$$

$$= -\frac{p'(a_1)}{q(a_1)}(s - a_1) + \text{const} \cdot (s - a_1)^2 + \cdots$$

Again we invert:

$$s - a_1 = -\frac{q(a_1)}{p'(a_1)}K + \text{const} \cdot K^2 + \cdots$$

Thus the tangent to the locus at $s = a_1$ is the ray

$$s = a_1 - \frac{q(a_1)}{p'(a_1)} K \quad (0 < K < \infty).$$

If we write $q(s) = B_0(s - b_1) \cdots (s - b_m)$, $p(s) = A_0(s - a_1) \cdots (s - a_n)$, we find

$$q(a_1) = B_0(a_1 - b_1) \cdots (a_1 - b_m),$$

$$p'(a_1) = A_0(a_1 - a_2) \cdots (a_1 - a_n),$$

so that the tangent has angle of inclination

$$\arg \frac{-q(a_1)}{p'(a_1)} = \pi + \sum_{l=1}^{m} \arg (a_1 - b_l) - \sum_{j=2}^{n} \arg (a_1 - a_j) = \gamma_1,$$

as asserted. The case of a simple zero b_l of $q(s)$ is similar.

Now let a_1 be a root of $p(s)$ of multiplicity $\mu > 1$, so that $p(s) = (s - a_1)^\mu p_1(s)$, $p_1(a_1) \neq 0$. Then we have a series expansion near a_1:

$$K = - \frac{p(s)}{q(s)}$$

$$= -(s - a_1)^\mu \frac{p_1(s)}{p(s)}$$

$$= -(s - a_1)^\mu \left[\frac{p_1(a_1)}{q(a_1)} + \left(\frac{p_1}{q} \right)' \bigg|_{a_1} (s - a_1) + \cdots \right].$$

As in the analysis of the asymptotes above, we write $t = K^{1/\mu}$, $(0 < t < \infty)$, so that $t^\mu = K$, and our equation reads

$$t^\mu = - \frac{p_1(a_1)}{q(a_1)} (s - a_1)^\mu [1 + \text{const} \cdot (s - a_1) + \cdots].$$

Again we take the μth root to obtain the μ branches

$$t = \frac{s - a_1}{\epsilon_k} [1 + \text{const} \cdot (s - a_1) + \cdots],$$

where ϵ_k varies over the μth roots of $-q(a_1)/p_1(a_1)$, $(k = 0, 1, \ldots, \mu - 1)$. Inverting, we find

$$s = a_1 + \epsilon_k t + \text{const} \cdot t^2 + \cdots,$$

so that the tangent lines are the lines $s = a_1 + \epsilon_k t$. We verify that $\arg \epsilon_k = (\gamma_1 + 2k\pi)/\mu$, as asserted. A similar analysis applies to the zeros of $q(s)$.

PROBLEMS

1. Verify the properties of Theorem 5 on Example 1 of Section 7–9 (Fig. 7–8) as follows:

(a) Property I. Decompose the locus into two branches, one from -1 to -3 and one from -2 to ∞.

(b) Property II. Show that $pq'' - qp'' \neq 0$ at the points where the branches of part (a) meet and show in what sense the branches meet at right angles.

(c) Property III.

(d) Property IV.

(e) Property V. Show that there is one asymptote, which is the negative real axis.

(f) Property VI. Choose $a_1 = -1$, $a_2 = -2$, $b_1 = -3$. Evaluate γ_1, γ_2, η_1 and verify that there is one branch at the proper angle from each of the three points.

2. Verify the properties of Theorem 5 on Example 2 of Section 7–9 (Fig. 7–9) as follows:

(a) Property I.

(b) Property II. Show that $pq' - qp' \neq 0$ on the locus.

(c) Property III.

(d) Property IV.

(e) Property V. Obtain the three asymptotes required.

(f) Property VI. Let $a_1 = -1 + i$, $a_2 = -1 - i$, $a_3 = -2$, and compute $\gamma_i = \pi - \sum_{i \neq j} \arg (a_i - a_j)$. Show that one branch starts at a_i at angle of inclination γ_i.

3. Comparing with Fig. 7–10 (next page), graph the following root loci:

(a) $s + 1 + K(s + 2) = 0$ (b) $s^2 + 2s + 2 + K = 0$
(c) $s^2 + 2s + 1 + K = 0$ (d) $s^2 + 3s + 2 + K = 0$
(e) $s^2 + 2s + 1 + K(s + 2) = 0$ (f) $(s + 1)(s + 3) + K(s + 2) = 0$
(g) $(s + 1)(s + 2)(s + 3) + K = 0$
(h) $(s + 1)(s + 2)(s + 4) + K(s + 3) = 0$
(i) $(s + 1)(s + 2)(s + 3) + K(s + 4) = 0$
(j) $(s^2 + 2s + 2)(s + 2) + K(s^2 + 1) = 0$

4. Show that if $A_0 > 0$ and $B_0 > 0$, then validity of (7–103) for $0 < K < \infty$ implies $|s| < 1 + R$, where R is the same as in Property V of Theorem 5. [*Hint:* Consider the graph of $(A_1 + KB_1)/(A_0 + KB_0)$ as a function of K, $0 < K < \infty$, and conclude that this function has values between A_1/A_0 and B_1/B_0, so that its absolute value cannot exceed the larger of $|A_1|/A_0$, $|B_1|/B_0$. Reason similarly for the other ratios in (7–103).]

5. Let n be a positive integer. Let $f(z) = 1 + c_1 z + c_2 z^2 + \cdots$ be analytic for $|z| < R$, so that an analytic branch of $w = [f(z)]^{1/n}$ can be defined such that $w = 1$ when $z = 0$ (see Theorem 9 in Section 3–4 and the definition of z^a in Section 3–5). Show that for this branch,

$$w = 1 + \frac{c_1}{n} z + \frac{2nc_2 - (n - 1)c_1^2}{2n^2} z^2 + \cdots$$

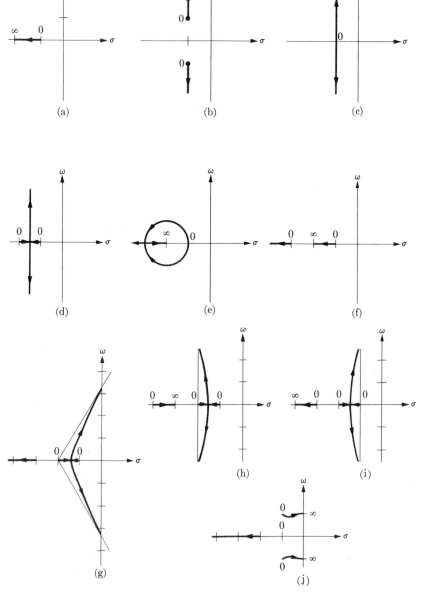

Fɪɢ. 7–10. Examples of root loci (Problem 3 following Section 7–10).

Suggested References

1. David K. Cheng, *Analysis of Linear Systems*. Reading, Mass.: Addison-Wesley, 1959.
2. L. E. Dickson, *First Course in the Theory of Equations*. New York: John Wiley, 1922.
3. K. Knopp, *Theory and Application of Infinite Series* (transl. by Miss R. C. Young). Glasgow: Blackie and Son, 1928.
4. K. Knopp, *Theory of Functions*, 2 Vols. (transl. by F. Bagemihl). New York: Dover, 1945.
5. James B. Scarborough, *Numerical Mathematical Analysis*, 2nd ed. Baltimore: Johns Hopkins Press, 1950.
6. J. H. Truxal, *Automatic Feedback Control System Synthesis*. New York: McGraw-Hill, 1955.
7. C. S. Wilts, *Principles of Feedback Control*. Reading, Mass.: Addison-Wesley, 1960.

CHAPTER 8

TIME-VARIANT LINEAR SYSTEMS

The transform methods of the preceding chapters have been applied thus far only to linear differential equations with constant coefficients. For many applications it is necessary to consider linear equations with variable coefficients. These equations describe *time-variant* linear systems, for which there is no property of stationarity, and for which the response to given input depends on the precise time at which the input is applied. Such systems arise naturally in a variety of physical problems. In particular, they arise in the study of the behavior of nonlinear systems in the neighborhood of a known nonconstant solution (steady state); if x measures the departure from the steady state at time t, then to first approximation x obeys a linear differential equation with variable coefficients (pp. 451–452 of Reference 10).

The general concepts of Chapter 2 were developed for equations with variable coefficients. However, the following concepts were restricted to equations with constant coefficients: the translation principle, Duhamel's integral, weighting function, characteristic function, transfer function, frequency response function. Of these concepts, the Duhamel integral and the weighting function will be extended here to time-variant linear systems. In Section 8–17 some consideration will be given to generalization of the transfer function and related ideas.

The major difficulty in the study of time-variant linear systems lies in the determination of the "transients," that is, the solutions of the related homogeneous differential equations. That which was a problem of algebra for the equations with constant coefficients becomes a problem demanding the most elaborate tools of mathematical analysis. In practical problems one is forced to make various approximations in order to obtain solutions in usable form. When the related homogeneous equation has been solved, one can easily give formulas for the solution of the nonhomogeneous equation. However, these are more complicated than the corresponding formulas for equations with constant coefficients, and a corresponding analysis of the transformation of input into output is much more involved.

8–1 The linear differential equation of order n; the Wronskian. We consider the equation

$$a_0(t)D^n x + a_1(t)D^{n-1}x + \cdots + a_n(t)x = f(t), \qquad (8\text{-}10)$$

as in Section 1–1. We assume the $a_j(t)$ are defined on an interval $A < t < B$ on which $a_0(t) \neq 0$. Furthermore we assume the $a_j(t)$ and $f(t)$ are continuous on the interval; this requirement will be generalized below. Related to Eq. (8–10) is the homogeneous equation

$$a_0(t)D^n x + \cdots + a_n(t)x = 0. \tag{8–11}$$

By the basic existence theorems (Theorems 1, 2, and 3 of Section 1–1), we know that Eq. (8–10) has a unique solution $x(t)$ satisfying prescribed initial conditions:

$$x(t_0) = x_0, \quad x'(t_0) = x_0', \ldots, \; x^{(n-1)}(t_0) = x_0^{(n-1)} \tag{8–12}$$

at a point t_0 of the interval. The general solution of Eq. (8–10) has the form

$$x = x^*(t) + c_1 x_1(t) + \cdots + c_n x_n(t), \tag{8–13}$$

where $x^*(t)$ is one solution and $x_1(t), \ldots, x_n(t)$ are linearly independent solutions of Eq. (8–11).

The linear independence of the solutions $x_1(t), \ldots, x_n(t)$ is equivalent to the condition that the Wronskian $w(t) \neq 0$ for $A < t < B$, where

$$w(t) = \begin{vmatrix} x_1(t) & \cdots & x_n(t) \\ x_1'(t) & \cdots & x_n'(t) \\ \vdots & & \vdots \\ x_1^{(n-1)}(t) & \cdots & x_n^{(n-1)}(t) \end{vmatrix} \tag{8–14}$$

(see Reference 10, p. 118). The Wronskian w will be important in the analysis that follows.

THEOREM 1. *Let $x_1(t), \ldots, x_n(t)$ be solutions of the homogeneous equation* (8–11) *for $A < t < B$, and let $w(t)$ be the corresponding Wronskian* (8–14). *Then*

$$a_0(t)Dw + a_1(t)w = 0, \quad (A < t < B), \tag{8–15}$$

so that, for some constant c,

$$w(t) = c \exp \int \left[\frac{-a_1(t)}{a_0(t)} \right] dt. \tag{8–16}$$

Proof. We compute Dw by using the rule of calculus (a consequence of the rule for differentiation of a product) that the derivative of a determinant of order n is the sum of n determinants, of which the first is obtained by differentiating the elements of the first row, the second by differentiating the elements of the second row, and so on. Thus

$$
Dw = \begin{vmatrix} x_1'(t) & \cdots & x_n'(t) \\ x_1'(t) & \cdots & x_n'(t) \\ \vdots & & \vdots \\ x_1^{(n-1)}(t) & \cdots & x_n^{(n-1)}(t) \end{vmatrix} + \begin{vmatrix} x_1(t) & \cdots & x_n(t) \\ x_1''(t) & \cdots & x_n''(t) \\ \vdots & & \vdots \\ x_1^{(n-1)}(t) & \cdots & x_n^{(n-1)}(t) \end{vmatrix} + \cdots
$$

The first determinant on the right is zero, since the first and second rows are identical. For a similar reason, the second, third, $\ldots$, $(n-1)$st determinants are all zero, and we conclude that Dw equals the determinant obtained from w by differentiating the last row alone. Hence

$$
a_0 Dw =
$$

$$
\begin{vmatrix} x_1(t) & \cdots & x_n(t) \\ x_1'(t) & \cdots & x_n'(t) \\ \vdots & & \vdots \\ x_1^{(n-2)}(t) & \cdots & x_n^{(n-2)}(t) \\ a_0 x_1^{(n)}(t) & \cdots & a_0 x_n^{(n)}(t) \end{vmatrix} = \begin{vmatrix} x_1(t) & & \cdots \\ x_1'(t) & & \cdots \\ \vdots & & \vdots \\ x_1^{(n-2)}(t) & & \cdots \\ -[a_1 x_1^{(n-1)} + a_2 x_1^{(n-2)} + \cdots] & & \cdots \end{vmatrix} ,
$$

by virtue of the differential equation (8–11). Accordingly,

$$
a_0 Dw = -a_1 \begin{vmatrix} x_1 & \cdots \\ x_1' & \cdots \\ \vdots & \vdots \\ x_1^{(n-2)} & \cdots \\ x_1^{(n-1)} & \cdots \end{vmatrix} - a_2 \begin{vmatrix} x_1 & \cdots \\ x_1' & \cdots \\ \vdots & \vdots \\ x_1^{(n-2)} & \cdots \\ x_1^{(n-2)} & \cdots \end{vmatrix} - a_3 \begin{vmatrix} & & \\ & & \\ & & \\ & & \\ & & \end{vmatrix} \cdots
$$

Here the second, third, $\ldots$ determinants are zero since two rows are identical in each case; the first determinant is w itself. Therefore $a_0 Dw + a_1 w = 0$, as asserted.

From Eq. (8–16) we see that, if $w(t) = 0$ for one value of t in the interval, then $w(t) \equiv 0$. Hence either $w(t) \neq 0$ (and $x_1(t), \ldots, x_n(t)$ are linearly independent) or $w(t) \equiv 0$ (and the functions are linearly dependent).

Accordingly, if the solutions $x_1(t), \ldots, x_n(t)$ are linearly independent in one subinterval (A', B') of the interval (A, B), then they are linearly independent in every other subinterval (A'', B''); for $w(t) \neq 0$ in (A', B') implies $w(t) \neq 0$ in (A, B), and hence $w(t) \neq 0$ in (A'', B'').

To obtain the particular solution $x^*(t)$ of Eq. (8–10) when the linearly independent solutions $x_1(t), \ldots, x_n(t)$ of the related homogeneous equation (8–11) are known, we use the method of variation of parameters. This is described in Problem 5 following Section 2–4; for a full discussion, refer to pp. 145–150 of Reference 10. The solution $x^*(t)$ is obtained in the form

$$x^*(t) = v_1(t)x_1(t) + \cdots + v_n(t)x_n(t), \qquad (8\text{–}17)$$

where $v_1(t), \ldots, v_n(t)$ are chosen to satisfy the equations

$$
\begin{aligned}
v_1'(t)x_1(t) \quad &+ \cdots + \quad v_n'(t)x_n(t) = 0, \\
v_1'(t)x_1'(t) \quad &+ \cdots + \quad v_n'(t)x_n'(t) = 0, \\
&\ \ \vdots \\
v_1'(t)x_1^{(n-2)}(t) &+ \cdots + v_n'(t)x_n^{(n-2)}(t) = 0, \\
v_1'(t)x_1^{(n-1)}(t) &+ \cdots + v_n'(t)x_n^{(n-1)}(t) = f(t)/a_0(t).
\end{aligned}
\qquad (8\text{–}18)
$$

These are n simultaneous linear (algebraic) equations for $v_1'(t), \ldots, v_n'(t)$, which can be solved by elimination or by determinants. When $v_1'(t), \ldots, v_n'(t)$ have been found, $v_1(t), \ldots, v_n(t)$ are obtained by integration. We remark that equations (8–18) can always be solved for $v_1'(t), \ldots, v_n'(t)$, since the determinant of coefficients is $w(t)$, and we know that $w(t) \neq 0$. In the final step of integrating to find $v_1(t)$ from $v_1'(t)$, $v_2(t)$ from $v_2'(t)$, $\ldots$, n arbitrary constants appear. These can be chosen so that prescribed initial conditions (8–12) are satisfied at a point t_0 of the interval.

8–2 The generalized weighting function; the kernel function.

THEOREM 2. *Let $x_1(t), \ldots, x_n(t)$ be linearly independent solutions of the homogeneous equation (8–11) for $A < t < B$, and let $w(t)$ be the corresponding Wronskian. Let*

$$
\phi(t, u) = [a_0(u)w(u)]^{-1}
\begin{vmatrix}
x_1(u) & \cdots & x_n(u) \\
x_1'(u) & \cdots & x_n'(u) \\
\vdots & & \vdots \\
x_1^{(n-2)}(u) & \cdots & x_n^{(n-2)}(u) \\
x_1(t) & \cdots & x_n(t)
\end{vmatrix}.
\qquad (8\text{–}20)
$$

Let $A < t_0 < B$. Then the particular solution of Eq. (8–10) such that $x(t_0) = 0$, $x'(t_0) = 0$, $\ldots$, $x^{(n-1)}(t_0) = 0$ is given by

$$x = \int_{t_0}^{t} \phi(t, u)f(u)\, du. \qquad (8\text{–}21)$$

Proof. We solve by variation of parameters. Solution of equations (8–18) by determinants gives

$$v_1' = \begin{vmatrix} 0 & x_2(t) & \cdots & x_n(t) \\ 0 & x_2'(t) & \cdots & x_n'(t) \\ \vdots & & & \vdots \\ f/a_0 & x_2^{(n-1)}(t) & \cdots & x_n^{(n-1)}(t) \end{vmatrix} \div w(t).$$

We choose $v_1(t)$ so that $v_1(t_0) = 0$. Then

$$v_1 = \int_{t_0}^{t} [w(u)]^{-1} \begin{vmatrix} 0 & x_2(u) & \cdots & x_n(u) \\ \vdots & & & \vdots \\ \dfrac{f(u)}{a_0(u)} & x_2^{(n-1)}(u) & \cdots & x_n^{(n-1)}(u) \end{vmatrix} du,$$

and similar expressions are obtained for $v_2(t), \ldots, v_n(t)$. Therefore,

$$x(t) = v_1 x_1 + \cdots + v_n x_n$$

$$= x_1(t) \int_{t_0}^{t} [w(u)]^{-1} \begin{vmatrix} 0 & x_2(u) & \cdots \\ \vdots & & \\ \dfrac{f(u)}{a_0(u)} & x_2^{(n-1)}(u) & \cdots \end{vmatrix} du + \cdots$$

$$+ x_n(t) \int_{t_0}^{t} [w(u)]^{-1} \begin{vmatrix} x_1(u) & \cdots & 0 \\ \vdots & & \vdots \\ x_1^{(n-1)}(u) & \cdots & \dfrac{f(u)}{a_0(u)} \end{vmatrix} du$$

$$= \int_{t_0}^{t} \dfrac{f(u)}{w(u)a_0(u)} \begin{vmatrix} x_1(u) & \cdots & x_n(u) \\ \vdots & & \vdots \\ x_1^{(n-2)}(u) & \cdots & x_n^{(n-2)}(u) \\ x_1(t) & \cdots & x_n(t) \end{vmatrix} du = \int_{t_0}^{t} \phi(t, u)f(u)\, du.$$

From the equation $x = v_1 x_1 + \cdots + v_n x_n$ and the fact that $v_1(t_0) = 0$, $\ldots, v_n(t_0) = 0$, we conclude that $x(t_0) = 0$. Now

$$x'(t) = v_1 x_1' + \cdots + v_n x_n' + v_1' x_1 + \cdots + v_n' x_n$$

$$= v_1(t)x_1'(t) + \cdots + v_n(t)x_n'(t)$$

by virtue of the first equation of (8–18). Similarly,

$$x''(t) = v_1(t)x_1''(t) + \cdots + v_n(t)x_n''(t),$$
$$\vdots$$
$$x^{(n-1)}(t) = v_1(t)x_1^{(n-1)}(t) + \cdots + v_n(t)x_n^{(n-1)}(t).$$

Accordingly, since $v_1(t_0) = 0, \ldots, v_n(t_0) = 0$, we have also $x'(t_0) = 0$, $\ldots, x^{(n-1)}(t_0) = 0$, as required.

The function $\phi(t, u)$ is clearly a generalization of the weighting function; the relationship will be made clear below. We call $\phi(t, u)$ the *kernel function* associated with Eq. (8–10) or Eq. (8–11). It will be seen that $\phi(t, u)$ is independent of the choice of the solutions $x_1(t), \ldots, x_n(t)$. From its definition (8–20), $\phi(t, u)$ is a continuous function of t and u.

THEOREM 3. *For each fixed u, $A < u < B$, the kernel function $\phi(t, u)$ is a solution of Eq. (8–11) and*

$$\phi(u, u) = 0, \; \phi_t(u, u) = 0, \ldots, \frac{\partial^{n-2}\phi}{\partial t^{n-2}}(u, u) = 0, \frac{\partial^{n-1}\phi}{\partial t^{n-1}}(u, u) = \frac{1}{a_0(u)}.$$

Proof. By (8–20), for fixed u, $\phi(t, u)$ is of form $\sum c_l x_l(t)$, where the c_l are constants. Hence $\phi(t, u)$ is a solution of Eq. (8–11). Furthermore,

$$\frac{\partial^k \phi}{\partial t^k} = [a_0(u)w(u)]^{-1} \begin{vmatrix} x_1(u) & \ldots & x_n(u) \\ x_1'(u) & \ldots & x_n'(u) \\ \vdots & & \vdots \\ x_1^{(n-2)}(u) & \ldots & x_n^{(n-2)}(u) \\ x_1^{(k)}(t) & \ldots & x_n^{(k)}(t) \end{vmatrix}.$$

If we now set $t = u$, then, for $k = 0, 1, \ldots, n - 2$, we obtain a determinant with two rows equal, while for $k = n - 1$ we obtain the determinant $w(u)$. Accordingly, Theorem 3 is established.

Remarks. When $n = 1$, all but the last of the equations $\phi(u, u) = 0$, $\phi_t(u, u) = 0, \ldots$ disappears, and the last one becomes $\phi(u, u) = 1/[a_0(u)]$. Equation (8–20) becomes $\phi(t, u) = x_1(t)/[a_0(u)w(u)]$, and $w(t) = x_1(t)$. (See Problem 2, following Section 8–3.)

Theorem 3 shows that $\phi(t, u)$ could be *defined* as the solution $x(t)$ of Eq. (8–11) such that $x(u) = 0$, $x'(u) = 0$, $\ldots$, $x^{(n-2)}(u) = 0$, but $x^{(n-1)}(u) = 1/a_0(u)$; thus $\phi(t, u)$ is uniquely determined and is independent of the choice of the functions $x_1(t), \ldots, x_n(t)$. In particular, let Eq. (8–11) have constant coefficients and let us choose $u = 0$; then $\phi(t, u)$ becomes $\phi(t, 0)$, a solution $x(t)$ of Eq. (8–11) for $-\infty < t < \infty$ with initial values $x(0) = 0$, $x'(0) = 0$, $\ldots$, $x^{(n-2)}(0) = 0$, $x^{(n-1)}(0) = 1/a_0$. By Lemma 1 of Section 6–7, we conclude that $\phi(t, 0)$ coincides with the weighting function $W(t)$ for $t \geqq 0$; since $W(t) \equiv 0$ for $t < 0$, we can write

$$W(t) = \phi(t, 0)h(t). \tag{8–22}$$

We note further that, with constant coefficients, $\phi(t, u)$ differs from $\phi(t, 0)$

in that the initial conditions are applied at $t = u$ rather than at $t = 0$. Hence by stationarity (Section 2–6),

$$\phi(t, u) = \phi(t - u, 0), \tag{8–23}$$

and from (8–22)

$$W(t - u) = \phi(t - u, 0)h(t - u) = \phi(t, u)h(t - u). \tag{8–24}$$

Accordingly, the particular solution (8–21) becomes, for $t_0 = 0$ and $t \geq 0$,

$$x = \int_0^t W(t - u)f(u)\, du = W * f.$$

Thus *the kernel function generalizes the weighting function. For the equation with constant coefficients, the kernel function $\phi(t, 0)$ coincides with the weighting function $W(t)$ for $t \geq 0$, and $\phi(t, u)$ coincides with $W(t - u)$ for $t \geq u$.*

EXAMPLE 1. $(D^2 + 3D + 2)x = f$. We choose $x_1(t)$ as e^{-t}, $x_2(t)$ as e^{-2t}. The Wronskian $w(t)$ is given by (8–14) as

$$w = \begin{vmatrix} e^{-t} & e^{-2t} \\ -e^{-t} & -2e^{-2t} \end{vmatrix} = -e^{-3t}$$

in agreement with (8–16). By (8–20)

$$\phi(t, u) = -e^{3u} \begin{vmatrix} e^{-u} & e^{-2u} \\ e^{-t} & e^{-2t} \end{vmatrix} = e^{u-t} - e^{2(u-t)},$$

while

$$W(t) = \mathcal{L}^{-1}[(s^2 + 3s + 2)^{-1}] = (e^{-t} - e^{-2t})h(t)$$

and

$$\phi(t, 0)h(t) = (e^{-t} - e^{-2t})h(t) = W(t),$$

as asserted. It should be remarked that $\phi(t, 0)$ satisfies the homogeneous equation for $-\infty < t < \infty$, while $W(t)$ satisfies the equation for $t > 0$ (and for $t < 0$), but not at $t = 0$. The discontinuity in $W(t)$ at $t = 0$ is related to the fact that $W(t)$ is the *impulse response*.

8–3 Green's function; impulse response.   In the study of boundary value problems, one often introduces a Green's function as a function satisfying a differential equation with special boundary conditions. Here we consider a special case, a Green's function for Eq. (8–10). We assume first that $n \geq 2$ and that the coefficients $a_k(t)$ and the right-hand member $f(t)$ are defined and continuous for $0 \leq t < \infty$. We define the corresponding Green's function as a function $G(t, u)$, $0 \leq t < \infty$, $0 \leq u < \infty$, satisfying the following conditions.

(a) For each fixed u, $G(t, u)$ satisfies Eq. (8–11) for $0 \leqq t < u$ and for $t > u$.

(b) $G(t, u)$ is continuous in t and u for $0 \leqq t < \infty$, $0 \leqq u < \infty$. For each fixed $u > 0$, $G(t, u)$ has continuous partial derivatives up to order $n - 2$ with respect to t for $t \geqq 0$, while $\partial^{n-1} G/\partial t^{n-1}$ is continuous except for a jump of $1/a_0(u)$ at $t = u$.

(c) For each fixed $u > 0$, one has $G(0, u) = 0$, $G_t(0, u) = 0, \ldots$, $(\partial^{n-1} G/\partial t^{n-1})(0, u) = 0$.

By (a) and (c), for $u > 0$, G satisfies a homogeneous linear differential equation in t for $0 \leqq t < u$ and has initial values zero at $t = 0$. Accordingly, $G(t, u) \equiv 0$ for $0 \leqq t < u$. By condition (b), $G(t, u) = 0$ also for $t = u$ and $G_t(u, u) = 0, \ldots, \partial^{n-2} G/\partial t^{n-2} = 0$ for $t = u$, while $\partial^{n-1} G/\partial t^{n-1}$ has limit 0 as $t \to u-$, limit $1/a_0(u)$ as $t \to u+$. Accordingly, by (a), for $t \geqq u$, $G(t, u)$ is the solution of the homogeneous equation (8–11) with all initial values zero at $t = u$, except for the $(n - 1)$st derivative, which equals $1/a_0(u)$. Thus $G(t, u)$ must coincide with $\phi(t, u)$ for $t \geqq u$, and we conclude that

$$G(t, u) = \phi(t, u)h(t - u). \tag{8–30}$$

This relation has been established for $u > 0$. However, since $\phi(t, u) = 0$ for $t = u$, the right-hand side is continuous for all t, u; by (b), $G(t, u)$ is also continuous, and we conclude that the relation is valid also for $u = 0$. From (8–24) and (8–30) we see that for an equation with constant coefficients, the Green's function $G(t, u)$ becomes the weighting function $W(t - u)$ in the region $t \geqq 0$, $u \geqq 0$.

The definition of the Green's function can be modified to fit the case $n = 1$. Condition (a) is unchanged, and condition (c) becomes $G(0, u) = 0$, $u > 0$; condition (b) becomes (b'): $G(t, u)$ is continuous for $0 \leqq t < \infty$, $0 \leqq u < \infty$ except along the line $t = u$; for each fixed $u > 0$, $G(t, u)$ jumps by $1/a_0(u)$ at $t = u$, and $G(0, 0) = 1/a_0(0) = \lim G(t, 0)$ as $t \to 0+$. We then verify that (8–30) is still valid.

Impulse response. Let us consider the response of a system, described by Eq. (8–10) and found to be initially at rest, when an impulse is applied. We take $t_0 = 0$ in Eq. (8–21) and $f(t) = \delta(t - c)$, $c > 0$. Accordingly,

$$x = \int_0^t \phi(t, u)\, \delta(u - c)\, du = \begin{cases} 0, & (t < c) \\ \phi(t, c) & (t > c) \end{cases}$$
$$= \phi(t, c)h(t - c);$$

the value at $t = c$ is ambiguous, but since (for $n \geqq 2$) $\phi(c, c) = 0$, we obtain a continuous x by considering this value to be zero. By (8–30), we therefore can *identify the Green's function with the impulse response.* In fact, by direct substitution we verify that $x = G(t, u)$ satisfies the differential equation (8–10) when $f = \delta(t - u)$. (See Problem 7 below.)

EXAMPLE 2. $(t^2 + 3t + 2)D^2x + (t^2 + 4t + 2)Dx + tx = f, t > -1$. Here $x_1 = e^{-t}$ and $x_2 = (t + 2)^{-1}$ are linearly independent solutions of the related homogeneous equation. Hence

$$w(t) = \begin{vmatrix} e^{-t} & (t + 2)^{-1} \\ -e^{-t} & -(t + 2)^{-2} \end{vmatrix} = e^{-t}(t + 2)^{-2}(t + 1) \neq 0,$$

$$\phi(t, u) = \begin{vmatrix} e^{-u} & (u + 2)^{-1} \\ e^{-t} & (t + 2)^{-1} \end{vmatrix} \div [w(u)(u^2 + 3u + 2)]$$

$$= \frac{u + 2}{(u + 1)^2(t + 2)} - \frac{e^{u-t}}{(u + 1)^2}.$$

We verify that $\phi(u, u) = 0$, $\phi_t(u, u) = (u^2 + 3u + 2)^{-1}$, in agreement with Theorem 3. Therefore our response with initial values zero at $t = 0$ is

$$x(t) = \int_0^t \phi(t, u)f(u)\, du = \int_0^t \left[\frac{u + 2}{(u + 1)^2(t + 2)} - \frac{e^{u-t}}{(u + 1)^2}\right]f(u)\, du.$$

In general, evaluation of this integral for particular f is awkward. However, special cases can be found for which the calculation is easy. For example, if $f(t) = 2(t + 1)^2$, then $x = t + 2e^{-t} - 4(t + 2)^{-1}$. If we write the equation, with this choice of f, in the form

$$D^2x + \frac{t^2 + 4t + 2}{t^2 + 3t + 2} Dx + \frac{t}{t^2 + 3t + 2} x = \frac{2(t + 1)}{t + 2},$$

then we see that for large t the equation has approximately constant coefficients. Furthermore, the solutions of the homogeneous equation are $c_1e^{-t} + c_2(t + 2)^{-1}$, which approach zero as $t \to +\infty$. However, the input is $2(t + 1)/(t + 2)$, which approaches the constant value 2 as $t \to +\infty$, and the output is $x = t$ plus transient, which becomes infinite as $t \to +\infty$. Thus an apparently stable system with approximately constant coefficients can have an unbounded response to a bounded input. This point is considered further in Section 8–16.

By Eq. (8–30) the Green's function, or impulse response, for this system is

$$G(t, u) = \left[\frac{u + 2}{(u + 1)^2(t + 2)} - \frac{e^{u-t}}{(u + 1)^2}\right]h(t - u).$$

In Fig. 8–1 the impulse response is graphed for three values of u: 0, 1, 2.

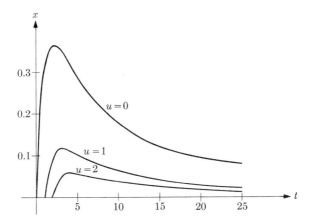

FIG. 8–1. Response of time-variant linear system to impulse applied at time u.

Because of time variance, *the impulse response depends on the time of application of the impulse.* It is striking that the later the impulse is applied, the weaker the response; however, each response also dies out very slowly as t becomes infinite.

PROBLEMS

1. Obtain the kernel function $\phi(t, u)$ and the Green's function $G(t, u)$ for each of the following equations:

(a) $(D^2 + 5D + 6)x = f(t)$
(b) $(D + \cos t)x = f(t)$
(c) $[(t^2 + 1)D^2 + (t - 1)^2 D - 2t]x = f(t)$

[*Hint for* (c): $x = e^{-t}$ and $x = t^2 - 2t + 3$ are solutions of the related homogeneous equation.]

2. For the general first order equation

$$[a_0(t) D + a_1(t)]x = f(t),$$

let $p(t) = \int_{t_0}^{t} [a_1(u)/a_0(u)] \, du$, $w(t) = e^{-p(t)}$.

(a) Show that $w(t)$ can be considered as the Wronskian determinant of a solution of the related homogeneous equation.

(b) Show that the kernel function is $\phi(t, u) = w(t)/[a_0(u)w(u)]$.

(c) Show that if $a_0(t)$ and $a_1(t)$ have period τ, then $p(t)$ has form $kt + q(t)$, where k is a constant and $q(t)$ has period τ, and that $\phi(t + \tau, u + \tau) = \phi(t, u)$.

3. Use the results of Problem 2 to find $\phi(t, u)$, $G(t, u)$ for each of the following equations, and graph $G(t, 0)$, $G(t, 1)$, $G(t, 2)$:

(a) $(D + 1)x = f(t)$
(b) $[D + (2 + \cos t)]x = f(t)$
(c) $(D + 2t)x = f(t)$

4. For the equation $\{D + [1 + 2t(t^2 + 1)^{-1}]\}x = f(t)$ the coefficients approach constant values 1, 1 as $t \to \infty$. Show that the replacement of $\phi(t, u)$ by the value $e^{-(t-u)}$ corresponding to these constant coefficients leads to a good approximation to the output for large t.

5. Show that the kernel function $\phi(t, u)$ for Eq. (8–10) has the properties

$$\frac{\partial^{l+k}\phi}{\partial t^l\, \partial u^k} = 0 \quad \text{for } t = u, \, k = 0, 1, \ldots, n - 2, \, l < n - k - 1,$$

and

$$\frac{\partial^{n-1}\phi}{\partial t^{n-k-1}\, \partial u^k} \neq 0 \quad \text{for } t = u, \, k = 0, 1, \ldots, n - 1.$$

[*Hint:* For $n \geqq 2$, differentiate all but the last of the relations

$$\phi(u, u) = 0, \, \phi_t(u, u) = 0, \ldots, \frac{\partial^{n-2}\phi}{\partial t^{n-2}} (u, u) = 0, \, \frac{\partial^{n-1}\phi}{\partial t^{n-1}} (u, u) \neq 0,$$

with respect to u, to obtain

$$\phi_t(u, u) + \phi_u(u, u) = 0, \ldots, \frac{\partial^{n-1}\phi}{\partial t^{n-1}} (u, u) + \frac{\partial^{n-1}\phi}{\partial t^{n-2}\, \partial u} (u, u) = 0,$$

and conclude that

$$\phi_u(u, u) = 0, \, \phi_{tu}(u, u) = 0, \ldots, \frac{\partial^{n-2}\phi}{\partial t^{n-3}\, \partial u} (u, u) = 0, \, \frac{\partial^{n-1}\phi}{\partial t^{n-2}\, \partial u} (u, u) \neq 0.$$

Repeat inductively to obtain the desired conclusion.]

Remark. We assume here that the partial derivatives of ϕ which appear are continuous. See the Corollary to Theorem 5 in Section 8–4.

6. In the notation of Problem 5, show that for each u the functions $\phi(t, u)$, $\phi_u(t, u)$, $\ldots$, $(\partial^{n-1}\phi/\partial u^{n-1})(t, u)$ are solutions of the homogeneous equation (8–11) and are moreover linearly independent, so that the general solution of Eq. (8–10) can be written (for $t_0 = 0$)

$$x = \int_0^t \phi(t, u)f(u)\, du + c_1\phi(t, 0) + c_2\phi_u(t, 0) + \cdots + c_n \frac{\partial^{n-1}\phi}{\partial u^{n-1}} (t, 0).$$

This generalizes Eq. (6–159).

[*Hint:* The fact that the functions $\phi(t, u)$, $\phi_u(t, u)$, $\ldots$ are solutions of the homogeneous equation follows from the form of Eq. (8–20). The linear independence follows from Problem 5. If for some u

$$C_1\phi(t, u) + C_2\phi_u(t, u) + \cdots + C_n \frac{\partial^{n-1}\phi}{\partial u^{n-1}} (t, u) \equiv 0$$

with constant $C_1, \ldots, C_n$, then let $t = u$ to conclude that $C_n = 0$; differentiate with respect to t and let $t = u$ to conclude that $C_{n-1} = 0$, and so on.]

7. *Response to generalized functions.* (a) Show that $x = G(t, u) = \phi(t, u)h(t - u)$ satisfies Eq. (8–10) when $f = \delta(t - u)$. [*Hint:* Verify that, for fixed u, $D^k[\phi(t, u)h(t - u)] = D^k\phi(t, u) \cdot h(t - u)$ for $k = 0, 1, \ldots, n - 1$, $= [1/a_0(u)]\, \delta(t - u) + D^n\phi \cdot h(t - u)$ for $k = n$. See Section 1–13.]

(b) Show that $x = (-1)^k(\partial^k/\partial u^k)G(t, u)$ satisfies Eq. (8–10) when $f = \delta^{(k)}(t - u)$, $k = 1, 2, \ldots$ Assume all differentiability needed; see the Corollary in Section 8–4. [*Hint:* Let $D_t = \partial/\partial t$, $D_u = \partial/\partial u$. Verify that if $\psi(t, u)$ has continuous partial derivatives,

$$D_t D_u[\psi(t, u)h(t - u)] = D_u D_t[\psi(t, u)h(t - u)],$$

$$D_t D_u[\psi(t, u)\, \delta^{(l)}(t - u)] = D_u D_t[\psi(t, u)\, \delta^{(l)}(t - u)],$$

and hence that

$$L[D_u^k\{\phi(t, u)h(t - u)\}] = D_u^k L[\phi(t, u)h(t - u)],$$

where L is the operator $a_0(t) D_t^n + a_1(t) D_t^{n-1} + \cdots$ In the proof make use of the relation $\delta^{(k)}(a - t) = (-1)^k\, \delta^{(k)}(t - a)$. This can be regarded as a definition or can be deduced from a sufficiently complete theory of generalized functions.]

8. Apply the results of Problem 7 to obtain a response $x(t)$ for each of the following equations, with given choice of $f(t)$. Verify that $x(t)$ is a solution by substitution in the differential equation.

(a) $(D^2 + 3D + 2)x = \delta(t - 1)$ (See Example 1, Section 8–2)

(b) $(D^2 + 3D + 2)x = \delta'(t)$

(c) $(D^2 + 3D + 2)x = \delta''(t)$

(d) $(D + \cos t)x = \delta(t - 2)$ (See Problem 1(b))

(e) $(D + \cos t)x = \delta'(t)$

(f) $(D + \cos t)x = 2\, \delta(t) + 3\, \delta'\left(t - \dfrac{\pi}{2}\right) + 5\, \delta''(t - \pi)$

9. *Simultaneous differential equations.* Apply variation of parameters to show that the solution, with initial values zero for $t = 0$, of the equations

$$Dx = a_1(t)x + b_1(t)y + f_1(t), \qquad Dy = a_2(t)x + b_2(t)y + f_2(t)$$

can be written in the form

$$x = \int_0^t \phi_{11}(t, u)f_1(u)\, du + \int_0^t \phi_{12}(t, u)f_2(u)\, du,$$

$$y = \int_0^t \phi_{21}(t, u)f_1(u)\, du + \int_0^t \phi_{22}(t, u)f_2(u)\, du.$$

[*Hint:* See Problem 7 following Section 2–4. The discussion of simultaneous equations is simplified by the use of matrices; see Section 8–13.]

10. Let $x_1(t), \ldots, x_n(t)$ be linearly independent solutions of Eq. (8–11) for $A < t < B$. Show that n distinct values $t_1, \ldots, t_n$ can be chosen in the interval

(A, B) such that

$$\begin{vmatrix} x_1(t_1) & \cdots & x_n(t_1) \\ x_1(t_2) & \cdots & x_n(t_2) \\ \vdots & & \vdots \\ x_1(t_n) & \cdots & x_n(t_n) \end{vmatrix} \neq 0.$$

Hence show that there exists one and only one solution $x(t)$ such that $x(t_j)$ has a given value k_j for $j = 1, \ldots, n$.

[*Hint:* We can choose an interval I_{11} in which $x_1(t) \neq 0$. Then we can choose an interval I_{12}, contained in I_{11}, in which $x_2(t) \neq 0$. Continuing thus, we finally obtain t_1, such that all $x_j(t_1)$ are different from zero. Next there is an interval I_{21} in which $\Delta_{12}(t) \neq 0$, where $\Delta_{ij}(t)$ is the determinant

$$\begin{vmatrix} x_i(t_1) & x_j(t_1) \\ x_i(t) & x_j(t) \end{vmatrix}.$$

There is an interval I_{22} contained in I_{21} in which $\Delta_{13}(t) \neq 0$. Proceeding thus, finally choose t_2 so that $\Delta_{ij}(t_2) \neq 0$ for $1 \leq i < j \leq n$. Now proceed inductively to obtain $t_1, \ldots, t_n$ as desired.]

Answers

1. (a) $\phi = e^{2(u-t)} - e^{3(u-t)}$　　　　(b) $\phi = \exp(\sin u - \sin t)$
   (c) $\phi = (u^2 + 1)^{-2}[t^2 - 2t + 3 - e^{u-t}(u^2 - 2u + 3)]$

In all cases $G = \phi(t, u)h(t - u)$.

3. (a) $\phi = e^{u-t}$　　　　(b) $\phi = e^{2u-2t+\sin u - \sin t}$
   (c) $\phi = \exp(u^2 - t^2)$

8. (a) $(e^{1-t} - e^{2(1-t)})h(t - 1)$　　　　(b) $(2e^{-2t} - e^{-t})h(t)$
   (c) $(e^{-t} - 4e^{-2t})h(t) + \delta(t)$　　　　(d) $\exp[\sin 2 - \sin t]h(t - 2)$
   (e) $\delta(t) - e^{-\sin t}h(t)$
   (f) $e^{-\sin t}[2h(t) + 5h(t - \pi)] + 3\,\delta(t - \tfrac{1}{2}\pi) + 5\,\delta(t - \pi) + 5\,\delta'(t - \pi)$

8-4 Adjoint equation.

To each operator

$$L = a_0(t)D^n + \cdots + a_n(t)$$

with variable coefficients, there is associated a second such operator

$$L^+ = (-1)^n D^n a_0 + (-1)^{n-1}D^{n-1}a_1 + \cdots + a_n; \qquad (8\text{--}40)$$

that is,

$$L^+[x] = (-1)^n D^n(a_0 x) + (-1)^{n-1}D^{n-1}(a_1 x) + \cdots$$

We call L^+ the *adjoint* of L. For L^+ to have meaning it is necessary that

$a_k(t)$ have derivatives through order $n - k$ on the interval considered, $k = 0, 1, \ldots, n$. The differential equation $L^+[x] = 0$ is said to be the *adjoint* of the equation $L[x] = 0$. For example, the adjoint of

$$(1 + t^2)D^2x + 3tDx + e^tx = 0$$

is

$$D^2[(1 + t^2)x] - D(3tx) + e^tx = 0;$$

that is,

$$(1 + t^2)D^2x + tDx + (e^t - 1)x = 0.$$

The adjoint equation arises naturally in searching for an integrating factor $v(t)$ of the equation $L[x] = 0$, that is, a function $v(t)$ such that $v(t)L[x] \equiv dF/dt$ for some function

$$F = b_0(t)D^{n-1}x + b_1(t)D^{n-2}x + \cdots + b_{n-1}(t)x.$$

One can verify (Problem 2 below) that $v(t)$ is such an integrating factor if and only if $L^+[v] = 0$. If an integrating factor $v(t)$ has been found, one can lower the order of the equation $L[x] = 0$ by integrating once to get the equation

$$F \equiv b_0(t)D^{n-1}x + \cdots + b_{n-1}(t)x = c_1,$$

where c_1 is a constant.

THEOREM 4 (Lagrange's identity). *Let L, L^+ be as above. Then for each pair of functions $x(t)$, $y(t)$ having derivatives through order n on the interval considered,*

$$yL[x] - xL^+[y] \equiv \frac{d}{dt} B(x, y), \qquad (8\text{-}41)$$

where

$$B(x, y) = \sum_{m=1}^{n} [(D^{m-1}x)a_{n-m}y - (D^{m-2}x)D(a_{n-m}y) + \cdots$$
$$+ (-1)^{m-1}xD^{m-1}(a_{n-m}y)]. \qquad (8\text{-}42)$$

The expression $B(x, y)$ is called the *bilinear concomitant* of $L[x]$. The verification of (8-41) is left as Problem 4 below. With the aid of the Lagrange identity one can relate the kernel function of the given equation to that of the adjoint equation.

THEOREM 5. *Let $a_k(t)$ have continuous derivatives through order $n - k$ for $\alpha < t < \beta$, $k = 0, 1, \ldots, n$. Let $\phi(t, u)$ be the kernel function for the differential equation $L[x] = f$; let $\phi^+(t, u)$ be the kernel function for the equation $L^+[x] = f$. Then*

$$\phi^+(t, u) = -\phi(u, t). \qquad (8\text{-}43)$$

Proof. We apply the Lagrange identity to $x(t) = \phi(t, u)$, $y(t) = \phi^+(t, v)$; here u and v are regarded as constants, $\alpha < u < \beta$, $\alpha < v < \beta$. We obtain

$$\phi^+(t, v)L[\phi(t, u)] - \phi(t, u)L^+[\phi(t, v)] = \frac{d}{dt}\{B[\phi(t, u), \phi(t, v)]\}.$$

But for each fixed u and v, $x = \phi(t, u)$ is a solution of $L[x] = 0$, and $y = \phi^+(t, v)$ is a solution of $L[y] = 0$. Thus the left-hand side is identically zero, and $B[\phi(t, u), \phi^+(t, v)]$ is constant (as a function of t). In particular, its value at $t = u$ equals its value at $t = v$:

$$B[\phi(t, u), \phi^+(t, v)]|_{t=u} = B[\phi(t, u), \phi^+(t, v)]|_{t=v}. \tag{8–44}$$

When $t = u$, then $\phi(t, u) = 0$, $\phi_t(t, u) = 0$, ..., $D^{n-2}\phi(t, u) = 0$, $D^{n-1}\phi(t, u) = 1/a_0(u)$; when $t = v$, then $\phi^+(t, v) = 0$, $\phi_t^+(t, v) = 0,\ldots,$ $D^{n-2}\phi^+(t, v) = 0$, $D^{n-1}\phi^+(t, v) = (-1)^n/a_0(v)$, since the leading term of $L^+[y]$ is $(-1)^n a_0(t)D^n y$. Therefore, substitution of these values in both sides of Eq. (8–44), in accordance with Eq. (8–42), leads to the relation $\phi^+(u, v) = (-1)^{n-1}\phi(v, u)(-1)^n$, from which the desired conclusion (8–43) follows at once.

COROLLARY. *Under the hypotheses of Theorem 5, the kernel function $\phi(t, u)$ can be written as follows:*

$$\phi(t, u) = \sum_{i=1}^{n} x_i(t)y_i(u), \tag{8–45}$$

where $x_1(t), \ldots, x_n(t)$ are linearly independent solutions of the homogeneous equation $L[x] = 0$, and $y_1(t), \ldots, y_n(t)$ are linearly independent solutions of the adjoint $L^+[y] = 0$. Furthermore, the partial derivatives $\partial^{k+l}\phi/\partial t^k \partial u^l$ exist and are continuous for $0 \leq k \leq n$, $0 \leq l \leq n$, and

$$\frac{\partial^{k+l}\phi}{\partial t^k \partial u^l} = \sum_{i=1}^{n} x_i^{(k)}(t)y_i^{(l)}(u). \tag{8–46}$$

Proof. We choose a fixed t_0, $\alpha < t_0 < \beta$, and solutions $x_1(t), \ldots, x_n(t)$ of Eq. (8–11) such that the $x_i(t)$ have all initial values zero at t_0 except for $x_1(t_0) = -1$, $x_2'(t_0) = -1, \ldots, x_n^{(n-1)}(t_0) = -1$. For the corresponding Wronskian $w(t)$ we have $w(t_0) = (-1)^n$, so that $x_1(t), \ldots, x_n(t)$ are linearly independent. As in the proof of Theorem 2, we form $\phi(t, u)$ using the functions $x_1(t), \ldots, x_n(t)$. By (8–20), $\phi(t, u)$ has the form (8–45). By Theorem 5, the adjoint has kernel function $\phi^+(t, u) = -\sum_1^n x_i(u)y_i(t)$. But by (8–20) we know that $\phi^+(t, u)$ has the form $\sum_1^n x_i^+(t)v_i(u)$, where

$x_1^+(t), \ldots, x_n^+(t)$ are linearly independent solutions of $L^+[x] = 0$. Hence

$$\sum_{i=1}^{n} x_i(u)y_i(t) \equiv - \sum_{i=1}^{n} x_i^+(t)v_i(u).$$

Now we can choose values $t_1, \ldots, t_n$ such that

$$\Delta = \begin{vmatrix} x_1(t_1) & \cdots & x_n(t_1) \\ \vdots & & \vdots \\ x_1(t_n) & \cdots & x_n(t_n) \end{vmatrix} \neq 0.$$

(See Problem 10 following Section 8–3.) Thus the equations

$$\sum_{i=1}^{n} x_i(t_j)y_i(t) = - \sum_{i=1}^{n} x_i^+(t)v_i(t_j) \quad (j = 1, \ldots, n)$$

can be solved for $y_1(t), \ldots, y_n(t)$. We conclude that each $y_i(t)$ is a linear combination of $x_1^+(t), \ldots, x_n^+(t)$ and is hence a solution of the adjoint. Since the $x_i(t)$ and $y_i(t)$ have continuous derivatives through order n, it follows that Eq. (8–45) can be differentiated to yield Eq. (8–46) and that the partial derivatives of ϕ are continuous as stated.

Now for each u, $\phi^+(t, u), \phi_u^+(t, u), \ldots, (\partial^{n-1}/\partial u^{n-1})\phi^+(t, u)$ are linearly independent solutions of the adjoint (Problem 6 following Section 8–3). Hence for each u, the functions

$$g_k(t) = - \sum_{i=1}^{n} x_i^{(k-1)}(u)y_i(t) \quad (k = 1, \ldots, n)$$

are linearly independent for $\alpha < t < \beta$. If we take $u = t_0$, then $g_k(t) = y_k(t)$, so that $y_1(t), \ldots, y_n(t)$ are linearly independent.

EXAMPLE. $D^2x + e^t Dx + e^t x = 0$. The adjoint is $D^2x - e^t Dx = 0$. Therefore $x = 1$ is a solution of the adjoint and also serves as an integrating factor for the given equation; that is, the given equation is exact. Indeed,

$$D^2x + e^t Dx + e^t x \equiv \frac{d}{dt}[Dx + e^t x].$$

Hence, upon integrating,

$$Dx + e^t x = c_1.$$

This is a first order equation, whose general solution is

$$x = \exp{(-e^t)}\int \exp{(e^t)}c_1 \, dt + c_2 \exp{(-e^t)},$$

and we have the desired general solution of the given homogeneous equation;

we can write this as $x = c_1 x_1(t) + c_2 x_2(t)$, where

$$x_1(t) = \exp(-e^t) \int_0^t \exp(e^v)\, dv, \qquad x_2(t) = \exp(-e^t).$$

Accordingly, the kernel function for the corresponding nonhomogeneous equation is found to be

$$\phi(t, u) = \exp(-e^t) \int_u^t \exp(e^v)\, dv.$$

The adjoint has particular solutions

$$x_1^+(t) = 1, \qquad x_2^+(t) = \int_0^t \exp(e^v)\, dv,$$

and the associated kernel function is (see Problem 3 below)

$$\phi^+(t, u) = \exp(-e^u) \int_u^t \exp(e^v)\, dv = -\phi(u, t).$$

Remark 1. We can seek an interpretation of the relation $\phi^+(t, u) = -\phi(u, t)$ in terms of the corresponding Green's functions

$$G(t, u) = \phi(t, u) h(t - u), \qquad G^+(t, u) = \phi^+(t, u) h(t - u).$$

We can write

$$\begin{aligned} G^+(t, u) &= -\phi(u, t) h(t - u) = \phi(u, t)[h(u - t) - 1] \\ &= G(u, t) - \phi(u, t). \end{aligned}$$

This relation is more complicated. In the literature one meets the relation $G^+(t, u) = G(u, t)$. However, $G^+(t, u)$ is then defined with respect to the adjoint differential equation with *adjoint boundary conditions*. Our Green's function can be considered as one for an interval $0 \leq t \leq \tau$ for some fixed $\tau > 0$, with initial values zero at $t = 0$. We can verify that $G(u, t)$ is a Green's function for the adjoint equation with initial values zero for $t = \tau$. (These are the adjoint boundary conditions.)

Remark 2. One can verify (Problem 5 below) that the adjoint of the adjoint is the original differential equation. Hence the quest for integrating factors effectively terminates at the adjoint equation.

PROBLEMS

1. For each of the following equations, find the adjoint, find a solution of the adjoint and obtain $\phi(t, u)$:

(a) $(t - 1) D^2 x + (t + 2) Dx + 2x = 0$
(b) $(e^t + 1) D^2 x + (3e^t + e^{2t} - 1) Dx + e^{2t} x = 0$

2. (a) Show that

$$\frac{d}{dt}\,[b_0(t)\,D^{n-1}x + \cdots + b_{n-1}(t)x] \equiv v(t)[a_0(t)\,D^n x + \cdots + a_n(t)x] \qquad (*)$$

implies $b_0 = va_0$, $b_1 + b_0' = va_1$, $b_2 + b_1' = va_2$, $\ldots$, $b_{n-1} + b_{n-2}' = va_{n-1}$, $b_{n-1}' = va_n$, and hence $b_1 = va_1 - (va_0)'$, $b_2 = va_2 - (va_1)' + (va_0)''$, $\ldots$, and finally

$$0 = va_n - (va_{n-1})' + \cdots + (-1)^n(va_0)^{(n)}.$$

Hence conclude that if $v(t)$ is an integrating factor for $L[x] = 0$, then $L^+[v] = 0$. Conversely, show that if $L^+[v] = 0$, then the preceding equations define b_0, b_1, $\ldots$, b_{n-1} so that $(*)$ holds and v is an integrating factor.

(b) Show that if $L^+[v] = 0$, then $vL[x] = (d/dt)B(x, v)$, and hence conclude that v is an integrating factor.

3. Verify the expressions for $\phi(t, u)$, $\phi^+(t, u)$ in the example of the text.

4. Prove the Lagrange identity (8–41). [*Hint:* Differentiate $B(x, y)$, and verify that in each term of the sum all terms cancel except two, so that

$$\frac{d}{dt}\,B(x, y) = \sum_{m=1}^{n} \{D^m x \cdot a_{n-m}y + (-1)^{m-1}x D^m(a_{n-m}y)\}.]$$

5. Prove that the adjoint of the operator L^+ defined by Eq. (8–40) is the operator L.

ANSWERS

1. (a) $(t - 1)D^2 x - tDx + x = 0$; t, e^t are solutions;

$$\phi(t, u) = (t - 1)^{-2}(u - te^{u-t})$$

(b) $(e^t + 1)D^2 x + (1 - e^t - e^{2t})Dx - (2e^t + e^{2t})x = 0$; e^{-t}, $\exp(e^t)$ are solutions; $\phi(t, u) = (e^t + 1)^{-2}[e^{t-u} - \exp(e^u - e^t)]$

8–5 Solution of linear differential equations by infinite series. Determination of the solutions and hence of the kernel function $\phi(t, u)$ for a specific linear differential equation is ordinarily an awkward problem. For some equations particular solutions or integrating factors may happen to be easily obtainable, so that the order can be reduced as in Section 8–4 (or as in pp. 309–310 of Reference 10); when the order has been reduced to one, solution is reduced to integration. We could of course work backwards from the solutions to form the differential equations; thus a catalogue of equations with known solutions could be formed. However, it appears to be very difficult to organize such a catalogue in a fashion which would make it useful. The equations with constant coefficients appear to form the only class of reasonable generality for which the solutions are known elementary functions. This class can be somewhat enlarged by addition of equations reducible to ones with constant coefficients by appropriate substitutions. For example, the *Cauchy equation*

$$(a_0 t^n D^n + a_1 t^{n-1} D^{n-1} + \cdots + a_{n-1} tD + a_n)x = 0 \qquad (8\text{–}50)$$

is reducible to an equation with constant coefficients by a substitution $t = e^v$. (See Problem 1, following Section 8–6.)

For equations with polynomial coefficients or, more generally, coefficients which are analytic in t (expandable in power series), solutions are obtainable in the form of power series. The theory and application of such solutions have been very highly developed, and we here merely summarize the main results.

Let the equation be

$$[D^n + a_1(t) D^{n-1} + \cdots + a_n(t)]x = 0. \qquad (8\text{–}51)$$

Let the coefficients $a_k(t)$ be expressible in power series in powers of $(t - t_0)$ for $|t - t_0| < \rho$; thus the $a_k(t)$ can be considered as analytic functions of the complex variable t (Chapter 3). Then the general solution of Eq. (8–51) for $|t - t_0| < \rho$ is of form $x = c_1 x_1(t) + \cdots + c_n x_n(t)$, where $x_\alpha(t)$ is analytic for $|t - t_0| < \rho$, so that

$$x_\alpha(t) = \sum_{\beta=0}^{\infty} b_{\alpha\beta}(t - t_0)^\beta \quad (|t - t_0| < \rho, \ \alpha = 1, \ldots, n). \qquad (8\text{–}52)$$

In particular, if all $a_k(t)$ are analytic for all complex t, then the solutions are valid for all t (real or complex).

EXAMPLE 1. $(D^2 + t^2 D + t)x = 0$. We choose $t_0 = 0$. The solutions $x_1(t)$, $x_2(t)$ with initial values $x_1(0) = 1$, $x_1'(0) = 0$, $x_2(0) = 0$, $x_2'(0) = 1$ are obtainable as series

$$x_1(t) = 1 + \sum_{\beta=1}^{\infty} b_{1\beta} t^\beta, \qquad x_2(t) = t + \sum_{\beta=1}^{\infty} b_{2\beta} t^\beta,$$

valid and linearly independent for all t. Upon substituting these expressions in the differential equation and equating to zero the coefficient of t^β for $\beta = 0, 1, 2, \ldots$, we find the solutions to be (Problem 2(a), following Section 8–6)

$$x_1(t) = 1 + \sum_{\beta=1}^{\infty} (-1)^\beta \frac{1^2 4^2 \cdots (3\beta - 2)^2}{(3\beta)!} t^{3\beta},$$

$$x_2(t) = t + t \sum_{\beta=1}^{\infty} (-1)^\beta \frac{2^2 5^2 \cdots (3\beta - 1)^2}{(3\beta + 1)!} t^{3\beta}.$$

Equation with regular singular point. If in the equation

$$b_0(t) D^n x + \cdots + b_n(t) x = 0 \qquad (8\text{–}53)$$

the $b_k(t)$ are analytic for $|t - t_0| < \rho$, but $b_0(t_0) = 0$, then t_0 is said to be a *singular point* of Eq. (8–53). We assume $b_0(t) \not\equiv 0$, so that (see Section 3–11) $b_0(t)$ has a zero of some order N at $t = t_0$. Upon division of the equation by $b_0(t)$, we obtain a new equation of form (8–51), but in which each $a_k(t)$ may have a pole at t_0. If the pole is of order at most k (or if there is no pole) for each $k = 1, \ldots, n$, then t_0 is said to be a *regular singular point* of Eq. (8–53), or of the equivalent Eq. (8–51). (If each a_k has no pole at t_0, then the singular point of Eq. (8–53) is *removable*. It will be convenient to consider this as a degenerate case of a regular singular point; the methods to be described remain applicable.) For example, the equation

$$[2t^2 D^2 + (t^2 - 5t) D + (3 + t)] x = 0 \qquad (8\text{–}54)$$

has a regular singular point at $t = 0$. Here $b_0(t)$ has a zero of order 2 at $t = 0$, and, after division of the equation by $2t^2$, the coefficients of Dx, x become

$$a_1(t) = \frac{t - 5}{2t}, \qquad a_2(t) = \frac{3 + t}{2t^2},$$

with poles of orders 1 and 2 respectively at $t = 0$.

Let the equation (8–53) have a regular singular point at $t = t_0$, and let the equation be written as $L[x] = 0$. If we expand the $b_k(t)$ in powers of $t - t_0$ and replace x by $(t - t_0)^\alpha$ in the left-hand side, we obtain series which can be combined to yield one series in increasing powers of $t - t_0$:

$$L[(t - t_0)^\alpha] = P(\alpha)(t - t_0)^\gamma + P_1(\alpha)(t - t_0)^{\gamma+1} + \cdots$$

The coefficient $P(\alpha)$ is a polynomial in α of degree n, and the equation $P(\alpha) = 0$ is called the *indicial equation* associated with Eq. (8–53) at t_0.

We now seek solutions of Eq. (8–53) of form

$$x = (t - t_0)^\alpha \sum_{\beta=0}^{\infty} c_\beta (t - t_0)^\beta. \qquad (8\text{–}55)$$

We find that such solutions are obtainable if α is a root of the indicial equation. In particular, if the indicial equation has n distinct real roots α_1, $\ldots, \alpha_n$, no two of which differ by an integer, then n solutions of form (8–55) are obtained, with $\alpha = \alpha_1, \alpha_2, \ldots, \alpha_n$, respectively. The series converge and the solutions are valid and linearly independent for $0 < |t - t_0| < \rho$, where ρ is the largest radius within which the coefficients $a_k(t)$ of the

corresponding Eq. (8–51) have no singularities other than at t_0. The procedures can be modified to take care of complex roots, coincident roots, and roots differing by an integer. For proofs and details refer to References 4, 8, and 19.

EXAMPLE 2. We consider Eq. (8–54) at $t = 0$. We find

$$L[t^\alpha] = 2t^2\alpha(\alpha - 1)t^{\alpha-2} + (t^2 - 5t)\alpha t^{\alpha-1} + (3 + t)t^\alpha$$
$$= (2\alpha^2 - 7\alpha + 3)t^\alpha + (\alpha + 1)t^{\alpha+1}$$

The indicial equation is
$$2\alpha^2 - 7\alpha + 3 = 0;$$

the roots are $\alpha = \frac{1}{2}$, $\alpha = 3$. We set

$$x_1 = t^{1/2} \sum_{\beta=0}^\infty c_\beta t^\beta$$

in Eq. (8–54) and combine terms into one series, finding

$$\sum_{\beta=1}^\infty [c_\beta(2\beta^2 - 5\beta) + c_{\beta-1}(\beta + \tfrac{1}{2})]t^\beta = 0.$$

Accordingly,
$$c_\beta = \frac{2\beta + 1}{2\beta(5 - 2\beta)} c_{\beta-1} \quad (\beta = 1, 2, 3, \ldots).$$

The coefficient c_0 is arbitrary. We choose $c_0 = 1$ and obtain the solution

$$x_1(t) = t^{1/2}\left[1 + \sum_{\beta=1}^\infty \frac{3 \cdot 5 \cdots (2\beta + 1)}{2^\beta\beta!3 \cdot 1 \cdot (-1)(-3) \cdots (5 - 2\beta)} t^\beta\right], \quad (8\text{–}56)$$

valid in particular for $t > 0$. Similarly, with $\alpha = 3$,

$$x_2(t) = t^3\left[1 + \sum_{\beta=1}^\infty \frac{(-1)^\beta(\beta + 1)(\beta + 2)(\beta + 3)}{6[7 \cdot 9 \cdots (2\beta + 5)]} t^\beta\right] \quad (8\text{–}57)$$

for all t. The solutions are linearly independent and, for proper interpretation of $t^{1/2}$, are also valid for complex t $(t \neq 0$ for $x_1(t))$.

8–6 Equations with coefficients asymptotic to constants. The systems problems leading to time-variant equations with regular singular points are rather rare. But other equations are reducible to such equations by appropriate substitutions.

EXAMPLE. $[2D^2 + (7 - e^{-t})D + (3 + e^{-t})]x = 0$. Here there is no singular point, and we can obtain power series solutions $\sum c_\beta t^\beta$ converging for all t. However, computing the solutions numerically from the series is effective only for small $|t|$, since the number of terms needed for given accuracy increases as $|t|$ increases. Instead of using power series in the given equation, we make the substitution $z = e^{-t}$. Since

$$\frac{dx}{dt} = \frac{dx}{dz}\frac{dz}{dt} = -e^{-t}\frac{dx}{dz} = -z\frac{dx}{dz},$$

$$\frac{d^2x}{dt^2} = -z\frac{d}{dz}\left(-z\frac{dx}{dz}\right) = z^2\frac{d^2x}{dz^2} + z\frac{dx}{dz},$$

we obtain the equation

$$2\left(z^2\frac{d^2x}{dz^2} + z\frac{dx}{dz}\right) + (7 - z)\left(-z\frac{dx}{dz}\right) + (3 + z)x = 0;$$

that is,

$$2z^2\frac{d^2x}{dz^2} + (z^2 - 5z)\frac{dx}{dz} + (3 + z)x = 0. \tag{8-60}$$

This is the same as Eq. (8-54), except that t is replaced by z. Hence the functions $x_1(z)$, $x_2(z)$ obtained from (8-56) and (8-57) are linearly independent solutions for all z except zero. Finally, $x_1(e^{-t})$, $x_2(e^{-t})$ are solutions of the given equation in t, valid for all t (since $e^{-t} \neq 0$):

$$x_1 = e^{-(1/2)t}\left[1 + \sum_{\beta=1}^{\infty} \frac{3 \cdot 5 \cdots (2\beta + 1)}{2^\beta \beta! 3 \cdot 1 \cdot (-1) \cdots (5 - 2\beta)} e^{-\beta t}\right],$$

$$x_2 = e^{-3t}\left[1 + \sum_{\beta=1}^{\infty} \frac{(-1)^\beta(\beta + 1)(\beta + 2)(\beta + 3)}{6[7 \cdot 9 \cdots (2\beta + 5)]} e^{-\beta t}\right]. \tag{8-61}$$

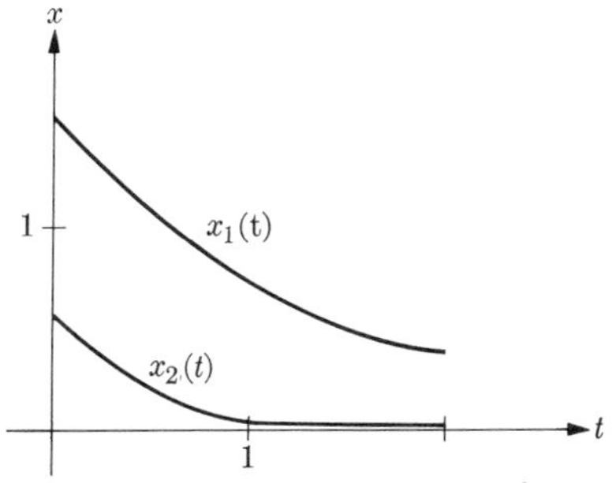

FIG. 8-2. Solutions of the equation $[2D^2 + (7 - e^{-t})D + (3 + e^{-t})]x = 0$.

These series converge very rapidly for large t and can be used to compute the solutions for all $t \geqq 0$. The results are graphed in Fig. 8–2. We remark that $x_1(t)$ and $x_2(t)$ approach zero as $t \to +\infty$, so that the corresponding system is *stable*. Except for small values of t (between zero and one) we obtain very good accuracy (error less than 0.05) by taking the first two terms of the series:

$$x_1(t) \sim e^{-t/2} + \tfrac{1}{2}e^{-3t/2}, \qquad x_2(t) \sim e^{-3t} - \tfrac{4}{7}e^{-4t};$$

even at $t = 0$, the error is less than 0.2 with these approximations. The Wronskian of the solutions can be obtained from the series solutions, or more easily from (8–16):

$$w(t) = c \exp\left[-\int \frac{7 - e^{-t}}{2} \, dt\right] = c \exp \frac{-7t - e^{-t}}{2}. \qquad (8\text{–}62)$$

The constant c can be found from the series (8–61) as follows: For large t only the first terms are needed, and we find

$$w(t) \sim \begin{vmatrix} e^{-(1/2)t} & e^{-3t} \\ -\tfrac{1}{2}e^{-(1/2)t} & -3e^{-3t} \end{vmatrix} = -\tfrac{5}{2}e^{-7t/2}. \qquad (8\text{–}63)$$

From (8–62), $w \sim ce^{-7t/2}$ for large t, so that $c = -\tfrac{5}{2}$:

$$w(t) = -\tfrac{5}{2} \exp \frac{-7t - e^{-t}}{2}. \qquad (8\text{–}62')$$

From this expression we obtain a series expansion:

$$w(t) = -\tfrac{5}{2}e^{-7t/2} \exp\left(-e^{-t}/2\right)$$
$$= -\tfrac{5}{2}e^{-7t/2}\left(1 - \frac{e^{-t}}{2} + \frac{e^{-2t}}{8} + \cdots + \frac{(-1)^k e^{-kt}}{2^k k!} + \cdots\right);$$

a good two-term approximation for $t > 0$ is

$$w(t) \sim -\tfrac{5}{2}e^{-7t/2} + \tfrac{5}{4}e^{-9t/2}.$$

In the same way,

$$\frac{1}{w(t)} = -\frac{2}{5} \exp \frac{7t + e^{-t}}{2}$$
$$= -\tfrac{2}{5}e^{7t/2}(1 + \tfrac{1}{2}e^{-t} + \tfrac{1}{8}e^{-2t} + \cdots) \sim -\tfrac{2}{5}e^{7t/2} - \tfrac{1}{5}e^{5t/2}. \qquad (8\text{–}64)$$

From (8–61) and (8–64) we obtain a series representation for $\phi(t, u)$. With the aid of Eq. (8–20) and the two-term approximations, we find

$$\phi(t,u) \sim -\tfrac{1}{140} \begin{vmatrix} 28e^{3u} + 28e^{2u} + 7e^{u} & 28e^{u/2} - 2e^{-u/2} - 8e^{-3u/2} \\ e^{-t/2} + \tfrac{1}{2}e^{-3t/2} & e^{-3t} - \tfrac{4}{7}e^{-4t} \end{vmatrix}.$$

(8–65)

We remark that $\phi(t,u)$ has a special form:

$$\phi(t,u) = \sum_{\beta,\gamma} A_{\beta\gamma} e^{\beta t} e^{\gamma u}.$$

(8–66)

Accordingly, computation of $T_0[f]$ is simple if f has simple form, in particular, if f is itself a sum of exponentials: $f(t) = \sum B_\eta e^{\eta t}$. We then obtain

$$x = T_0[f] = \int_0^t \phi(t,u) f(u)\, du$$

$$= \sum_{\beta,\gamma,\eta} A_{\beta\gamma} B_\eta e^{\beta t} \int_0^t e^{\gamma u} e^{\eta u}\, du$$

$$= \sum_{\beta,\gamma,\eta} A_{\beta\gamma} B_\eta e^{\beta t} \frac{e^{(\gamma+\eta)t} - 1}{\gamma + \eta}.$$

(8–67)

(If $\gamma = -\eta$, the integral is equal to t.)

We can also write

$$\phi(t,u) = \sum_{\beta,\gamma} A_{\beta\gamma} e^{(\beta+\gamma)t} e^{\gamma(u-t)}$$

and hence (at least formally)

$$x = T_0[f] = \sum_{\beta,\gamma} A_{\beta\gamma} e^{(\beta+\gamma)t} \int_0^t e^{-\gamma(t-u)} f(u)\, du$$

$$= \sum_{\beta,\gamma} A_{\beta\gamma} e^{(\beta+\gamma)t} (e^{-\gamma t} * f(t)).$$

We can employ Laplace transforms: if $\mathcal{L}[f] = F(s)$, then

$$\mathcal{L}[e^{-\gamma t} * f] = \frac{F(s)}{s + \gamma},$$

$$\mathcal{L}[e^{(\beta+\gamma)t}(e^{-\gamma t} * f)] = \frac{F(s - \beta - \gamma)}{s - \beta},$$

$$\mathcal{L}[x] = \sum_{\beta,\gamma} A_{\beta\gamma} \frac{F(s - \beta - \gamma)}{s - \beta}.$$

In our example, the coefficients approach constant values as $t \to +\infty$, and it is natural to compare the results obtained with those for the equation with constant coefficients: $(2D^2 + 7D + 3)x = 0$. The characteristic roots are $-\frac{1}{2}$, -3, two linearly independent solutions are $e^{-t/2}$, e^{-3t}, their Wronskian is $w_1 = -\frac{5}{2}e^{-7t/2}$, and the kernel function is $\phi_1 = -\frac{1}{5}(e^{3u-3t} - e^{(u-t)/2})$. These can all be considered as principal terms in the corresponding series expansions. In particular, $\phi_1(t, u)$ corresponds to the two terms in (8–66) for which $\beta = -\gamma$; for every other term, $\beta < 0$ and $\beta + \gamma < 0$, and the corresponding weighted average has total weight

$$A_{\beta\gamma}\int_0^t e^{\beta t}e^{\gamma u}\,du = A_{\beta\gamma}e^{\beta t}\frac{e^{\gamma t} - 1}{\gamma},$$

which approaches zero as $t \to +\infty$. Thus $\phi_1(t, u)$ gives the principal terms of the kernel function.

Our example suggests a broad generalization. Let an equation of order n be given, in which the coefficients approach constant values b_k as $t \to +\infty$. We then try to represent the coefficients by series in powers of e^{-t} to obtain an equation

$$[b_0D^n + (b_1 + b_{11}e^{-t} + b_{12}e^{-2t} + \cdots)D^{n-1} + \cdots$$
$$+ (b_n + b_{n1}e^{-t} + \cdots)]x = f(t). \qquad (8\text{–}68)$$

We then set $z = e^{-t}$ and hope to obtain an equation with regular singular point at $z = 0$, so that we can follow the procedure of the example. There are general theorems (p. 92 of Reference 4) to the effect that when the coefficients approach constant values as $t \to +\infty$, the solutions behave, for large t, like those of the corresponding equation with constant coefficients:

$$(b_0D^n + \cdots + b_n)x = f(t). \qquad (8\text{–}69)$$

We can verify this behavior for the case at hand.

THEOREM 6. *Let Eq. (8–68) be given, in which $b_0 \neq 0$, $f(t)$ is continuous, and the coefficient series converge absolutely for $t \geq 0$. The substitution $z = e^{-t}$ converts the corresponding homogeneous equation into one with a regular singular point at $z = 0$. The corresponding indicial equation is*

$$b_0(-\alpha)^n + b_1(-\alpha)^{n-1} + \cdots + b_n = 0,$$

which is the characteristic equation of Eq. (8–69) with s replaced by $-\alpha$. If the characteristic roots are real and distinct and no two roots differ by an integer, then n linearly independent solutions are obtained for $t > 0$:

$$x_k(t) = z^{\alpha_k}\left[1 + \sum_{\beta=1}^{\infty} c_{k\beta}z^\beta\right] = e^{-\alpha_k t}\left[1 + \sum_{\beta=1}^{\infty} c_{k\beta}e^{-\beta t}\right] \quad (k = 1, \ldots, n).$$

These solutions are asymptotic to corresponding solutions $e^{-\alpha_k t}$ of the homogeneous equation related to Eq. (8–69): $x_k(t)/e^{-\alpha_k t} \to 1$ as $t \to +\infty$. The corresponding Wronskian has the form

$$w(t) = ce^{-b_1 t/b_0}\left(1 + \sum_{\beta=1}^{\infty} w_\beta e^{-\beta t}\right) \quad (t > 0),$$

where

$$c = \begin{vmatrix} 1 & \cdots & 1 \\ s_1 & \cdots & s_n \\ \vdots & & \vdots \\ s_1^{n-1} & \cdots & s_n^{n-1} \end{vmatrix} = \prod_{i>j}(s_i - s_j)$$

$(s_k = -\alpha_k)$. *The kernel function can be written in the form*

$$\phi(t, u) = \sum_{k=1}^{n} e^{s_k(t-u)} R_k(u) Q_k(t),$$

where

$$R_k(u) = \sum_{\beta=0}^{\infty} r_{k\beta} e^{-\beta u} \quad (u > 0),$$

$$Q_k(t) = \sum_{\beta=0}^{\infty} q_{k\beta} e^{-\beta t} \quad (t > 0).$$

The response, with initial values zero at $t = 0$, is

$$x = T_0[f] = \int_0^t \phi(t, u) f(u)\, du$$

$$= \sum_{k=1}^{n} Q_k(t)[e^{s_k t} * \{R_k(t)f(t)\}].$$

Proof. As above, we set $z = e^{-t}$ and write $\mathfrak{D}x = dx/dz$. Then

$$Dx = -z\mathfrak{D}x, \quad D^2 x = (-z\mathfrak{D})^2 x = (z^2\mathfrak{D}^2 + z\mathfrak{D})x, \ldots$$

The homogeneous equation corresponding to Eq. (8–68) becomes

$$b_0[(-1)^n z^n \mathfrak{D}^n + \cdots]x$$
$$+ (b_1 + b_{11}z + \cdots)[(-1)^{n-1}z^{n-1}\mathfrak{D}^{n-1} + \cdots]x + \cdots = 0$$

or

$$b_0(-1)^n z^n \mathfrak{D}^n x + \psi_{n-1}(z)\mathfrak{D}^{n-1}x + \cdots + \psi_0(z)x = 0,$$

where the $\psi_k(z)$ are power series converging absolutely for $|z| \leq 1$, the term of lowest degree in $\psi_k(z)$ having degree l, $l \geq k$. Hence we obtain an equation with a regular singular point at $z = 0$. The indicial equation is obtained by replacing x by z^α in the left-hand side and finding the coefficient of the term of lowest degree; equivalently, we replace x by $e^{-\alpha t}$ in the left-hand side of Eq. (8–68) and find the term of lowest degree in e^{-t}:

$$P(\alpha)e^{-\alpha t} = [b_0(-\alpha)^n + b_1(-\alpha)^{n-1} + \cdots]e^{-\alpha t}.$$

Hence the indicial equation is as stated in the theorem. From the general theory of equations with regular singular points (Section 8–5), we obtain n linearly independent solutions as stated. The series solutions are valid for $|z| < 1$, that is, for $|e^{-t}| < 1$, or $t > 0$. Under the hypotheses made, the coefficients in Eq. (8–68) are continuous for $t \geqq 0$, so that the solutions $x_k(t)$ (if properly defined at $t = 0$) must remain continuous at $t = 0$. From the form of the solutions, we see that as $t \to \infty$, $x_k(t)e^{\alpha_k t} \to 1$, as asserted. We may also differentiate the series term by term and conclude that, for $m = 1, 2, \ldots,$

$$\frac{x_k^{(m)}(t)}{(-\alpha_k)^m e^{-\alpha_k t}} \to 1 \quad \text{as } t \to +\infty.$$

From (8–16) we obtain an expression for the Wronskian $w(t)$:

$$w(t) = c \exp\left[-\frac{1}{b_0}\int (b_1 + b_{11}e^{-t} + \cdots)\, dt\right]$$

$$= ce^{-b_1 t/b_0}e^{\rho(t)},$$

where

$$\rho(t) = \frac{1}{b_0}[b_{11}e^{-t} + b_{12}(e^{-2t}/2) + \cdots] \to 0 \quad \text{as } t \to +\infty.$$

Hence

$$\lim_{t\to\infty} e^{b_1 t/b_0}w(t) = c.$$

Now from (8–14)

$$w(t) = \begin{vmatrix} x_1(t) & \cdots \\ x_1'(t) & \cdots \\ \vdots & \vdots \end{vmatrix}$$

$$= \begin{vmatrix} e^{-\alpha_1 t}\omega_{11}(t) & e^{-\alpha_2 t}\omega_{12}(t) & \cdots \\ -\alpha_1 e^{-\alpha_1 t}\omega_{21}(t) & -\alpha_2 e^{-\alpha_2 t}\omega_{22}(t) & \cdots \\ \vdots & \vdots \end{vmatrix},$$

where $\omega_{ij}(t) \to 1$ as $t \to +\infty$. Now $\alpha_1 + \cdots + \alpha_n = b_1/b_0$ by the rule of algebra for the sum of the roots of an algebraic equation. Hence

$$e^{b_1 t/b_0} w(t) = e^{(\alpha_1 + \cdots + \alpha_n)t} w(t)$$

$$= \begin{vmatrix} \omega_{11}(t) & \omega_{12}(t) & \cdots \\ -\alpha_1 \omega_{21}(t) & -\alpha_2 \omega_{22}(t) & \cdots \\ \vdots & \vdots & \end{vmatrix}$$

$$\rightarrow \begin{vmatrix} 1 & 1 & \cdots & 1 \\ -\alpha_1 & -\alpha_2 & \cdots & -\alpha_n \\ \vdots & & & \vdots \\ (-\alpha_1)^{n-1} & & \cdots & (-\alpha_n)^{n-1} \end{vmatrix} = c$$

as $t \rightarrow +\infty$. Since $s_k = -\alpha_k$, we obtain the expression for c as given in the theorem.

Since $\rho(t)$ is represented by a convergent power series in $z = e^{-t}$ for $|z| < 1$, we conclude that $e^{\rho(t)}$ is also representable by such a series, and we can write

$$w(t) = c e^{-b_1 t/b_0} \left(1 + \sum_{\beta=1}^{\infty} w_\beta e^{-\beta t} \right) \quad (t > 0),$$

where the w_β are constants. Similarly,

$$\frac{1}{w(t)} = \frac{1}{c} e^{b_1 t/b_0} e^{-\rho(t)} = \frac{1}{c} e^{b_1 t/b_0} \left(1 + \sum_{\beta=1}^{\infty} v_\beta e^{-\beta t} \right) \quad (t > 0).$$

By (8–20) the kernel function can now be represented as follows:

$$\phi(t, u) = \frac{e^{b_1 u/b_0}}{c} (1 + \textstyle\sum v_\beta e^{-\beta u}) \begin{vmatrix} e^{s_1 u} \omega_{11}(u) & \cdots \\ s_1 e^{s_1 u} \omega_{21}(u) & \cdots \\ \vdots & \vdots \\ e^{s_1 t} \omega_{11}(t) & \cdots \end{vmatrix}$$

$$= \frac{1 + \sum v_\beta e^{-\beta u}}{c} \begin{vmatrix} \omega_{11}(u) & \cdots \\ s_1 \omega_{21}(u) & \cdots \\ \vdots & \vdots \\ e^{s_1(t-u)} \omega_{11}(t) & \cdots \end{vmatrix},$$

since $s_1 + \cdots + s_n = -(\alpha_1 + \cdots + \alpha_n) = -b_1/b_0$. Now $\omega_{ij}(u)$ is a power series in powers of e^{-u} and hence, by expanding the determinant and multiplying the series, we finally obtain the desired expression

$$\phi(t, u) = \sum_{k=1}^{n} e^{s_k(t-u)} R_k(u) Q_k(t),$$

where $R_k(u)$ and $Q_k(t)$ are as stated in the theorem. In fact, $Q_k(t) = \omega_{1k}(t) = x_k(t)e^{-s_k t}$. Since the $x_k(t)$ can be considered as solutions of the homogeneous equation for $t \geq 0$, $\phi(t, u)$ is continuous for $t \geq 0$, $u \geq 0$, and the response to $f(t)$ is obtained in the usual manner:

$$x(t) = \int_0^t \phi(t, u) f(u) \, du$$

$$= \sum_{k=1}^{n} Q_k(t) \int_0^t e^{s_k(t-u)} R_k(u) f(u) \, du$$

$$= \sum_{k=1}^{n} Q_k(t) \{ e^{s_k t} * [R_k(t) f(t)] \}.$$

If $f(t) = e^{at}$, we can write the response (apart from a solution of the homogeneous equation) as e^{at} times a power series in powers of e^{-t}; the leading term of this series gives the response for the equation with constant coefficients (8–69).

Remark. In practical problems having coefficients which approach constants as $t \to +\infty$, we can proceed as follows: Set $z = e^{-t}$, or $t = -\log z$, in each coefficient, so that each becomes a function of z on the interval $0 \leq z \leq 1$. Now approximate each coefficient by a polynomial in z, for example, with the aid of an expansion in Legendre polynomials (Section 4–12). With $z^k = e^{-kt}$, the desired representation of the coefficients has been obtained. We can even extend the method to equations whose coefficients do not approach constants. For example, in the nonhomogeneous Mathieu equation

$$D^2 x + (a + b \cos t) x = f(t),$$

one can insert a factor $e^{-\sigma t}$ ($\sigma > 0$) before the cosine:

$$D^2 x + (a + b e^{-\sigma t} \cos t) x = f(t);$$

the method just described can then be applied for each $\sigma > 0$. By letting $\sigma \to 0$, we then obtain the response for the given equation. For σ close to 0, the approximation will be good over a large interval of t.

Problems

1. (a) Show that the substitution $t = e^v$ reduces a Cauchy equation (8–50) to one with constant coefficients.
   (b) Find the general solution: $(t^2 D^2 + 6tD + 4)x = 0$.

2. Find two linearly independent solutions of form $\sum c_n t^n$:
   (a) $(D^2 + t^2 D + t)x = 0$ (Example 1 of Section 8–5)
   (b) $[(t^2 + 1)D^2 + 5tD + 4]x = 0$

3. Find two linearly independent solutions of form $\sum c_n t^n$ (obtain terms through t^3): $D^2 x + (3 + e^{-t}) Dx + 2x = 0$.

4. For each of the following equations verify that there is a regular singular point at $t = 0$, and obtain two linearly independent solutions of form $t^\alpha \sum c_\beta t^\beta$; in part (b) treat b as a parameter:
   (a) $6t^2 D^2 x + tDx + (1 - t)x = 0$
   (b) $2t^2 D^2 x - 3tDx + (3 - bt)x = 0$

5. With the aid of the solution to Problem 4(a) obtain the kernel function $\phi(t, u)$ for the equation $6D^2 x + 5Dx + (1 - e^{-t})x = f(t)$.

6. With the aid of the solution to Problem 4(b) obtain the kernel function for the equation
$$2D^2 x + 5Dx + (3 - be^{-t})x = f(t).$$

Find $T_0[f]$ for $f = e^t$ and for $b = 0$, $b = 0.1$, $b = 1$ respectively. Graph the results to show the influence of the time-varying coefficient.

Answers

1. (b) $x = c_1 t^{-1} + c_2 t^{-4}$

2. (b) $1 + \sum_{\beta=1}^{\infty} (-1)^\beta t^{2\beta} \dfrac{2 \cdot 4 \cdots (2\beta)}{1 \cdot 3 \cdots (2\beta - 1)}$, $t + t\sum_{\beta=1}^{\infty} (-1)^\beta t^{2\beta} \dfrac{3 \cdot 5 \cdots (2\beta + 1)}{2 \cdot 4 \cdots (2\beta)}$

3. $1 - t^2 + (4t^3/3) + \cdots$, $t - 2t^2 + (5t^3/2) + \cdots$

4. (a) $t^{1/2}\left(1 + \sum_{\beta=1}^{\infty} \{t^\beta/[\beta! 7 \cdot 13 \cdots (6\beta + 1)]\}\right)$,

   $t^{1/3}\left(1 + \sum_{\beta=1}^{\infty} [t^\beta/\{\beta! 5 \cdot 11 \cdots (6\beta - 1)\}]\right)$

   (b) $t\left(1 + \sum_{\beta=1}^{\infty} [t^\beta b^\beta/\{\beta! 1 \cdot 3 \cdots (2\beta - 1)\}]\right)$,

   $t^{3/2}\left(1 + \sum_{\beta=1}^{\infty} [t^\beta b^\beta/\{\beta! 3 \cdot 5 \cdots (2\beta + 1)\}]\right)$

5. $\phi(t, u) = A(t, u)e^{-(t-u)/3} - B(t, u)e^{-(t-u)/2}$, where

$$A(t, u) = \left(1 + \sum[e^{-\beta u}/\{\beta!7 \cdot 13 \cdots (6\beta + 1)\}]\right)$$
$$\times \left(1 + \sum[e^{-\beta t}/\{\beta!5 \cdot 11 \cdots (6\beta - 1)\}]\right),$$

$$B(t, u) = \left(1 + \sum[e^{-\beta u}/\{\beta!5 \cdot 11 \cdots (6\beta - 1)\}]\right)$$
$$\times \left(1 + \sum[e^{-\beta t}/\{\beta!7 \cdot 13 \cdots (6\beta + 1)\}]\right)$$

6. $\phi(t, u) = A(t, u)e^{-(t-u)} - B(t, u)e^{-3(t-u)/2}$, where

$$A(t, u) = \left(1 + \sum[e^{-\beta u}b^\beta/\{\beta!3 \cdot 5 \cdots (2\beta + 1)\}]\right)$$
$$\times \left(1 + \sum[e^{-\beta t}b^\beta/\{\beta!1 \cdot 3 \cdots (2\beta - 1)\}]\right),$$

$$B(t, u) = \left(1 + \sum[e^{-\beta u}b^\beta/\{\beta!1 \cdot 3 \cdots (2\beta - 1)\}]\right)$$
$$\times \left(1 + \sum[e^{-\beta t}b^\beta/\{\beta!3 \cdot 5 \cdots (2\beta + 1)\}]\right);$$

$$T_0[f] = (e^t/30)[3 - 15e^{-2t} + 12e^{-5t/2} + b(e^{-t} - 10e^{-2t}$$
$$+ 20e^{-5t/2} + 4e^{-7t/2} - 15e^{-3t}) + b^2(\ldots) + \cdots].$$

8–7 Perturbation method. We consider our differential equation in the form

$$b_0 D^n x + [b_1 + \epsilon p_1(t)]D^{n-1}x + \cdots + [b_n + \epsilon p_n(t)]x = f(t). \quad (8\text{–}70)$$

We assume that $b_0, b_1, \ldots, b_n$ are constants, that $b_0 \neq 0$, and that $f(t)$, $p_1(t), \ldots, p_n(t)$ are continuous for $t \geqq 0$. We regard ϵ as a parameter; for each fixed value of ϵ, we can solve Eq. (8–70) with given initial values $x^{(l)}(t_0)$, $l = 0, 1, \ldots, n - 1$. The solution $x(t)$ then depends on ϵ. It can be shown (Chapter 1 of Reference 4) that x is continuous in t, ϵ, and that, for each t, x is an analytic function of ϵ for all complex ϵ. Thus we can write

$$x(t) = \sum_{k=0}^{\infty} q_k(t)\epsilon^k. \quad (8\text{–}71)$$

Here we are permitted to differentiate term by term to obtain $x'(t), \ldots,$ $x^{(n)}(t)$:

$$x^{(l)}(t) = \sum_{k=0}^{\infty} q_k^{(l)}(t)\epsilon^k \quad (l = 0, 1, \ldots, n).$$

In particular, at $t = t_0$, for all ϵ,

$$x^{(l)}(t_0) = \sum_{k=0}^{\infty} q_k^{(l)}(t_0)\epsilon^k.$$

Accordingly, by Theorem 22 of Section 3–8, we conclude that, for $l = 0$, $1, \ldots, n - 1$,

$$q_0^{(l)}(t_0) = x^{(l)}(t_0), \quad q_k^{(l)}(t_0) = 0 \quad \text{for } k = 1, 2, \ldots \quad (8\text{–}72)$$

We now rewrite the differential equation:

$$L_0[x] + \epsilon L_1[x] = f(t),$$

$$L_0 = b_0 D^n + b_1 D^{n-1} + \cdots + b_n,$$

$$L_1 = p_1 D^{n-1} + p_2 D^{n-2} + \cdots + p_n.$$

We then substitute the expression (8–71) for x and collect terms of same degree in ϵ:

$$L_0[q_0] + \sum_{k=1}^{\infty} \{L_0[q_k] + L_1[q_{k-1}]\} \epsilon^k = f(t).$$

Since this equation holds for all ϵ, we conclude as above that

$$L_0[q_0] = f(t), \quad (8\text{–}73)$$

$$L_0[q_k] = -L_1[q_{k-1}] \quad (k = 1, 2, \ldots). \quad (8\text{–}74)$$

Accordingly, $q_0(t)$ is the solution of the differential equation with constant coefficients $(b_0 D^n + \cdots + b_n)q = f$, with given initial values at $t = t_0$ as in (8–72), while $q_k(t)$ can be regarded as the solution of the differential equation with constant coefficients

$$(b_0 D^n + \cdots + b_n)q = f_k(t), \quad f_k = -L_1[q_{k-1}],$$

with initial values zero at $t = t_0$ as in (8–72). In particular, if $t_0 = 0$ and all initial values are zero, then

$$q_0 = W * f, \ q_1 = -W * L_1[q_0], \ q_2 = -W * L_1[q_1], \ldots, \quad (8\text{–}75)$$

where $W = \mathcal{L}^{-1}[1/(b_0 s^n + \cdots + b_n)]$, and $x = \sum q_k(t) \epsilon^k = T_0[f]$. If $|\epsilon|$ is small, a few terms of the series will give a good approximation to $x(t)$.

In general, if a differential equation has variable coefficients remaining close to constant values, then the differential equation can be considered as a *perturbation* of an equation with constant coefficients and, with the aid of an appropriate parameter ϵ, a solution can be found as indicated.

EXAMPLE. $Dx + (1 + \cos t)x = 2e^t$. Here we insert ϵ as follows:

$$Dx + (1 + \epsilon \cos t)x = 2e^t,$$

so that we must later set $\epsilon = 1$. The operator L_0 is $D + 1$, and the operator L_1 is $\cos t$ (as multiplier). We seek the solution with initial value one at $t = 0$. Hence $(D + 1)q_0 = 2e^t$, $q_0(0) = 1$, so that $q_0(t) = e^t$. Next $(D + 1)q_1 = -(\cos t)q_0 = -e^t \cos t$, $q_1(0) = 0$. Thus, with $W = \mathcal{L}^{-1}[1/(s + 1)] = e^{-t}$,

$$q_1 = -e^{-t} * (e^t \cos t) = \tfrac{2}{5}e^{-t} - \tfrac{1}{5}e^t(2 \cos t + \sin t),$$

$$x = e^t + \frac{\epsilon}{5} [2e^{-t} - e^t(2 \cos t + \sin t)] + \cdots$$

For $\epsilon = 1$ we obtain the desired solution of the given equation.

We can apply the perturbation method to the general equation (8–70) to obtain a series for the kernel function $\phi(t, u)$:

$$\phi(t, u) = \sum_{k=0}^{\infty} \phi_k(t, u)\epsilon^k. \qquad (8\text{--}76)$$

In accordance with Theorem 3 (Section 8–2), $\phi(t, u)$ is the solution of the homogeneous equation related to Eq. (8–70) such that $x(u) = 0$, $x'(u) = 0$, $\ldots$, $x^{(n-2)}(u) = 0$, $x^{(n-1)}(u) = 1/b_0$. By the procedure described above, $\phi_0(t, u)$ is then the solution of the homogeneous equation with constant coefficients

$$L_0[x] = b_0 D^n x + \cdots + b_n x = 0$$

with these initial values at $t = u$. Accordingly, $\phi_0(t, u)$ is the kernel function for the equation $L_0[x] = f$ and $\phi_0(t, u) = W(t - u)$ for $t \geq u$, where $W = \mathcal{L}^{-1}[1/(b_0 s^n + \cdots + b_n)]$. Next $\phi_1(t, u)$ is the solution of

$$L_0[x] = -L_1[\phi_0(t, u)]$$

with initial values zero at $t = u$. Accordingly, by Theorem 2 (Section 8–2)

$$\phi_1(t, u) = -\int_u^t L_1[\phi_0(t, u)]\Big|_{t=v} \phi_0(t, v) \, dv,$$

and in general

$$\phi_k(t, u) = -\int_u^t L_1[\phi_{k-1}(t, u)]\Big|_{t=v} \phi_0(t, v) \, dv.$$

8–8 Equations with coefficients which are piecewise constant. In our differential equation

$$a_0(t) D^n x + \cdots + a_n(t) x = f(t) \qquad (8\text{--}80)$$

we have up to this point assumed $f(t)$ and the coefficients $a_0(t), \ldots, a_n(t)$ to be continuous functions on an interval. We now relax this restriction by requiring only that these functions be piecewise continuous. In addition, we require that $a_0(t) \neq 0$ and that $a_0(t)$ not have zero as a limiting value to the left or right at jump points. We can then proceed exactly as in Section 1–12 to define solutions of Eq. (8–80); a solution $x(t)$ is continuous, has continuous derivatives up to order $n - 1$, and satisfies Eq. (8–80) except perhaps at the jump points. The existence theorem continues to apply: there is a unique solution $x(t)$ with given initial values at a point t_0 of the interval. If for example the basic interval is $0 \leq t < \infty$ and $t_0 = 0$, let the first jump point be t_1 ($t_1 > 0$). We then solve as usual in the interval $0 \leq t \leq t_1$, in which the coefficients can be considered to be continuous. Then $x(t)$, $x'(t)$, ..., $x^{(n-1)}(t)$ all have limits as $t \to t_1-$; these limits serve as initial values for the interval $t_1 \leq t \leq t_2$, and so on.

If the coefficients have particularly simple forms in each interval of continuity, it may be possible to solve explicitly. This is so if the coefficients are piecewise constant.

EXAMPLE. Let the equation be given in two stages:

$$(D^2 + 3D + 2)x = 0 \quad (0 \leq t < 1);$$

$$(D^2 + 4D + 3)x = 0 \quad (1 \leq t < \infty).$$

The solution $x_1(t)$, such that $x_1(0) = 1$, $x_1'(0) = -1$, will then equal e^{-t} for $0 \leq t \leq 1$. At $t = 1-$ we obtain the limiting values $x_1 = e^{-1}$, $x_1' = -e^{-1}$. The solution $x_1(t)$ of the appropriate equation for the interval $1 \leq t < \infty$ with initial values $x_1(1) = e^{-1}$, $x_1'(1) = -e^{-1}$ is easily found to be e^{-t}, so that in this case $x_1(t) = e^{-t}$ for $0 \leq t < \infty$. Next the solution $x_2(t)$ with initial values $x_2(0) = 1$, $x_2'(0) = -2$ is equal to e^{-2t} for $0 \leq t \leq 1$ and to $\frac{1}{2}(e^{-1-t} + e^{1-3t})$ for $t \geq 1$. The Wronskian of the two solutions is $w(t)$, equal to $-e^{-3t}$ for $0 \leq t \leq 1$, and to $-e^{1-4t}$ for $1 \leq t < \infty$. Furthermore, $w(t)$ is continuous and satisfies the equation $w' + 3w = 0$ for $0 < t < 1$, and the equation $w' + 4w = 0$ for $t > 1$. We can now form the kernel function $\phi(t, u)$ in the usual manner.

If one deals with equations with piecewise constant coefficients over many intervals $0 < t < t_1$, $t_1 < t < t_2$, ..., the procedure described can lead to quite involved algebraic manipulation. The work can be systematized with the aid of matrices (Reference 6 and pp. 4–12 of Reference 18). For more general coefficients, one can *approximate* the coefficients by piecewise constant functions to obtain a procedure for explicit solution. This is, in essence, a numerical integration of the differential equation and should be considered as an alternative to one of the many known methods for numerical solution (Reference 14).

8–9 Application of Laplace transforms. If the coefficients in Eq. (8–80) are polynomials in t, the Laplace transform can be applied; the differential in x becomes a new differential equation in $X(s) = \mathcal{L}[x]$. The crucial property of Laplace transforms needed here is the rule

$$\mathcal{L}[tf] = -F'(s), \qquad \mathcal{L}[t^k]f = (-1)^k F^{(k)}(s) \quad (k = 2, 3, \ldots), \quad (8\text{–}90)$$

where $F(s) = \mathcal{L}[f]$ (Theorem 7, Section 6–5). Accordingly, if m is the highest power to which t appears, then we will obtain a differential equation of order m for $X(s)$. If m is 1, the new differential equation is of first order and can be solved easily in terms of indefinite integrals. Hence a useful form of the solution may result. Throughout we assume $t \geq 0$.

EXAMPLE 1. $[tD^2 + (1 - t)D + 1]x = 0$. The equation has a singular point at $t = 0$, so that the existence theorem is inapplicable. However, we try to obtain a solution $x(t)$ such that $x(0) = a$, $x'(0) = b$. If $\mathcal{L}[x] = X(s)$, then $\mathcal{L}[x'] = sX(s) - a$, $\mathcal{L}[x''] = s^2X(s) - as - b$, and by the rule (8–90) our equation becomes

$$-\frac{d}{ds}[s^2X - as - b] + \left(1 + \frac{d}{ds}\right)(sX - a) + X = 0.$$

Upon simplification we find

$$(s - s^2)X' + (2 - s)X = 0.$$

(Thus the initial values have disappeared from the equation.) We can now solve:

$$X = c\,\frac{s - 1}{s^2}, \qquad x = c(1 - t).$$

Thus we obtain a family of solutions depending on only one arbitrary constant. It can be verified that the other solutions have a logarithmic singularity at $t = 0$ and are not obtainable by Laplace transforms.

In general, the transform method leads to a differential equation of order m for $X(s)$; if we carry along arbitrary initial values for x, we obtain as general solution for $X(s)$ an expression

$$X(s) = X^*(s) + c_1X_1(s) + c_2X_2(s) + \cdots + c_mX_m(s)$$
$$+ x(0)G_1(s) + x'(0)G_2(s) + \cdots + x^{(n-1)}(0)G_n(s). \quad (8\text{–}91)$$

Here some or all of the $G_k(s)$ may be zero, as in the example. Furthermore, not all the functions of s on the right may be Laplace transforms of functions of t. In general, (8–91) will not provide solutions depending on more than n arbitrary constants; in fact, it will usually provide solutions depending on less than n arbitrary constants, as in the example. The complications can be attributed to singular points, either on the interval $0 \leq t < \infty$ or at $t = \infty$.

EXAMPLE 2. $tDx - x = 0$. We find $X = cs^{-2}$, $x = ct$. Thus all solutions are obtained by transforms, but necessarily $x_0 = x(0) = 0$.

EXAMPLE 3. $Dx + tx = 0$. We find

$$X(s) = -x_0 e^{s^2/2} \int e^{-s^2/2} \, ds + c e^{s^2/2}.$$

We make a particular choice of the indefinite integral by writing

$$X = x_0 e^{s^2/2} \int_s^\infty e^{-\zeta^2/2} \, d\zeta + c e^{s^2/2},$$

where the path of integration is on the line $\zeta = s + t$, $0 \leq t < \infty$. For by uniform convergence

$$\frac{d}{ds} \int_s^\infty e^{-\zeta^2/2} \, d\zeta = \frac{d}{ds} \int_0^\infty e^{-(s+t)^2/2} \, dt$$

$$= -\int_0^\infty e^{-(s+t)^2/2}(s + t) \, dt = -e^{-s^2/2},$$

so that minus the integral from s to ∞ does represent an indefinite integral of $e^{-s^2/2}$. With $t = \zeta - s$ on the path, we find

$$X = x_0 \int_0^\infty e^{-t^2/2} e^{-st} \, dt + c e^{s^2/2}. \tag{8–92}$$

Now the general solution of the given differential equation is easily found to be $x = x_0 e^{-t^2/2}$. Thus the first term on the right of (8–92) gives the Laplace transform of the solutions. The second term is not a Laplace transform unless $c = 0$ (p. 49 of Reference 5). Hence the solutions provided by (8–92) depend on only one arbitrary constant.

EXAMPLE 4. $tDx + 2x = 0$. We find that $X = cs$ is not a transform unless $c = 0$. The solutions are $x = c/t^2$; $x(0)$ cannot be prescribed.

Verification of the missing details in these examples is left as an exercise (Problem 3 below).

PROBLEMS

1. Given the equation $[2D^2 + (7 - e^{-t})D + (3 + e^{-t})]x = 0$ of the example of Section 8–6, introduce a parameter ϵ as follows:

$$[2D^2 + (7 - \epsilon e^{-t})D + (3 + \epsilon e^{-t})]x = 0.$$

Apply the perturbation method to obtain the solution such that $x(0) = 1$, $x'(0) = 0$. Exhibit explicitly the terms through ϵ^2.

2. In the Example of Section 8–6, approximate the coefficient $7 - e^{-t}$ by the function $6[h(t) - h(t - 1)] + 7h(t - 1)$, and the coefficient $3 + e^{-t}$ by $4[h(t) - h(t - 1)] + 3h(t - 1)$; then find the solution of the approximating equation such that $x(0) = 1$, $x'(0) = 0$.

3. Verify the solution by Laplace transforms of the following examples in Section 8–9:

(a) Example 2   (b) Example 3   (c) Example 4

4. Given the equation $[(t - 1)D^2 - tD + 1]x = 0$, apply Laplace transforms to obtain the solution such that $x(0) = a$, $x'(0) = b$.

ANSWERS

1. $\frac{1}{5}(6e^{-t/2} - e^{-3t}) + (\epsilon/35)(-11e^{-t/2} + 21e^{-3t/2} - 14e^{-3t} + 4e^{-4t}) + (\epsilon^2/1260)(13e^{-t/2} - 198e^{-3t/2} + 945e^{-5t/2} - 1008e^{-3t} + 288e^{-4t} - 40e^{-5t}) + \cdots$

2. $(2e^{-t} - e^{-2t})[h(t) - h(t - 1)] + \frac{1}{5}[(8e^{-1/2} - 2e^{-3/2})e^{-t/2} + (2e^2 - 3e) \times e^{-3t}]h(t - 1)$

4. $a(e^t - t) + bt + c(e^{t-1} - t)h(t - 1)$. Solutions depend on two arbitrary constants for $0 \leq t \leq 1$ and for $1 \leq t < \infty$; $t = 1$ is a singular point.

8–10 Equations with periodic coefficients. We first consider an equation of first order:

$$Dx + a(t)x = f(t);$$ (8–100)

we assume that $a(t)$ and $f(t)$ are continuous for all t and that $a(t)$ has period $\tau > 0$: $a(t + \tau) = a(t)$. Then as in Chapter 4 we can form the Fourier series of $a(t)$:

$$p_0 + \sum_{n=1}^{\infty} (p_n \cos n\omega t + q_n \sin n\omega t) \quad (\omega = 2\pi/\tau).$$ (8–101)

We can integrate (Section 4–9):

$$b(t) = \int_0^t a(u) \, du$$

$$= p_0 t + \sum_{n=1}^{\infty} \frac{p_n \sin n\omega t - q_n \cos n\omega t}{n\omega} + \text{const.}$$

Then $e^{b(t)}$ is an integrating factor of Eq. (8–100), and the solutions are given by

$$x = e^{-b(t)} \left[\int f(t) e^{b(t)} \, dt + c \right].$$

Here the solutions of the related homogeneous equation are the functions

$$x_c(t) = c_1 e^{-b(t)} = c e^{-p_0 t} g(t),$$

$$g(t) = \exp \sum_{n=1}^{\infty} \frac{q_n \cos n\omega t - p_n \sin n\omega t}{n\omega}.$$

Thus $g(t)$ has period τ, and $g(t) \neq 0$. However, $x_c(t)$ is not periodic for $c \neq 0$ unless p_0 (assumed real) is zero. If $p_0 > 0$, then $x_c(t) \to 0$ as $t \to \infty$, so that the equation is stable.

The kernel function (see Problem 2 following Section 8–3) is

$$\phi(t, u) = e^{b(u)-b(t)} = e^{p_0(u-t)} \frac{g(t)}{g(u)},$$

and the response to f, with initial value zero, is

$$x = g(t) \int_0^t e^{p_0(u-t)} \frac{f(u)}{g(u)} \, du. \tag{8–102}$$

Since $g(t)$ is periodic, has a continuous derivative, and is never zero, $1/g(t)$ has the same properties and can also be expanded in a Fourier series (which we write in complex form):

$$\frac{1}{g(t)} = \sum_{n=-\infty}^{\infty} c_n e^{in\omega t}.$$

If now $f(t) = e^{\alpha t}$, then we can compute the response (8–102):

$$x = g(t) \int_0^t e^{p_0(u-t)} e^{\alpha u} \sum c_n e^{in\omega u} \, du$$

$$= g(t) e^{-p_0 t} \sum_{-\infty}^{\infty} c_n \frac{e^{(p_0+\alpha+in\omega) t} - 1}{p_0 + \alpha + in\omega}$$

$$= e^{\alpha t} P(t) + k e^{-p_0 t} g(t),$$

$$P(t) = g(t) \sum [c_n e^{in\omega t} (p_0 + \alpha + in\omega)^{-1}] \quad (k = \text{const}).$$

The term-by-term integration is justified, since the series for $1/g$ is uniformly convergent (Section 4–6). If $p_0 + \alpha + in\omega = 0$ for some n, the corresponding term in the summation is to be replaced by $c_n t$. If this does not occur, then $P(t)$ is the product of two functions of period τ and hence also has period τ; if in addition $p_0 > 0$, then the term $k e^{-p_0 t} g(t)$ represents a transient. Thus, in general, the *steady-state response to an exponential forcing function* $e^{\alpha t}$ *has form* $e^{\alpha t} P(t)$, *where* $P(t)$ *has period* τ. In particular, if $\alpha = 0$, so that the input is constant, the steady state output is the periodic function $P(t)$.

Now let α be pure imaginary, $\alpha = i\omega_1$, so that $f(t)$ has period $2\pi/\omega_1$. If $p_0 > 0$, the conclusions of the previous paragraph hold, and the steady state is $e^{i\omega_1 t}P(t)$. Unless the frequencies ω_1 and ω are commensurable, we cannot conclude that the response is periodic. If ω_1 is a rational multiple of ω, $\omega_1 = l\omega/m$ (l, m positive integers without common factor), then the factor $e^{i\omega_1 t}$ has period $2\pi/\omega_1 = 2\pi m/(l\omega) = (m/l)\tau$, and hence also has period $m\tau = l(m/l)\tau$. Thus both factors $e^{i\omega_1 t}$, $P(t)$ have period $m\tau$, and the output has this period (it may have a smaller one). Since the output period is in general a multiple of the input period, the output should be considered a *subharmonic response*.

Now let $p_0 = 0$ and $\omega_1 = N\omega$ for some positive integer N. Unless $c_{-N} = 0$, the response includes a term $c_{-N} t g(t)$ which is unbounded as $t \to +\infty$. Thus the phenomenon of *resonance* can occur.

The number p_0 plays a crucial role in the preceding discussion. Its negative, $-p_0$, is called the *characteristic exponent* of the differential equation.

We now turn to the equation of order n and formulate a general result:

THEOREM 7 (Floquet-Poincaré). *Let the functions $a_j(t)$ ($j = 0, 1, \ldots, n$) be continuous for all t and have period $\tau > 0$; let $a_0(t) \neq 0$. There exist n constants $\rho_1, \ldots, \rho_n$ such that the differential equation*

$$a_0(t)D^n x + \cdots + a_n(t)x = 0 \qquad (8\text{--}103)$$

has n linearly independent solutions

$$x_k(t) = e^{\rho_k t}\psi_k(t) \quad (-\infty < t < \infty, k = 1, \ldots, n); \quad (8\text{--}104)$$

here each $\psi_k(t)$ has period τ, when $e^{\rho_1\tau}, \ldots, e^{\rho_n\tau}$ are distinct, and otherwise $\psi_k(t)$ equals a periodic function of period τ times a polynomial in t (of degree less than the number of times $e^{\rho_k\tau}$ is repeated).

For a proof see pp. 78–81 of Reference 4. The main idea of the proof is indicated in the discussion below.

The numbers $\rho_1, \ldots, \rho_n$ are the *characteristic exponents*. They are not unique, but the numbers $\lambda_k = e^{\rho_k\tau}$ (often called *characteristic multipliers*) are unique; thus the ρ_k are unique up to addition of integral multiples of $i\omega$, $\omega = 2\pi/\tau$; in particular, Re (ρ_k) is uniquely determined. In general, the ρ_k are complex and [the coefficients in (8–103) being real] come in complex conjugate pairs. Accordingly, the solutions (8–104) are complex; real solutions are obtained by taking real or imaginary parts (Section 1–9).

From the form of the solutions (8–104) it is clear that Eq. (8–103) *is stable precisely when* Re $(\rho_k) < 0$ *for* $k = 1, \ldots, n$.

When the λ_k are distinct, each of the solutions (8–104) has the property that

$$x_k(t + \tau) = \lambda_k x_k(t). \tag{8–105}$$

For $x_k(t + \tau) = e^{\rho_k(t+\tau)}\psi_k(t + \tau) = e^{\rho_k t}\lambda_k\psi_k(t)$. If the general solution of Eq. (8–103) is known, we can use the property (8–105) to find the characteristic exponents. Let the general solution be

$$x = c_1 g_1(t) + \cdots + c_n g_n(t). \tag{8–106}$$

Since the coefficients in (8–103) have period τ, not only $g_k(t)$ but also $g_k(t + \tau)$ is a solution. Hence

$$g_k(t + \tau) = \sum_{\beta=1}^{n} b_{k\beta} g_\beta(t)$$

for some constants $b_{k\beta}$. We now require that $x(t)$, as defined by (8–106), satisfies the condition $x(t + \tau) = \lambda x(t)$ for some λ, different from zero. We obtain the equations

$$\sum_{k=1}^{n} c_k g_k(t + \tau) = \lambda \sum_{k=1}^{n} c_k g_k(t),$$

$$\sum_{k=1}^{n} \sum_{\beta=1}^{n} c_k b_{k\beta} g_\beta(t) = \lambda \sum_{k=1}^{n} c_k g_k(t).$$

Since the $g_k(t)$ are linearly independent, we can compare coefficients of the $g_k(t)$ on left and right:

$$\sum_{\alpha=1}^{n} c_\alpha b_{\alpha k} = \lambda c_k \quad (k = 1, \ldots, n).$$

These are n homogeneous linear equations for $c_1, \ldots, c_n$. We obtain a nontrivial solution when

$$\begin{vmatrix} b_{11} - \lambda & b_{21} & \cdots & \cdots \\ b_{12} & b_{22} - \lambda & \cdots & \cdots \\ \vdots & & & \vdots \\ b_{1n} & & \cdots & b_{nn} - \lambda \end{vmatrix} = 0. \tag{8–107}$$

We note that $\lambda = 0$ cannot be a root of Eq. (8–107); for if it were, then one could choose $c_1, \ldots, c_n$, not all zero, so that $\sum_k c_k g_k(t + \tau) \equiv 0$, in contradiction to the linear independence of $g_1(t), \ldots, g_n(t)$. If Eq. (8–107) has n distinct roots $\lambda_1, \ldots, \lambda_n$, then we obtain n corresponding solutions $x_1(t), \ldots, x_n(t)$. We can verify that they are linearly independent (Problem 2, following Section 8–12) and that (8–105) holds. Now set $\psi_k(t) = x_k(t)e^{-\rho_k t}$, where $\rho_k = \tau^{-1} \log \lambda_k$. Then (8–104) holds, and

$$\psi_k(t + \tau) = x_k(t + \tau)e^{-\rho_k(t+\tau)} = \lambda_k x_k(t)e^{-\rho_k t}e^{-\rho_k \tau}$$

$$= x_k(t)e^{-\rho_k t} = \psi_k(t).$$

Thus ψ_k is periodic. We have in effect proved Theorem 7 for the case in which Eq. (8–107) has distinct roots. It should be noted that

$$\rho_k = \frac{1}{\tau} \log \lambda_k = \frac{1}{\tau} [\log |\lambda_k| + i\,(\arg \lambda_k + 2m\pi)]$$

(Section 3–5); hence the ρ_k are determined only up to multiples of $2\pi i/\tau$.

THEOREM 8. *The Wronskian of the n solutions of Theorem 7 has the form*

$$w = e^{(\rho_1 + \cdots + \rho_n)t}\chi(t),$$

where $\chi(t)$ has period τ. Furthermore, for proper choice of the ρ_k,

$$\rho_1 + \cdots + \rho_n = -\frac{1}{\tau}\int_0^\tau \frac{a_1(t)}{a_0(t)}\,dt. \tag{8–108}$$

THEOREM 9. *The kernel function for Eq. (8–103) is obtainable from the n functions (8–104) in the form*

$$\phi(t, u) = \sum_{k=1}^n e^{\rho_k(t-u)}\psi_k(t)\theta_k(u), \tag{8–109}$$

where, if the λ_k are distinct, the $\psi_k(t)$ and $\theta_k(u)$ have period τ. Furthermore,

$$\phi(t + \tau, u + \tau) \equiv \phi(t, u).$$

The proofs of Theorems 8 and 9 are left as exercises (Problems 3 and 4, following Section 8–12).

Response to given inputs. By Theorem 9, the response to $f(t)$ with initial values zero at $t = 0$ is

$$x(t) = \sum_{k=1}^n \psi_k(t)\int_0^t e^{\rho_k(t-u)}\theta_k(u)f(u)\,du.$$

If all ρ_k have negative real parts, so that the equation is stable, then $x(t)$ can be regarded as the steady-state solution. When the λ_k are distinct, then ψ_k and θ_k have period τ, and $x(t)$ is a sum of terms of the same form as that considered above for the first order equation. Thus by taking $f = e^{\alpha t}$, we again conclude that the steady-state response has form $e^{\alpha t}P(t)$, where $P(t)$ has period τ; if $\alpha = i(l/m)\omega$, then f has a subharmonic response.

8-11 Evaluation of characteristic exponents. As in Section 8-10, we first seek n linearly independent solutions $g_1(t), \ldots, g_n(t)$. We then set

$$x = c_1 g_1(t) + \cdots + c_n g_n(t)$$

and try to choose $c_1, \ldots, c_n$ so that $x(t + \tau) = \lambda x(t)$. It is sufficient to make

$$x(\tau) = \lambda x(0), \, x'(\tau) = \lambda x'(0), \, \ldots, \, x^{(n-1)}(\tau) = \lambda x^{(n-1)}(0).$$

For then $y(t) = x(t + \tau) - \lambda x(t)$ is a solution of Eq. (8-103) such that $y(0) = 0$, $y'(0) = 0$, $\ldots$, $y^{(n-1)}(0) = 0$; hence $y(t) \equiv 0$, so that $x(t + \tau) \equiv \lambda x(t)$. We are thus led to the n equations

$$\sum_{m=1}^{n} c_m g_m(\tau) = \lambda \sum_{m=1}^{n} c_m g_m(0),$$

$$\sum_{m=1}^{n} c_m g_m'(\tau) = \lambda \sum_{m=1}^{n} c_m g_m'(0), \qquad (8\text{-}110)$$

$$\vdots$$

$$\sum_{m=1}^{n} c_m g_m^{(n-1)}(\tau) = \lambda \sum_{m=1}^{n} c_m g_m^{(n-1)}(0).$$

These are n homogeneous linear equations for $c_1, \ldots, c_n$. They have a nontrivial solution when

$$\begin{vmatrix} g_1(\tau) - \lambda g_1(0) & g_2(\tau) - \lambda g_2(0) & \cdots & & \cdots \\ g_1'(\tau) - \lambda g_1'(0) & g_2'(\tau) - \lambda g_2'(0) & \cdots & & \cdots \\ \vdots & & & & \vdots \\ g_1^{(n-1)}(\tau) - \lambda g_1^{(n-1)}(0) & & & \cdots & g_n^{(n-1)}(\tau) - \lambda g_n^{(n-1)}(0) \end{vmatrix} = 0.$$

$$(8\text{-}111)$$

If in particular we choose the $g_k(t)$ so that $g_1(0) = 1$, $g_1'(0) = 0$, $\ldots$, $g_2(0) = 0$, $g_2'(0) = 1$, $g_2''(0) = 0$, $\ldots$, and in general

$$g_m^{(l-1)}(0) = \delta_{lm} = \begin{cases} 1 & (l = m) \\ 0 & (l \neq m) \end{cases}$$

for $l = 1, \ldots, n$, $m = 1, \ldots, n$, then Eq. (8–111) becomes

$$\begin{vmatrix} g_1(\tau) - \lambda & g_2(\tau) & \cdots & g_n(\tau) \\ g_1'(\tau) & g_2'(\tau) - \lambda & \cdots & g_n'(\tau) \\ \vdots & & & \vdots \\ \cdots & & \cdots & g_n^{(n-1)}(\tau) - \lambda \end{vmatrix} = 0. \quad (8\text{–}111')$$

In either case, we have an equation of degree n for λ; from its roots λ_1, $\ldots, \lambda_n$ we obtain the desired characteristic exponents $\rho_k = (1/\tau) \log \lambda_k$. In addition, solution for the $c_1, \ldots, c_n$ corresponding to each λ_k yields the corresponding solution $x_k(t) = \sum c_m g_m(t) = e^{\rho_k t} \psi_k(t)$.

Use of power series. If after division of Eq. (8–103) by $a_0(t)$, the coefficients are expandable in power series for $t_0 - \rho < t < t_0 + \rho$, $\rho > \tau$, then the $g_m(t)$ can be obtained as such series. It is more convenient to expand in powers of t for $-\tau/2 \leqq t \leqq \tau/2$. Then in Eqs. (8–110), (8–111), and (8–111'), τ is to be replaced by $\tau/2$, and zero (as value of t) by $-\tau/2$.

EXAMPLE 1. $20D^2x + (9 + \cos t)Dx + x = 0$, $\tau = 2\pi$. We replace $\cos t$ by its power series $1 - (t^2/2) + \cdots$, and set $x = \sum c_n t^n$. Proceeding as in Section 8–5, we obtain equations for the coefficients:

$$40c_2 + 10c_1 + c_0 = 0, \; 120c_3 + 20c_2 + c_1 = 0, \ldots$$

We obtain linearly independent solutions $g_1(t)$, $g_2(t)$ by setting $c_0 = 1$, $c_1 = 0$ and then $c_0 = 0$, $c_1 = 1$. We find

$$g_1(t) = 1 - \frac{t^2}{40} + \frac{t^3}{240} - \frac{t^4}{2400} - \frac{t^5}{32{,}000}$$
$$+ \; 0.000014t^6 + 0.000003t^7 + \cdots,$$

$$g_2(t) = t - \frac{t^2}{4} + \frac{t^3}{30} - \frac{t^4}{960} - 0.000604t^5$$
$$+ \; 0.000066t^6 + 0.000018t^7 - 0.000005t^8 + \cdots$$

Here, in accordance with Section 8–5, the series converge for all t. We now evaluate $g_k(\pm \pi)$, $g_k'(\pm \pi)$ from the indicated terms of the series; it appears that the results are accurate to at least two significant figures. (An upper bound for the error can be found as in pp. 360–364 and 519–523 of Reference 10.) The results are shown in Table 8–1.

Equation (8–111) becomes

$$\begin{vmatrix} 0.855 - 0.597\lambda & 1.49 + 6.60\lambda \\ -0.055 - 0.311\lambda & 0.114 - 3.51\lambda \end{vmatrix} = 0,$$

TABLE 8–1

$x(t)$	$x(\pi)$	$x(-\pi)$	$x'(\pi)$	$x'(-\pi)$
$g_1(t)$	0.855	0.597	−0.055	0.311
$g_2(t)$	1.49	−6.60	0.114	3.51

or $4.15\lambda^2 - 2.24\lambda + 0.179 = 0$, so that we obtain $\lambda_1 = 0.443$, $\lambda_2 = 0.098$. The characteristic exponents can be chosen to be real:

$$\rho_1 = \frac{\log \lambda_1}{2\pi} = -0.130, \qquad \rho_2 = \frac{\log \lambda_2}{2\pi} = -0.370.$$

Hence $\rho_1 + \rho_2 = -0.500$. We can employ Eq. (8–108) as a check:

$$-\frac{1}{2\pi}\int_0^{2\pi} \frac{9 + \cos t}{20}\, dt = -\frac{9}{20} = -0.45.$$

Hence the accuracy appears to be only fair.

To obtain the solutions $x_1(t)$, $x_2(t)$ of Theorem 7, we form the equations (8–110). For $\lambda = \lambda_1$, we are led to the relation $0.591c_1 + 4.41c_2 = 0$, so that we can choose $c_1 = 1$, $c_2 = -0.134$, and

$$x_1(t) = g_1(t) - 0.134g_2(t)$$
$$= 1 - 0.134t + 0.0085t^2 - 0.00030t^3 + \cdots$$

For $\lambda = \lambda_2$, we find $0.796c_1 + 2.14c_2 = 0$, $c_1 = 1$, $c_2 = -0.372$,

$$x_2(t) = g_1(t) - 0.372g_2(t)$$
$$= 1 - 0.372t + 0.068t^2 - 0.0082t^3 + \cdots$$

Then

$$\psi_1(t) = x_1(t)e^{-\rho_1 t} = x_1(t)e^{0.130t} = x_1(t)(1 + 0.130t + 0.0085t^2 + \cdots)$$
$$= 1 - 0.004t + 0.0004t^2 + \cdots$$

Similarly, $\psi_2(t)$ can be found.

By (8–16), the Wronskian $w(t)$ of the solutions $x_1(t)$, $x_2(t)$ has the form

$$w = c \exp\left[-\int \frac{9 + \cos t}{20}\, dt\right] = c \exp \frac{-9t - \sin t}{20}.$$

To evaluate c, we write

$$c = w(0) = \begin{vmatrix} 1 & 1 \\ -0.134 & -0.372 \end{vmatrix} = -0.238.$$

The kernel function is then

$$\phi(t, u) = \frac{1}{20w(u)} \begin{vmatrix} e^{\rho_1 u}\psi_1(u) & e^{\rho_2 u}\psi_2(u) \\ e^{\rho_1 t}\psi_1(t) & e^{\rho_2 t}\psi_2(t) \end{vmatrix}$$

$$= 0.210 e^{(\sin u)/20}[e^{\rho_1(t-u)}\psi_1(t)\psi_2(u) - e^{\rho_2(t-u)}\psi_2(t)\psi_1(u)].$$

EXAMPLE 2. *Solution by perturbation method.* $20D^2 x + (9 + \epsilon \cos t)Dx + x = 0$. We solve by the perturbation method of Section 8–7. If we set $\epsilon = 1$, we obtain Example 1 again. Here $L_0 = 20D^2 + 9D + 1$, $L_1 = (\cos t)D$. We seek $g_1(t) = \sum_0^\infty q_k(t)\epsilon^k$ such that $g_1(0) = 1$, $g_1'(0) = 0$. We find

$$q_0(t) = 5e^{-t/5} - 4e^{-t/4},$$

$$q_1(t) = \tfrac{1}{401}[20(e^{-t/4} - e^{-t/5})(1 + \cos t)$$

$$+ (e^{-t/4} + e^{-t/5})\sin t], \dots$$

For $\epsilon = 1$, $g_1(t) = q_0(t) + q_1(t) + \cdots = 1 - 0.025t^2 + 0.004166t^3 - 0.00645t^4 + \cdots$; here the power series has been obtained from q_0 and q_1 alone. The result is in fair agreement with that found above for Example 1. We find $g_1(\pi) = 0.844$, $g_1(-\pi) = 0.598$, $g_1'(\pi) = -0.080$, $g_1'(-\pi) = 0.317$, in agreement with Table 8–1. The computation of $g_2(t)$ such that $g_2(0) = 0$, $g_2'(0) = 1$ is similar, and from $g_1(t)$, $g_2(t)$ we obtain the characteristic exponents as we did for Example 1.

Remark. For the general equation

$$D^n x + [b_1 + \epsilon p_1(t)]D^{n-1}x + \cdots + [b_n + \epsilon p_n(t)]x = 0,$$

if we choose $g_1(t), \dots, g_n(t)$ as linearly independent solutions satisfying fixed initial conditions, independent of ϵ, then the $g_m(t)$ are expressible as power series in powers of ϵ. The same then applies to the quantities $g_m^{(l)}(\tau)$ in Eq. (8–111′), so that λ satisfies an equation

$$\lambda^n + G_1(\epsilon)\lambda^{n-1} + \cdots + G_{n-1}(\epsilon)\lambda + G_n(\epsilon) = 0,$$

where the $G_k(\epsilon)$ are analytic for all ϵ. Accordingly, the roots $\lambda_1, \dots, \lambda_n$ depend algebraically on analytic functions of ϵ. We can deduce from this that the λ_k are analytic functions of ϵ, though they are in general multiple-valued and have algebraic branch points. (See Volume II, Chapter 5 of Reference 12.) If for $\epsilon = 0$ we obtain distinct roots $\lambda_1, \dots, \lambda_n$, then for $|\epsilon|$ sufficiently small the λ_k remain distinct and can be represented by convergent power series in powers of ϵ.

Solution by piecewise constant approximation. We illustrate the procedure by approximating the coefficients in Example 1 by piecewise constant functions. We replace the coefficient $9 + \cos t$ by $a_1(t)$, a function equal to 10 for $-\pi/2 < t < \pi/2$, equal to 8 for $\pi/2 < t < 3\pi/2$, and having period 2π. We write a for $-\pi/2$, b for $\pi/2$, and c for $3\pi/2$. In the interval $a < t < b$ our equation is $(20D^2 + 10D + 1)x = 0$. The characteristic roots are $s_1 = -0.361$, $s_2 = -0.139$. In the interval $b < t < c$ the equation is $(20D^2 + 8D + 1)x = 0$; the characteristic roots are $\sigma_1 = -0.2 - 0.1i$, $\sigma_2 = -0.2 + 0.1i$. Let $g_j(t)$ be the solution which equals $\exp(s_j t)$ for $a \leqq t \leqq b$. In the interval $b \leqq t \leqq c$, $g_j(t)$ has the form $d_{j1}e^{\sigma_1 t} + d_{j2}e^{\sigma_2 t}$. Continuity of $g_j(t)$ and $g_j'(t)$ at $t = b$ leads to the equations

$$d_{j1}e^{\sigma_1 b} + d_{j2}e^{\sigma_2 b} = e^{s_j b},$$

$$\sigma_1 d_{j1}e^{\sigma_1 b} + \sigma_2 d_{j2}e^{\sigma_2 b} = s_j e^{s_j b},$$

from which we find

$$d_{j1} = e^{-\sigma_1 b}\,\frac{\sigma_2 - s_j}{\sigma_2 - \sigma_1}\,e^{s_j b}, \qquad d_{j2} = e^{-\sigma_2 b}\,\frac{s_j - \sigma_1}{\sigma_2 - \sigma_1}\,e^{s_j b}.$$

We seek a solution $x(t) = c_1 g_1(t) + c_2 g_2(t)$ such that $x(c) = \lambda x(a)$, $x'(c) = \lambda x'(a)$. By analogy with Eq. (8–111), we obtain the equation

$$\begin{vmatrix} g_1(c) - \lambda g_1(a) & g_2(c) - \lambda g_2(a) \\ g_1'(c) - \lambda g_1'(a) & g_2'(c) - \lambda g_2'(a) \end{vmatrix} = 0$$

for the characteristic multipliers λ. After carrying out the algebra, we find $\lambda_1 = 0.31$, $\lambda_2 = 0.19$, and hence $\rho_1 = -0.17$, $\rho_2 = -0.26$. In view of the roughness of the approximation of the time-varying term, the result is surprisingly close to that obtained above.

8–12 The equation of second order with periodic coefficients

THEOREM 10. *Let an equation of second order*

$$[D^2 + a_1(t)D + a_2(t)]x = f(t) \qquad (8\text{–}120)$$

be given, in which $a_1(t)$, $a_1'(t)$, $a_2(t)$, $f(t)$ *are continuous for* $-\infty < t < \infty$. *The substitution*

$$x = yp(t), \qquad p(t) = \exp\left[-\tfrac{1}{2}\int_0^t a_1(v)\,dv\right] \qquad (8\text{–}121)$$

converts Eq. (8–120) into the equation

$$D^2 y + \tfrac{1}{4}(4a_2 - a_1^2 - 2a_1')y = f(t)/p(t). \qquad (8\text{–}120')$$

If Eq. (8–120) has coefficients of period τ, *then Eq. (8–120′) has coefficients of period* τ. *If the characteristic exponents of Eq. (8–120) are* ρ_1, ρ_2, *then the characteristic exponents of Eq. (8–120′) are* $\sigma_1 = \rho_1 + \frac{1}{2}c_0$, $\sigma_2 = \rho_2 + \frac{1}{2}c_0$, *where*

$$c_0 = \frac{1}{\tau} \int_0^\tau a_1(t)\, dt.$$

The proof of Theorem 10 is left as an exercise (Problem 10 below). The effect of this theorem is to permit us to reduce the study of a general second order equation to the study of one in which the first derivative term is absent. For the homogeneous case, one is thus led to an equation

$$D^2x + r(t)x = 0. \tag{8–122}$$

We shall assume for the remainder of this section that $r(t)$ *is continuous for all t and has period* π. *If* $r(t)$ *is even, Eq. (8–122) is called Hill's equation;* this term is sometimes used for Eq. (8–122) when $r(t)$ is periodic but not necessarily even. A Hill's equation in which $r(t)$ is representable by its Fourier series can be written

$$D^2x + r(t)x \equiv D^2x + \left(\sum_{n=0}^\infty c_n \cos 2nt\right)x = 0; \tag{8–122′}$$

a special case is the *Mathieu equation:*

$$D^2x + (a - 2q \cos 2t)x = 0. \tag{8–122″}$$

THEOREM 11. *For the equation (8–122), in which* $r(t)$ *has period* π, *the characteristic multipliers* λ_1, λ_2 *satisfy the relation* $\lambda_1\lambda_2 = 1$; *accordingly, the characteristic exponents* ρ_1, ρ_2 *can be chosen so that* $\rho_1 + \rho_2 = 0$. *For the special case of the Hill equation* [$r(t)$ *even*], *if* $\lambda_1 \neq \lambda_2$, *there are two linearly independent solutions* $x_1(t) = e^{\rho t}\psi(t)$, $x_2(t) = e^{-\rho t}\psi(-t)$.

Proof. By Theorem 8 (Section 8–10), since $a_1(t) \equiv 0$, we can choose ρ_1, ρ_2 so that $\rho_1 + \rho_2 = 0$. Then $\lambda_1\lambda_2 = e^{\rho_1\pi}e^{\rho_2\pi} = 1$. Now let $r(t)$ be even and let $\lambda_1 \neq \lambda_2$, so that $\lambda_1 \neq \pm 1$. Set $\rho = \rho_1$, so that $\rho_2 = -\rho$, $\rho \neq 0$. Let the solution $x_1(t) = e^{\rho t}\psi(t)$ be chosen as in Theorem 7 [Eq. (8–104)]. Then because $r(t)$ is even, $x_2(t) = e^{-\rho t}\psi(-t)$ is also a solution. We assert that the solutions $x_1(t)$, $x_2(t)$ are linearly independent. Suppose, to the contrary, that there is a relation

$$c_1 e^{\rho t}\psi(t) + c_2 e^{-\rho t}\psi(-t) \equiv 0,$$

where c_1 and c_2 are constants, not both zero. If, for example, $c_1 \neq 0$, then we choose a t such that $\psi(t) \neq 0$ [$\psi(t)$ cannot be identically zero], and we write

$$e^{2\rho t} = -\frac{c_2}{c_1}\frac{\psi(-t)}{\psi(t)}.$$

Accordingly, for this value of t,

$$e^{2\rho(t+\pi)} = -\frac{c_2}{c_1}\frac{\psi(-t-\pi)}{\psi(t+\pi)} = -\frac{c_2}{c_1}\frac{\psi(-t)}{\psi(t)} = e^{2\rho t}$$

by the periodicity of $\psi(t)$. Therefore $e^{2\rho\pi} = 1$ and $\lambda_1 = e^{\rho\pi} = \pm 1$, so that $\lambda_2 = \lambda_1$, contrary to hypothesis. Hence the two solutions $x_1(t)$, $x_2(t)$ are linearly independent and the theorem is proved.

Remark. If (as we normally assume) $r(t)$ is real, then the equation (8–111) for the characteristic multipliers λ has real coefficients. Since $\lambda_1\lambda_2 = 1$, either λ_1, λ_2 are both real or they are both complex, and $\lambda_2 = \bar{\lambda}_1$, with $\lambda_1\bar{\lambda}_1 = 1$; in the latter case, $\lambda_1 = e^{i\beta}$, $\lambda_2 = e^{-i\beta}$ for some real β, $0 < \beta < \pi$. Accordingly, we have the three cases of Fig. 8–3. When $\lambda_1 = \lambda_2 = 1$, $\rho_1 = \rho_2 = 0$; when $\lambda_1 = \lambda_2 = -1$, we can choose $\rho_1 = -\rho_2 = i$. It should be remarked that ρ_1 and ρ_2 cannot both have negative real parts, so that the equation is never stable in the strict sense. However, when $\lambda_1 = e^{i\beta}$, $0 < \beta < \pi$, then ρ_1 and ρ_2 are pure imaginary; the solutions $e^{\rho_j t}\psi_j(t)$ are *bounded* and there is a *neutral stability*. (In the mathematical literature, what we term stability for linear differential equations is usually called *asymptotic stability*; what we term neutral stability is usually called *stability*. See References 3, 4, 11.)

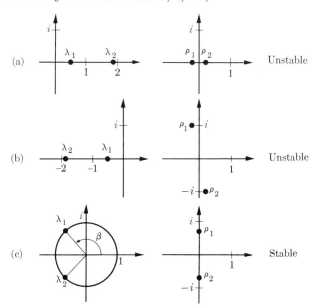

Fig. 8–3. Characteristic multipliers λ and exponents ρ for the equation $D^2 x + r(t)x = 0$, $r(t + \pi) = r(t)$; ρ is determined only up to multiples of i.

THEOREM 12. *Let Eq. (8–122) be a Hill equation* [$r(t)$ *even, with period* π]. *Let* $g_1(t)$, $g_2(t)$ *be the solutions such that* $g_1(0) = g_2'(0) = 1$, $g_1'(0) = g_2(0) = 0$. *Then* $g_1(t)$ *and* $g_2'(t)$ *are even;* $g_1'(t)$ *and* $g_2(t)$ *are odd. Let* $a_{11} = g_1(\pi)$, $a_{12} = g_1'(\pi)$, $a_{21} = g_2(\pi)$, $a_{22} = g_2'(\pi)$. *Then*

$$g_1(t + \pi) = a_{11}g_1(t) + a_{12}g_2(t),$$

$$g_2(t + \pi) = a_{21}g_1(t) + a_{22}g_2(t),$$

$$\text{(8–123)}$$

and

$$a_{11} = a_{22}, \qquad a_{11}a_{22} - a_{12}a_{21} = 1. \qquad \text{(8–124)}$$

Furthermore, for all t,

$$g_1(t)g_2'(t) - g_2(t)g_1'(t) = 1. \qquad \text{(8–125)}$$

The characteristic multipliers are the roots of the equation

$$\lambda^2 - 2a_{11}\lambda + 1 = 0, \qquad \text{(8–126)}$$

and the characteristic exponents are the solutions of the equation

$$\cosh \pi\rho \equiv \cos \pi i\rho = a_{11}. \qquad \text{(8–127)}$$

Proof. From the form of the differential equation, if $x(t)$ is a solution, so is $x(-t)$. Hence $g_1(t)$, $g_1(-t)$ are both solutions. But both have value one and derivative zero at $t = 0$. Therefore, by the uniqueness of solutions with given initial values, $g_1(t) \equiv g_1(-t)$, so that $g_1(t)$ is even. Differentiating, we conclude that $g_1'(t) = -g_1'(-t)$, so that $g_1'(t)$ is odd. Similar arguments show that $g_2(t)$ is odd, $g_2'(t)$ is even.

The Wronskian $w(t)$ of the solutions $g_1(t)$, $g_2(t)$ satisfies the equation $w' + 0w = 0$. Hence $w(t) \equiv g_1(t)g_2'(t) - g_2(t)g_1'(t) = \text{const} = w(0) = 1$, by the given initial values. Thus (8–125) is established.

Since $g_1(t + \pi)$, $g_2(t + \pi)$ are solutions and $g_1(t)$, $g_2(t)$ are linearly independent solutions, equations (8–123) must hold for some constants a_{ij}. Differentiating, we obtain the relations

$$g_1'(t + \pi) = a_{11}g_1'(t) + a_{12}g_2'(t),$$

$$g_2'(t + \pi) = a_{21}g_1'(t) + a_{22}g_2'(t).$$

$$\text{(8–123')}$$

If we now set $t = 0$ in (8–123), (8–123') and substitute the given initial values, we conclude that $g_1(\pi) = a_{11}, g_2(\pi) = a_{21}, g_1'(\pi) = a_{12}, g_2'(\pi) = a_{22}$. If we then put $t = -\pi$ in (8–123) and (8–123') and use the oddness or evenness of g_1, g_2, g_1', g_2', we conclude that

$$a_{11}^2 - a_{12}a_{21} = 1, \quad a_{21}(a_{11} - a_{22}) = 0,$$

$$a_{12}(a_{11} - a_{22}) = 0, \quad a_{22}^2 - a_{21}a_{12} = 1.$$

From (8–125) for $t = \pi$, we obtain

$$a_{11}a_{22} - a_{21}a_{12} = 1.$$

From the last five equations we deduce that $a_{11} = a_{22}$ (Problem 10 below). Hence (8–124) is established.

We now apply Eq. (8–111'), with $\tau = \pi$, to obtain the equation for the characteristic multipliers λ_1, λ_2:

$$\begin{vmatrix} g_1(\pi) - \lambda & g_2(\pi) \\ g_1'(\pi) & g_2'(\pi) - \lambda \end{vmatrix} = \begin{vmatrix} a_{11} - \lambda & a_{21} \\ a_{12} & a_{22} - \lambda \end{vmatrix} = 0.$$

By Eq. (8–124), this equation reduces to Eq. (8–126). If λ satisfies Eq. (8–126), then

$$a_{11} = \frac{1}{2}\left(\lambda + \frac{1}{\lambda}\right) = \tfrac{1}{2}(e^{\pi\rho} + e^{-\pi\rho}) = \cos \pi i\rho = \cosh \pi\rho.$$

Conversely, if ρ satisfies Eq. (8–127), then $\lambda = e^{\pi\rho}$ satisfies Eq. (8–126). Thus Theorem 12 is proved.

COROLLARY.   *A Hill equation* (8–122) *with* $r(t)$ *real is unstable if* $|g_1(\pi)| > 1$, *and is neutrally stable if* $|g_1(\pi)| < 1$.

For from (8–126) at least one of the multipliers λ_1, λ_2 is real and has absolute value greater than one, when $|a_{11}| = |g_1(\pi)| > 1$. Both multipliers are complex, of form $e^{\pm i\beta}$ ($0 < \beta < \pi$), when $|a_{11}| = |g_1(\pi)| < 1$. In the latter case, $\rho = \pm i\beta/\pi$, so that $\cos \pi i\rho = \cos \beta = a_{11}$.

Accordingly, to study the stability of a particular Hill equation, it is sufficient to obtain the one solution $g_1(t)$ and evaluate $g_1(\pi)$. The solution $g_1(t)$ can be found analytically, for example, by power series; it can be obtained by numerical solution of the differential equation; it can also be found by means of an analogue or digital computer (Reference 7).

The Mathieu equation. This is the special case (8–122'') listed above. For $q = 0$, $a > 0$, we find $g_1(t)$ to be $\cos \sqrt{a}\, t$, so that $g_1(\pi) = \cos \sqrt{a}\, \pi$. Thus $|g_1(\pi)| < 1$ unless $\sqrt{a}$ is an integer; that is, unless $a = 1, 4, 9, 16, \ldots$ Hence, by continuity, the equation will be stable for sufficiently small $|q|$ for a not equal to the square of an integer. If $q = 0$, $a < 0$, then $g_1(t)$ is $\cosh \sqrt{-a}\, t$, so that $|g_1(\pi)| > 1$ for all $a < 0$; hence again for sufficiently small $|q|$ the equation will be unstable for negative a.

For more precise results, one is forced to use series or some numerical method. Typical conclusions are shown in Fig. 8–4. The curves shown are obtained from tables computed by Ince (p. 363 of Reference 9). In his article, Ince describes in detail the methods used to obtain the tables. The replacement of q by $-q$ is equivalent to a phase shift in the cosine and hence has no influence on stability; thus the graph is symmetrical in the a-axis.

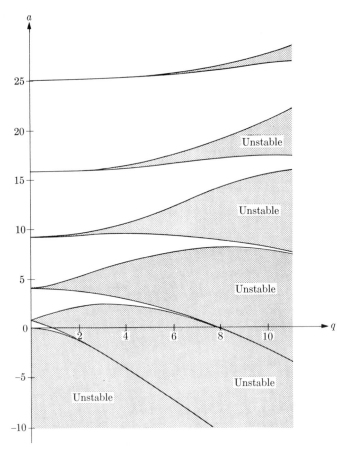

Fig. 8-4.   Stable and unstable zones for the Mathieu equation $D^2x + (a - 2q \cos 2t)x = 0$. Unstable zones are shaded.

THEOREM 13 (Liapounoff). *In Eq. (8–122) let $r(t)$ be continuous and have period π, $r(t) \not\equiv 0$. Let $r(t)$ be real and satisfy the two inequalities*

$$\int_0^\pi r(t)\, dt \geqq 0, \tag{8–128}$$

$$\int_0^\pi |r(t)|\, dt \leqq \frac{4}{\pi}. \tag{8–129}$$

Then the characteristic exponents ρ_1, ρ_2 can be chosen to be pure imaginary, and all solutions of Eq. (8–122) are bounded for $-\infty < t < \infty$.

Proof. We first establish an auxiliary theorem on real functions.

LEMMA. *Let $x(t)$ be defined for $a \leqq t \leqq b$, where $a < b$. Let x have continuous derivatives through the second order, and let $x(a) = x(b) = 0$, $x(t) \neq 0$ for $a < t < b$. Then*

$$\int_a^b \left| \frac{x''(t)}{x(t)} \right| dt > \frac{4}{b-a} .$$

Proof. We first remark that, since $x(t) \neq 0$ for $a < t < b$, the integral in question has meaning as an improper integral:

$$\int_a^b |\ldots| \, dt = \lim_{\substack{\epsilon \to 0+ \\ \delta \to 0+}} \int_{a+\delta}^{b-\epsilon} |\ldots| \, dt.$$

Since the integrand is positive or zero, the value of the integral is between zero and $+\infty$, inclusive.

We can assume without loss of generality that $x(t) > 0$ for $a < t < b$. Let $M = \max x(t)$ for $a \leqq t \leqq b$, so that $M > 0$. Let $x(c) = M$, so that $a < c < b$. Let $c - a = u(b - a)$, so that $b - c = (1 - u)(b-a)$ and $0 < u < 1$. Then since

$$\frac{1}{u} + \frac{1}{1-u} \geqq 4 \quad \text{for } 0 < u < 1,$$

we conclude that

$$\frac{b-a}{c-a} + \frac{b-a}{b-c} \geqq 4.$$

Thus

$$\frac{4}{b-a} \leqq \frac{1}{c-a} + \frac{1}{b-c} = \frac{1}{x(c)} \left[\frac{x(c) - x(a)}{c-a} - \frac{x(b) - x(c)}{b-c} \right]$$

$$= \frac{1}{M} [x'(t_1) - x'(t_2)] \quad (a < t_1 < c < t_2 < b)$$

$$= \frac{1}{M} \int_{t_1}^{t_2} [-x''(t)] \, dt \leqq \frac{1}{M} \int_{t_1}^{t_2} |x''(t)| \, dt$$

$$\leqq \frac{1}{M} \int_a^b |x''(t)| \, dt < \int_a^b \left| \frac{x''(t)}{x(t)} \right| dt.$$

Here we have used the equations $x(a) = x(b) = 0$ to obtain the first equality, and have applied the law of the mean for the second equality. The last inequality follows from the fact that $|x(t)| \leqq M$ for $a \leqq t \leqq b$ and, in particular, $|x(t)| < M$ for t sufficiently close to a or b. Thus the lemma is established.

We now return to the proof of the theorem. We know from Theorem 11 that $\lambda_1 \lambda_2 = 1$ and that instability occurs only if the λ's are real. Hence if the equation is unstable, there is a real solution $x(t)$ (not identically zero) such that $x(t + \pi) = \lambda x(t)$, λ real, $\lambda \neq 0$. Hence either $x(t)$ has no zeros or $x(t)$ has infinitely many zeros for $-\infty < t < \infty$.

If $x(t)$ is never zero, then by (8–122) $x''(t) + r(t)x(t) = 0$, and

$$0 = \int_0^\pi \frac{x''(t)}{x(t)}\, dt + \int_0^\pi r(t)\, dt$$

$$= \frac{x'(t)}{x(t)}\bigg|_0^\pi + \int_0^\pi \left[\frac{x'(t)}{x(t)}\right]^2 dt + \int_0^\pi r(t)\, dt,$$

by integration by parts. In the last expression, the first term is zero since $x(t + \pi) = \lambda x(t)$, $x'(t + \pi) = \lambda x'(t)$, so that $x'(t)/x(t)$ has period π. The second is positive unless $x'(t) \equiv 0$, so that $x(t) \equiv$ const; but then $x''(t) \equiv 0$, and the differential equation gives $r(t)x(t) \equiv 0$, which is impossible since $x(t) \neq 0$ and $r(t) \not\equiv 0$. Hence the second term is positive; the third is positive or zero by hypothesis. Hence the sum of the three terms cannot be zero, and we have a contradiction.

Accordingly, $x(t)$ has infinitely many zeros. At each zero t_0, $x'(t_0) \neq 0$; for otherwise, $x(t) \equiv 0$ by uniqueness of solutions of the differential equation. Hence $x(t) \neq 0$ in some deleted neighborhood of t_0. Since $x(t + \pi) = \lambda x(t)$, successive zeros differ by at most π. Let a, b be two successive zeros, $a < b$, so that $b - a \leqq \pi$. We apply the lemma:

$$\frac{4}{b - a} < \int_a^b \left|\frac{x''(t)}{x(t)}\right| dt = \int_a^b |r(t)|\, dt \leqq \frac{4}{\pi}.$$

Hence $b - a > \pi$, a contradiction. Therefore the characteristic multipliers cannot be real, and the theorem is proved.

Problems

1. Find the characteristic exponent and kernel function of the following:

(a) $(2 + \cos t)Dx - (\sin t)x = f$

(b) $Dx + (1 + \cos t)x = f$

2. Prove that if $\tau > 0$, $\lambda_1, \ldots, \lambda_n$ are distinct, $x_1(t), \ldots, x_n(t)$ are defined for all t, and $x_k(t + \tau) = \lambda_k x_k(t)$ for all t, no $x_k(t)$ being identically zero, then $x_1(t), \ldots, x_n(t)$ are linearly independent for all t. [*Hint:* Suppose the $x_k(t)$ are linearly dependent. Then, for proper numbering, there is a relation: $c_1 x_1(t) + \cdots + c_m x_m(t) \equiv 0$, none of the c_k being zero, $m \geqq 1$. Hence also $c_1 \lambda_1 x_1(t) + \cdots + c_m \lambda_m x_m(t) \equiv 0$. If now $m > 1$, we can eliminate $x_m(t)$ between these two relations to obtain one of form $d_1 x_1(t) + \cdots + d_{m-1} x_{m-1}(t) \equiv 0$, with none of the d_k equal to zero. Repeating the process, we eventually conclude that $x_1(t) \equiv 0$, contrary to assumption.]

3. Prove Theorem 8 of Section 8–10. [*Hint:* Apply Eq. (8–16) to show that $w(t) = e^{kt}\Phi_1(t)$, where $\Phi_1(t)$ has period τ and k is the constant on the right of Eq. (8–108). Apply Theorem 7 to show that

$$w(t) = \exp\left[(\rho_1 + \cdots + \rho_n)t\right]\Phi_2(t),$$

where $\Phi_2(t)$ is a sum of products of polynomials by functions of period τ. Now show that $\Phi_2(t + \tau) = C\Phi_2(t)$ for an appropriate constant C and that, from the known form of $\Phi_2(t)$, C must be 1, so that $\Phi_2(t)$ is the desired function $X(t)$. Show finally that one can choose $\rho_1, \ldots, \rho_n$, so that $\rho_1 + \cdots + \rho_n = k$.]

4. Prove Theorem 9 of Section 8–10. [*Hint:* To show that $\phi(t + \tau, u + \tau) = \phi(t, u)$, verify that $\phi_1(t, u) = \phi(t + \tau, u + \tau)$ satisfies the given differential equation (8–103) for each fixed u, and that, when $t = u$, $\phi_1 = 0, \ldots,$ $D^{n-2}\phi_1 = 0$, $D^{n-1}\phi_1 = 1/a_0(u)$.]

5. Verify the calculations leading to the characteristic exponents ρ_1, ρ_2 in Example 1 of Section 8–11.

6. Verify the calculations leading to the solution $g_1(t)$ in Example 2 of Section 8–11.

7. Consider a homogeneous differential equation of second order in which the coefficients have period τ and are piecewise constant for $-\tau/4 < t < \tau/4$ and for $\tau/4 < t < 3\tau/4$. Let the characteristic roots for the first interval be s_1, s_2, and those for the second interval σ_1, σ_2; in both cases let the roots be distinct.

(a) Show that the equation for the characteristic multipliers is $\lambda^2 + B\lambda + C = 0$, where

$$
\begin{aligned}
B = {}&(\sigma_2 - \sigma_1)^{-1}(s_2 - s_1)^{-1}[S_2\{\Sigma_1(\sigma_2 - s_2)(s_1 - \sigma_1) \\
&+ \Sigma_2(s_2 - \sigma_1)(s_1 - \sigma_2)\} + S_1\{\Sigma_1(\sigma_2 - s_1)(\sigma_1 - s_2) \\
&+ \Sigma_2(s_1 - \sigma_1)(\sigma_2 - s_2)\}],
\end{aligned}
$$

$$C = S_1 S_2 \Sigma_1 \Sigma_2, \qquad S_k = e^{s_k \tau/2}, \qquad \Sigma_k = e^{\sigma_k \tau/2} \quad (k = 1, 2).$$

(b) Show that if the characteristic roots $s_1 = a + bi$, $s_2 = a - bi$, $\sigma_1 = \alpha + \beta i$, $\sigma_2 = \alpha - \beta i$ are complex, then the equation for λ becomes $\lambda^2 + B\lambda + e^{(a+\alpha)\tau} = 0$, where

$$
\begin{aligned}
B = {}&(b\beta)^{-1}e^{(a+\alpha)\tau/2}[\{(a - \alpha)^2 + (b - \beta)^2\}\sin(b\tau/2)\sin(\beta\tau/2) \\
&- 2b\beta\cos(b + \beta)\tau/2].
\end{aligned}
$$

8. Apply the rule of part (a) of Problem 7 to verify the conclusions obtained in the last paragraph of Section 8–11.

9. Prove Theorem 10 in Section 8–12.

10. Complete the proof that $a_{11} = a_{22}$, as suggested in the proof of Theorem 12 (Section 8–12).

11. Solve the Mathieu equation by the perturbation method (Section 8–7), with $\epsilon = 2q$, $a > 0$, to obtain the solution $g_1(t)$ as in Theorem 12. Show that retention of terms through ϵ^2 leads to the value

$$g_1(\pi) = \cos b\pi + \frac{q^2}{4(b^2 - 1)}\frac{\pi \sin b\pi}{b} \quad (b = \sqrt{a}).$$

Compare the loci $g_1(\pi) = \pm 1$ with the boundary of the stability regions of Fig. 8–4.

12. Apply Theorem 13 to the Mathieu equation to obtain a sufficient condition for neutral stability. In particular, show that the equation is stable for $a = 0$, $|q| \leqq \pi^{-1}$.

13. *The WKB method*, introduced by Wentzel, Kramers, and Brillouin for problems of quantum mechanics, provides approximate solutions for the equation $D^2x + r(t)x = 0$. Let $r(t) > 0$, and set $s(t) = [r(t)]^{1/4}$. Then the approximate solutions are given by $s^{-1} \cos \int s^2 \, dt$, $s^{-1} \sin \int s^2 \, dt$ or in complex form by $s^{-1} \exp [\pm i \int s^2 \, dt]$.

(a) Show that if $s(t)$ is very large compared to 1, while $|s'/s|$, $|s''/s|$ are very small compared to 1, then the differential equation is approximately satisfied by $x = s^{-1} \exp [i \int s^2 \, dt]$.

(b) Show that for a Mathieu equation with $a > 0$, $|q| < a/2$, the *WKB* method yields the characteristic exponents $\rho = \pm (i/\pi) \int_0^\pi s^2 \, dt$, so that only stability is predicted.

14. Prove that if $q(t) \leqq 0$, then the equation $[D^2 + q(t)]x = 0$ is unstable. [*Hint:* Show that the solution for which $x(0) = 0$, $x'(0) = 1$ has $D^2x \geqq 0$, $Dx \geqq 1$ for $t \geqq 0$, and hence $x(t) \to +\infty$ as $t \to +\infty$.]

Answers

  1. (a) $\rho = 0, \phi = 1/(1 + \cos t)$
     (b) $\rho = -1, \phi = \exp [u - t + \sin u - \sin t]$

8–13 Matrix formulation of differential equations. Up to this point we have treated differential equations of order n and simultaneous differential equations without the aid of matrices. It is well known that matrices are a very valuable tool in simplifying both the description and the study of differential equations; hence our disregard of them may appear strange. The justification of their omission is simply that all the essential ideas and techniques of the preceding chapters could be adequately presented without them. Furthermore, in most applications the common practice is to reduce each problem to the form of a single equation of order n: $\phi_1(D)x = \phi_2(D)f$. (See the discussion of elimination in Section 1–5.)

It would, however, be a mistake to conclude that matrices are unimportant for the problems at hand. A more profound theory is essentially forced to employ matrices, and a variety of ideas become transparent only when framed in terms of matrices. An illustration is the theory of stability of differential equations, of which a few theorems are presented in the following sections. The important text of Coddington and Levinson (Reference 4) is formulated almost completely in terms of matrices.

We first summarize very briefly the necessary properties of matrices. For more information, refer to References 4, 13, and 15.

By an $m \times n$ matrix A we mean a rectangular array of complex numbers having m rows and n columns:

$$A = \begin{bmatrix} a_{11} & a_{12} & \ldots & a_{1n} \\ a_{21} & a_{22} & \ldots & a_{2n} \\ \vdots & & & \vdots \\ a_{m1} & a_{m2} & \ldots & a_{mn} \end{bmatrix}.$$

We write $A = (a_{ij})$ as abbreviations for the matrix and call the a_{ij} the *elements* of the matrix. When $m = 1$, A is a *row vector*: $[b_1, \ldots, b_n]$, $b_k = a_{1k}$; when $n = 1$, A is a *column vector*, which we shall denote by col $(b_1, \ldots, b_m)$, $b_k = a_{k1}$. Two $m \times n$ matrices can be added:

$$A + B = (a_{ij}) + (b_{ij}) = (a_{ij} + b_{ij}) = (c_{ij}) = C.$$

A matrix can be multiplied by a scalar k (complex number): $kA = k(a_{ij}) = (ka_{ij})$. An $m \times p$ matrix A can be multiplied by a $p \times n$ matrix B:

$$AB = A \cdot B = (a_{ij})(b_{ij}) = (c_{ij}) = C,$$

$$c_{ij} = \sum_{k=1}^{p} a_{ik}b_{kj} \quad (i = 1, \ldots, m, j = 1, \ldots, n).$$

In general, $A \cdot B \neq B \cdot A$ (even when both are defined). However, $A + B = B + A$, $A(B + C) = AB + AC$, $A(BC) = (AB)C$ always (all products and sums being assumed to have meaning). We write zero for a matrix in which all the elements are zero; the context indicates the number of rows and columns. Hence $A + 0 = A$, $A \cdot 0 = 0$.

An $n \times n$ matrix is called a *square matrix of order n*. In particular, the square matrix

$$I = (\delta_{ij}) = \begin{bmatrix} 1 & 0 & 0 & \ldots & & 0 \\ 0 & 1 & 0 & \ldots & & 0 \\ \vdots & & & & & \vdots \\ 0 & 0 & 0 & \ldots & 0 & 1 \end{bmatrix}$$

is the *unit matrix* of order n; again the context indicates the number of rows and columns. We verify that $AI = IA = A$ for every square matrix A of order n. We write det A for the determinant corresponding to A:

$$\det A = \det (a_{ij}) = \begin{vmatrix} a_{11} & \ldots & a_{1n} \\ \vdots & & \vdots \\ a_{n1} & \ldots & a_{nn} \end{vmatrix}.$$

If det $A \neq 0$, A is called *nonsingular*. If A is nonsingular, then there is a unique *inverse matrix* of A, denoted by A^{-1}, such that $AA^{-1} = I$, $A^{-1}A = I$. We also define $A^0 = I$, $A^1 = A$, $A^2 = A \cdot A$, $\ldots$ The roots of the equation det $(A - \lambda I) = 0$ are called *characteristic values* (eigenvalues) of A.

We define the *norm* of matrix $A = (a_{ij})$, denoted by $\|A\|$, to be the non-negative real number $\sum |a_{ij}|$, where the sum is for $i = 1, \ldots, m$, $j = 1,$ $\ldots, n$. We can verify that $\|A + B\| \leq \|A\| + \|B\|$, $\|kA\| = |k| \|A\|$, $\|AB\| \leq \|A\| \cdot \|B\|$.

If the elements of A are functions of t, over some interval, then A becomes a function of t: $A = A(t) = (a_{ij}(t))$. If $a_{ij}(t) \to c_{ij}$ as $t \to t_0$, then $A(t)$ is said to have limit $C = (c_{ij})$ as $t \to t_0$; if the $a_{ij}(t)$ are continuous (at t_0), then $A(t)$ is said to be continuous (at t_0). If $\|A(t)\| < k$ for some constant k, $\alpha < t < \beta$, then $A(t)$ is said to be bounded for $\alpha < t < \beta$. We can define the derivative and integral of a matrix function of t:

$$\frac{dA}{dt} = \frac{d}{dt} \left(a_{ij}(t) \right) = \left(\frac{da_{ij}}{dt} \right),$$

$$\int_\alpha^\beta A(t) \, dt = \left(\int_\alpha^\beta a_{ij}(t) \, dt \right).$$

Thus $A(t)$ is differentiable if all $a_{ij}(t)$ are differentiable, and $A(t)$ is integrable if $A(t)$ is continuous. We can also verify rules of calculus:

$$\frac{d}{dt} (A + B) = \frac{dA}{dt} + \frac{dB}{dt}, \qquad \frac{d}{dt} [k(t)A] = k(t) \frac{dA}{dt} + k'(t)A,$$

$$\frac{d}{dt} (A \cdot B) = A \cdot \frac{dB}{dt} + \frac{dA}{dt} \cdot B,$$

$$\int_\alpha^\beta [A(t) + B(t)] \, dt = \int_\alpha^\beta A(t) \, dt + \int_\alpha^\beta B(t) \, dt,$$

$$\left\| \int_\alpha^\beta A(t) \, dt \right\| \leq \int_\alpha^\beta \|A(t)\| \, dt \quad (\alpha < \beta).$$

If A is a square matrix, we define the exponential function $\exp A = e^A$ by a series:

$$e^A = I + A + \frac{1}{2!} A^2 + \cdots + \frac{1}{m!} A^m + \cdots$$

It can be shown that the series converges in the sense that the m-th partial sum, which is a square matrix of order n (the order of A), has elements which approach limits as $m \to +\infty$. The rule $e^{A+B} = e^A \cdot e^B$ is valid if A and B commute ($AB = BA$).

A set of simultaneous linear differential equations

$$\frac{dx_i}{dt} = \sum_{j=1}^n a_{ij}x_j + f_i(t) \quad (i = 1, \ldots, n)$$

is equivalent to a matrix equation:

$$\frac{dx}{dt} = Ax + f(t), \tag{8–130}$$

where $x = \text{col}\,(x_1, \ldots, x_n)$, $A = (a_{ij})$, in general depending on t, $f(t) = \text{col}\,(f_1(t), \ldots, f_n(t))$. The related homogeneous system is

$$\frac{dx}{dt} = Ax. \tag{8–130'}$$

An equation of order n,

$$D^n y + a_1 D^{n-1} y + \cdots + a_n y = F(t),$$

is equivalent to an equation (8–130) in which $x = \text{col}\,(y, y', \ldots, y^{(n-1)})$, $f = \text{col}\,(0, \ldots, 0, F)$,

$$A = \begin{bmatrix} 0 & 1 & 0 & \cdots & & 0 \\ 0 & 0 & 1 & \cdots & & 0 \\ \vdots & & & & & \vdots \\ 0 & & & \cdots & 0 & 1 \\ -a_n & -a_{n-1} & & \cdots & & -a_1 \end{bmatrix} \tag{8–131}$$

The basic existence theorem (Section 1–4) can be reformulated in terms of matrices:

THEOREM 14. *In the matrix differential equation* (8–130) *let the matrices* A, f *be continuous for* $\alpha < t < \beta$, *and let* $\alpha < t_0 < \beta$. *There exist* n *solutions* $x = x^{(i)}(t)$ $(i = 1, \ldots, n)$ *of the related homogeneous equation* (8–130') *which are linearly independent for* $\alpha < t < \beta$; *that is,*

$$c_1 x^{(1)}(t) + \cdots + c_n x^{(n)}(t) \equiv 0 \quad (\alpha < t < \beta),$$

for constant scalars $c_1, \ldots, c_n$, *implies* $c_1 = 0, \ldots, c_n = 0$. *The general solution of Eq.* (8–130') *is*

$$x = c_1 x^{(1)}(t) + \cdots + c_n x^{(n)}(t),$$

where $c_1, \ldots, c_n$ *are arbitrary constant scalars, or, equivalently,*

$$x = X(t)c,$$

where $c = \text{col}\,(c_1, \ldots, c_n)$ *and*

$$X = (x_i^{(j)}(t)) = \begin{bmatrix} x_1^{(1)} & \cdots & x_1^{(n)} \\ \vdots & & \vdots \\ x_n^{(1)} & \cdots & x_n^{(n)} \end{bmatrix}.$$

The general solution of Eq. (8–130) is given by

$$x = X(t)c + x^*(t),$$

where $x^*(t)$ is a particular solution. The solution can be chosen to satisfy given initial conditions:

$$x(t_0) = \xi = \text{col}\,(\xi_1, \ldots, \xi_n).$$

In the remainder of this section we assume that $A(t)$ and $f(t)$ satisfy the hypotheses of Theorem 14.

By a *fundamental set of solutions* of Eq. (8–130') we mean a set of n linearly independent solutions $x^{(1)}(t), \ldots, x^{(n)}(t)$, for $\alpha < t < \beta$. The matrix $X(t)$, formed as above with the column vectors $x^{(j)}(t)$ as its columns, is called a *fundamental matrix* for Eq. (8–130'). We then have

$$\frac{dX}{dt} = AX. \tag{8–132}$$

THEOREM 15.   *Let $x^{(1)}(t), \ldots, x^{(n)}(t)$ be solutions of Eq. (8–130') for $\alpha < t < \beta$. Then they are linearly independent for $\alpha < t < \beta$ if and only if the Wronskian determinant*

$$w(t) = \det X = \begin{vmatrix} x_1^{(1)}(t) & \cdots & x_1^{(n)}(t) \\ \vdots & & \vdots \\ x_n^{(1)}(t) & \cdots & x_n^{(n)}(t) \end{vmatrix}$$

is never zero for $\alpha < t < \beta$. In particular, if $X(t)$ is a fundamental matrix for Eq. (8–130'), then $X(t)$ is nonsingular.

The requirement that $x = X(t)c$ satisfy given initial conditions leads to the equation $X(t_0)c = \xi$; this matrix equation is equivalent to n simultaneous linear equations for $c_1, \ldots, c_n$. The equations are solvable for $c_1, \ldots, c_n$ for arbitrary ξ, precisely when $\det X(t_0) \neq 0$. Thus Theorem 15 is a consequence of Theorem 14. Furthermore, $c = X^{-1}(t_0)\xi$, so that the solution of Eq. (8–130') with initial value ξ at t_0 is $x = X(t)X^{-1}(t_0)\xi$.

THEOREM 16.   *If $w(t)$ is as in Theorem 15, then one has*

$$\frac{dw}{dt} = (a_{11} + a_{22} + \cdots + a_{nn})w,$$

so that

$$w = c \exp \int_{t_0}^{t} (a_{11} + \cdots + a_{nn})\, dt,$$

where c is a constant scalar.

THEOREM 17. *Let $X(t)$ be a fundamental matrix for Eq. (8–130′). Then for each fixed u the matrix $\Phi(t, u) = X(t)X^{-1}(u)$ is the unique solution of Eq. (8–132) such that $\Phi = I$ when $t = u$; equivalently, the j-th column of $\Phi(t, u)$ is the unique solution of Eq. (8–130′) such that $x_i(u) = \delta_{ij}$. The unique solution of Eq. (8–130) such that $x(t_0) = \xi$ is given by*

$$x = \Phi(t, t_0)\xi + \int_{t_0}^t \Phi(t, u)f(u)\, du$$
$$= X(t)X^{-1}(t_0)\xi + X(t)\int_{t_0}^t X^{-1}(u)f(u)\, du. \qquad (8\text{--}133)$$

Thus $\Phi(t, u)$ is our *matrix kernel function.* If, as above, an equation of order n is replaced by an equivalent matrix differential equation, we verify that $\phi(t, u)$ is the element ϕ_{1n} of $\Phi(t, u) = (\phi_{ij})$ (Problem 9, following Section 8–16).

THEOREM 18. *Let A be a constant matrix. Then the solution of Eq. (8–132) such that $X(0) = I$ is*

$$X = e^{tA} = I + tA + \frac{t^2}{2!}A^2 + \cdots$$

Correspondingly, the solution of Eq. (8–130′) such that $x(0) = \xi$ is given by $X(t)\xi = e^{tA}\xi$. Let $\lambda_1, \ldots, \lambda_n$ be the characteristic values of A. If the λ_j are distinct, then

$$X(t) = C \begin{bmatrix} e^{\lambda_1 t} & 0 & & \cdots & 0 \\ 0 & e^{\lambda_2 t} & 0 & \cdots & 0 \\ \vdots & & & & \vdots \\ 0 & & & \cdots & e^{\lambda_n t} \end{bmatrix} C^{-1},$$

where $C = (c_{ij})$ is a constant matrix. If $\mathrm{Re}\,(\lambda_j) < -a < 0$ for $j = 1, \ldots, n$, then for some constant $b > 0$

$$\|X(t)\| \leqq be^{-at} \to 0 \quad as\ t \to +\infty.$$

Furthermore,

$$X^{-1}(t) = X(-t), \qquad X(t)X^{-1}(u) = X(t - u),$$

and for $t_0 = 0$, $\xi = 0$, Eq. (8–133) becomes

$$x = \int_0^t X(t - u)f(u)\, du. \qquad (8\text{--}134)$$

Accordingly, $X(t) = e^{tA}$ is the *matrix weighting function* for the system with constant coefficients $x' = Ax + f$.

8-14 Stability theorems. We first establish an auxiliary inequality:

THEOREM 19. *Let $u(t)$, $v(t)$ be real functions, defined and continuous for $0 \leq t \leq t_1, t_1 > 0$; let $v(t) \geq 0$; let c be a constant. If*

$$u(t) \leq c + \int_0^t u(s)v(s)\, ds \quad (0 \leq t \leq t_1),$$

then $u(t) \leq c \exp \int_0^t v(s)\, ds$.

Proof. We can write $u(t) = c + \int_0^t u(s)v(s)\, ds - w(t)$, where $w(t) \geq 0$. Let $x(t) = u(t) + w(t)$, so that $x(0) = u(0) + w(0) = c$. Now $x'(t) = [u(t) + w(t)]' = u(t)v(t) = [x(t) - w(t)]v(t)$, or $x' - vx = -vw$. This is a linear differential equation for x. Accordingly,

$$x(t)q(t) = \int_0^t [-v(s)w(s)q(s)]\, ds + x(0),$$

where $q(t) = \exp\left[-\int_0^t v(s)\, ds\right]$, so that $q(t) > 0$. Since $t \geq 0$ and the integrand $-v(s)w(s)q(s)$ is not greater than zero for $0 \leq s \leq t_1$, we conclude that the corresponding integral is not greater than zero. Thus $x(t)q(t) \leq x(0) = c$, $x(t) \leq cq^{-1} = c \exp \int_0^t v(s)\, ds$. Since $u(t) \leq u(t) + w(t) = x(t)$, the theorem is established.

THEOREM 20. *Let A be a constant $n \times n$ matrix. Let $B(t)$ be a continuous $n \times n$ matrix function of t, $0 \leq t < \infty$, and let $\int_0^\infty \|B(t)\|\, dt < \infty$. If all solutions of the equation*

$$\frac{dx}{dt} = Ax \tag{8-140}$$

are bounded for $0 \leq t < \infty$, then so also are all solutions of the equation

$$\frac{dy}{dt} = [A + B(t)]y. \tag{8-141}$$

Proof. Let $y(t)$ be a solution of Eq. (8-141) for $0 \leq t < \infty$. Then $dy/dt = Ay + f$, where $f(t) = B(t)y(t)$. Hence by (8-133)

$$y(t) = x(t) + \int_0^t X(t-s)B(s)y(s)\, ds,$$

where $X(t-s)$ is the matrix kernel function for Eq. (8-140) and $x(t)$ is the solution of Eq. (8-140) such that $x(0) = y(0)$. Since the columns of $X(t)$ are solutions of Eq. (8-140) and all solutions are bounded, we conclude that

$$\|x(t)\| < k, \qquad \|X(t-s)\| < k \quad (0 \leq t < \infty, 0 \leq s \leq t),$$

for some constant k. Now

$$\|y(t)\| \leq \|x(t)\| + \left\| \int_0^t X(t-s)B(s)y(s)\, ds \right\|$$

$$\leq \|x(t)\| + \int_0^t \|X(t-s)B(s)y(s)\|\, ds$$

$$\leq k + k\int_0^t \|B(s)\|\, \|y(s)\|\, ds.$$

Therefore Theorem 19 is applicable, with $u(t) = \|y(t)\|$, $v(t) = k\|B(t)\|$, $c = k$. Accordingly, for each finite t

$$\|y(t)\| \leq k \exp\int_0^t k\|B(s)\|\, ds.$$

Since $\int_0^\infty \|B(s)\|\, ds$ is finite, the right-hand member is bounded for $0 \leq t < \infty$, and we conclude that $y(t)$ is bounded.

Remark 1. We can apply the theorem, for example, to an equation

$$D^2x + [a - 2qb(t)\cos 2t]x = 0.$$

If $b(t) \equiv 1$, we obtain a Mathieu equation. We assume $a > 0$, so that the equation $D^2x + ax = 0$ is stable (that is, neutrally stable—all solutions bounded). Then, if $\int_0^\infty |b(t)|\, dt < \infty$, the given equation is also neutrally stable. For example, the conclusion is valid if $b(t) = 1/(1+t^2)$.

Remark 2. If we assume that the characteristic values λ_j of A all have negative real parts, then the solutions of Eq. (8–140) are bounded (in fact, by Theorem 18, they all approach zero as $t \to \infty$), and the theorem is applicable. If, in addition, the λ_j are distinct, then one can show that the solutions of Eq. (8–141), under the hypotheses stated, behave like the solutions of Eq. (8–140) for large t. Thus there are n linearly independent solutions of Eq. (8–140) having form $e^{\lambda_j t}p_j$ $(j = 1, \ldots, n)$, where p_j is a constant vector. One can establish the existence of n linearly independent solutions $y_j(t)$ of Eq. (8–141) such that $y_j(t) = e^{\lambda_j t}q_j(t)$, where $q_j(t) \to p_j$ as $t \to \infty$. For a proof of this result (in an even stronger form) see p. 92 of Reference 4. The equations considered in Section 8–6 provide an illustration.

THEOREM 21. *Let A be a constant $n \times n$ matrix, and let the characteristic values of A all have negative real parts, so that every solution of Eq. (8–140) approaches zero as $t \to \infty$. There exists a constant k depending on A such that, if $B(t)$ is continuous for $t \geq 0$ and $\|B(t)\| < k$, then all solutions of Eq. (8–141) also approach zero as $t \to \infty$.*

Proof. We choose $X(t)$ as in Theorem 18. Under the hypotheses stated, $\|X(t)\| \leqq be^{-at}$ for $t \geqq 0$, for some constants $b > 0$, $a > 0$. Let $y(t)$ be a solution of Eq. (8–141) such that $\|y(0)\| = 1$. Then as above

$$y(t) = x(t) + \int_0^t X(t - u)B(u)y(u)\,du,$$

where $x(t) = X(t)y(0)$. Accordingly,

$$\|x(t)\| \leqq \|X(t)\|\,\|y(0)\| = \|X(t)\| \leqq be^{-at}$$

for $t \geqq 0$. Therefore, for $t \geqq 0$, $\|B(t)\| < k$,

$$\|y(t)\| \leqq be^{-at} + b\int_0^t \|X(t - u)\|\,\|B(u)\|\,\|y(u)\|\,du$$

$$\leqq be^{-at} + b\int_0^t e^{-a(t-u)}k\|y(u)\|\,du;$$

$$\|y(t)\|e^{at} \leqq b + \int_0^t e^{au}\|y(u)\|kb\,du.$$

We now apply Theorem 19, with $u(t) = \|y(t)\|e^{at}$, $v(t) = kb$; accordingly,

$$\|y(t)\|e^{at} \leqq be^{kbt},$$

$$\|y(t)\| \leqq be^{(kb-a)t}.$$

If we thus choose k to be $a/(2b)$ (or any positive number less than a/b), then $y(t)$ approaches zero as $t \to \infty$. The constant k depends on A, since a and b are determined by A. In detail: a is any number such that $0 > -a > \max \operatorname{Re} (\lambda_j)$, where $\lambda_1, \ldots, \lambda_n$ are the characteristic values of A, and $b = \max (\|e^{tA}\|e^{at})$, $t \geqq 0$. If the λ_j are distinct, a can be chosen as $-\max \operatorname{Re} (\lambda_j)$ and b as the "least upper bound" of $\|e^{tA}\|e^{at}$.

If $y(t)$ is a solution of Eq. (8–141) such that $\|y(0)\| \neq 1$ and $y(0) \neq 0$, then $y_1(t) = y(t)/\|y(0)\|$ is a solution of Eq. (8–141) such that $\|y_1(0)\| = 1$ and $\|y(t)\| = \|y(0)\| \cdot \|y_1(t)\| \to 0$ as above. If $y(0) = 0$, then $y(t) \equiv 0$, by the uniqueness of solutions of the differential equation.

Remark 3. Again the equations of Section 8–6 provide an illustration. One can also apply the theorem to the perturbation process of Section 8–7. If an equation with constant coefficients $\sum_0^n a_k D^{n-k}x = 0$ is stable and the functions $b_k(t)$ are continuous and bounded for $0 \leqq t < \infty$, then for $|\epsilon|$ sufficiently small the equation $\sum_0^n [a_k + \epsilon b_k(t)]D^{n-k}x = 0$ is also stable.

8–15 Application of Hermitian matrices. A square matrix $A = (a_{ij})$ of order n is said to be Hermitian if $a_{ji} = \bar{a}_{ij}$ (the bar denoting complex conjugate) for $i = 1, \ldots, n$, $j = 1, \ldots, n$. In particular, $a_{ii} = \bar{a}_{ii}$, so that the diagonal elements of A are real. One can show that the characteristic values of A are also real (p. 26 of Reference 13). Associated with a Hermitian matrix A is the *Hermitian form*

$$Q(x) = Q(x_1, \ldots, x_n) = \sum_{i,j=1}^{n} a_{ij} x_i \bar{x}_j. \qquad (8\text{–}150)$$

Here Q is a function of the n complex variables $x_1, \ldots, x_n$. However, Q takes on only real values, since

$$\bar{Q} = \sum_{i,j} \bar{a}_{ij} \bar{x}_i x_j = \sum_{i,j} a_{ji} \bar{x}_i x_j = \sum_{i,j} a_{ij} x_i \bar{x}_j = Q.$$

A square matrix $T = (t_{ij})$ of order n is said to be *unitary* if, for every column vector $y = \text{col}\,(y_1, \ldots, y_n)$, the column vector $x = Ty$ satisfies the equation $\sum_1^n |x_i|^2 = \sum_1^n |y_i|^2$. A unitary matrix is necessarily nonsingular. For if $\det\,(t_{ij}) = 0$, then the equations $\sum_j t_{ij} y_j = 0$ $(i = 1, \ldots, n)$ have a nontrivial solution $(y_1, \ldots, y_n)$, so that $\sum_i |y_i|^2 \neq 0$; but $0 = x_i = \sum_j t_{ij} y_j$ $(i = 1, \ldots, n)$, and hence $\sum_i |x_i|^2 \neq \sum_i |y_i|^2$, contrary to hypothesis. Since T is nonsingular, the equations

$$x_i = \sum_{j=1}^{n} t_{ij} y_j \quad (i = 1, \ldots, n) \qquad (8\text{–}151)$$

can be solved uniquely for $y_1, \ldots, y_n$ in terms of $x_1, \ldots, x_n$ ($x = Ty$ and $y = T^{-1}x$), so that T describes a one-to-one correspondence between the set of all $(x_1, \ldots, x_n)$ and the set of all $(y_1, \ldots, y_n)$.

LEMMA 1. *For every Hermitian form (8–150) there is a unitary matrix* $T = (t_{ij})$ *such that the substitution (8–151) reduces $Q(x)$ to the form* $Q_1(y) = \sum_1^n \lambda_i |y_i|^2$, *where $\lambda_1, \ldots, \lambda_n$ are the characteristic values of* $A = (a_{ij})$.

For a proof, see p. 63 of Reference 13.

LEMMA 2. *Let the Hermitian matrix A have characteristic values $\lambda_1, \ldots, \lambda_n$ (necessarily real). Let $m = \min\,(\lambda_1, \ldots, \lambda_n)$, $M = \max\,(\lambda_1, \ldots, \lambda_n)$. Then for all $(x_1, \ldots, x_n)$ the Hermitian form $Q(x)$ defined by (8–150) satisfies the inequalities*

$$m \sum_{i=1}^{n} |x_i|^2 \leq Q(x) \leq M \sum_{i=1}^{n} |x_i|^2. \qquad (8\text{–}152)$$

Proof. We choose T as in Lemma 1. Then to each $x = (x_1, \ldots, x_n)$ there is a corresponding $y = (y_1, \ldots, y_n)$ and, by definition, $Q(x) = Q_1(y)$. Now T describes a one-to-one correspondence between the loci $\sum_i |x_i|^2 = k$ and $\sum_i |y_i|^2 = k$, for each $k \geq 0$. Therefore

$$\max_{\sum_i |x_i|^2 = k} Q(x) = \max_{\sum_i |y_i|^2 = k} Q_1(y) = \max_{\sum_i |y_i|^2 = k} [\lambda_1 |y_1|^2 + \cdots + \lambda_n |y_n|^2]$$

$$\leq \max_{\sum_i |y_i|^2 = k} M \sum_i |y_i|^2 = M \cdot k.$$

(The $\leq$ sign can be replaced by an equals sign; if, for example, $\lambda_1 = M$, then the maximum value Mk is attained for $y_1 = \sqrt{k}$, $y_2 = y_3 = \cdots = 0$). For any particular $x^0 = (x_1^0, \ldots, x_n^0)$, $Q(x^0)$ is at most equal to the maximum of Q among all x such that $\sum_i |x_i|^2 = k = \sum_i |x_i^0|^2$; hence

$$Q(x^0) \leq M \cdot k = M \sum_i |x_i^0|^2.$$

Thus one half of (8–152) is established. The other half is proved in similar fashion.

Remark. Since $Q_1(y)$ does attain its maximum Mk in the set of y such that $\sum_i |y_i|^2 = k$, $Q(x)$ also attains its maximum Mk in the set of x such that $\sum_i |x_i|^2 = k$. Similar statements apply to the minimum. Thus in (8–152) the constants m, M are "best possible."

We are now prepared to establish a theorem useful in analysis of stability. (See pp. 48–49 of Reference 3.)

THEOREM 22. *Let $A(t) = (a_{ij}(t))$ be a square matrix of order n depending continuously on t for $0 \leq t < \infty$. Let $C(t)$ be the matrix $(c_{ij}(t))$, where*

$$c_{ij} = \frac{a_{ji} + \bar{a}_{ij}}{2} \quad (i, j = 1, \ldots, n).$$

Then, for each t, C is a Hermitian matrix. Let $m(t)$, $M(t)$ be the minimum and maximum characteristic values of $C(t)$. Then for every solution $x(t)$ of the differential equation

$$\frac{dx}{dt} = Ax \tag{8–153}$$

one has

$$u(0) \exp\left[2\int_0^t m(s)\, ds \right] \leq u(t) \leq u(0) \exp\left[2\int_0^t M(s)\, ds \right], \tag{8–154}$$

where

$$u(t) = \sum_{i=1}^n |x_i(t)|^2.$$

Proof. Since $\bar{c}_{ij} = c_{ji}$, C is a Hermitian matrix. We can write $u = \sum_i x_i \bar{x}_i$. Therefore

$$
\begin{aligned}
\frac{du}{dt} &= \sum_i \left(x_i \frac{d\bar{x}_i}{dt} + \bar{x}_i \frac{dx_i}{dt} \right) \\
&= \sum_{i=1}^n \left[x_i \sum_{j=1}^n \bar{a}_{ij}\bar{x}_j + \bar{x}_i \sum_{j=1}^n a_{ij}x_j \right] \\
&= \sum_{i=1}^n \sum_{j=1}^n (\bar{a}_{ij} + a_{ji}) x_i \bar{x}_j \\
&= 2 \sum_{i,j=1}^n c_{ij} x_i \bar{x}_j = 2Q(x),
\end{aligned}
$$

where $Q(x)$ is the Hermitian form $\sum c_{ij}x_i\bar{x}_j$. Accordingly, by Lemma 2, the inequalities (8–152) are valid; that is,

$$ m(t)u(t) \leq Q(x(t)) \leq M(t)u(t); $$

thus for each t,

$$ 2mu \leq \frac{du}{dt} \leq 2Mu. $$

From the inequality on the right we deduce that

$$ \exp\left\{ -2\int_0^t M(s)\,ds \right\} \left[\frac{du}{dt} - 2M(t)u \right] \leq 0, $$

$$ \frac{d}{dt} \left[u \exp\left\{ -2\int_0^t M(s)\,ds \right\} \right] \leq 0, $$

$$ u(t) \exp\left\{ -2\int_0^t M(s)\,ds \right\} \leq u(0) \quad (0 \leq t < \infty). $$

Thus one half of (8–154) follows. The other half is proved in the same way.

Remark. Since $M(t)$ and $m(t)$ are the maximum and minimum, respectively, of the continuous function $Q(x(t))$ for $\sum_i |x_i|^2 = 1$, $M(t)$ and $m(t)$ are continuous functions of t.

We note that, if $u(t) \to 0$, then all $x_i(t) \to 0$, so that $\|x\| \to 0$; if $u(t)$ is bounded, then $x(t)$ is bounded; if $u(t) \to +\infty$, then, since

$$ \sum_{i=1}^n |x_i|^2 \leq \left(\sum_{i=1}^n |x_i| \right)^2, $$

$\|x(t)\| \rightarrow \infty$ also. Accordingly, we deduce:

COROLLARY TO THEOREM 22. *If $\int_0^\infty M(t)\,dt = -\infty$, then the differential equation (8–153) is stable; that is, every solution $x = x(t)$ approaches zero as $t \rightarrow \infty$. If $\int_0^\infty M(t)\,dt$ is finite, then Eq. (8–153) is neutrally stable; that is, all solutions are bounded. If $\int_0^\infty m(t)\,dt$ is finite, Eq. (8–153) is not stable; more precisely, $x(t)$ cannot approach zero as $t \rightarrow \infty$ unless $x(t) \equiv 0$. If $\int_0^\infty m(t)\,dt = +\infty$, all solutions of Eq. (8–153), other than the solution $x(t) \equiv 0$, are unbounded.*

EXAMPLE. $[D^2 + 3D + (2 + a \cos t)]x = 0$, $a =$ real const. We replace this second order equation by an equivalent pair of first order equations, in order to apply the matrix technique. However, the replacement can be carried out in many different ways. We shall see that the information obtained varies accordingly.

We first form the equivalent pair:

$$Dx_1 = x_2, \qquad Dx_2 = -(2 + a \cos t)x_1 - 3x_2 \quad (x_1 = x).$$

The matrices A and C become

$$A = \begin{bmatrix} 0 & 1 \\ -2 - a \cos t & -3 \end{bmatrix},$$

$$C = \begin{bmatrix} 0 & \dfrac{-1 - a \cos t}{2} \\ \dfrac{-1 - a \cos t}{2} & -3 \end{bmatrix}.$$

The characteristic equation associated with C is

$$\lambda^2 + 3\lambda - \tfrac{1}{4}(1 + a \cos t)^2 = 0.$$

Hence the maximum root $M(t)$ is always ≥ 0, and in fact $\int_0^\infty M(t)\,dt = \infty$; the minimum root $m(t)$ is at most -3, so that $\int_0^\infty m(t)\,dt = -\infty$. Thus no conclusions can be reached on the basis of the corollary.

We now form another equivalent pair. We set

$$y_1 = 2x_1 + x_2 = 2x + Dx, \qquad y_2 = -x_1 - x_2 = -x - Dx,$$

and obtain the equations:

$$Dy_1 = -y_1 - (a \cos t)(y_1 + y_2),$$
$$Dy_2 = -2y_2 + (a \cos t)(y_1 + y_2).$$

Thus the new system is chosen to yield a diagonal form when $a = 0$. We note that $x = y_1 + y_2$, so that stability for the y system implies stability for the equation in x. The matrices A, C become

$$A = \begin{bmatrix} -1 - a\cos t & -a\cos t \\ a\cos t & -2 + a\cos t \end{bmatrix},$$

$$C = \begin{bmatrix} -1 - a\cos t & 0 \\ 0 & -2 + a\cos t \end{bmatrix}.$$

The characteristic values for C are $\lambda_1 = -1 - a\cos t$, $\lambda_2 = -2 + a\cos t$. Accordingly, if $|a| < 1$, then $M(t) \leqq -1 + |a| < 0$ and $\int_0^\infty M(t)\,dt = -\infty$, so that the equation is stable. (See also Problem 5 below.)

8–16 Response to bounded inputs. In Section 2–10 we have introduced the concept of a *resonant* system as one for which some bounded input yields an unbounded output. There we showed that nonresonance was equivalent to stability for a system described by a differential equation with constant coefficients. We now proceed to establish an analogous result for equations with variable coefficients. However, it will be seen that stability alone is not equivalent to nonresonance; one must also require that the transients die out *exponentially* (as they do automatically for a stable system described by differential equations with constant coefficients).

THEOREM 23. *Let the square matrix $A(t)$ be continuous for $0 \leqq t < \infty$. Let constants $a > 0$, $b > 0$ exist such that for every solution of the homogeneous differential equation $dx/dt = Ax$ one has*

$$\|x(t)\| \leqq b\|x(u)\|e^{-a(t-u)} \quad (0 \leqq u < t < \infty). \tag{8–160}$$

Then for each $f(t)$, bounded and continuous for $0 \leqq t < \infty$, every solution of the equation

$$\frac{dx}{dt} = A(t)x + f(t) \tag{8–161}$$

is also bounded for $0 \leqq t < \infty$.

Proof. We form the matrix kernel function $\Phi(t, u) = X(t)X^{-1}(u)$ as in Theorem 17 (Section 8–13). Then $x = \Phi(t, u)\xi$ (where ξ is a column vector) is the solution of the homogeneous equation such that $x(u) = \Phi(u, u)\xi = \xi$. Accordingly, by (8–160),

$$\|\Phi(t, u)\xi\| \leqq b\|\xi\|e^{-a(t-u)} \quad (0 \leqq u \leqq t < \infty); \tag{8–160'}$$

validity of (8–160′) for all ξ is in fact equivalent to validity of (8–160) for all solutions $x(t)$ of the homogeneous equation.

Now let $x(t)$ be a solution of the nonhomogeneous equation (8–161), where $\|f(t)\| < K = \text{const}$ for $0 \leq t < \infty$. Then by Theorem 17

$$x(t) = \Phi(t, 0)x(0) + \int_0^t \Phi(t, u)f(u) \, du;$$

the first term provides the solution of the homogeneous equation with given initial value $x(0)$, the second provides the solution of the nonhomogeneous equation with initial value zero. Since the solutions of the homogeneous equation approach zero exponentially as $t \to \infty$, the first term approaches zero and is therefore bounded for $0 \leq t < \infty$. For the second term we apply (8–160′):

$$\left\| \int_0^t \Phi(t, u)f(u) \, du \right\| \leq \int_0^t \|\Phi(t, u)f(u)\| \, du$$

$$\leq \int_0^t b\|f(u)\| e^{-a(t-u)} \, du$$

$$\leq bKe^{-at} \int_0^t e^{au} \, du = \frac{bK}{a}(1 - e^{-at}) < \frac{bK}{a}.$$

Hence this second term is bounded, and the theorem is proved.

Remark 1. By the proof above, the ratio b/a can be considered as an upper estimate for the amplification of the "disturbance" $f(t)$. For a stable equation with constant coefficients we have, by Theorem 18, $\Phi(t, u) = X(t - u)$, where $\|X(t)\| \leq be^{-at}$, and hence

$$\|\Phi(t, u)\| = \|X(t - u)\| \leq be^{-a(t-u)} \quad (t > u),$$

$$\|\Phi(t, u)\xi\| \leq \|\Phi(t, u)\| \, \|\xi\| \leq b\|\xi\|e^{-a(t-u)}.$$

Thus Eq. (8–160′) is satisfied and hence so also is (8–160). As in the proof of Theorem 21, a can be chosen to be any number such that $\max \text{Re}(\lambda_j) < -a < 0$ and b as $\max(e^{at}\|e^{tA}\|)$ for $t \geq 0$.

Remark 2. Another inequality equivalent to (8–160) and (8–160′) is the following:

$$\|\Phi(t, u)\| \leq b_1 e^{-a(t-u)} \quad (0 \leq u \leq t < \infty). \qquad (8\text{–}160'')$$

For if (8–160″) holds, then $\|\Phi(t, u)\xi\| \leq \|\Phi(t, u)\| \, \|\xi\| \leq b_1\|\xi\|e^{-a(t-u)}$, so that (8–160) and (8–160′) hold with $b = b_1$. Conversely, let (8–160) hold; then each column of $\Phi(t, u)$ is a solution of the equation $x' = Ax$ with norm 1 when $t = u$, and hence each column satisfies an inequality (8–160) with $\|x(u)\| = 1$. Since the norm of Φ is the sum of the norms of its columns, we conclude that

$$\|\Phi(t, u)\| \leq bne^{-a(t-u)} \quad (0 \leq u \leq t < \infty),$$

so that (8–160″) holds with $b_1 = bn$.

Remark 3. An analogous theorem holds for the equation of order n: if the kernel function $\phi(t, u)$ satisfies an inequality $|\phi(t, u)| \leq be^{-a(t-u)}$, $0 \leq u \leq t < \infty$, then the response to each bounded input f is bounded. The proof is similar to that of Theorem 23. (The inequality imposed on ϕ also implies that the solutions of the homogeneous equation decay exponentially; see Theorem 25.) An illustration is provided by the equations of Section 8–6. If the characteristic roots s_k in Theorem 6 have negative real parts, then Re $s_k < -a < 0$ for some a, so that by Theorem 6,

$$|\phi(t, u)| \leq e^{-a(t-u)} \sum_{k=1}^{n} |R_k(u)|\,|Q_k(t)| \quad (0 \leq u < t).$$

Since $R_k(u) \to r_{k0}$ as $u \to +\infty$, and $Q_k(t) \to q_{k0}$ as $t \to +\infty$, these functions are bounded for $0 \leq u < \infty$, $0 \leq t < \infty$, and hence $|\phi(t, u)| \leq be^{-a(t-u)}$ for some constant b.

THEOREM 24. *Let $A(t)$ be a continuous square matrix of order n for $0 \leq t < \infty$, and let $\|A(t)\| < c_1 = \text{const for } 0 \leq t < \infty$. For every $f(t)$, bounded and continuous for $0 \leq t < \infty$, let every solution $x(t)$ of the differential equation $x' = A(t)x + f$ be bounded for $0 \leq t < \infty$. Then there exist constants $a > 0$, $b > 0$, such that*

$$\|x(t)\| \leq b\|x(t_0)\|e^{-a(t-t_0)} \quad (0 \leq t_0 \leq t < \infty),$$

for every solution $x(t)$ of the differential equation $dx/dt = A(t)x$.

Proof. The proof will be given with the aid of several lemmas.

LEMMA 1. *Let $g(u)$ be a continuous real-valued function of u for $a \leq u \leq b$ and let $\int_a^b |g(u)|\,du > K > 0$. Then there exists a continuous real-valued function $F(u)$, $a \leq u \leq b$, such that $F(a) = F(b) = 0$, $|F(u)| \leq 1$, and $\int_a^b F(u)g(u)\,du > K$.*

Proof. Let $\epsilon = \int_a^b |g(u)|\,du - K$, so that $\epsilon > 0$. We choose a polynomial $P(u)$ so that $|P(u) - g(u)| < \epsilon/[8(b - a)]$; this is possible by the

Weierstrass approximation theorem (pp. 131–132 of Reference 17). Then

$$\int_a^b |P(u)|\, du = \int_a^b |g(u) + P(u) - g(u)|\, du$$

$$\geq \int_a^b |g(u)|\, du - \int_a^b |P(u) - g(u)|\, du$$

$$\geq K + \epsilon - \frac{(b-a)\epsilon}{8(b-a)} = K + \frac{7}{8}\,\epsilon. \qquad (8\text{-}162)$$

Now let $M = \max |P(u)|$ for $a \leq u \leq b$, so that necessarily $M > 0$, and let

$$\delta = \min\left(\frac{b-a}{3},\ \frac{\epsilon}{8(b-a)},\ \frac{\epsilon}{16M}\right).$$

Let B_δ denote the set formed of the intervals $[a, a + \delta]$ and $[b - \delta, b]$; I_δ the set formed of the intervals in which $|P(u)| \geq \delta$, for $a + \delta \leq u \leq b - \delta$; E_δ that formed of the intervals in which $|P(u)| \leq \delta$ for $a + \delta \leq u \leq b - \delta$. Since $P(u) = \pm\delta$ only at a finite number of points, each set is formed of a finite number of intervals. We now define $F(u)$ as follows: $F(a) = F(b) = 0$; $F(u) = 1$ where $P(u) \geq \delta$ in I_δ; $F(u) = -1$ where $P(u) \leq -\delta$ in I_δ; $F(u)$ varies linearly on the intervals remaining. Thus $F(u)$ is continuous for $a \leq u \leq b$ and $|F(u)| \leq 1$. Now

$$\int_a^b F(u)P(u)\, du = \int_{I_\delta} F(u)P(u)\, du + \int_{E_\delta} \ldots du + \int_{B_\delta} \ldots du.$$

Within I_δ, $F(u)P(u) = |P(u)|$. Hence

$$\left| \int_a^b F(u)P(u)\, du - \int_a^b |P(u)|\, du \right|$$

$$= \left| \int_{E_\delta} [F(u)P(u) - |P(u)|]\, du + \int_{B_\delta} [F(u)P(u) - |P(u)|]\, du \right|$$

$$\leq \int_{E_\delta} |P(u)|\, [|F(u)| + 1]\, du + \int_{B_\delta} |P(u)|\, [|F(u)| + 1]\, du$$

$$\leq 2\delta(b-a) + 4\delta M,$$

since $|P(u)| \leq \delta$ in E_δ, $|P(u)| \leq M$ in B_δ, $|F(u)| \leq 1$. By the definition of δ, $2\delta(b-a) < \epsilon/4$, $4\delta M < \epsilon/4$. Hence

$$\left| \int_a^b F(u)P(u)\, du - \int_a^b |P(u)|\, du \right| < \frac{\epsilon}{4} + \frac{\epsilon}{4} = \frac{\epsilon}{2}. \qquad (8\text{-}163)$$

Furthermore,

$$\left| \int_a^b F(u)g(u) \, du - \int_a^b F(u)P(u) \, du \right|$$

$$= \left| \int_a^b F(u)[g(u) - P(u)] \, du \right| \leq \frac{\epsilon}{8(b-a)} \, (b-a) = \frac{\epsilon}{8}, \quad (8\text{–}164)$$

since

$$|F(u)| \leq 1 \quad \text{and} \quad |g(u) - P(u)| < \epsilon/[8(b-a)].$$

Combining (8–163) and (8–164), we conclude

$$\left| \int_a^b F(u)g(u) \, du - \int_a^b |P(u)| \, du \right| < \frac{\epsilon}{2} + \frac{\epsilon}{8} = \frac{5}{8} \, \epsilon. \quad (8\text{–}165)$$

By (8–162) and (8–165), finally,

$$\int_a^b F(u)g(u) \, du > K + \frac{7}{8} \, \epsilon - \frac{5}{8} \, \epsilon = K + \frac{\epsilon}{4} > K.$$

Thus $F(u)$ has all the asserted properties.

LEMMA 2. *Let $\phi(t, u)$ be a real-valued continuous function of t and u for $0 \leq u \leq t < \infty$. For each fixed s let $\int_0^s |\phi(t, u)| \, du$ be bounded for $s \leq t < \infty$. For every $F(u)$, bounded and continuous for $0 \leq u < \infty$, let $\int_0^t \phi(t, u)F(u) \, du$ be bounded for $0 \leq t < \infty$. Then there exists a constant α such that $\int_s^t |\phi(t, u)| \, du < \alpha$ for all s, t such that $0 \leq s \leq t < \infty$.*

Proof. We shall prove the existence of α such that $\int_0^t |\phi(t, u)| \, du < \alpha$ for all t. Since for $s \leq t$

$$\int_s^t |\phi(t, u)| \, du \leq \int_0^t |\phi(t, u)| \, du,$$

the desired conclusion will follow. Let us suppose the contrary, that no such α exists. Then we can choose a t (as large as desired) so that $\int_0^t |\phi(t, u)| \, du$ is as large as desired. In particular, we choose s_1 so that $\int_0^{s_1} |\phi(s_1, u)| \, du > 1$. By hypothesis, there is a constant k_1 such that

$$\int_0^{s_1} |\phi(t, u)| \, du < k_1 \quad (s_1 \leq t < \infty). \quad (8\text{–}166)$$

We now choose s_2 so large, and greater than s_1, that

$$\int_0^{s_2} |\phi(s_2, u)| \, du > 2 + 2k_1.$$

By (8–166), $\int_0^{s_1} |\phi(s_2, u)| \, du < k_1$. Hence necessarily

$$\int_{s_1}^{s_2} |\phi(s_2, u)| \, du > 2 + k_1.$$

Proceeding in this manner, we obtain sequences $\{k_n\}$, $\{s_n\}$, such that $0 < s_1 < s_2 < \cdots < s_n < s_{n+1} < \cdots, s_n \to \infty$ as $n \to \infty$, and

$$\int_0^{s_1} |\phi(s_1, u)| \, du > 1, \quad \int_{s_{n-1}}^{s_n} |\phi(s_n, u)| \, du > n + k_{n-1} \quad \text{for } n = 2, 3, \ldots,$$

$$\int_0^{s_n} |\phi(t, u)| \, du < k_n \quad \text{for } s_n \leqq t < \infty.$$

We now define $F(u)$ in accordance with Lemma 1 in each of the intervals $[0, s_1], [s_1, s_2], \ldots$, so that $F(0) = 0$, $F(s_n) = 0$, $|F(u)| \leqq 1$, $F(u)$ is continuous for $0 \leqq u \leqq s_1, \ldots, s_{n-1} \leqq u \leqq s_n, \ldots$, and hence for $0 \leqq u < \infty$, and

$$\int_0^{s_1} \phi(s_1, u)F(u) \, du > 1, \qquad \int_{s_{n-1}}^{s_n} \phi(s_n, u)F(u) \, du > n + k_{n-1}.$$

Now for $n = 2, 3, \ldots$, since $|F(u)| \leqq 1$,

$$\left| \int_0^{s_{n-1}} \phi(s_n, u)F(u) \, du \right| \leqq \int_0^{s_{n-1}} |\phi(s_n, u)| \, du < k_{n-1},$$

$$\int_0^{s_n} \phi(s_n, u)F(u) \, du$$

$$= \int_{s_{n-1}}^{s_n} \phi(s_n, u)F(u) \, du + \int_0^{s_{n-1}} \phi(s_n, u)F(u) \, du > n + k_{n-1} - k_{n-1} = n.$$

Thus $F(u)$ is bounded, but $\int_0^t \phi(t, u)F(u) \, du$ is not bounded, contrary to hypothesis. Therefore the constant α must exist, and Lemma 2 is proved.

LEMMA 3. *Under the hypotheses for Theorem 24, there exists a constant c_2 such that the matrix kernel function $\Phi(t, u)$ for the differential equation $dx/dt = Ax + f$ satisfies the inequality*

$$\int_{t_0}^t \|\Phi(t, u)\| \, du < c_2 \quad (t_0 \leqq t < \infty).$$

Proof. Let $\Phi(t, u) = (\phi_{ij}(t, u))$. The response, with initial values zero, to a forcing function (column vector) f with all elements zero except f_j $[f = \text{col} (0, \ldots, 0, f_j, 0, \ldots, 0)]$ is $x(t) = \text{col} (x_1(t), \ldots, x_n(t))$, where

$$x_i(t) = \int_0^t \phi_{ij}(t, u)f_j(u) \, du \quad (i = 1, \ldots, n).$$

If $f_j(u)$ is bounded and continuous for $0 \leq u < \infty$, then, by hypothesis, $\|x(t)\|$ is bounded for $0 \leq t < \infty$, so that also $|x_i(t)|$ is bounded for each i. Thus for fixed i, j, $\phi(t, u) \equiv \phi_{ij}(t, u)$ satisfies the condition that $\int_0^t \phi(t, u) F(u) \, du$ is bounded for every bounded $F(u)$.

Furthermore, for each fixed s, $\int_0^s |\phi_{ij}(t, u)| \, du$ is bounded for $0 \leq t < \infty$. To show this, we remark that $\Phi = X(t)X^{-1}(u)$, so that $\phi_{ij}(t, u) = \sum_k x_{ik}(t) x_{kj}^{(-1)}(u)$, where $X = (x_{ij})$, $X^{-1} = (x_{ij}^{(-1)})$. Since the columns of $X(t)$ are solutions of the homogeneous differential equation (response to input zero), each column is bounded for $0 \leq t < \infty$, so that each $|x_{ik}(t)|$ is bounded and we can write

$$|\phi_{ij}(t, u)| \leq K \sum_{k=1}^{n} |x_{kj}^{(-1)}(u)| = K\psi(u),$$

where K is a constant and $\psi(u)$ is continuous for $0 \leq u < \infty$. Therefore, for each fixed s,

$$\int_0^s |\phi_{ij}(t, u)| \, du \leq K \int_0^s \psi(u) \, du = \text{const.}$$

Hence $\int_0^s |\phi_{ij}(t, u)| \, du$ is bounded, as asserted.

Thus $\phi(t, u) \equiv \phi_{ij}(t, u)$ satisfies all the hypotheses of Lemma 2, and we conclude that there exists a constant α_{ij} such that

$$\int_{t_0}^t |\phi_{ij}(t, u)| \, du < \alpha_{ij} \quad \text{for } t_0 \leq t < \infty.$$

If we set $c_2 = \sum_{i,j} |\alpha_{ij}|$, we have the desired conclusion:

$$\int_{t_0}^t \|\Phi(t, u)\| \, du < c_2 \quad \text{for } t_0 \leq t < \infty.$$

LEMMA 4. *The matrix kernel function $\Phi(t, u)$ satisfies the relation*

$$\Phi(t, u)\Phi(u, s) = \Phi(t, s) \tag{8–167}$$

$$(0 \leq t < \infty, 0 \leq s < \infty, 0 \leq u < \infty).$$

In particular,

$$\Phi(t, u)\Phi(u, t) = \Phi(t, t) = I, \tag{8–168}$$

so that $\Phi(u, t) = [\Phi(t, u)]^{-1}$. Furthermore,

$$\frac{d\Phi(t, u)}{du} = -\Phi(t, u)A(u) \quad (t = \text{const.}) \tag{8–169}$$

The proof is left as an exercise (Problem 7 below).

LEMMA 5. *Under the hypotheses of Theorem 24, the matrix kernel function* $\Phi(t, u)$ *is bounded for all* $t \geq 0$, $u \geq 0$: $\|\Phi(t, u)\| < c_3 = const.$

Proof. We have, with the aid of Lemmas 3 and 4,

$$
\begin{aligned}
\|\Phi(t, s)\| &= \|\Phi(t, t) - \Phi(t, s) - \Phi(t, t)\| \\
&\leq \|\Phi(t, t) - \Phi(t, s)\| + \|\Phi(t, t)\| \\
&\leq \left\| \int_s^t \frac{d\Phi(t, u)}{du} \, du \right\| + \|I\| \\
&\leq \left\| \int_s^t \Phi(t, u) A(u) \, du \right\| + n \\
&\leq \int_s^t \|\Phi(t, u)\| \, \|A(u)\| \, du + n \\
&\leq c_1 \int_s^t \|\Phi(t, u)\| \, du + n \\
&< c_1 c_2 + n = c_3.
\end{aligned}
$$

(We have assumed $s \leq t$; if $t \leq s$, s and t must be interchanged in the last two integrals.) Thus Lemma 5 is proved.

We now turn to the proof of Theorem 24. We have, for $t_0 < t$, by Lemmas 3, 4, and 5,

$$
\|(t - t_0)\Phi(t, t_0)\| = \left\| \int_{t_0}^t \Phi(t, t_0) \, du \right\| = \left\| \int_{t_0}^t \Phi(t, u)\Phi(u, t_0) \, du \right\|
$$

$$
\leq \int_{t_0}^t \|\Phi(t, u)\| \, \|\Phi(u, t_0)\| \, du \leq c_3 \int_{t_0}^t \|\Phi(t, u)\| \, du < c_2 c_3.
$$

Therefore $\|\Phi(t, t_0)\| < c_2 c_3/(t - t_0)$, and for $T = 2c_2 c_3$,

$$
\|\Phi(t_0 + T, t_0)\| < \frac{c_2 c_3}{T} = \frac{1}{2},
$$

$$
\|\Phi(t_0 + 2T, t_0)\| = \|\Phi(t_0 + 2T, t_0 + T) \cdot \Phi(t_0 + T, t_0)\| < \frac{c_2 c_3}{T} \cdot \frac{c_2 c_3}{T} = \frac{1}{4},
$$

and in general

$$
\|\Phi(t_0 + kT, t_0)\| \leq n2^{-k} \quad (k = 0, 1, 2, \ldots)
$$

(the factor n being inserted to take care of the case $k = 0$, when

$\|\Phi(t_0, t_0)\| = \|I\| = n$). We can also write

$$\|\Phi(t_k, t_0)\| \leq ne^{-c_4(t_k-t_0)},$$

where $t_k = t_0 + kT$, $c_4 = T^{-1}\log 2$.   Now let $0 \leq \tau < T$.   Then

$$\|\Phi(t_k + \tau, t_0)\| = \|\Phi(t_k + \tau, t_k)\Phi(t_k, t_0)\|$$
$$\leq \|\Phi(t_k + \tau, t_k)\|ne^{-c_4(t_k-t_0)}$$
$$< nc_3e^{c_4\tau}e^{-c_4(t_k+\tau-t_0)}$$
$$< c_5e^{-c_4(t_k+\tau-t_0)},$$

where $c_5 = nc_3e^{c_4 T}$. Hence, if $t = t_k + \tau$ (and every $t \geq t_0$ can be so represented for some k and τ),

$$\|\Phi(t, t_0)\| < be^{-a(t-t_0)},$$

where $b = c_5 > 0$, $a = c_4 > 0$. Thus, for every solution $x(t)$ of the homogeneous differential equation,

$$\|x(t)\| = \|\Phi(t, t_0)x(t_0)\| \leq b\|x(t_0)\|e^{-a(t-t_0)},$$

as asserted.

For further information on this topic, see References 1 and 16. In the proof of the theorem we have implicitly assumed all quantities to be real. The theorem remains valid in the complex case, as one can deduce by taking real and imaginary parts and applying the result for the real case.

EXAMPLE. $Dx + (t + 2)^{-1}x = f(t)$. Here $n = 1$ and the matrix $A(t)$ has one element, $-(t + 2)^{-1}$, so that $\|A(t)\| < c_1 = 1$ for $0 \leq t < \infty$. However, the solutions of the homogeneous equation are $x = c(t + 2)^{-1}$, which approach zero as $t \to \infty$, but not exponentially. Resonance occurs, as we illustrate by taking $f(t) = 1$. A particular solution is $x = \frac{1}{2}(t + 2)$, which is unbounded for $0 \leq t < \infty$.

Remarks. Theorem 24 can be applied to an equation of order n, $L[x] \equiv (D^n + a_1D^{n-1} + \cdots + a_n)x = f$, by rewriting the equation in matrix form or, equivalently, as n simultaneous first order equations:

$$Dx_1 = x_2, Dx_2 = x_3, \ldots, Dx_n = -a_nx_1 - \cdots - a_1x_n + f,$$

where $x_1 = x$. The theorem then asserts that, if all $a_j(t)$ are bounded and continuous for $t \geq 0$ and if for every set of bounded continuous disturbances $f_1(t), \ldots, f_n(t)$ $(f_n = f)$, all solutions of the equations

$$Dx_1 = x_2 + f_1, Dx_2 = x_3 + f_2, \ldots, Dx_n = -a_nx_n - \cdots - a_1x_n + f_n$$

are bounded for $t \geq 0$, then there are constants $a > 0$, $b > 0$, such that each solution $\{x_i(t)\}$ of the related homogeneous system satisfies the inequalities

$$\sum_{i=1}^{n} |x_i(t)| \leq be^{-a(t-t_0)} \sum_{i=1}^{n} |x_i(t_0)| \quad (t \geq t_0)$$

or, equivalently, the matrix kernel function $\Phi(t, u)$ is similarly bounded:

$$\|\Phi(t, u)\| \leq b_1 e^{-a(t-u)} \quad (0 \leq u \leq t < \infty).$$

Thus for the equation of order n, for each solution $x(t)$ of the related homogeneous equation, $x(t)$, $x'(t)$, $\ldots$, $x^{(n-1)}(t)$ all approach zero exponentially; since the kernel function $\phi(t, u)$ is an element of the matrix $\Phi(t, u)$ (Problem 9 below), one has also

$$|\phi(t, u)| \leq b_1 e^{-a(t-u)} \quad (0 \leq u < t < \infty).$$

As in Remark 3 following the proof of Theorem 23, the last condition is sufficient to guarantee that for each bounded scalar $f(t)$, each solution $x(t)$ of the n-th order equation $L[x] = f$ is bounded. However, the condition is much less stringent than the matrix condition $\|\Phi(t, u)\| < b_1 e^{-a(t-u)}$, which implies that $x(t)$, $x'(t)$, $\ldots$, $x^{(n-1)}(t)$ all approach zero exponentially for every solution of the homogeneous equation. Nevertheless, the inequality imposed on ϕ implies that at least the solutions $x(t)$ of the homogeneous equation approach zero exponentially:

THEOREM 25. *Let $a_1(t)$, $\ldots$, $a_n(t)$ be continuous for $0 \leq t < \infty$ and let $|a_j(t)| < k_j = \text{const for } 0 \leq t < \infty$. Let $\phi(t, u)$ be the kernel function for the equation $L[x] \equiv [D^n + a_1(t)D^{n-1} + \cdots + a_n(t)]x = f(t)$, and let constants a, b exist such that*

$$|\phi(t, u)| \leq be^{-a(t-u)} \quad (0 \leq u \leq t < \infty).$$

Then there is a constant K such that for every solution $x(t)$ of the related homogeneous equation $L[x] = 0$ one has

$$|x(t)| \leq K\|x(u)\|e^{-a(t-u)} \quad (1 \leq u \leq t < \infty),$$

where $\|x(t)\| = |x(t)| + |x'(t)| + \cdots + |x^{(n-1)}(t)|$.

Proof. Let $x(t)$ be a solution of the homogeneous equation and let u be fixed, $u \geq 1$. We shall then show that for $t \geq u$, $x(t)$ can be expressed in the form

$$x(t) = \int_{u-1}^{u} f(v)\phi(t, v) \, dv;$$

that is, $x(t)$ is the response to a forcing function which acts only for $u - 1 \leq t \leq u$.

In order to find $f(t)$, we first construct a function $\psi(t), 0 \leq t \leq 1$, having continuous derivatives through order n and such that $\psi(t) \equiv 0$ for $0 \leq t \leq \frac{1}{3}, \psi(t) \equiv 1$ for $\frac{2}{3} \leq t \leq 1$. Now let

$$g(t) = \left[x(u) + x'(u)(t - u) + \cdots + \frac{x^{(n-1)}(u)}{(n-1)!}(t - u)^{(n-1)} \right] \psi(t + 1 - u).$$

Then $g(t)$ is defined for $u - 1 \leq t \leq u$, and has continuous derivatives through order n. For $u - \frac{1}{3} \leq t \leq u$, $g(t)$ coincides with the polynomial in brackets; hence $g(u) = x(u), g'(u) = x'(u), \ldots, g^{(n-1)}(u) = x^{(n-1)}(u)$, where the derivatives of g are evaluated to the left. Now let $f(t) = L[g(t)]$ for $u - 1 \leq t \leq u$, let $f(t) = 0$ for $t > u$. Then for $t \geq u - 1$, $\int_{u-1}^{t} f(v)\phi(t, v)\, dv$ describes the solution of the equation $L[x] = f$ with initial values zero at $t = u - 1$. But $L[g] = f$ and g has initial values zero at $t = u - 1$. Hence

$$\int_{u-1}^{t} f(v)\phi(t, v)\, dv = g(t) \quad (u - 1 \leq t \leq u).$$

For $t \geq u$, the integral gives a solution of the equation $L[x] = 0$; by continuity (Section 1–12) this solution has at $t = u$ the initial values $g(u), g'(u), \ldots, g^{(n-1)}(u)$; but these values are the same, respectively, as $x(u), x'(u), \ldots, x^{(n-1)}(u)$. Hence

$$\int_{u-1}^{t} f(v)\phi(t, v)\, dv = x(t) \quad (t \geq u).$$

Here the upper limit can be replaced by u, since $f(v) = 0$ for $v > u$.

We now proceed to estimate $|x(t)|$. First of all, for $u - 1 \leq t \leq u$,

$$|g(t)| \leq \left\{ |x(u)| + |x'(u)|\, |t - u| + \cdots \right.$$

$$\left. + |x^{(n-1)}(u)|\frac{|t - u|^{n-1}}{(n-1)!} \right\} |\psi(t + 1 - u)|$$

$$\leq \{|x(u)| + |x'(u)| + \cdots + |x^{(n-1)}(u)|\} |\psi(t + 1 - u)|$$

$$= \|x(u)\|\, |\psi(t + 1 - u)|.$$

But $|\psi(t)| < \text{const}$ for $0 \leq t \leq 1$, and hence $|g(t)| < \text{const} \cdot \|x(u)\|$, where the constant is independent of u. If we differentiate $g(t)$ successively, similar estimates are obtained for $g'(t), g''(t), \ldots$ Thus we can choose one constant C, independent of u, such that

$$|g^{(l)}(t)| \leq C\|x(u)\| \quad (u - 1 \leq t \leq u, l = 0, 1, \ldots, n).$$

Hence also

$$|f| = |L[g]| = |g^{(n)}(t) + a_1(t)g^{(n-1)}(t) + \cdots + a_n g(t)|$$
$$\leqq C\|x(u)\|(1 + k_1 + \cdots + k_n) = C_1\|x(u)\|.$$

Therefore, for $t \geqq u$,

$$|x(t)| = \left| \int_{u-1}^u f(v)\phi(t, v) \, dv \right| \leqq C_1\|x(u)\| \int_{u-1}^u |\phi(t, v)| \, dv$$
$$\leqq C_1\|x(u)\| \int_{u-1}^u be^{-a(t-v)} \, dv = K\|x(u)\|e^{-a(t-u)},$$

where $K = C_1 b(1 - e^{-a})/a$ for $a \neq 0$, $K = C_1 b$ for $a = 0$. Thus the theorem is established.

Problems

1. (a) Find the fundamental matrix $X = e^{tA}$ of Theorem 18 (Section 8–13) for the system: $Dx = -7x - 4y$, $Dy = 5x + 2y$.

(b) Find constants $a > 0$, $b > 0$ such that $\|X(t)\| \leqq be^{-at}$ for $t \geqq 0$.

2. For each of the following apply Theorem 20 or Theorem 21 (Section 8–14) to show either that all solutions are bounded or that all solutions approach zero as $t \to +\infty$:

(a) $Dx + (3 + 2e^{-t})x = 0$

(b) $D^2x + [5 + (1 + t^2)^{-1}]x = 0$

(c) $D^2x + [7 + (4 + \cos t)^{-1}]Dx + 6x = 0$

(d) $Dx = (-7 + 0.05 \cos t)x - 4y$, $Dy = 5x + (2 - 0.05 \cos t)y$

(See Problem 1.)

3. Apply Theorem 22 (Section 8–15) to the following to show stability: $Dx = (-3 - 10t^2 - t^4)x + (4 - t + 2t^2)y$, $Dy = (6 + t)x - 5y$.

4. Prove that every solution of the equation $[D^2 + (4 + b \cos t)]x = 0$ ($b = $ positive const) satisfies a relation of the form

$$|x(t)| < \text{const} \cdot \exp\left[\frac{b}{4} \int_0^t |\cos s| \, ds\right].$$

5. Consider the equation $[D^2 + 3D + (2 + a \cos t)]x = 0$ of the Example in Section 8–15.

(a) Show that in the second matrix formulation $M(t) = \frac{1}{2}(|1 - 2a \cos t| - 3)$ and that for $|a| \leqq 2$, $\int_0^\infty M(t) \, dt = -\infty$, so that the equation is stable. [Hint: Show first that, for any two real numbers p, q, we have

$$\max (p, q) = \frac{1}{2}(|p - q| + p + q).]$$

(b) Show more precisely that $\int_0^\infty M(t) \, dt = -\infty$ if and only if $|a| < \frac{1}{2} \sec \beta$, where β (approximately 1.35) is the root of the equation $\tan \beta - \beta - \pi = 0$ between zero and $\pi/2$; thus the equation is stable for $|a| < 2.27$.

6. Verify that if $f(t)$ is bounded and continuous for $0 \leqq t < \infty$, then every solution of the equation $Dx + tx = f$ is bounded for $0 \leqq t < \infty$. Show that the conclusion could be obtained from Theorem 23, but that the equation is not covered by Theorem 24 (Section 8–16).

7. Prove Lemma 4 of Section 8–16. [*Hint:* Use the representation $\Phi(t, u) = X(t)X^{-1}(u)$ to obtain (8–167) and (8–168). Differentiate (8–168) with respect to u, with t held constant, to obtain (8–169).]

8. *Equations with periodic coefficients in matrix form.* Let $A(t)$ be a square matrix, continuous for all t and having period $\tau > 0$: $A(t + \tau) = A(t)$. Show that if $X(t)$ is a fundamental matrix for the differential equation $dx/dt = A(t)x$, then $X(t)$ can be represented in the form $X(t) = P(t)e^{tB}$, where $P(t)$ has period τ and B is a constant matrix. (The eigenvalues of B are the characteristic exponents of Section 8–10.) [*Hint:* It is known that if C is a nonsingular square matrix, then C can be expressed as $e^{\tau B}$ for some matrix B. (See Chapter 1 of Reference 2.) Let $X(t)$ be a fundamental matrix. Show that $X(t + \tau)$ is also a fundamental matrix and that $X(t + \tau) = X(t)C$ for some nonsingular constant matrix C. Write $C = e^{\tau B}$, set $P(t) = X(t)e^{-tB}$ and show that $P(t)$ is periodic.]

9. *Kernel function of equation of order n.* Show that if $\Phi(t, u) = (\phi_{ij}(t, u))$ is the matrix kernel function for the equation $x' = A(t)x + f(t)$, where $A(t)$ is the matrix (8–131), then $\phi_{1n}(t, u)$ is the kernel function for the equation of order n: $(D^n + a_1 D^{n-1} + \cdots)x = f$.

ANSWERS

1. (a) $X = \begin{bmatrix} 5e^{-3t} - 4e^{-2t} & 4e^{-3t} - 4e^{-2t} \\ -5e^{-3t} + 5e^{-2t} & -4e^{-3t} + 5e^{-2t} \end{bmatrix}$

  (b) $a = 2, b = 18$

2. (a) solutions approach zero (b) solutions are bounded (c) solutions approach zero (d) solutions approach zero

8–17 Operational methods.
In a series of papers (see, for example, References 20 and 21), Lotfi Zadeh has developed generalizations of the transform methods of Chapters 5 and 6 to equations with variable coefficients. We present here some of the central ideas of his theory.

We first recall that for an equation with constant coefficients $(a_0 D^n + \cdots + a_n)x = f(t)$ the transfer function $Y(s) = 1/(a_0 s^n + \cdots + a_n)$ has the crucial property that, when $f = e^{st}$, a particular solution is $Y(s)e^{st}$. It is now important to ask: *which* particular solution? A proper characterization of this solution is essential for the generalization to be presented to equations with variable coefficients.

Let us assume that the given equation is *stable*. Then we can characterize the solution $x = Y(s)e^{st}$, for Re $s > 0$, as the solution with *initial values zero at* $t = -\infty$. (See Section 5–17.) Indeed, $x(t) \to 0$, $x'(t) \to 0, \ldots,$ $x^{(n-1)}(t) \to 0$ as $t \to -\infty$. By the stability assumption, every solution of the homogeneous equation (except the trivial one, $x \equiv 0$) becomes infinite as $t \to -\infty$. Therefore, $x = Y(s)e^{st}$ is the only solution with initial values zero at $t = -\infty$.

As in Chapter 5, we can introduce the weighting function $W(t) = \Phi^{-1}[Y(i\omega)]$ (Φ^{-1} here denoting the inverse Fourier transform). Then for proper restrictions on f,

$$x = \int_{-\infty}^{t} W(t - u)f(u)\,du$$

gives the solution with "small" initial values at $t = -\infty$. In particular, for $f = e^{st}$, Re $s > 0$, $x = Y(s)e^{st}$. We remark that the weighting function W can also be obtained by Laplace transforms: $W = \mathcal{L}^{-1}[Y(s)]$.

We now turn to an equation with variable coefficients, which we write in the form

$$L[x] \equiv [D^n + a_1(t)D^{n-1} + \cdots + a_n(t)]x = f(t). \qquad (8\text{--}170)$$

We assume the $a_j(t)$ and $f(t)$ to be continuous for $-\infty < t < \infty$ and that each $a_j(t)$ has bounded and continuous derivatives through order $n - j$, so that the adjoint $L^+[x]$ is well-defined (Section 8–4). Furthermore we postulate that Eq. (8–170) is *stable*. In detail, we require a strong form of stability; namely, that the kernel function $\phi(t, u)$ satisfy an equality

$$|\phi(t, u)| \leqq be^{-a(t-u)} \quad (-\infty < u \leqq t < \infty, a > 0, b > 0). \qquad (8\text{--}171)$$

We have seen (Theorem 25, Section 8–16) that this condition implies that all solutions of the homogeneous equation approach zero exponentially as $t \to \infty$. In (8–171) we permit t and u to be *negative;* the effect of this is that certain integrals with lower limit $-\infty$ have meaning. In particular, under appropriate restrictions on $f(t)$,

$$x = \int_{-\infty}^{t} \phi(t, u)f(u)\,du \qquad (8\text{--}172)$$

defines a solution of Eq. (8–170) and can be considered as the solution "with small initial values at $t = -\infty$."

We now obtain our generalized transfer function, called *system function* by Zadeh, by choosing $f = e^{st}$:

DEFINITION. The system function $H(s, t)$ is defined by the equation

$$H(s, t)e^{st} = \int_{-\infty}^{t} \phi(t, u)e^{su}\,du. \qquad (8\text{--}173)$$

THEOREM 26. *Under the hypotheses stated, the system function $H(s, t)$ is defined for Re $s > -a$, $-\infty < t < \infty$, and $H(s, t)e^{st}$ is a solution of*

Eq. (8–170) when $f = e^{st}$. The system function is the Laplace transform, with respect to v, of $\phi(t, t - v)$:

$$H(s, t) = \mathcal{L}_v[\phi(t, t - v)] = \int_0^\infty \phi(t, t - v)e^{-sv}\, dv. \qquad (8\text{–}174)$$

Hence, for each fixed t, $H(s, t)$ is an analytic function of s for Re $s > -a$. If the coefficients in Eq. (8–170) are analytic for all t, so that $\phi(t, u)$ is analytic in t and u, and if the inequality

$$|\phi(t, t - v)| \leqq be^{|a||v|} \qquad (8\text{–}175)$$

holds for all complex v, then for each fixed t, $H(s, t)$ is analytic at $s = \infty$:

$$H(s, t) = \frac{c_0(t)}{s} + \frac{c_1(t)}{s^2} + \cdots + \frac{c_n(t)}{s^{n+1}} + \cdots \qquad (|s| > |a|),$$

where

$$\phi(t, t - v) = c_0(t) + c_1(t)v + \cdots + \frac{c_n(t)}{n!} v^n + \cdots$$

Proof. By (8–171), if Re $s = \sigma$,

$$|\phi(t, u)e^{su}| \leqq be^{-a(t-u)}e^{\sigma u} = be^{-at}e^{(a+\sigma)u}.$$

Since $\int_{-\infty}^t e^{(a+\sigma)u}\, du$ exists for $\sigma > -a$, the integral on the right of (8–173) is absolutely convergent and $H(s, t)$ is well-defined for $\sigma > -a$.

We next prove that $H(s, t)e^{st}$ is a solution of the equation $L[x] = e^{st}$. To this end we form the kernel function $\phi^+(t, u)$ of the adjoint equation. By Theorem 5 (Section 8–4), $\phi^+(t, u) = -\phi(u, t)$. Hence by (8–171),

$$|\phi^+(t, u)| \leqq be^{a(t-u)} \quad (t \leqq u, a > 0, b > 0).$$

This is an inequality analogous to (8–171) for the "past" as opposed to the future, with a replaced by $-a$. Exactly as in the proof of Theorem 25, we now conclude that each solution $y(t)$ of the adjoint equation satisfies a similar inequality:

$$|y(t)| \leqq K\|y(u)\|e^{a(t-u)} \quad (t \leqq u),$$

or, upon interchanging t and u,

$$|y(u)| \leqq K\|y(t)\|e^{a(u-t)} \quad (u \leqq t).$$

Hence for $u \leqq t$,

$$|y(u)e^{su}| \leqq K\|y(t)\|e^{-at}e^{(a+\sigma)u}.$$

Since $\int_{-\infty}^0 e^{(a+\sigma)u}\, du$ exists for $\sigma > -a$, we conclude that $\int_{-\infty}^0 y(u)e^{su}\, du$ also exists. Now by the Corollary to Theorem 5, $\phi(t, u)$ can be expressed as $\sum_1^n x_i(t)y_i(u)$, where $L[x_i] = 0$, $L^+[y_i] = 0$. Therefore for each t

$$H(s, t)e^{st} = \int_{-\infty}^0 \phi(t, u)e^{su}\, du = \int_{-\infty}^0 \phi(t, u)e^{su}\, du + \int_0^t \phi(t, u)e^{su}\, du$$

$$= \int_{-\infty}^0 \sum_{i=1}^n x_i(t)y_i(u)e^{su}\, du + \int_0^t \phi(t, u)e^{su}\, du$$

$$= \sum_{i=1}^n x_i(t)\int_{-\infty}^0 y_i(u)e^{su}\, du + \int_0^t \phi(t, u)e^{su}\, du.$$

Here the first term on the right is a linear combination of $x_1(t), \ldots, x_n(t)$ and is hence a solution of the homogeneous equation. The second term is $T_0[e^{st}]$, a particular solution of the nonhomogeneous equation. Hence $H(s, t)e^{st}$ is a solution of the equation $L[x] = e^{st}$, as asserted.

If we now set $v = t - u$, (8–173) becomes

$$H(s, t)e^{st} = \int_0^\infty \phi(t, t - v)e^{s(t-v)}\, dv,$$

from which (8–174) follows at once. We note that Eq. (8–172) can be written similarly:

$$x = \int_0^\infty f(t - v)\phi(t, t - v)\, dv;$$

if $f(t) = 0$ for $t < 0$, this becomes

$$x = \int_0^t f(t - v)\phi(t, t - v)\, dv.$$

If we compare this relation with that for an equation with constant coefficients:

$$x = \int_0^t f(t - v)W(v)\, dv,$$

we see that $\phi(t, t - v)$ should be considered as the *generalized weighting function;* its Laplace transform, with respect to v, is the generalized transfer function, or system function, $H(s, t)$.

The statements about the analyticity of $H(s, t)$ are consequences of properties of Laplace transforms (Sections 6–6 and 6–9). Thus Theorem 26 is established.

Remarks. If we modify the assumption (8–171) to permit a to be negative or zero, then we of course lose the implication of stability and the interpretation of $H(s, t)e^{st}$ as the unique solution with small initial values at $-\infty$. However, Theorem 26 and its proof remain valid without change. The system function $H(s, t)$ is defined and analytic for $\operatorname{Re} s > -a$.

In (8–171) the requirement that t and u vary from $-\infty$ to ∞ is a restriction on the "past" of the system considered. For most practical problems, the form of the equation in the remote past is of no significance, and one may wish to modify a given form to ensure that (8–171) is satisfied. This can usually be achieved most simply by assuming that the coefficients in the differential equation become constant for t sufficiently large negative, and that the equation with constant coefficients is stable.

The series expansion of $H(s, t)$ at $s = \infty$ permits one to obtain an approximation of $H(s, t)$ by a rational function of s:

$$H(s, t) \sim \frac{c_0(t)}{s} + \frac{c_1(t)}{s^2} + \cdots + \frac{c_n(t)}{s^{n+1}}.$$

Another such approximation can be obtained for the equations of Section 8–6, for which

$$\phi(t, u) = \sum_{k=1}^{n} e^{s_k(t-u)} R_k(u) Q_k(t).$$

We approximate $R_k(u)$ by two terms of its series:

$$R_k(u) \sim r_{k0} + r_{k1} e^{-u},$$

and obtain the approximations

$$\phi(t, t - v) \sim \sum_{k=1}^{n} e^{s_k v}(r_{k0} + r_{k1} e^{-(t-v)}) Q_k(t),$$

$$H(s, t) \sim \sum_{k=1}^{n} Q_k(t) \left[\frac{r_{k0}}{s - s_k} + \frac{r_{k1} e^{-t}}{s - (s_k + 1)} \right].$$

It is natural to ask whether $H(s, t)$ can exactly coincide with a rational function of s. A fairly complete answer is provided by the following theorem:

THEOREM 27. *Let the system function of Theorem 26 be expressible in the form*

$$H(s, t) = \sum_{k=1}^{m} \frac{A_k(t)}{s - s_k(t)},$$

where the functions $s_k(t)$, $A_k(t)$ are continuous for all t, the $A_k(t)$ have continuous derivatives through order $n - 1$, and no $A_k(t)$ is identically zero in an interval. Then $m \geq n$ and $s_k(t) \equiv$ const. If $m = n$, then Eq. (8–170) has constant coefficients. If $m > n$, then both the homogeneous equation $L[x] = 0$, and its adjoint can be written in the form

$$(b_{01} e^{q_1 t} + \cdots + b_{0\mu} e^{q_\mu t}) D^n x + \cdots + (b_{n1} e^{q_1 t} + \cdots + b_{n\mu} e^{q_\mu t}) x = 0,$$

where the b_{ij} and q_j are constants. The general solution of the adjoint equation has the form

$$x = c_1 \sum_{k=1}^{m} d_{1k}e^{s_k t} + \cdots + c_n \sum_{k=1}^{m} d_{nk}e^{s_k t} \quad (d_{ik} = \text{const}),$$

while the general solution of the equation $L[x] = 0$ has the form

$$x = c_1 \sum_{k=1}^{m} A_k(t)e^{s_k t} - c_2 \sum_{k=1}^{m} s_k A_k(t)e^{s_k t} + \cdots$$

$$+ c_n(-1)^{n+1} \sum_{k=1}^{m} s_k^{n-1} A_k(t)e^{s_k t};$$

each $A_k(t)$ is a rational function of $e^{s_1 t}, \ldots, e^{s_m t}$. The kernel function of the given equation and that of the adjoint are given by

$$\phi(t, u) = \sum_{k=1}^{m} A_k(t)e^{s_k(t-u)}, \qquad \phi^+(t, u) = - \sum_{k=1}^{m} A_k(u)e^{s_k(u-t)}.$$

Proof. Since $H(s, t) = \mathcal{L}_v[\phi(t, t - v)]$, we obtain

$$\phi(t, t - v) = \sum_{k=1}^{m} A_k(t)e^{s_k(t)v},$$

$$\phi(t, u) = \sum_{k=1}^{m} A_k(t)e^{s_k(t)(t-u)}.$$

By Theorem 5 (Section 8–4) the kernel function for the adjoint equation is $\phi^+(t, u) = -\phi(u, t)$. Hence

$$\phi^+(t, u) = \sum_{k=1}^{m} B_k(u)e^{-s_k(u)t}, \qquad B_k(u) = -A_k(u)e^{u s_k(u)}.$$

Now for each fixed u, $\phi^+(t, u)$ is a solution of the adjoint equation. If the $s_k(u)$ are not all identically constant, then by varying u we could obtain solutions of the adjoint: $\phi^+(t, u_1), \phi^+(t, u_2), \ldots, \phi^+(t, u_N)$, each involving an exponential function $\exp[-s_k(u_l)t]$ not occurring in any of the other solutions. For $N > n$, this would yield more than n linearly independent solutions of the adjoint, which is impossible. Thus the $s_k(u)$ must be constant, and we can assume they are distinct. Now, as shown in Problem 6 following Section 8–3, the functions $\phi^+(t, u), \partial\phi^+/\partial u, \ldots, \partial^{n-1}\phi^+/\partial u^{n-1}$ are, for each fixed u, n linearly independent solutions of the adjoint equation. However, for fixed u, these n functions are all linear combinations of the m functions $e^{-s_1 t}, \ldots, e^{-s_m t}$. If $m < n$, they cannot be n linearly independent functions (Problem 8 below); hence this case cannot arise. If $m = n$, the functions form n linearly independent solutions of the equation with constant coefficients $(D - s_1)(D - s_2) \cdots (D - s_n)x = 0$; therefore the adjoint equation and, accordingly, the original equation, must have constant coefficients. If $m > n$, one obtains the general solution of the adjoint in the form stated in the theorem, where, for example, $d_{1l} = B_l(0) = -A_l(0)$, $d_{2l} = B_l'(0)$, ..., $d_{nl} = B_l^{(n-1)}(0)$. If we form the linear differential equation of order n whose general solution has this form (pp. 27–28 of Reference 10), we obtain an equation whose coefficients are linear combinations of exponential functions as stated in the theorem; the q_j are sums of certain of the s_k. The adjoint of such an equation has similar form.

From the indicated general solution of the adjoint, we can again form $\phi^+(t, u)$. The Wronskian $w^+(u)$ is a linear combination of exponentials: $w^+(u) = \sum_1^\mu w_j e^{q_j t}$, where the q_j are as above. Accordingly, by (8–20), $\phi^+(t, u)$ has the form $\sum B_k(u) e^{-s_k t}$, where the $B_k(u)$ are rational functions of $e^{s_1 u}$, ..., $e^{s_m u}$. Therefore, $\phi(t, u) = -\phi^+(u, t)$ has the form stated in the theorem. Reasoning as above, we obtain n linearly independent solutions of the given equation by evaluating $\phi(t, u)$, $\phi_u(t, u)$, ..., $\partial^{n-1}\phi/u^{n-1}$ at a fixed u, for example, $u = 0$. Hence the general solution has the form stated.

EXAMPLE 1. We form the equation of order 1 whose adjoint has solution $x = e^{at} + e^{bt}$, $a \neq b$. The adjoint can be written in the form

$$(e^{at} + e^{bt})Dx - (ae^{at} + be^{bt})x = 0,$$

and the original equation (nonhomogeneous, with leading coefficient 1) is

$$Dx + \frac{ae^{at} + be^{bt}}{e^{at} + e^{bt}} x = f(t).$$

We find the complementary function to be $c(e^{at} + e^{bt})^{-1}$ and the kernel function to be

$$\phi(t, u) = \frac{e^{au} + e^{bu}}{e^{at} + e^{bt}} = -\phi^+(u, t).$$

Hence the generalized weighting function and system function are

$$\phi(t, t - v) = \frac{e^{a(t-v)} + e^{b(t-v)}}{e^{at} + e^{bt}},$$

$$H(s, t) = \frac{1}{e^{at} + e^{bt}} \left(\frac{e^{at}}{s + a} + \frac{e^{bt}}{s + b} \right).$$

Thus $s_1 = -a$, $s_2 = -b$, $A_1(t) = e^{at}/(e^{at} + e^{bt})$, $A_2(t) = e^{bt}/(e^{at} + e^{bt})$. Thus all the relations described in the theorem are satisfied.

We note that if $0 < a < b$, then the original equation is stable and $H(s, t)$ is analytic for Re $s > -a$. Furthermore, we verify (Problem 1 below) that

$$|\phi(t, u)| \leq e^{-a(t-u)} \quad (t \geq u),$$

so that condition (8–171) is satisfied. For large negative t, the equation approaches the form $Dx + ax = f$, and the complementary function is approximated by ce^{-at}; for large positive t, a must be replaced by b in these expressions. Thus in this example it is the behavior in the past which determines the crucial constant a.

EXAMPLE 2. The equation is

$$D^2x + \frac{3 + 8e^t}{1 + 2e^t} Dx + \frac{12e^{2t} + 12e^t + 2}{(1 + 2e^t)^2} x = f(t).$$

We verify (Problem 2 below) that

$$H(s, t) = \frac{1}{1 + 2e^t} \left(\frac{1 + e^t}{s + 1} - \frac{1}{s + 2} - \frac{e^t}{s + 3} \right).$$

Here $H(s, t)$ is analytic for Re $s > -1$, and the equation is stable.

THEOREM 28. *Let* $f(t)$ *be piecewise continuous for* $-\infty < t < \infty$, *let* $f(t) = 0$ *for* $t < 0$, *let* $\mathcal{L}[f]$ *be absolutely convergent for* $\sigma > \sigma_0$, *and let* $\mathcal{L}[f] = F(s) = As^{-1} + G(s)$, *where* $|G(s)| < B/|s|^2$ *for* $\sigma > \sigma_0$ *and* A, B *are constants. Let the kernel function* $\phi(t, u)$ *for Eq.* (8–170) *satisfy an inequality:* $|\phi(t, u)| \leq be^{-a(t-u)}$ *for* $t \geq u$, *where* a, b *are constants* $(b > 0)$. *Then the solution of Eq.* (8–170) *for* $t \geq 0$ *with initial values zero at* $t = 0$ *is given by*

$$x = T_0[f] = \frac{1}{2\pi} (P) \int_{-\infty}^{\infty} H(s, t)F(s)e^{st} \, d\omega \quad (s = \sigma + i\omega), \quad (8\text{–}176)$$

where σ *is chosen so that* $\sigma > 0$, $\sigma > -a$, $\sigma > \sigma_0$.

Proof. We first assume that $A = 0$. We can write

$$x = \int_0^t \phi(t, u)f(u) \, du = \int_{-\infty}^t \phi(t, u)f(u) \, du,$$

since $f(u) = 0$ for $u < 0$. Since $\mathcal{L}[f] = F(s)$, we have

$$f(u) = \frac{1}{2\pi} \int_{-\infty}^{\infty} F(s)e^{su} \, d\omega \quad s = \sigma + i\omega, \quad \sigma > \sigma_0, \quad \sigma > 0,$$

no principal value being needed since $|F(s)| < B/|s|^2$. Accordingly,

$$x = \frac{1}{2\pi} \int_{-\infty}^{t} \int_{-\infty}^{\infty} F(s)\phi(t,\, u)e^{su} \, d\omega \, du.$$

Now $|F(s)| = |G(s)| < B/|s|^2 = B/(\sigma^2 + \omega^2)$, $\sigma > \sigma_0$. Accordingly, for each fixed t and for $u < t$,

$$|F(s)\phi(t,\, u)e^{su}| \leqq \frac{B}{\sigma^2 + \omega^2} \, be^{-a(t-u)}e^{\sigma u} \leqq \text{const} \cdot \frac{e^{(a+\sigma)u}}{\sigma^2 + \omega^2}.$$

Since for $\sigma \neq 0$, $\sigma > -a$, the function $e^{(a+\sigma)u}/(\sigma^2 + \omega^2)$ is integrable over the region $-\infty < u \leqq t$, $-\infty < \omega < \infty$, we conclude that the given integral of $F(s)\phi(t,\, u)e^{su}$ over this region is absolutely convergent, and hence we can interchange the order of integration (see the argument in Section 5–11):

$$x(t) = \frac{1}{2\pi} \int_{-\infty}^{\infty} \int_{-\infty}^{t} F(s)\phi(t,\, u)e^{su} \, du \, d\omega$$

$$= \frac{1}{2\pi} \int_{-\infty}^{\infty} F(s)H(s,\, t) \, d\omega = \frac{1}{2\pi i} \int_{\sigma - i\infty}^{\sigma + i\infty} F(s)H(s,\, t) \, ds.$$

Thus the theorem is established for the case $A = 0$.

If $A \neq 0$, we consider the term A/s separately. By adding the results, we obtain the desired conclusion. Now for $F = A/s$, $f(t) = Ah(t)$, and the asserted formula defines x thus:

$$x(t) = \frac{1}{2\pi} \, (P) \int_{-\infty}^{\infty} H(s,\, t)F(s)e^{st} \, d\omega = \frac{A}{2\pi} \, (P) \int_{-\infty}^{\infty} \frac{H(s,\, t)}{s} \, e^{st} \, d\omega.$$

Now we can write $H(s,\, u) = \mathcal{L}_t[\phi(u,\, u - t)]$. Hence, for $\sigma > 0$, $\sigma > -a$,

$$\frac{1}{s} H(s,\, u) = \mathcal{L}_t\left[\int_{0}^{t} \phi(u,\, u - v) \, dv \right].$$

Therefore, with a principal value understood,

$$\frac{A}{2\pi} \int_{-\infty}^{\infty} \frac{H(s,\, u)}{s} \, e^{st} \, d\omega = A \int_{0}^{t} \phi(u,\, u - v) \, dv$$

and, for $u = t \geqq 0$,

$$x = \frac{A}{2\pi} \int_{-\infty}^{\infty} \frac{H(s,\, t)}{s} \, e^{st} \, d\omega = \int_{0}^{t} A\phi(t,\, t - v) \, dv = \int_{0}^{t} \phi(t,\, u)Ah(u) \, du.$$

Thus x is the response to $f(t) = Ah(t)$, as asserted, and Theorem 28 follows.

Remark. The formula (8–176) does not give $x(t)$ strictly as an inverse Laplace transform. However, as above, we can write

$$g(t, u) = \frac{1}{2\pi} \int_{-\infty}^{\infty} H(s, u)F(s)e^{st} \, d\omega$$

$$= \mathcal{L}^{-1}[H(s, u)F(s)] = \phi(u, u - t) * f(t),$$

and then $x(t) = g(t, t)$. Thus in Example 1 above,

$$g(t, u) = \mathcal{L}^{-1}\left[\frac{1}{e^{au} + e^{bu}} \left(\frac{e^{au}}{s + a} + \frac{e^{bu}}{s + b} \right) F(s) \right].$$

If

$$f(t) = e^{\alpha t}h(t) \quad (\alpha \neq -a, \, \alpha \neq -b),$$

then $F(s) = 1/(s - \alpha)$ and

$$g(t, u) = \mathcal{L}^{-1}\left[\frac{1}{e^{au} + e^{bu}} \left\{ \left(\frac{e^{au}}{a + \alpha} + \frac{e^{bu}}{b + \alpha} \right) \frac{1}{s - \alpha} \right.\right.$$

$$\left.\left. - \frac{e^{au}}{(a + \alpha)(s + a)} - \frac{e^{bu}}{(b + \alpha)(s + b)} \right\} \right]$$

$$= \frac{1}{e^{au} + e^{bu}} \left[\left(\frac{e^{au}}{a + \alpha} + \frac{e^{bu}}{b + \alpha} \right) e^{\alpha t} - \frac{e^{au}}{a + \alpha} e^{-at} - \frac{e^{bu}}{b + \alpha} e^{-bt} \right],$$

$$x(t) = g(t, t) = \frac{1}{e^{at} + e^{bt}} \left\{ \frac{e^{(a+\alpha)t}}{a + \alpha} + \frac{e^{(b+\alpha)t}}{b + \alpha} - \frac{1}{a + \alpha} - \frac{1}{b + \alpha} \right\}.$$

We can at once verify that $x(t)$ is the desired solution. The last two terms provide a solution of the related homogeneous equation.

We can summarize the procedure indicated in the one formula:

$$x(t) = \mathcal{L}^{-1}[H(s, u)F(s)]|_{u=t}.$$

The result makes possible the application of Laplace transform methods to a wide variety of equations, in general, quite different from those considered in Section 8–9. By restricting s to be pure imaginary, a similar application of Fourier transforms is made possible.

PROBLEMS

1. Prove that, in Example 1 above, if $0 < a < b$, then $|\phi(t, u)| \leq e^{-a(t-u)}$ for $t \geq u$.

2. Show that the expression given for the system function in Example 2 is correct.

3. Find $H(s, t)$ for the equation $Dx + 2t(1 + t^2)^{-1}x = f$.

4. Let the following equation be given:

$$D^2x + \frac{3 + e^{-t}}{1 + e^{-t}} \, Dx + \frac{2}{1 + e^{-t}} \, x = f.$$

(a) Find $\phi(t, u)$ and $H(s, t)$. [*Hint:* Show that $x_1(t) = e^{-t}$ and $x_2(t) = (1 + e^t)^{-1}$ are solutions of the related homogeneous equation.]

(b) Show that $|\phi(t, u)| < be^{-a(t-u)}$ with $a = 1$ for the range $0 \leq u < t < \infty$, and with $a = 0$ for the range $-\infty < u < t < \infty$. Show that $H(s, t)$ is analytic for Re $s > \sigma_0$ only if $\sigma_0 \geq 0$. (Thus the behavior in the past, when the equation approaches the form $D^2x + Dx = f$, affects the domain of analyticity of $H(s, t)$.)

(c) For large positive t the differential equation approaches the form $(D^2 + 3D + 2)x = f$. Thus one expects two solutions of the related homogeneous equation approaching the form e^{-t}, e^{-2t} as $t \to \infty$. Show that $x_1(t)$, $x_2(t)$ do not provide these two solutions, but that certain linear combinations of $x_1(t)$, $x_2(t)$ do.

5. Find a particular solution of the following:

(a) $Dx + (1 + 2e^t)(1 + e^t)^{-1}x = \sin t$ (see Example 1 above)

(b) $D^2x + (3 + 8e^t)(1 + 2e^t)^{-1}Dx + (12e^{2t} + 12e^t + 2)(1 + 2e^t)^{-2}x = 1$ (see Example 2 above)

6. Prove that the system function $H(s, t)$ for Eq. (8–170) satisfies the differential equation

$$[(D + s)^n + a_1(t)(D + s)^{n-1} + \cdots + a_n(t)]H(s, t) = 1.$$

[*Hint:* Use the fact that $x = H(s, t)e^{st}$ satisfies the equation

$$\{D^n + a_1(t)D^{n-1} + \cdots\}x = e^{st}.]$$

7. Prove that if the coefficients in Eq. (8–170) have period τ, then for each fixed s, $H(s, t)$ has period τ in t. [*Hint:* Use Theorem 9, Section 8–10.]

8. Prove that if $m < n$ and $g_i(t) = \sum_{j=1}^{m} a_{ij}e^{s_j t}$, $i = 1, \ldots, n$, where the a_{ij} and s_j are constants, then $g_1(t), \ldots, g_n(t)$ are linearly dependent for all t. [*Hint:* Show that there cannot be more than n linearly independent solutions of a homogeneous linear differential equation of order m, and that $g_1(t), \ldots, g_n(t)$ all satisfy such an equation.]

Answers

3. $s^{-1} + (1 + t^2)^{-1}(2s^{-3} - 2ts^{-2})$

4. (a) $\phi(t, u) = -e^{-t}(e^u + e^{2u}) + (1 + e^t)^{-1}(1 + e^u)^2$,

   $H(s, t) = (1 + e^t)^{-1}[(s + 1)^{-1}(1 - e^t) - (s + 2)^{-1}e^t + s^{-1}]$

   (c) $x_1(t) - x_2(t)$ and $x_1(t)$ or $x_2(t)$

5. (a) $[10(1 + e^t)]^{-1}[5 \sin t - 5 \cos t + e^t(4 \sin t - 2 \cos t)]$

   (b) $[6(1 + 2e^t)]^{-1}(4e^t - 3 - 6e^{-t} + 5e^{-2t})$

SUGGESTED REFERENCES

1. RICHARD BELLMAN, "On an Application of a Banach-Steinhaus Theorem to the Study of the Boundedness of Solutions of Nonlinear Differential and Difference Equations," *Annals of Mathematics*, 2nd Series, vol. 49, pp. 515–522. Princeton, N. J.: Princeton University Press, 1948.

2. RICHARD BELLMAN, *Stability Theory of Differential Equations*. New York: McGraw-Hill, 1953.

3. LAMBERTO CESARI, *Asymptotic Behavior and Stability Problems in Ordinary Differential Equations*. Berlin: Springer, 1959.

4. EARL A. CODDINGTON and NORMAN LEVINSON, *Theory of Ordinary Differential Equations*. New York: McGraw-Hill, 1955.

5. G. DOETSCH, *Theorie und Anwendung der Laplace Transformation*. Berlin: Springer, 1937.

6. EDWARD O. GILBERT, "A Method for the Symbolic Representation and Analysis of Linear Periodic Feedback Systems," *Transactions of American Institute of Electrical Engineers*, vol. 78, part II (1959) pp. 512–523.

7. H. J. GRAY, R. MERWIN, and J. G. BRAINERD, "Solutions of the Mathieu Equation." *Transactions of American Institute of Electrical Engineers*, vol. 67, part I (1948) pp. 429–441.

8. E. L. INCE, *Ordinary Differential Equations*. New York: Dover, 1956.

9. E. L. INCE, "Tables of the Elliptic Cylinder Functions." *Proceedings of the Royal Society of Edinburgh*, vol. 52 (1932) pp. 355–423.

10. WILFRED KAPLAN, *Ordinary Differential Equations*. Reading, Mass.: Addison-Wesley, 1958.

11. WILFRED KAPLAN, "Stability Theory." *Proceedings of the Symposium on Nonlinear Circuit Analysis*, vol. VI (1957), pp. 3–21. New York: Polytechnic Institute of Brooklyn.

12. KONRAD KNOPP, *Theory of Functions* (2 vols.), transl. by F. Bagemihl. New York: Dover, 1945.

13. C. C. MacDUFFEE, *The Theory of Matrices*. New York: Chelsea Publishing Co., 1946.

14. W. E. MILNE, *Numerical Solution of Differential Equations*. New York: John Wiley, 1953.

15. SAM PERLIS, *Theory of Matrices*. Reading, Mass.: Addison-Wesley, 1952.

16. O. PERRON, "Die Stabilitätsfragen bei Differenzialgleichungen." *Mathematische Zeitschrift*, vol. 32 (1930) pp. 703–728.

17. W. RUDIN, *Principles of Mathematical Analysis*. New York: McGraw-Hill, 1953.

18. *Transactions, Professional Group on Circuit Theory*, Institute of Radio Engineers, New York, vol. CT–2, No. 1, March, 1955. (special issue on time-variant linear systems).

19. E. T. WHITTAKER and G. N. WATSON, *A Course of Modern Analysis*, 4th ed. Cambridge, Eng.: Cambridge University Press, 1940.

20. LOTFI A. ZADEH, "Circuit Analysis of Linear Varying-parameter Networks," *Journal of Applied Physics*, vol. 21 (1950) pp. 1171–1177.

21. LOTFI A. ZADEH, "Initial Conditions in Linear Varying-parameter Systems," *Journal of Applied Physics*, vol. 22 (1951) pp. 782–786.

Appendixes

APPENDIX I

THE OPERATIONAL CALCULUS OF MIKUSINSKI

A-1 Introduction. In a series of writings from 1950 on, the mathematician Jan Mikusiński has evolved an operational calculus quite independent of Fourier or Laplace transforms. The formal procedures are very similar to those based on Laplace transforms, but considerably greater generality is achieved. Most remarkably, ordinary functions, generalized functions, and operators are all treated on the same basis as objects of one algebraic structure α, the field of "operators," as Mikusiński calls them. Furthermore, these objects can be added, subtracted, multiplied, and divided, whereby the multiplication corresponds to *convolution:*

$$fg = \int_0^t f(u)g(t-u)\,du,$$

and *not* to usual multiplication. In fact, usual multiplication of ordinary functions or of ordinary functions by generalized functions is *excluded* from the theory.

It is our purpose here to give a brief introduction to Mikusiński's theory. For a complete exposition, refer to References 1, 2, and 3.

A-2 Convolution as multiplication in the linear space $\mathcal{C}$. We consider continuous complex-valued functions $f(t)$, $g(t)$, ... on the interval $0 \leq t < \infty$. It will be convenient to consider each function to be zero for $t < 0$. To avoid confusion, we shall write $\{f(t)\}$, in braces, for the function whose value at each t is $f(t)$; in particular, $\{1\}$ denotes the function always equal to 1 for $t \geq 0$. As usual, we shall also abbreviate $\{f(t)\}$, $\{g(t)\}$, ... by $f, g, \ldots$

As noted in Section 2–2, our functions $f, g, \ldots$ form a linear space, which we denote by $\mathcal{C}$. Thus the operations of addition, $f + g$, and multiplication by a constant, αf, are defined and obey the familiar laws. Within $\mathcal{C}$ we also have the operation of convolution. *We regard convolution as a kind of multiplication and write $f \cdot g$ or fg for the convolution of f and g: $fg = p$,* where

$$\{p(t)\} = \left\{ \int_0^t f(u)g(t-u)\,du \right\}. \tag{A-20}$$

The notation $f * g$ will not be used, since the shorter notation presents various advantages and since multiplication in the familiar sense will not occur. We verify as in Section 2–8 that in $\mathbb{C}$ the following rules hold:

$$f + g = g + f, \quad f + (p + q) = (f + p) + q, \quad f + 0 = f,$$
$$\text{(A–21)}$$
$$fg = gf, \quad f(pq) = (fp)q, \quad f(p + q) = fp + fq, \quad f \cdot 0 = 0,$$

where $0 = \{0\}$. Furthermore, the equation $f + \phi = g$ has a unique solution for ϕ: $\phi = g - f$. There is one further remarkable property, not previously mentioned:

THEOREM OF TITCHMARSH. *If f, g are in $\mathbb{C}$ and $fg = 0$, then $f = 0$ or $g = 0$.*

This property is easy to deduce if f and g have Laplace transforms $F(s)$, $G(s)$. For then fg has transform $F(s)G(s)$. But $fg = 0$, so that $F(s)G(s) \equiv 0$. Since $F(s)$, $G(s)$ are analytic functions, we conclude that $F(s) \equiv 0$ or $G(s) \equiv 0$, and accordingly that $f = 0$ or $g = 0$. Since it is an essential feature of Mikusiński's theory to allow for functions f, g not having Laplace transforms [for example, e^{t^2}, $\exp(e^t)$], another proof is needed. Proofs have been given by various writers, beginning with Titchmarsh; one demonstration, along with bibliographical references, is included in Chapter II of Reference 3.

COROLLARY (Cancellation law). *If f, g_1, g_2 are in $\mathbb{C}$, and $f \neq 0$, then $fg_1 = fg_2$ implies $g_1 = g_2$.*

For $fg_1 = fg_2$ implies $fg_1 - fg_2 = 0$ or $f(g_1 - g_2) = 0$, so that $f = 0$ or $g_1 - g_2 = 0$. Since $f \neq 0$, we conclude that $g_1 - g_2 = 0$, or $g_1 = g_2$.

It follows, therefore, that if p, q are given functions in $\mathbb{C}$, then there is at most one function g in $\mathbb{C}$ such that

$$pg = q.$$

For example, if $p = \{t\}$, $q = \{t^4\}$, then *necessarily* $g = \{12t^2\}$. We can write

$$g = \frac{q}{p}.$$

Thus our "multiplication" (convolution) leads to a corresponding "division." However, division is not always possible, even when the denominator is different from zero. In fact, p/p is not defined in $\mathbb{C}$. Indeed, if $p \neq 0$ and $g = p/p$, so that $pg = p$, then for each t

$$p(t) = \int_0^t p(u)g(t - u) \, du.$$

Let t_1 be fixed and let M be the maximum of $\{g(t)\}$ for $0 \leq t \leq t_1$. Then for $0 \leq t \leq t_1$,

$$|p(t)| \leq \int_0^t M|p(u)|\,du.$$

We now apply Theorem 19 of Section 8–14, with $u = \{|p(t)|\}$, $v = \{M\}$, $c = 0$. Hence $|p(t)| \leq 0$ for $0 \leq t \leq t_1$, so that $p(t) \equiv 0$ for $0 \leq t \leq t_1$. Since t_1 is arbitrary, we conclude that $\{p(t)\} = 0$, contrary to hypothesis. Thus p/p has no meaning in $\mathcal{C}$. We remark that p/p can be interpreted as a generalized function, namely, $\delta = \{\delta(t)\}$; for $p\,\delta$ [the convolution of $\{p(t)\}$ with $\{\delta(t)\}$] is $\{p(t)\}$, as in Section 6–14. It is precisely in this manner that generalized functions enter the theory.

A–3 The operator field $\mathcal{Q}$. By the term *operator* we mean an indicated quotient q/p of two functions q, p in $\mathcal{C}$, where $p \neq 0$; we write $a = q/p$. Two operators $a_1 = q_1/p_1$, $a_2 = q_2/p_2$ are considered to be equal if $q_1 p_2 = q_2 p_1$:

$$\frac{q_1}{p_1} = \frac{q_2}{p_2} \quad \text{if } q_1 p_2 = q_2 p_1. \tag{A–30}$$

The sum $a_1 + a_2$ and product $a_1 \cdot a_2$ or $a_1 a_2$ of two operators $a_1 = q_1/p_1$, $a_2 = q_2/p_2$ are defined as follows:

$$a_1 + a_2 = \frac{q_1}{p_1} + \frac{q_2}{p_2} = \frac{q_1 p_2 + q_2 p_1}{p_1 p_2},$$

$$a_1 \cdot a_2 = \frac{q_1}{p_1} \cdot \frac{q_2}{p_2} = \frac{q_1 q_2}{p_1 p_2}. \tag{A–31}$$

It can be verified that these definitions are consistent with the definition of equality; that is, $a_1 + a_2$, $a_1 \cdot a_2$ are uniquely defined as operators. The class of operators, with the indicated operations of addition and multiplication, will be called the *operator field* $\mathcal{Q}$.

One can verify that the properties (A–21) hold for operators, where zero denotes the operator $\{0\}/\{1\}$. Furthermore, subtraction is uniquely defined. Thus $\mathcal{Q}$ has the same basic algebraic properties as $\mathcal{C}$. Also, division, other than by zero, is uniquely defined in $\mathcal{Q}$: if $a_1 = q_1/p_1 \neq 0$, $a_2 = q_2/p_2$, then the equation $a_1 a = a_2$ has a unique solution $a = a_2/a_1$ in $\mathcal{Q}$, namely,

$$a = \frac{q_2}{p_2} \div \frac{q_1}{p_1} = \frac{p_1 q_2}{p_2 q_1}. \tag{A–32}$$

In particular, if $a_1 \neq 0$, then a_1/a_1 is a uniquely defined operator, independent of a_1, which we denote by 1. The operator 1 has the basic property

$$a \cdot 1 = 1 \cdot a = a \quad \text{for every operator } a \text{ in } \mathcal{Q}. \tag{A–33}$$

The verification of the properties stated is straightforward. In fact, the passage from $\mathcal{C}$ to $\mathcal{Q}$ by forming quotients is in strict analogy with the

passage from integers to rational numbers (ratios of integers) in ordinary
algebra. The structure of e is that of a *ring* (addition, subtraction, and
multiplication, but no division), whereas the structure of a is that of a
field (all four operations), to employ the terminology of modern algebra.

Just as the integers are particular rational numbers $(1 = \frac{2}{2}, 2 = \frac{4}{2}, \ldots)$,
so can the functions $f, g, \ldots$ of e be considered to be particular operators
in a. In detail, we assign to each f in e the operator

$$a = \frac{\{1\}f}{\{1\}}.$$ (A-34)

It can be verified that $f = g$ in e if and only if the corresponding operators
in a are equal, and that the sum and product of two functions in e corresponds to the sum and product, respectively, of the corresponding
operators. Thus e can be considered as "imbedded" in a; that is, as a part
of a. Accordingly, a includes all the continuous functions on the interval
$0 \leqq t < \infty$; every function is a particular "operator."

However, a also includes genuine operators. Let

$$l = \{1\},$$ (A-35)

so that l is also in e. Then for every function f in e we have

$$lf = \left\{\int_0^t f(u)\, du\right\}.$$ (A-36)

Hence l is an *integration operator* (the operator D_0^{-1} of Section 1-6).

Now let

$$s = \frac{1}{l} = \frac{\text{operator } 1}{\text{operator } l}.$$ (A-37)

Let $f = \{f(t)\}$ be a function in e having a continuous derivative $f' = \{f'(t)\}$ and let $f(0) = 0$. Then $sf = f'$. Indeed, $f = \{\int_0^t f'(u)\, du\} = lf'$,
so that $sf = slf' = 1 f' = f'$. Thus, at least for the functions f considered,
s is a *differentiation operator*: $sf = Df$.

The operator 1 must be distinguished from the function $l = \{1\}$. Indeed,
1 can be written as p/p, where p is a function, but $1 = p/p$ cannot itself be
a function, as was shown above. We can in fact consider 1 to be the generalized function $\{\delta(t)\}$. As an operator, 1 acts very simply as the *identity operator*:

$$1f = f = \{f(t)\}.$$ (A-38)

We can introduce general *numerical operators* α, where α is a complex
number, by writing

$$\alpha = \frac{\{\alpha\}}{\{1\}}.$$ (A-39)

Then for every f in $\mathcal{C}$, α operates on f to yield

$$\alpha f = \{\alpha f(t)\}. \tag{A-38'}$$

Indeed,

$$\alpha f = \frac{\{\alpha\}}{\{1\}} \frac{\{1\}f}{\{1\}} = \frac{\{\alpha\}f}{\{1\}} = \frac{\{\alpha\}f}{l}$$

$$= s\{\alpha\}f = s\left\{\int_0^t \alpha f(u)\, du\right\} = \{\alpha f(t)\}.$$

Thus the scalars α (complex numbers) are included in $\mathcal{C}$, and we can verify that equality, sum, and product of scalars correspond to equality, sum, and product of the corresponding operators. We stress that each α, as an object in $\mathcal{C}$, is an operator, and not a function in $\mathcal{C}$. In fact, α can be identified with the generalized function $\{\alpha\,\delta(t)\}$. Exceptionally, for $\alpha = 0$,

$$\alpha = \frac{\{0\}}{\{1\}} = \{0\} = 0,$$

a function in $\mathcal{C}$.

A–4 Differential operators and linear differential equations. We have seen that $sf = Df$, when f has a continuous derivative and $f(0) = 0$. If $f(0) \neq 0$, we can write

$$\{f(t)\} = \left\{\int_0^t f'(u)\, du\right\} + \{f(0)\};$$

that is, $f = lf' + f(0)l$. Hence, since $sl = 1$,

$$sf = slf' + slf(0) = f' + f(0), \tag{A-40}$$

where $f(0)$ is, as above, a numerical operator. The formula (A–40) is equivalent to the familiar one

$$Df = f'(t) + f(0)\,\delta(t)$$

for differentiation in the context of generalized functions (Section 1–13). Thus s can still be interpreted as the operator of differentiation.

If f'' is in $\mathcal{C}$, we have (with $s^2 = s \cdot s$)

$$s^2 f = sf' + sf(0) = f'' + f'(0) + sf(0), \tag{A-41}$$

and in general, if $f^{(n)}$ is in $\mathcal{C}$,

$$s^n f = f^{(n)} + f^{(n-1)}(0) + sf^{(n-2)}(0) + \cdots + s^{n-1}f(0), \tag{A-42}$$

where $s^n = s \cdot s \cdots s$ (n times) and all derivatives at zero are derivatives to the right. Equivalently, we can write

$$f^{(n)} = s^n f - [f^{(n-1)}(0) + s f^{(n-2)}(0) + \cdots + s^{n-1} f(0)]. \quad \text{(A-42$'$)}$$

This formula parallels the familiar one for the Laplace transform of $D^n f$ (Section 6–5).

We can now form general *polynomial differential operators*

$$\alpha_0 s^n + \alpha_1 s^{n-1} + \cdots + \alpha_n \quad \text{(A-43)}$$

and *rational operators*

$$\frac{\alpha_0 s^n + \alpha_1 s^{n-1} + \cdots + \alpha_n}{\beta_0 s^m + \beta_1 s^{m-1} + \cdots + \beta_m}. \quad \text{(A-44)}$$

Here $\alpha_0, \alpha_1, \ldots, \beta_0, \beta_1, \ldots$ are numerical operators, $\beta_0 \neq 0$. The familiar algebraic operations can now be applied to these expressions: the polynomials can be factored, and the proper rational "functions" of s can be decomposed into partial fractions.

From the equation

$$s\{e^{\alpha t}\} = \alpha\{e^{\alpha t}\} + 1$$

we conclude that $(s - \alpha)\{e^{\alpha t}\} = 1$ or that

$$\frac{1}{s - \alpha} = \{e^{\alpha t}\}. \quad \text{(A-45)}$$

Accordingly,

$$\frac{1}{(s - \alpha)} \cdot \frac{1}{(s - \alpha)} = \frac{1}{(s - \alpha)^2} = \{e^{\alpha t}\}\{e^{\alpha t}\}$$

$$= \left\{ \int_0^t e^{\alpha u} e^{\alpha(t-u)}\, du \right\} = \{t e^{\alpha t}\},$$

and, in general,

$$\frac{1}{(s - \alpha)^n} = \left\{ \frac{t^{n-1} e^{\alpha t}}{(n-1)!} \right\} \quad (n = 1, 2, \ldots). \quad \text{(A-46)}$$

Thus each proper rational function of the operator s can be written as a function in $\mathcal{C}$.

The solutions of a linear differential equation with constant coefficients can now be found. For example, to find $\{x(t)\}$ such that

$$x'' + 3x' + 2x = f,$$

$$x(0) = 1, \qquad x'(0) = 0,$$

where f is in $\mathcal{C}$, we apply (A–42′) and rewrite the equation:

$$s^2 x - s + 3(sx - 1) + 2x = f,$$

$$x = \frac{s+3}{s^2 + 3s + 2} + \frac{f}{s^2 + 3s + 2}$$

$$= \frac{2}{s+1} - \frac{1}{s+2} + \left(\frac{1}{s+1} - \frac{1}{s+2}\right) f$$

$$= \{2e^{-t}\} - \{e^{-2t}\} + \{e^{-t} - e^{-2t}\}\{f(t)\}.$$

The result is the same as that obtained by Laplace transforms:

$$x = 2e^{-t} - e^{-2t} + W * f,$$

$$W = \mathcal{L}^{-1}[(s^2 + 3s + 2)^{-1}] = e^{-t} - e^{-2t}.$$

Similar techniques are applicable to simultaneous differential equations

A–5 Discontinuous functions and generalized functions in $\mathcal{C}$. Thus far we have identified only the continuous functions and the generalized functions $\{\alpha\delta(t)\}$ as members of $\mathcal{C}$. The discontinuous function $\{h(t - c)\}$, $c > 0$, can be identified with the derivative of the continuous function $\{(t - c)h(t - c)\}$, that is, with

$$s\{(t - c)h(t - c)\} = \frac{\{(t - c)h(t - c)\}}{l} = \frac{\{(t - c)h(t - c)\}}{\{1\}},$$

a ratio of two functions of $\mathcal{C}$. Thus we can consider $\{h(t - c)\}$ to be a member of $\mathcal{C}$.

Similarly, each piecewise continuous function $\{f(t)\}$ on the interval $0 \leq t < \infty$ can be represented as

$$\frac{\left\{\int_0^t f(u) \, du\right\}}{l}.$$

Thus $\mathcal{C}$ can be considered to include all piecewise continuous functions. It should be noted that the value of the piecewise continuous function at each discontinuity is ignored in this representation; that is, two piecewise continuous functions which are equal except at jump points should be considered to be the same function. The sum and convolution of two piecewise continuous functions can be shown to be represented by the sum and product, respectively, of the corresponding objects in $\mathcal{C}$.

To the derivative of the discontinuous function $\{h(t - c)\}$, that is, to $\{\delta(t - c)\}$, we assign the operator

$$e^{-cs} = s\{h(t - c)\} \quad (c > 0). \tag{A-50}$$

This is a *definition* of the symbol e^{-cs}, which has otherwise no meaning (s is not a complex variable, but an operator). The notation is justified by the property

$$e^{-c_1 s} e^{-c_2 s} = e^{-(c_1 + c_2)s}, \tag{A-51}$$

which can be immediately verified. We can also define e^{0s} to be 1, since $s\{h(t)\} = sl = 1$, and $\{\delta(t - c)\}$ now corresponds to e^{-cs} for $c \geq 0$.

The "exponential function" thus introduced is a *translation operator;* that is,

$$e^{-cs}\{f(t)\} = \{f(t - c)\} \quad (c \geq 0), \tag{A-52}$$

for each f in $\mathcal{C}$, and indeed for each piecewise continuous function f. For

$$e^{-cs}f = s\{h(t - c)\}f = s\left\{\int_0^t h(u - c)f(t - u)\,du\right\}$$

$$= s\left\{\int_c^t f(t - u)\,du\,h(t - c)\right\}$$

$$= s\left\{\int_0^{t-c} f(v)\,dv\,h(t - c)\right\} = \{f(t - c)\}.$$

In particular, $e^{-cs}\{h(t)\} = e^{-cs}l = \{h(t - c)\}$, so that

$$\{h(t - c)\} = \frac{e^{-cs}}{s}. \tag{A-53}$$

The derivatives $\{\delta'(t - c)\}$, $\{\delta''(t - c)\}$, ... can also be considered to be included in $\mathcal{C}$, the corresponding operators being se^{-cs}, $s^2 e^{-cs}$, ...:

$$\{\delta^{(k)}(t - c)\} = s^k e^{-cs} \quad (k = 0, 1, \ldots, c \geq 0). \tag{A-54}$$

The operator relations

$$s^m e^{-\alpha s} \cdot s^n e^{-\beta s} = s^{m+n} e^{-(\alpha + \beta)s},$$

$$s^k e^{-cs}\{f(t)\} = \{f^{(k)}(t - c)\}$$

are then equivalent to the convolution formulas (6–148) in Section 6–14.

It follows that $\mathcal{C}$ includes all continuous and piecewise continuous functions on the interval $0 \leq t < \infty$, all generalized functions which are linear combinations of the functions $\{\delta^{(k)}(t - c)\}$, $c \geq 0$, and integration and differentiation operators. Since the operator s corresponds to differentiation in the context of generalized functions, we can say that $\mathcal{Q}$ includes so many generalized functions that every continuous function has derivatives of every order within $\mathcal{Q}$.

Furthermore, we have already identified certain continuous and piecewise continuous functions as "functions" of the operator s:

$$\left\{\frac{t^{n-1}e^{\alpha t}}{n!}\right\} = \frac{1}{(s-\alpha)^n}, \qquad \{h(t-c)\} = \frac{e^{-cs}}{s}.$$

By superposition and translation (multiplication by e^{-cs}) we obtain a larger collection of functions with known representation in terms of s:

$$\{h(t) - h(t-c)\} = \frac{1 - e^{-cs}}{s},$$

$$\{t[h(t) - h(t-c)]\} = \frac{1 - e^{-cs} - cse^{-cs}}{s^2}, \qquad (A-55)$$

$$\{\cos \beta t\} = \left\{\frac{e^{\beta it} + e^{-\beta it}}{2}\right\} = \frac{s}{s^2 + \beta^2}, \qquad \{\sin \beta t\} = \frac{\beta}{s^2 + \beta^2}.$$

If $f(t)$ is periodic for $t \geqq 0$, with period τ, and $f(t) = g(t)$ for $0 \leqq t < \tau$, $g(t)$ being zero outside this interval, then $\{f(t)\} = \{g(t)\} + \{f(t-\tau)\}$, so that in $\mathcal{Q}$ we have $f = g + e^{-\tau s}f$ or

$$f = \frac{g}{1 - e^{-\tau s}}. \qquad (A-56)$$

For example, a *square wave f* of period $\tau = 2c$ for which $\{g(t)\} = \{h(t) - h(t-c)\}$ has the representation

$$f = \frac{1 - e^{-cs}}{s(1 - e^{-2cs})} = \frac{1}{s(1 + e^{-cs})}. \qquad (A-57)$$

A *sawtooth wave f* of period c for which $\{g(t)\} = \{t[h(t) - h(t-c)]\}$ has the representation

$$f = \frac{1 - e^{-cs} - cse^{-cs}}{s^2(1 - e^{-cs})}. \qquad (A-58)$$

It is evident that each representation given thus far matches the corresponding formula for the Laplace transform. The parallel can be further extended by giving a meaning to a series $\sum_0^\infty c_n n! s^{-n-1}$ corresponding to a function $F(s)$ analytic at ∞, and then showing that $F(s)$ as operator in $\mathcal{Q}$ is equal to the function $\{f(t)\} = \{\sum_0^\infty c_n t^n\}$.

Suggested References

1. A. Erdélyi, "From Delta Functions to Distributions," in *Modern Mathematics for the Engineer*, Second Series, E. F. Beckenbach, ed., pp. 5–50. New York: McGraw-Hill, 1961.

2. A. Erdélyi, *Operational Calculus and Generalized Functions.* New York: Holt, Rinehart and Winston, 1962.

3. Jan Mikusiński, *Operational Calculus*, 5th ed. London: Pergamon Press, 1959.

Appendix II

APPENDIX II

RECAPITULATION OF PRINCIPAL TABLES

In this appendix, we repeat the principal tables which occur in the text. These are Table 4–1 from p. 194; Table 4–2 from pp. 196–197; Table 4–3 from pp. 204–205; Table 4–4 from p. 210; Table 5–1 from pp. 256–257; Table 5–2 from pp. 266–267; Table 5–3 from pp. 282–283; Table 6–1 from pp. 306–308; Table 6–2 from pp. 320–321; Table 6–4 from pp. 344–345; Table 6–5 from pp. 378–379.

TABLE 4–1

TRUNCATED LAPLACE TRANSFORMS

No.	$f(t)$	$\mathcal{L}_\tau[f] = \int_0^\tau f(t)e^{-st}\,dt$ $= F_\tau(s) \quad (\tau > 0)$
1	$h(t)$	$\dfrac{1 - e^{-s\tau}}{s}$
2	t	$\dfrac{1 - e^{-s\tau}(1 + s\tau)}{s^2}$
3	$t^k \quad (k = 1, 2, \ldots)$	$\dfrac{k! - e^{-s\tau}g_k(s\tau)}{s^{k+1}}$ (See Note)
4	e^{at}	$\dfrac{1 - e^{(a-s)\tau}}{s - a}$
5	$t^k e^{at} \quad (k = 1, 2, \ldots)$	$\dfrac{k! - e^{(a-s)\tau}g_k((s - a)\tau)}{(s - a)^{k+1}}$
6	$h(t - c) \quad (c > 0)$	$\dfrac{e^{-sc} - e^{-s\tau}}{s}h(\tau - c)$

541

Table 4-2
Finite Fourier Transforms

No.	$f(t)$	$\phi(n) = \Phi_\tau[f] = \int_0^\tau f(t)e^{-in\omega t}\,dt$	Remarks
1	1	$\dfrac{1 - e^{-2\pi in}}{in\omega}$	$\phi(0) = \tau$ $\phi(n) = 0$ for $n \neq 0$
2	t	$\dfrac{1 - e^{-2\pi in}(1 + 2\pi in)}{(in\omega)^2}$	$\phi(0) = \tau^2/2$ $\phi(n) = 2\pi i/(n\omega^2)$ for $n \neq 0$
3	t^k, $k = 1, 2, \ldots$	$\dfrac{k! - e^{-2\pi in} g_k(2\pi in)}{(in\omega)^{k+1}}$	$\phi(0) = \tau^{k+1}/(k+1)$ $\phi(n) = [k! - g_k(2\pi in)]/(in\omega)^{k+1}$ for $n \neq 0$
4	e^{at}, $a - in\omega \neq 0$	$\dfrac{e^{(a-in\omega)\tau} - 1}{a - in\omega}$	$\phi(n) = \dfrac{e^{a\tau} - 1}{a - in\omega}$
5	$t^k e^{at}$, $a - in\omega \neq 0$, $k = 1, 2, \ldots$	$\dfrac{k! - e^{(a-in\omega)\tau} g_k[(in\omega - a)\tau]}{(in\omega - a)^{k+1}}$	$\phi(n) = \dfrac{k! - e^{a\tau} g_k(2\pi in - a\tau)}{(in\omega - a)^{k+1}}$
6	$h(t - c)$, $0 < c < \tau$	$\dfrac{e^{-in\omega c} - e^{-2\pi in}}{in\omega}$	$\phi(0) = \tau - c$ $\phi(n) = \dfrac{e^{-in\omega c} - 1}{in\omega}$ for $n \neq 0$

7	$(at+b)[h(t-\alpha) - h(t-\beta)]$, $0 \leq \alpha < \beta \leq \tau$	$(n^2\omega^2)^{-1}\{e^{-\beta in\omega}[(a\beta + b)in\omega + a] - e^{-\alpha in\omega}[(a\alpha + b)in\omega + a]\}$	$\phi(0) = \frac{1}{2}a(\beta^2 - \alpha^2) + b(\beta - \alpha)$
8	$t[h(t) - h(t - \frac{1}{2}\tau)]$ $+ (\tau - t)[h(t - \frac{1}{2}\tau) - h(t - \tau)]$	$\left(\dfrac{1 - e^{-\pi in}}{in\omega}\right)^2$	$\phi(0) = \tau^2/4$ $\phi(n) = \dfrac{2[(-1)^n - 1]}{n^2\omega^2}$ for $n \neq 0$
9	$t[1 - h(t-\alpha)] + \alpha h(t-\alpha)$, $0 \leq \alpha \leq \tau$	$\dfrac{e^{-\alpha in\omega} + \alpha in\omega e^{-2\pi in} - 1}{n^2\omega^2}$	$\phi(0) = \alpha\tau - \frac{1}{2}\alpha^2$ $\phi(n) = \dfrac{e^{-\alpha in\omega} + \alpha in\omega - 1}{n^2\omega^2}$ for $n \neq 0$
10	$\tau - \alpha + (\alpha - t)h(t-\alpha)$, $0 \leq \alpha \leq \tau$	$\dfrac{e^{-in\omega\alpha} - e^{-2\pi in} - in\omega(\tau - \alpha)}{n^2\omega^2}$	$\phi(0) = \dfrac{\tau^2 - \alpha^2}{2}$ $\phi(n) = \dfrac{e^{-in\omega\alpha} - 1 - 2\pi in + in\omega\alpha}{n^2\omega^2}$ for $n \neq 0$
11	$[h(t - \alpha) - h(t - \beta)] \cdot$ $\sin(at + b)$, $0 \leq \alpha < \beta \leq \tau$, a, b real, $a - n\omega \neq 0$	$(a^2 - n^2\omega^2)^{-1}[e^{-in\omega\alpha}(a\cos\theta + in\omega\sin\theta) - e^{-in\omega\beta}(a\cos\mu + in\omega\sin\mu)]$, $\theta = a\alpha + b$, $\mu = a\beta + b$	
12	$\left[h(t) - h\left(t - \dfrac{\pi}{a}\right)\right]\sin at$, $a \neq n\omega$, $a > \pi/\tau$	$\dfrac{a}{a^2 - n^2\omega^2}(1 + e^{-in\omega\pi/a})$	

TABLE 4-3

f-Convolutions

No.	$f(t), \; 0 \leqq t < \tau$	$\gamma_a e^{at} \, \Delta f, \; 0 \leqq t < \tau$
1	$h(t)$	$-a^{-1}h(t)$
2	t	$a^{-2}(-1 - at - \gamma_a a\tau e^{at})$
3	$t^k \;\; (k = 1, 2, \ldots)$	$a^{-k-1}[k!e^{at} - g_k(at) + \gamma_a e^{at}\{k!e^{at} - g_k(a\tau)\}]$
4	$e^{bt} \;\; (b \neq a)$	$(b - a)^{-1}[e^{bt} - e^{at} + \gamma_a e^{at}(e^{b\tau} - e^{a\tau})]$
5	$t^k e^{bt} \;\; (k = 1, 2, \ldots)$	$(a - b)^{-k-1}(k!e^{at} - e^{bt}g_k[(a - b)t] + \gamma_a e^{at}\{k!e^{a\tau} - e^{b\tau}g_k[(a - b)\tau]\})$
6	$h(t - c), \;\; 0 \leqq c \leqq \tau$	$a^{-1}[(e^{a(t-c)} - 1)h(t - c) - \gamma_a e^{at}(1 - e^{a(\tau-c)})]$
7	$(At + B)[h(t - \alpha) - h(t - \beta)],$ $0 \leqq \alpha < \beta \leqq \tau$	$Aa^{-2}\{e^{a(t-\alpha)}(1 + a\alpha)h(t - \alpha) - e^{a(t-\beta)}(1 + a\beta)h(t - \beta)$ $- (1 + at)[h(t - \alpha) - h(t - \beta)] + \gamma_a e^{at}[e^{a(\tau-\alpha)}(1 + a\alpha)$ $- e^{a(\tau-\beta)}(1 + a\beta)]\} + Ba^{-1}\{[e^{a(t-\alpha)} - 1]h(t - \alpha)$ $- [e^{a(t-\beta)} - 1]h(t - \beta) + \gamma_a e^{at}[e^{a(\tau-\alpha)} - e^{a(\tau-\beta)}]\}$

8	$t[h(t) - h(t - \frac{1}{2}\tau)]$ $+ (\tau - t)h(t - \frac{1}{2}\tau)]$	$a^{-2}\{e^{at} - 1 - at + [2 + 2at - a\tau - 2e^{a(2t-\tau)/2}]h(t - \frac{1}{2}\tau)$ $+ \gamma_a e^{at}[e^{a\tau} - 2e^{a\tau/2} + 1]\}$
9	$t[1 - h(t - \alpha)] + \alpha h(t - \alpha)$, $0 \leqq \alpha \leqq \tau$	$a^{-2}\{(1 + at)[h(t - \alpha) - 1] - [e^{a(t-\alpha)} + a\alpha]h(t - \alpha)$ $- \gamma_a e^{at}[e^{a(\tau-\alpha)} - 1 + a\alpha]\}$
10	$\tau - \alpha - (t - \alpha)h(t - \alpha)$, $0 \leqq \alpha \leqq \tau$	$-a^{-2}\{[e^{a(t-\alpha)} - 1 - at + a\alpha]h(t - \alpha) + \gamma_a e^{at}[e^{a(\tau-\alpha)} - 1]$ $+ (a\tau - a\alpha)(1 - \gamma_a e^{at})\}$
11	$[h(t - \alpha) - h(t - \beta)]\sin(At + B)$, A, B real, $0 \leqq \alpha < \beta \leqq \tau$	$(a^2 + A^2)^{-1}e^{at}([e^{-a\alpha}\{a\sin\eta + A\cos\eta\} - e^{-at}\{a\sin(At + B)$ $+ A\cos(At + B)\}][h(t - \alpha) - h(t - \beta)] + [e^{-a\alpha}\{a\sin\eta + A\cos\eta\}$ $- e^{-a\beta}\{a\sin\zeta + A\cos\zeta\}][\gamma_a e^{a\tau} + h(t - \beta)]),$ $\eta = A\alpha + B, \quad \zeta = A\beta + B$
12	$\left[h(t) - h\left(t - \frac{\pi}{A}\right)\right]\sin At$, $A > \pi/\tau$	$(a^2 + A^2)^{-1}\left\{[ke^{at} - a\sin At - A\cos At]\left[h(t) - h\left(t - \frac{\pi}{A}\right)\right]\right.$ $\left. + e^{at}[Ae^{-a\pi/A} + k]h\left(t - \frac{\pi}{A}\right)\right\}, \quad k = A\gamma_a(1 + e^{a\tau}e^{-a\pi/A})$

TABLE 4–4

INVERSE FINITE FOURIER TRANSFORMS

No.	$\phi(n)$	$\Phi_\tau^{-1}[\phi] = f = f(t), \quad 0 \leqq t < \tau$
1	$1, \quad n = 0$ $0, \quad n \neq 0$	$\dfrac{1}{\tau} h(t) = \dfrac{1}{\tau}$
2	$0, \quad n = 0; \quad \dfrac{1}{n}, \quad n \neq 0$	$\dfrac{2\pi i}{\tau}\left(\dfrac{1}{2} - \dfrac{t}{\tau}\right)$
3	$0, \quad n = 0; \quad \dfrac{1}{n^2}, \quad n \neq 0$	$\dfrac{(2\pi i)^2}{\tau}\left(-\dfrac{1}{12} + \dfrac{1}{2}\dfrac{t}{\tau} - \dfrac{1}{2}\dfrac{t^2}{\tau^2}\right)$
4	$0, \quad n = 0; \quad \dfrac{1}{n^3}, \quad n \neq 0$	$\dfrac{(2\pi i)^3}{\tau}\left(-\dfrac{1}{12}\dfrac{t}{\tau} + \dfrac{1}{4}\dfrac{t^2}{\tau^2} - \dfrac{1}{6}\dfrac{t^3}{\tau^3}\right)$
5	$0, \quad n = k; \quad \dfrac{1}{n-k}, \quad n \neq k$	$e^{ik\omega t}\dfrac{2\pi i}{\tau}\left(\dfrac{1}{2} - \dfrac{t}{\tau}\right)$
6	$\dfrac{1}{n-b}, \quad b \neq 0, \pm 1, \ldots$	$\dfrac{i\omega e^{bi\omega t}}{1-q}, \quad q = e^{2\pi bi}$
7	$\dfrac{1}{(n-b)^2}, \quad b \neq 0, \pm 1, \ldots$	$\dfrac{e^{bi\omega t}[(q-1)\omega^2 t - 2\pi\omega q]}{(q-1)^2},$ $q = e^{2\pi bi}$
8	$\dfrac{1}{in\omega - a}, \quad \dfrac{ai}{\omega} \neq 0, \pm 1, \ldots$	$\gamma_a e^{at}, \quad \gamma_a = (1 - e^{a\tau})^{-1}$
9	$\dfrac{1}{(in\omega - a)^2}, \quad \dfrac{ai}{\omega} \neq 0, \pm 1, \ldots$	$\gamma_a e^{at}[t + \tau(\gamma_a - 1)]$
10	$\dfrac{1}{(in\omega - a)^k}, \quad \dfrac{ai}{\omega} \neq 0, \pm 1, \ldots$	$\dfrac{1}{(k-1)!}\dfrac{\partial^{k-1}}{\partial a^{k-1}}(\gamma_a e^{at}), \quad k = 1, 2, \ldots$
11	$\dfrac{1}{(in\omega - \alpha)^2 + \beta^2},$ α, β real, $\beta \neq 0,$ $\dfrac{(\alpha i - \beta)}{\omega} \neq 0, \pm 1, \ldots,$	$\dfrac{e^{\alpha t}[\sin \beta t - e^{\alpha\tau}\sin\beta(t-\tau)]}{\beta(1 + e^{2\alpha\tau} - 2e^{\alpha\tau}\cos\beta\tau)}$

TABLE 5–1

FOURIER TRANSFORMS

No.	$f(t)$	$\Phi[f] = \int_{-\infty}^{\infty} f(t)e^{-i\omega t}\, dt = \phi(\omega)$
1	$h(t - c_1) - h(t - c_2)$, $c_1 < c_2$	$\dfrac{e^{-i\omega c_1} - e^{-i\omega c_2}}{i\omega}$
2	$h(t + c) - h(t - c)$, $c > 0$	$\dfrac{2\sin c\omega}{\omega}$
3	$e^{at}[h(t - c_1) - h(t - c_2)]$, $c_1 < c_2$	$\dfrac{e^{(a-i\omega)c_2} - e^{(a-i\omega)c_1}}{a - i\omega}$
4	$e^{at}h(t)$, $\operatorname{Re}(a) < 0$	$\dfrac{1}{i\omega - a}$
5	$e^{at}h(t - c)$, $\operatorname{Re}(a) < 0$	$\dfrac{e^{(a-i\omega)c}}{i\omega - a}$
6	$e^{-at}h(-t)$, $\operatorname{Re}(a) < 0$	$\dfrac{-1}{i\omega + a}$
7	$e^{a|t|}$, $\operatorname{Re}(a) < 0$	$\dfrac{-2a}{\omega^2 + a^2}$
8	$e^{at}h(t) - e^{-at}h(-t)$, $\operatorname{Re}(a) < 0$	$\dfrac{-2i\omega}{\omega^2 + a^2}$
9	$t^k e^{at}h(t)$, $k = 1, 2, \ldots$, $\operatorname{Re}(a) < 0$	$\dfrac{k!}{(i\omega - a)^{k+1}}$
10	$e^{ibt}[h(t + c) - h(t - c)]$, $c > 0$	$\dfrac{2\sin c(\omega - b)}{\omega - b}$
11	$\dfrac{1}{a^2 + t^2}$, $\operatorname{Re}(a) < 0$	$-\dfrac{\pi}{a}e^{a|\omega|}$
12	$\dfrac{t}{(a^2 + t^2)^2}$, $\operatorname{Re}(a) < 0$	$\dfrac{i\omega\pi}{2a}e^{a|\omega|}$
13	$\dfrac{e^{ibt}}{a^2 + t^2}$, $\operatorname{Re}(a) < 0$, b real	$-\dfrac{\pi}{a}e^{a|\omega - b|}$

(Continued)

TABLE 5–1 *Continued*

No.	$f(t)$	$\Phi[f] = \int_{-\infty}^{\infty} f(t)e^{-i\omega t}\, dt = \phi(\omega)$
14	$\dfrac{\cos bt}{a^2 + t^2},\quad \operatorname{Re}(a) < 0,\quad b \text{ real}$	$-\dfrac{\pi}{2a}[e^{a|\omega - b|} + e^{a|\omega + b|}]$
15	$\dfrac{\sin bt}{a^2 + t^2},\quad \operatorname{Re}(a) < 0,\quad b \text{ real}$	$-\dfrac{\pi}{2ai}[e^{a|\omega - b|} - e^{a|\omega + b|}]$
16	$t[h(t) - h(t - c)]$	$\dfrac{1 - e^{-i\omega c}(1 + i\omega c)}{-\omega^2}$
17	$t^k[h(t) - h(t - c)],$ $k = 1, 2, \ldots$	$\dfrac{k! - e^{-i\omega c}g_k(i\omega c)}{(i\omega)^{k+1}},$ $g_k(x) = x^k + kx^{k-1} + \cdots + k!$
18	$t[h(t) - h(t - c)] + (2c - t)$ $\times\, [h(t - c) - h(t - 2c)]$	$\dfrac{1 - 2e^{-i\omega c} + e^{-2i\omega c}}{-\omega^2}$
19	$e^{-at^2},\quad a > 0$	$\sqrt{\dfrac{\pi}{a}}\, e^{-\omega^2/(4a)}$

TABLE 5–2

CONVOLUTIONS

No.	$f(t)$, $-\infty < t < \infty$	$[e^{st}h(t)] * f(t)$, $\operatorname{Re}(s) < 0$
1	1	$-\dfrac{1}{s}$
2	t	$\dfrac{-1 - st}{s^2}$,
3	t^k, $k = 1, 2, \ldots$	$\dfrac{-g_k(st)}{s^{k+1}}$, $\quad g_k(x) = x^k + kx^{k-1} + \cdots$
4	e^{at}, $\operatorname{Re}(a) \geqq 0$	$\dfrac{e^{at}}{a - s}$
5	$t^k e^{at}$, $\operatorname{Re}(a) \geqq 0$	$\dfrac{-e^{at} g_k[(s - a)t]}{(s - a)^{k+1}}$
6	$\cos bt$, b real	$\dfrac{b \sin bt - s \cos bt}{s^2 + b^2}$
7	$\sin bt$, b real	$\dfrac{-b \cos bt - s \sin bt}{s^2 + b^2}$
8	$h(t)$	$\dfrac{e^{st} - 1}{s} h(t)$
9	$h(t - c)$	$\dfrac{e^{s(t-c)} - 1}{s} h(t - c)$
10	$h(t) - h(t - c)$	$\dfrac{e^{st} - 1}{s}[h(t) - h(t - c)]$ $\quad - \dfrac{e^{s(t-c)} - e^{st}}{s} h(t - c)$
11	$th(t)$	$\dfrac{e^{st} - 1 - st}{s^2} h(t)$
12	$(t - c)h(t - c)$	$\dfrac{e^{s(t-c)} - 1 - s(t - c)}{s^2} h(t - c)$

(Continued)

TABLE 5-2 *Continued*

No.	$f(t), \quad -\infty < t < \infty$	$[e^{st}h(t)] * f(t), \quad \text{Re}\,(s) < 0$
13	$t[h(t) - h(t - c)]$	$\dfrac{e^{st} - 1 - st}{s^2}[h(t) - h(t - c)]$ $+ \dfrac{e^{st}[1 - (1 + cs)e^{-sc}]}{s^2} h(t - c)$
14	$t[h(t) - h(t - c)] + (2c - t)$ $\times [h(t - c) - h(t - 2c)]$	$\dfrac{e^{st} - 1 - st}{s^2}[h(t) - h(t - c)]$ $+ \dfrac{e^{st} - 2e^{s(t-c)} + 1 + st - 2cs}{s^2}$ $\times [h(t - c) - h(t - 2c)]$ $+ \dfrac{e^{st}(1 - 2e^{-cs} + e^{-2cs})}{s^2}$ $\times h(t - 2c)$
15	$e^{at}h(t)$	$\dfrac{e^{at} - e^{st}}{a - s} h(t), \quad s \neq a$ $te^{st}h(t) \text{ for } s = a$
16	$t^k e^{at}h(t), \quad k = 1, 2, \ldots$	$\dfrac{k!e^{st} - e^{at}g_k[(s - a)t]}{(s - a)^{k+1}} h(t),$ $s \neq a,$ $\dfrac{e^{st}t^{k+1}}{k + 1} \text{ for } s = a$
17	$e^{at}h(-t), \quad \text{Re}\,(a) \geq 0$	$\dfrac{e^{at}[1 - h(t)] + e^{st}h(t)}{a - s}$

TABLE 5–3

FOURIER TRANSFORMS OF GENERALIZED FUNCTIONS*

No.	$f(t)$	$\phi(\omega) = \Phi[f]$
1	$\delta(t)$	1
2	$\delta(t - c)$	$e^{-ic\omega}$
3	$\delta^{(n)}(t)$	$(i\omega)^n$
4	$\delta^{(n)}(t - c)$	$e^{-ic\omega}(i\omega)^n$
5	1	$2\pi\,\delta(\omega)$
6	t	$2\pi i\,\delta'(\omega)$
7	t^n	$2\pi i^n\,\delta^{(n)}(\omega)$
8	$e^{i\alpha t}$	$2\pi\,\delta(\omega - \alpha)$
9	$t^n e^{i\alpha t}$	$2\pi i^n\,\delta^{(n)}(\omega - \alpha)$
10	$h(t)$	$\dfrac{1}{i\omega} + \pi\,\delta(\omega)$
11	$t^n h(t)$	$\dfrac{n!}{(i\omega)^{n+1}} + \pi i^n\,\delta^{(n)}(\omega)$
12	$h(t - c)$	$\dfrac{e^{-i\omega c}}{i\omega} + \pi\,\delta(\omega)$
13	$(t - c)^n h(t - c)$	$\dfrac{n!e^{-ic\omega}}{(i\omega)^{n+1}} + \pi i^n \displaystyle\sum_{r=0}^{n} \binom{n}{r}(ic)^{n-r}\,\delta^{(r)}(\omega)$
14	$e^{i\alpha t}h(t - c)$	$\dfrac{e^{-ic(\omega-\alpha)}}{i(\omega - \alpha)} + \pi\,\delta(\omega - \alpha)$
15	$e^{i\alpha t}t^n h(t)$	$\dfrac{n!}{[i(\omega - \alpha)]^{n+1}} + \pi i^n\,\delta^{(n)}(\omega - \alpha)$
16	$e^{i\alpha t}(t - c)^n h(t - c)$	$\dfrac{n!e^{-ic(\omega-\alpha)}}{[i(\omega - \alpha)]^{n+1}} + \pi i^n \displaystyle\sum_{r=0}^{n} \binom{n}{r}$ $\times (ic)^{n-r}\,\delta^{(r)}(\omega - \alpha)$
17	$\dfrac{1}{t}$	$\pi i - 2\pi i h(\omega)$

* *Note:* Throughout, $n = 1, 2, 3, \ldots$, and c and α are real constants.

(Continued)

TABLE 5–3 *Continued*

No.	$f(t)$	$\phi(\omega) = \Phi[f]$
18	$\dfrac{1}{t^n}$	$\dfrac{(-i\omega)^{n-1}}{(n-1)!}\,[\pi i - 2\pi i h(\omega)]$
19	$\dfrac{1}{t-c}$	$e^{-ic\omega}[\pi i - 2\pi i h(\omega)]$
20	$\dfrac{1}{(t-c)^n}$	$\dfrac{e^{-ic\omega}(-i\omega)^{n-1}}{(n-1)!}\,[\pi i - 2\pi i h(\omega)]$
21	$\dfrac{e^{i\alpha t}}{t-c}$	$e^{-ic(\omega-\alpha)}[\pi i - 2\pi i h(\omega-\alpha)]$
22	$\dfrac{e^{i\alpha t}}{(t-c)^n}$	$\dfrac{e^{-ic(\omega-\alpha)}[-i(\omega-\alpha)]^{n-1}}{(n-1)!}\,[\pi i - 2\pi i h(\omega-\alpha)]$

TABLE 6–1 LAPLACE TRANSFORMS

No.	$f(t)$ for $t \geqq 0$	$\mathcal{L}[f] = \phi(s) = \int_0^\infty f(t)e^{-st}\,dt$	Range of σ
1	1	$\dfrac{1}{s}$	$\sigma > 0$
2	e^{at}	$\dfrac{1}{s-a}$	$\sigma > \mathrm{Re}\,(a)$
3	$t^n,\ n > -1$	$\dfrac{n!}{s^{n+1}},\ n = 0, 1, \ldots,$ $\dfrac{\Gamma(n+1)}{s^{n+1}},$ any $n > -1$	$\sigma > 0$
4	$t^n e^{at},\ n > -1$	$\dfrac{n!}{(s-a)^{n+1}},\ n = 0, 1, \ldots,$ $\dfrac{\Gamma(n+1)}{(s-a)^{n+1}},$ any $n > -1$	$\sigma > \mathrm{Re}\,(a)$
5	$\cosh at$	$\dfrac{s}{s^2 - a^2}$	$\sigma > |\mathrm{Re}\,(a)|$
6	$\sinh at$	$\dfrac{a}{s^2 - a^2}$	$\sigma > |\mathrm{Re}\,(a)|$
7	$\cos at$	$\dfrac{s}{s^2 + a^2}$	$\sigma > |\mathrm{Im}\,(a)|$
8	$\sin at$	$\dfrac{a}{s^2 + a^2}$	$\sigma > |\mathrm{Im}\,(a)|$
9	$t^n \cos at,\ n > -1$	$\dfrac{\Gamma(n+1)}{2(s^2+a^2)^{n+1}}[(s+ai)^{n+1} + (s-ai)^{n+1}]$	$\sigma > |\mathrm{Im}\,(a)|$

(Continued)

TABLE 6–1 LAPLACE TRANSFORMS (*Continued*)

No.	$f(t)$ for $t \geqq 0$	$\mathcal{L}[f] = \phi(s) = \int_0^\infty f(t)e^{-st}\,dt$	Range of σ
10	$t^n \sin at, \quad n > -1$	$\dfrac{\Gamma(n+1)}{2i(s^2+a^2)^{n+1}}\left[(s+ai)^{n+1} - (s-ai)^{n+1}\right]$	$\sigma > \lvert \mathrm{Im}\,(a)\rvert$
11	$\cos^2 t$	$\dfrac{1}{2}\left(\dfrac{1}{s} + \dfrac{s}{s^2+4}\right)$	$\sigma > 0$
12	$\sin^2 t$	$\dfrac{1}{2}\left(\dfrac{1}{s} - \dfrac{s}{s^2+4}\right)$	$\sigma > 0$
13	$\sin at \sin bt$	$\dfrac{2abs}{[s^2+(a+b)^2][s^2+(a-b)^2]}$	$\sigma > \lvert \mathrm{Im}\,(a+b)\rvert$ and $\sigma > \lvert \mathrm{Im}\,(a-b)\rvert$
14	$e^{at}\sin(bt+c), \quad a, b, c$ real	$\dfrac{(s-a)\sin c + b\cos c}{(s-a)^2+b^2}$	$\sigma > a$
15	1 for $2nc \leq t < (2n+1)c$, 0 for $(2n+1)c \leq t < (2n+2)c$, $n = 0, 1, 2, \ldots, c > 0$ (square wave)	$\dfrac{1}{s(1+e^{-cs})}$	$\sigma > 0$
16	$h(t) - h(t-c), \quad c > 0$	$\dfrac{1-e^{-cs}}{s}$	$-\infty < \sigma < \infty$
17	$h(t-c), \quad c > 0$	$\dfrac{e^{-cs}}{s}$	$\sigma > 0$
18	$h(t-c_1) - h(t-c_2), \quad 0 < c_1 < c_2$	$\dfrac{e^{-c_1 s} - e^{-c_2 s}}{s}$	$-\infty < \sigma < \infty$

	$f(t)$	$F(s)$	
19	$e^{at}[h(t-c_1) - h(t-c_2)]$, $0 < c_1 < c_2$	$\dfrac{e^{-c_1(s-a)} - e^{c_2(s-a)}}{s-a}$	$-\infty < \sigma < \infty$
20	$e^{at}h(t-c)$, $c > 0$	$\dfrac{e^{-c(s-a)}}{s-a}$	$\sigma > \text{Re}(a)$
21	$t^n h(t-c)$, $c > 0$, $n = 1, 2, \ldots$	$\dfrac{e^{-cs}[(cs)^n + n(cs)^{n-1} + \cdots + n!]}{s^{n+1}}$	$\sigma > 0$
22	$t^n[h(t) - h(t-c)]$, $c > 0$, $n = 1, 2, \ldots$	$\dfrac{n! - e^{-cs}[(cs)^n + n(cs)^{n-1} + \cdots + n!]}{s^{n+1}}$	$-\infty < \sigma < \infty$
23	$t[h(t) - h(t-c)] + (2c-t)[h(t-c) - h(t-2c)]$	$\dfrac{1 - 2e^{-cs} + e^{-2cs}}{s^2}$	$-\infty < \sigma < \infty$
24	$\displaystyle\sum_{k=0}^{\infty}(t - 2kc)[h(t - 2kc) - h(t - 2kc - c)]$ $+ \displaystyle\sum_{k=0}^{\infty}(2kc + 2c - t)\cdot[h(t - 2kc - c)$ $- h(t - 2kc - 2c)]$ (triangular wave)	$\dfrac{1 - e^{-cs}}{s^2(1 + e^{-cs})}$	$\sigma > 0$
25	$a\displaystyle\sum_{k=0}^{\infty}(t - kc)\cdot[h(t - kc) - h(t - kc - c)]$ (sawtooth wave)	$\dfrac{a(1 + cs - e^{cs})}{s^2(1 - e^{cs})}$	$\sigma > 0$

TABLE 6–2

INVERSE LAPLACE TRANSFORMS

No.	$\phi(s)$	$f(t) = \mathcal{L}^{-1}[\phi]$ for $t \geqq 0$
1	$\dfrac{c}{as+b}, \quad a \neq 0$	$\dfrac{c}{a} e^{-(b/a)t}$
2	$\dfrac{1}{(s+\alpha)(s+\beta)}, \quad \alpha \neq \beta$	$\dfrac{e^{-\alpha t} - e^{-\beta t}}{\beta - \alpha}$
3	$\dfrac{ps+q}{(s+\alpha)(s+\beta)}, \quad \alpha \neq \beta$	$\dfrac{(q-p\alpha)e^{-\alpha t} - (q-p\beta)e^{-\beta t}}{\beta - \alpha}$
4	$\dfrac{1}{(s+\alpha)^2}$	$te^{-\alpha t}$
5	$\dfrac{ps+q}{(s+\alpha)^2}$	$e^{-\alpha t}[p + t(q - \alpha p)]$
6	$\dfrac{1}{as^2+bs+c},$ $b^2 - 4ac > 0, \quad a \neq 0$	$\dfrac{e^{-\beta t} - e^{-\alpha t}}{\mu}, \quad \alpha = \dfrac{b+\mu}{2a},$ $\beta = \dfrac{b-\mu}{2a}, \quad \mu = \sqrt{b^2 - 4ac}$
7	$\dfrac{ps+q}{as^2+bs+c},$ $b^2 - 4ac > 0, \quad a \neq 0$	$\dfrac{(q-p\beta)e^{-\beta t} - (q-p\alpha)e^{-\alpha t}}{\mu},$ $\alpha, \beta, \mu \quad \text{as in No. 6}$
8	$\dfrac{1}{as^2+bs+c},$ $b^2 - 4ac = 0, \quad a \neq 0$	$\dfrac{te^{-\alpha t}}{a}, \quad \alpha = \dfrac{b}{2a}$
9	$\dfrac{ps+q}{as^2+bs+c},$ $b^2 - 4ac = 0, \quad a \neq 0$	$\dfrac{e^{-\alpha t}[p + t(q - \alpha p)]}{a}, \quad \alpha = \dfrac{b}{2a}$
10	$\dfrac{1}{as^2+bs+c},$ $b^2 - 4ac < 0, \quad a \neq 0$	$\dfrac{2e^{-\alpha t}}{\mu} \sin \dfrac{\mu}{2a} t,$ $\mu = \sqrt{4ac - b^2}, \quad \alpha = \dfrac{b}{2a}$
11	$\dfrac{ps+q}{as^2+bs+c},$ $b^2 - 4ac < 0, \quad a \neq 0$	$e^{-\alpha t}\left[\dfrac{p}{a} \cos \dfrac{\mu}{2a} t + \dfrac{2q - 2\alpha p}{\mu} \sin \dfrac{\mu}{2a} t\right]$ $\alpha = \dfrac{b}{2a}, \quad \mu = \sqrt{4ac - b^2}$

TABLE 6-2 *Continued*

No.	$\phi(s)$	$f(t) = \mathcal{L}^{-1}[\phi]$ for $t \geqq 0$
12	$\dfrac{1}{(s+\alpha)(s+\beta)(s+\gamma)}$, α, β, γ distinct	$-\left(\dfrac{e^{-\alpha t}}{BC} + \dfrac{e^{-\beta t}}{AC} + \dfrac{e^{-\gamma t}}{AB}\right)$, $A = \beta - \gamma, \; B = \gamma - \alpha, \; C = \alpha - \beta$
13	$\dfrac{1}{(s+\alpha)^2(s+\beta)}$, $\quad \alpha \neq \beta$	$\dfrac{e^{-\beta t}}{(\beta-\alpha)^2} + \left[\dfrac{t}{\beta-\alpha} - \dfrac{1}{(\beta-\alpha)^2}\right]e^{-\alpha t}$
14	$\dfrac{1}{(s+\alpha)^3}$	$\dfrac{t^2}{2}e^{-\alpha t}$
15	$\dfrac{1}{(s+\alpha)(as^2+bs+c)}$, $N = a\alpha^2 - b\alpha + c \neq 0, \; a \neq 0$	$\dfrac{e^{-\alpha t}}{N} + \dfrac{1}{N}\mathcal{L}^{-1}\left[\dfrac{-as + a\alpha - b}{as^2+bs+c}\right]$
16	$\dfrac{1}{(as^2+bs+c)(As^2+Bs+C)}$, as^2+bs+c and As^2+Bs+C having no common roots, $a \neq 0, \; A \neq 0$.	$\mathcal{L}^{-1}\left[\dfrac{p_0 s + q_0}{as^2+bs+c}\right]$ $\quad + \mathcal{L}^{-1}\left[\dfrac{p_1 s + q_1}{As^2+Bs+C}\right];$ let $\beta = aB - bA, \; \gamma = aC - cA,$ $\delta_0 = a\gamma^2 - b\beta\gamma + c\beta^2,$ $\delta_1 = A\gamma^2 - B\beta\gamma + C\beta^2.$ Then $p_0 = \dfrac{-a^2\beta}{\delta_0}, \; q_0 = \dfrac{a^2\gamma - ab\beta}{\delta_0},$ $p_1 = \dfrac{A^2\beta}{\delta_1}, \quad q_1 = \dfrac{-A^2\gamma + AB\beta}{\delta_1}.$
17	$\dfrac{1}{(as^2+bs+c)^2}$, $a \neq 0, \; b^2 - 4ac < 0$	$\dfrac{e^{-\alpha t}}{2a^2\beta^3}(\sin\beta t - \beta t \cos\beta t),$ $\alpha = \dfrac{b}{2a}, \; \beta = \dfrac{\sqrt{4ac-b^2}}{2a}$
18	$\dfrac{e^{-cs}}{(s-a)^k}$, $c \geqq 0, \; k = 1, 2, \ldots$	$\dfrac{(t-c)^{k-1}e^{a(t-c)}h(t-c)}{(k-1)!}$

Table 6-4
Convolutions

No.	$f(t),\ t \geqq 0\ \ (f = 0\ \text{for}\ t < 0)$	$e^{st}h(t) * f(t),\quad t \geqq 0,\quad s = \text{const}$
1	$1,\ s \neq 0$	$\dfrac{e^{st} - 1}{s}$
2	$t,\ s \neq 0$	$\dfrac{e^{st} - 1 - st}{s^2}$
3	$t^n,\ n = 1, 2, \ldots,\ s \neq 0$	$\dfrac{n!e^{st} - g_n(st)}{s^{n+1}},\quad g_n(x) = x^n + nx^{n-1} + \cdots + n!$
4	$e^{at},\ a \neq s$	$\dfrac{e^{at} - e^{st}}{a - s}$
5	e^{st}	te^{st}
6	$t^n e^{at},\ a \neq s,\ n = 1, 2, \ldots$	$\dfrac{n!e^{st} - e^{at} g_n[(s - a)t]}{(s - a)^{n+1}}$
7	$t^n e^{st},\ n = 1, 2, \ldots$	$\dfrac{e^{st} t^{n+1}}{n + 1}$

8	$h(t - c)$, $c > 0$, $s \neq 0$	$\dfrac{e^{s(t-c)} - 1}{s} h(t - c)$
9	$(t - c)h(t - c)$, $c > 0$, $s \neq 0$	$\dfrac{e^{s(t-c)} - 1 - s(t - c)}{s^2} h(t - c)$
10	$t[h(t) - h(t - c)]$, $c > 0$, $s \neq 0$	$\dfrac{e^{st} - 1 - st}{s^2}[h(t) - h(t-c)] + \dfrac{e^{st}[1 - (1 + cs)e^{-sc}]}{s^2} h(t - c)$
11	$t[h(t) - h(t - c)] + (2c - t)[h(t - c) - h(t - 2c)]$, $c > 0$, $s \neq 0$	$\dfrac{e^{st} - 1 - st}{s^2}[h(t) - h(t-c)]$ $+ \dfrac{e^{st} - 2e^{s(t-c)} + 1 + st - 2cs}{s^2} \cdot [h(t-c) - h(t-2c)]$ $+ \dfrac{e^{st}(1 - 2e^{-cs} + e^{-2cs})}{s^2} h(t - 2c)$
12	$e^{at} \cos bt$, $s \neq a \pm bi$	$\dfrac{e^{at}[(a - s)\cos bt + b\sin bt] - (a - s)e^{st}}{(a - s)^2 + b^2}$
13	$e^{at} \sin bt$, $s \neq a \pm bi$	$\dfrac{e^{at}[(a - s)\sin bt - b\cos bt] + be^{st}}{(a - s)^2 + b^2}$

Table 6–5

z-Transforms*

$$Z[f] = \sum_{n=0}^{\infty} f(nT)z^{-n} = F(z), \quad |z| > R = \frac{1}{\rho}$$

No.	$f(t), \quad t = 0, T, 2T, \ldots$	$F(z)$	R
1	1	$\dfrac{z}{z-1}$	1
2	t	$\dfrac{zT}{(z-1)^2}$	1
3	t^2	$\dfrac{T^2 z(z+1)}{(z-1)^3}$	1
4	e^{ct}	$\dfrac{z}{z-e^{cT}}$	$|e^{cT}|$
5	$t^k e^{ct}$	$\dfrac{\partial^k}{\partial c^k}\dfrac{z}{z-e^{cT}}$	$|e^{cT}|$
6	$\cos bt$	$\dfrac{z(z-\cos bT)}{z^2 - 2z\cos bT + 1}$	1
7	$\sin bt$	$\dfrac{z\sin bT}{z^2 - 2z\cos bT + 1}$	1
8	$e^{ct}\cos bt$	$\dfrac{z(z-e^{cT}\cos bT)}{z^2 - 2ze^{cT}\cos bT + e^{2cT}}$	e^{cT}
9	$e^{ct}\sin bt$	$\dfrac{ze^{cT}\sin bT}{z^2 - 2ze^{cT}\cos bT + e^{2cT}}$	e^{cT}
10	$\cosh bt$	$\dfrac{z(z-\cosh bT)}{z^2 - 2z\cosh bT + 1}$	$e^{|b|T}$
11	$\sinh bt$	$\dfrac{z\sinh bT}{z^2 - 2z\cosh bT + 1}$	$e^{|b|T}$
12	$h(t-T)$	$\dfrac{1}{z-1}$	1
13	$h(t) - h(t-T)$	1	0

* *Note.* In Entries 6 through 11 the formulas are valid for complex b and c, but the range of z is given only for real b and c. In Entries 15 through 17, a is assumed real and $\neq 0$; however, the transforms remain valid for complex a (not 0), if we choose a definite value for the exponential a^z, for example, exp $(z \log a)$ as in Section 3–5.

TABLE 6–5 *Continued*

$$Z[f] \;=\; \sum_{n=0}^{\infty} f(nT)z^{-n} \;=\; F(z), \quad |z| \;>\; R \;=\; \frac{1}{\rho}$$

No.	$f(t), \quad t = 0, T, 2T, \ldots$	$F(z)$	R
14	$h(t - kT), \quad k = 1, 2, \ldots$	$\dfrac{1}{z^{k-1}(z-1)}$	1
15	$a^{(t-T)/T} h(t - T)$	$\dfrac{1}{z - a}$	$|a|$
16	$\dfrac{1}{T}(t - T)a^{(t-2T)/T} h(t - T)$	$\dfrac{1}{(z - a)^2}$	$|a|$
17	$\dfrac{(t - T)(t - 2T)\cdots(t - kT + T)}{(k-1)!\,T^{k-1}}$ $\times\, a^{(t-kT)/T} h(t - T)$	$\dfrac{1}{(z - a)^k}$	$|a|$
18	$h(t - T) - h(t - 2T)$	$\dfrac{1}{z}$	0
19	$h(t - kT) - h[t - (k+1)T]$	$\dfrac{1}{z^k}$	0
20	$\dfrac{1}{\Gamma\left(\dfrac{t}{T} + 1\right)}$	$e^{1/z}$	0
21	$\dfrac{T}{t} h(t - T)$	$\log \dfrac{z}{z-1} = \displaystyle\sum_{n=1}^{\infty} \dfrac{z^{-n}}{n}$	1
22	$f(nT) = \dfrac{k(k-1)\cdots(k-n+1)}{n!}$	$\left(1 + \dfrac{1}{z}\right)^k$	1
23	$\dfrac{(t + T)(t + 2T)\cdots(t + kT - T)}{(k-1)!\,T^{k-1}}$	$\left(1 - \dfrac{1}{z}\right)^{-k}, \quad k = 2, 3, \ldots$	1
24	f has period NT, $f = t$ for $0 \leqq t < NT$	$\dfrac{Tz}{z-1}\left(\dfrac{1}{z-1} - \dfrac{N}{z^N - 1}\right)$	1
25	f has period NT, $f = 1$ for $0 \leqq t \leqq MT, \quad M < N$	$\dfrac{z^{M+1} - 1}{z^{M-N}(z-1)(z^N - 1)}$	1
26	f has period NT, $f_1(t) = f(t)[1 - h(t - NT)]$	$\dfrac{z^N}{z^N - 1} F_1(z), \quad F_1 = Z[f_1]$	1

Appendix III

GLOSSARY OF SYMBOLS

Symbol	Significance	Section reference
$A = (a_{ij})$, $\|A\|$	matrix, norm of matrix	8–13
arg z	argument (polar angle) of complex number z	1–7
col $(x_1, \ldots, x_n)$	column matrix	8–13
$\mathcal{C}$	conjugation operator, $\mathcal{C}f(t) = \overline{f(t)}$	5–10
D, D^n	derivatives d/dt, d^n/dt^n	1–2
D^{-1}	indefinite integral $\int \ldots dt$	1–6
D^{-n}	n-fold indefinite integral	1–6
D_0^{-1}	integral from 0 to t	1–6
D_0^{-n}	n-fold integral, $D_0^{-2}f = D_0^{-1}[D_0^{-1}f]$, $\ldots$	1–6
det A	determinant of matrix A	8–13
e^A	exponential function of matrix A	8–13
$F^*(s)$	Laplace transform of sampled function	6–24
$f^*(t)$	sampled function	6–24
$f \triangle g$	f-convolution	4–17
$f * g$	convolution	2–8, 5–11, 6–12
$f *_T g$	z-convolution	6–22
$f'_+(t)$, $f'_-(t)$	special derivatives to right and left	4–7
$g_k(x)$	$x^k + kx^{k-1} + \cdots + k!$	4–15
$G(t, u)$	Green's function	8–3
$H(s, t)$	Zadeh system function	8–17
$\mathcal{H}[\alpha(\omega)]$	Hilbert transform	6–25
$h(t)$	Heaviside unit step function, $h(t) = 1$ for $t \geqq 0$, $h(t) = 0$ for $t < 0$	1–12
I	identity operator, unit matrix	2–2, 8–13
$L[x]$, $L^+[x]$	linear differential operator and adjoint operator	8–4
$L_n(t)$	n-th Laguerre polynomial	4–12, 6–10
$\mathcal{L}[f]$, $\mathcal{L}^{-1}[f]$	Laplace transform, inverse transform	4–15, 6–2, 6–7
$\mathcal{L}_\tau[f]$	truncated Laplace transform	4–15
log z, Log z	complex logarithm and principal value of log	3–5
Res $[f(z), z_0]$	residue of $f(z)$ at z_0	3–13, 3–14

Symbol	Significance	Section reference
$T, T[f]$	operator, operator applied to f	2–2
T^k	k-th power of operator T	2–2
T^{-1}	inverse of operator T	2–2
$T_0[f]$	response of linear system, initially at rest, to input f	2–3
$V(s)$	characteristic function	2–9
$W(t)$	weighting function	2–8, 5–18, 6–18, 8–2, 8–17
$W_\tau(t)$	weighting function for periodic inputs	4–23
$w(t)$	Wronskian determinant	8–1
$Y(s)$	transfer function	2–9
$Y(i\omega)$	frequency response function	2–9
$Z[f], Z^{-1}[f]$	z-transform of f, inverse z-transform	6–22
z^a, a^z	complex power function and general exponential function	3–5
$|z|$	absolute value of complex number z	1–7
$\Gamma(x)$	Gamma function	6–3
γ_a	$(1 - e^{a\tau})^{-1}$, constant used with finite Fourier transform	4–18
$\Delta_1, \ldots, \Delta_k$	Hurwitz determinants	7–4
$\Delta_c[f]$	translation operator	2–6
δ_{ij}	Kronecker delta	6–23, 8–13
$\delta(t)$	Dirac delta function, impulse	1–13
$\delta^{(k)}(t - c)$	k-th derivative of translated delta function	1–13
$\delta_\tau(t)$	periodic impulse	4–20
$\delta_\tau^{(k)}(t - c)$	k-th derivative of translated period impulse	4–20
Θ	reflection operator, $\Theta f(t) = f(-t)$	5–10
$\Phi[f], \Phi^{-1}[f]$	Fourier transform, inverse transform	5–3, 5–13
$\Phi_\tau[f], \Phi_\tau^{-1}[f]$	finite Fourier transform, inverse transform	4–14, 4–19
$\Phi(t, u)$	matrix kernel function	8–13
$\phi(t, u)$	kernel function	8–2
$\displaystyle\int_C f(z)\, dz$	complex line integral	3–3
$\displaystyle (P) \int_{-\infty}^{\infty} f(t)\, dt$	principal value of integral	1–8
$\displaystyle \binom{n}{k}$	binomial coefficient, $n(n - 1) \cdots (n - k + 1)/k!$	6–10

Index

INDEX

Randall Library – UNCW

QA432 .K3   NXWW
Kaplan / Operational methods for linear systems.

3049001765899